D1614960

IMPACT DYNAMICS

IMPACT DYNAMICS

JONAS A. ZUKAS
USA Ballistic Research Laboratory

THEODORE NICHOLAS
USAF Wright Aeronautical Laboratories

HALLOCK F. SWIFT
Physics Applications Inc.

LONGIN B. GRESZCZUK
McDonnell Douglas Astronautics Co.

DONALD R. CURRAN
SRI International

A Wiley-Interscience Publication
JOHN WILEY & SONS
New York Chichester Brisbane Toronto Singapore

Library of Congress Cataloging in Publication Data

Main entry under title:

Impact Dynamics.

"A Wiley-Interscience publication."
Includes bibliographical references and index.
1. Impact. 2. Structural dynamics. 3. Projectiles. I. Zukas, Jonas A.
TA418.34.I46 623'.516 81-11683
ISBN 0-471-08677-0 AACR2

Printed in the United States of America

10 9 8 7 6 5 4

PREFACE

This book is an outgrowth of a one-week short course on "Impact Dynamics" first taught by the authors in 1979. It was immediately apparent at the time that there was no single text available for the broad range of topics dealing with the various aspects of material behavior during high-speed impact. In fact, there was little available in published texts and reports as reference material for each of the topics covered. Rather, the information had to be gathered from scattered reports in journals, relatively obscure government and industrial reports, and personal knowledge of the field.

The primary emphasis of this book is on the material aspects of impact phenomenology as related to the general problem area of projectile impact against target materials. The treatment is aimed at the practicing engineer who wants to acquire a working knowledge of the field and the specialist who wants to obtain information on aspects of material behavior under impact outside his or her own area. No attempt is made to deal in any depth with other specific problems involving dynamic deformation of materials such as high-speed metal forming or machining. The approach is fundamental in nature and emphasizes basic principles. Thus it could be useful in the study of a wide variety of dynamic problems. The approach is a combination of a tutorial as well as a state-of-the-art review of the subject field. Numerous references are given for the reader to explore any of the topics covered in greater detail. In many cases, a historical perspective is given to illustrate the thinking and approaches used in reaching today's technology position. The subject matter covers both experimental and analytical approaches to material behavior under conditions varying from low-speed to hypervelocity impact. A basic understanding of the phenomena as well as a rational approach are emphasized; no emphasis is made on highly empirical approaches to model complex impact events.

JONAS A. ZUKAS
THEODORE NICHOLAS
HALLOCK F. SWIFT
LONGIN B. GRESZCZUK
DONALD R. CURRAN

Aberdeen Proving Ground, Maryland
Wright Patterson Air Force Base, Ohio
Huntington Beach, California
Dayton, Ohio
Menlo Park, California

September 1981

CONTENTS

1

STRESS WAVES IN SOLIDS

Jonas A. Zukas

The response of materials and structures to intense impulsive loading is quite complex. For conceptual clarity, the behavior of impacted solids may roughly be divided into three regimes. For loading conditions that result in stresses below the yield point, materials behave elastically. For metals, Hooke's law is applicable. A number of elegant mathematical solutions have been obtained for various loading conditions in this regime. Most, however, are for semiinfinite bodies. Practical impact problems often involve strikers and targets with finite boundaries which exert considerable influence on their behavior. As the intensity of the applied loading is increased, the material is driven into the plastic range. The behavior here involves large deformations, heating, and often failure of the colliding solids through a variety of mechanisms. With still further increases in loading intensity, pressures are generated that exceed the strength of the colliding solids by several orders of magnitude, which, in effect, behave hydrodynamically.

For low intensity excitations, both the geometry of the entire structure as well as the nature of the material from which it is made play a major role in resisting external forces. As loading intensity increases, the response tends to become highly localized and is more affected by the constitution of the material in the vicinity of load application than the geometry of the total structure. A description of the phenomena in terms of elastic, plastic, and shock wave propagation becomes appropriate.

This chapter provides a broad overview of stress wave effects in solids that are caused by impact or rapidly varying loading, together with fundamental results involving elastic and shock waves. The complex problem of elastoplastic wave propagation is treated in Chapter 4. It is not the aim here to be exhaustive. A rich literature on these subjects exists and will be cited as needed. Instead, results are presented that are useful in obtaining a qualitative appreciation of the behavior of solids under short duration loading, which may serve as precursors to more complete analyses.

1.1 RESPONSE OF MATERIALS TO IMPULSE LOADING

Elastic theory for isotropic solids indicates propagation of two types of waves. Dilatational (or longitudinal) waves are those in which the particle motion induced by the disturbance is normal to the wave front. Distortional (or transverse or shear) waves are those wherein material particles move in a plane at right angles to that in which the wave front propagates. Expressions for the velocities of propagation for the two types of disturbances, denoted by c_L and c_s respectively, are given in Table 1 for bounded and unbounded media. Under the proper loading conditions, torsional and flexural waves may also be generated. For a discussion of these, see the books by Kolsky (1963), Johnson (1972), and Achenbach (1975).

In addition to dilatational and distortional waves that travel through an extended solid medium, elastic waves may be propagated along the surface of a solid. Two types have been studied extensively. Rayleigh surface waves decay exponentially with depth from the surface to the medium interior. Rayleigh waves spread only in two dimensions, fall off more slowly with distance than the other types of elastic waves, and have an amplitude appreciable only near the surface of a body. The Love wave is another type of surface wave confined to a relatively shallow surface zone. It is a shear wave. A situation favorable to the establishment of a Love wave is that in which a layer of material that possesses one set of physical constants overlies a layer that possesses different physical constants.

Table 1 Wave Velocities in Isotropic Solids

Extended Media	Bounded Media
$c_L^2=\frac{\lambda+2\mu}{\rho}=\frac{E(1-\nu)}{\rho(1+\nu)(1-2\nu)}$	$c_L^2=\frac{E}{\rho}$
$c_s^2=\frac{\mu}{\rho}=\frac{G}{\rho}=\frac{E}{2\rho(1+\nu)}$	$c_s^2=\frac{G}{\rho}$

where E = Young's modulus
ν = Poisson's ratio
ρ = density
λ, μ = Lame parameters
G = shear modulus

It should be kept in mind that real solids never fit the idealized models developed to describe their behavior. When a wave disturbance is propagated through a real medium, part of the mechanical energy is converted into heat. The several different mechanisms by which this energy transformation occurs are collectively termed internal friction. Because of the complexity introduced into mathematical models, such phenomena are not generally considered for other than academic problems. Kolsky's book should be consulted for additional details. The texts by Achenbach and Whitham (1974) also deal with dissipative mechanisms in waves.

When either a dilatational or distortional wave impinges on a boundary of the solid, waves of both types are generated. The interaction of elastic waves with boundaries is treated lucidly in Achenbach's book. Of particular interest in impact situations is the normal impingement of a strong compressive pulse on a free surface. The pulse is reflected as a tensile wave and if its magnitude is greater than the tensile strength of the material, fractures occur causing material separation. Simple analyses predict reasonably well the location of the fracture plane, and the size and speed of the ejected material for high strength solids. If after fracture the magnitude of the stress pulse still exceeds the material's tensile strength, multiple fractures can occur.

As the intensity of the applied load increases, the material is driven beyond its elastic limit and becomes plastic. Two waves now propagate in the solid, an elastic wave (or precursor) followed by a much slower but more intense plastic wave. If the characteristics of the medium are such that the velocity of propagation of large disturbances is greater than the propagation velocity of smaller ones, the stress pulse develops a steeper and steeper front on passing through the medium, and the thickness of this front is ultimately determined by the molecular constitution of the medium. The shock wave (or steep pressure pulse) thus formed differs from the high pressures generated by conventional methods in that it relies on the inertial response of material to rapid acceleration rather than on static constraints.

Most of the work on plastic wave propagation has dealt with long rod geometries. Analytical work has assumed that only the longitudinal component of stress was nonzero because the cylindrical surface of the rod was free of traction. Although a lateral strain is associated with longitudinal stretch (the Poisson ratio effect), lateral inertia effects are commonly neglected in rod wave theories and only the longitudinal component of the equation of motion is included. Thermal effects are also ignored. For rod impacts, this seems reasonable since plastic flow limits the stress to the yield stress magnitude which increases only slowly with work hardening. The plasticity-limited stress holds down the work done for a given strain level to the extent that it produces a marginal thermal effect. Thus a purely mechanical stress-strain law is adequate, based on plasticity theory, and the thermodynamics of plastic deformation does not intrude.

The existing theories for one-dimensional plastic waves in rods are due to von Karman, Taylor, and Rakhmatulin who assume a rate-independent constitutive model and to Malvern who includes strain rate dependence. Arguments over the validity of both approaches have raged for years. Because of the different methods of testing and the wide range of materials used, comparison between experiments is most difficult. The article by Hopkins (1964) is an excellent review of the subject. The current trend seems to favor the strain rate-dependent model, although Wood (1963) has pointed out that both camps may be looking at different aspects of the same phenomenon and that the arguments may disappear once inertia effects are included in rod-wave theories. Lee (1971) has pointed out that one-dimensional rod theories which neglect effects of lateral inertia can only be expected to hold if wavelengths of the disturbances are long compared with the rod diameter. To assess the influence of strain rate on plastic flow, high strain gradients are needed and these tend to introduce a lateral inertia influence. Elastoplastic wave propagation is treated at length in Chapter 4.

If the intensity of the loading is so great or its duration so short that the material no longer possesses rigidity, it will behave as though it had the properties of a fluid. Transverse (shear) waves cannot exist within the body and only a longitudinal wave will be propagated with a velocity $c_B^2 = K/\rho$ where K is the bulk modulus and ρ the density. In this shock wave regime, extremely high pressures are generated which can lead to changes in density of materials such as steel as great as 30%.

Experiments in this regime typically involve the impact or explosive loading of plates. If one considers the situation away from the plate edges where nonuniform conditions exist, only the strain component normal to the wavefront is nonzero and because (due to symmetry) there is no lateral motion, this component of strain also expresses the volume change. The stress response is governed mainly by dilation, since pressures generated are typically measured in the hundreds of kilobars while material strengths are of the order of a few kilobars. This circumstance led to the development of hydrodynamic theories in which the material strength was neglected and the metal assumed to behave as a perfect fluid with resistance only to dilatation. There is a large expenditure of work because of the material's resistance to volume change. The scalar quantities hydrostatic pressure, volume ratio, and temperature express the state of the material, and wave propagation theory becomes a one-dimensional analysis with the only nontrivial equation of motion arising from the component normal to the wavefront.

Lee (1971) noted the historical separation of impulse loading studies to long rod and flat plate geometries and suggested that effects historically omitted need be accounted for if accurate characterization of the dynamic behavior of materials is desired. The need for inclusion of lateral inertia effects in rod theories has already been noted. Lee also suggests that hydrodynamic theories need to be modified to account for strength effects and finite deformations. Finite strains and strength effects can play a dominant role in determining stress-wave profile. For fracture problems, this profile is crucial for determining the location and type of failure. The first unloading of a steep pressure pulse is frequently of greatest interest in experiments. Inclusion of finite elastic and plastic strains in present hydrodynamic theories would lead to more rapid attenuation of the initial pressure pulse than is currently predicted and this in turn would affect failure predictions.

Because of the resulting one-dimensionality of the equations for the propagation of elastic waves and shock waves, solutions are possible for a variety of problems which, though oversimplified, greatly assist the study of impact phenomena. These topics are now considered.

1.2 ELASTIC WAVES

A number of texts have appeared lately that treat in detail various aspects of elastic wave propagation in solids, for example—Achenbach (1975), Whitham (1974), Wasley (1973), and Eringen and Suhubi (1975). Without repeating at length material in these treatises, we quote the necessary results and consider approximations for impact situations which, though quite simplistic, aid in visualizing the complex physical phenomena which take place. They prove to be useful in the design of experiments and assessment of the utility of more complex analytical schemes.

The system of equations governing the motion of a homogeneous, isotropic, linearly elastic body consists of the stress equations of motion, Hooke's Law, and the strain-displacement relations [Achenbach (1975)] given by:

$$\sigma_{ij,j} + \rho f_i = \rho \ddot{u}_i \tag{1}$$

$$\sigma_{ij} = \lambda \varepsilon_{kk} \delta_{ij} + 2\mu \varepsilon_{ij} \tag{2}$$

$$\varepsilon_{ij} = \tfrac{1}{2}(u_{i,j} + u_{j,i}) \tag{3}$$

In the above use of the indicial notation has been made so that repeated subscripts denote summation. Latin indices are assumed to take on the value 1, 2, and 3 whereas Greek subscripts take on the values 1 and 2. The quantities λ and μ are the Lame parameters, σ_{ij} the components of the stress tensor, ε_{ij} the components of the strain tensor, u_i the components of the displacement vector, ρ the density, and f_i body forces. A comma denotes partial differentiation with respect to coordinates whereas a dot indicates a time derivative. δ_{ij} is the well-known Kronecker delta which takes on the value 1 when $i=j$ and 0 when $i \neq j$.

Equations (1)–(3) may be combined to obtain the displacement equation of motion

$$\mu u_{i,jj} + (\lambda + \mu) u_{j,ji} + \rho f_i = \rho \ddot{u}_i \tag{4}$$

Equations (1)–(4) must be satisfied at every interior point of the undeformed body. With the specification of boundary conditions on the surface of the body as well as initial conditions, the statement of the elastodynamic problem is complete.

Equations (1)–(4), with appropriate boundary and initial conditions, are associated with problems of mechanical wave propagation and vibration effects in elastic media. The system of equations (4) has the disadvantageous feature of coupling the three displacement components. In application to specific problems, it is convenient to uncouple these equations and cast them into a standard form known as the wave equation. This can be done by various means, including expression of the displacements u_i in terms of vector and scalar potentials or elimination of two of the three displacement components through two of the three equations (4). Details are available in the texts mentioned previously. For the sake of brevity we will arrive at the basic form by first assuming that body forces are negligible in comparison to the other applied loads, which reduces (4) to

$$\rho \ddot{u}_i = (\lambda + \mu)\Delta,_i + \mu u_{i,jj} \tag{5}$$

where

$$\Delta = \varepsilon_{jj} = u_{j,j}$$

Differentiating each equation in (5) with respect to the spatial variables in respective order gives

$$\rho \ddot{u}_{i,i} = (\lambda + \mu)\Delta,_{ii} + \mu u_{i,jji}$$

or (6)

$$\frac{\rho \partial^2 \Delta}{\partial t^2} = (\lambda + 2\mu)\frac{\partial^2 \Delta}{\partial x_i \partial x_i}$$

since we treat λ, μ, and ρ as material constants. This equation is of the general form

$$\frac{\partial^2 \psi}{\partial t^2} = c^2 \frac{\partial^2 \psi}{\partial x_i \partial x_i} \tag{7}$$

where $\psi(x_i, t)$ is the dependent variable and is a measure of some property of the disturbance such as displacement or velocity and c is a physical constant. Equation (7) is the classical equation of wave motion.

There are two general alternate methods by which solutions to linear partial differential equations are obtained. The method due to Cauchy and Fourier consists of separation of the partial differential equation into ordinary differential equations and obtaining solutions to the ordinary differential equations through power series. The method due to Riemann and Hadamard requires transformation of the equations to a new set of independent variables by finding a set of natural coordinates that are related to the initial independent variables. The former method lends itself to a standing wave or normal mode solution to the wave equation appropriate to problems of elastic vibration in finite systems while the latter method lends itself to a traveling wave or progressive wave solution of that equation appropriate to problems in elastic wave propagation.

Let us limit the following discussion to one-dimensional wave propagation for the sake of conceptual clarity. Let the spatial coordinate in the direction of propagation be x.

It can be shown—for example, Wasley (1973)—that the most general solution of the one-dimensional wave equation is

$$\psi = f(x-ct) + g(x+ct) \tag{8}$$

where f and g are arbitrary functions of the arguments $(x-ct)$ and $(x+ct)$ respectively. The quantities f and g must be consistent with the requirements of continuity, small amplitude, and the various imposed boundary conditions. Two functions are necessary because the wave equation is of second order.

There is a simple interpretation for (8) in the one-dimensional situation. For purposes of illustration, consider only the function f. Then, at any time t_1, ψ is a function of x only, and can be represented by a certain curve as shown in Figure 1.

The specific shape of the curve is determined by the form of f. After time is increased by an amount Δt, the argument of the function becomes $x-c(t+\Delta t)$. Since we are dealing with a one-dimensional ideal elastic medium where c is a constant, the shape of the wave does not change during propagation. The function f will remain unchanged provided that simultaneously with the increase by Δt the points along x change by an amount $\Delta x = c\Delta t$. This means that the curve ψ as shown for the time t in the figure can also be used for the time $t+\Delta t$ if it is appropriately displaced in the x direction. Hence, f in the solution (8) represents a wave moving in the direction of the positive x axis with a constant velocity c. In the same manner, it can be shown that the function g represents a wave traveling in the direction of the negative x axis. Thus (8) physically represents two waves progressing along the x axis in opposite directions, each with constant velocity.

Solution of the field equations for small motions of homogeneous, perfectly elastic solids requires specification of boundary conditions. However, if the disturbance begins at an interior point far away from boundaries, geometric complications due to surface effects do not enter. Waves in unbounded and semiextended media (a semiinfinite solid with a plane boundary) are important for a variety of applications including the propagation of seismic disturbances. However, impact problems of practical interest involve either bounded media or impact velocities which mandate treatment above the elastic regime. This situation will therefore not be treated here. We refer the reader to the previously mentioned books by Achenbach, Wasley, Eringen and Suhubi

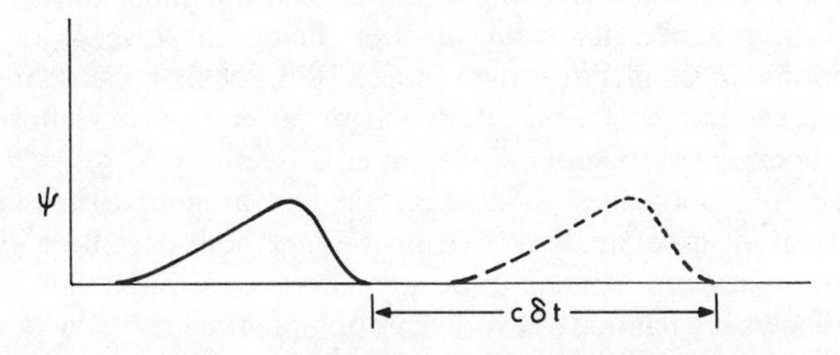

Figure 1 Propagation of an idealized one-dimensional elastic wave.

(1975) for additional information, as well as the book by Ewing, et al. (1975) for seismological problems, and pass to the consideration of wave propagation in bounded media.

The only difference between the propagation of elastic stress disturbances in bounded and unbounded media is geometrical. In theory, the transmission of such disturbances can be treated by solving the equations of small motion with the appropriate boundary conditions. In practice, however, addition of boundaries introduces immense complexities into the mathematical formulation of the problem so that very few "exact" analyses exist. An important analytical exact solution treats the problem of propagation of trains of progressive simple harmonic waves of infinite duration in isotropic circular cylinders of infinite length. This solution is due to Pochhammer and Chree and is described in detail in Wasley's book. We summarize briefly the basic features of the Pochhammer-Chree analysis and consider other approximate solutions before examining the characteristics of elementary solutions which are used for most work in impact situations.

In the Pochhammer-Chree approach, the governing equations in cylindrical coordinates are solved using expressions for sinusoidal waves of infinite duration propagating along an isotropic circular cylinder of infinite extent subject to the boundary conditions that the stresses σ_{rr}, $\sigma_{r\theta}$, and $\sigma_{\theta z}$ vanish at the cylinder surface. Dispersion equations—relations involving the frequency and wavelength of the vibrations, the radius of the cylinder, and its material properties—are derived for longitudinal, torsional, and flexural vibrations. For each of these modes (associated with a specific dispersion equation) the dispersion equation is found to have an infinite number of roots (called harmonics) each corresponding to a particular deformation pattern that exist independently of any other. The disturbances are transmitted at velocities that depend on both the mode and harmonic of transmission. For a given value of Poisson's ratio, the particular dispersion equation desired can be obtained involving only ratios of phase velocity to wave velocity ($\sqrt{E/\rho}$) and bar radius to wavelength. A numerical solution can then be computed, and the results for the various harmonics conveniently graphed.

From this type of analysis it is determined that the elementary wave theory can be expected to hold only when the ratio of bar radius to wavelength is much less than unity. However, for many impact situations this appears to be the case since the elementary theory agrees well with many aspects of impact experiments. Because of its complexity, the Pochhammer-Chree method is never used in practice, its main function being to serve as a check for approximations or simpler theories. Indeed, the Pochhammer-Chree analysis itself fails in the immediate neighborhood of an end cross section of a bar, although it is expected to apply at distant cross sections.

Two types of approaches have been taken to attempt to circumvent the complexities of the exact analysis—elementary methods have been extended or improved to obtain approximate dispersion curves or approximate solutions to the exact dispersion relations have been attempted, usually via power series. Most of the attempts fall in the former category.

Consider first longitudinal waves. Recall that the equation of motion is of the form

$$\frac{\partial^2 u}{\partial t^2} = c^2 \frac{\partial^2 u}{\partial x^2}; \qquad c^2 = \frac{E}{\rho} \tag{9}$$

Love reformulated the problem taking into account the inertia of lateral contraction while still assuming that plane sections remain plane and that the stress is uniform over a cross section, arriving at an equation of the form

$$\frac{\partial^2 u}{\partial t^2} = c^2 \frac{\partial^2 u}{\partial x^2} + \nu^2 K^2 \frac{\partial^4 u}{\partial x^2 \partial t^2} \tag{10}$$

where K is the radius of gyration. Further refinements were made by Mindlin and Herrmann who considered effects of radial shear as well as radial inertia. Their analysis, which includes two adjustable constants, gives rise to two dispersion curves which agree reasonably well with the fundamental and first harmonic of the Pochhammer-Chree analysis.

Approximate solutions for flexural waves were obtained by Rayleigh who considered the effects of rotary inertia and Love who took into account not only rotary inertia but distortion of the cross section during bending as well. The approach that most closely agrees with the exact solution is due to Timoshenko, which accounts for both rotary inertia and transverse shear effects.

1.3 ONE-DIMENSIONAL ELASTIC WAVES IN BARS

When a material is stressed with a suddenly applied load, the deformations and stresses are not immediately transmitted to all parts of the body, remote portions remaining undisturbed for some time. Deformations and stresses progress through the material in the form of one or more stress disturbances which travel at a finite velocity from the area of application of the load, this velocity being a characteristic of the material. Such a suddenly applied, or impulsive, load may be produced by a sharp mechanical blow, a detonating explosive, or by impact of a high velocity projectile. Regardless of the method of application, the consequent stress disturbances have identical properties.

In the elementary case we consider two types of stress pulses generated by an impulsive load. The first, the longitudinal wave, is also called a dilatational, irrotational, or primary (P) wave, the terms being synonymous. In a longitudinal pulse, the particle motion is parallel to the direction of propagation of the pulse and the strain is pure dilatation. In a transverse wave, otherwise called a distortional, rotational, secondary (S), or shear wave, the particle motion is normal to the direction of propagation of the pulse and the strain is a shearing strain.

For information on flexural waves, the reader is referred to the books by Achenbach and Wasley.

Representation of a pulse can be accomplished in any one of several ways:

1 Stress versus time.
2 Particle velocity versus time.
3 Stress versus distance.
4 Particle velocity versus distance.

Two velocities must be considered: the velocity of propagation c of the disturbance and the particle velocity v, the velocity with which a point in the material moves as the disturbance moves across it. These enter into the equations in distinctly different ways.

The equation of motion for the elementary theory is given by (9). Recall that this equation is valid for stress-pulse propagation in slender bars (a bar being defined as a long rod of circular cross section whose length is at least 10 times greater than its diameter) and implies neglect of:

1 Transverse strain.
2 Lateral inertia.
3 Body forces.
4 Internal friction (dissipative forces).

It can be shown that this equation should apply with reasonable success to problems where the wavelength of the propagating pulse is at least 6–10 times greater than the typical cross-sectional dimension of the bar [Wasley (1973), Achenbach (1975)].

1.3.1 Intensity of the Propagated Stress

The relationship between the longitudinal stress at any point in a body and the longitudinal particle velocity v_L at any point comes from Newton's second law

$$F_L\,dt=(mv_L) \tag{11}$$

where F_L is the longitudinal force acting on a given cross section, dt is the time the force acts, m is the mass it acts against, and v_L is the velocity imparted to m by F_L. Since

$$\sigma=\frac{F_L}{A}$$

$$m=\rho A\,dl$$

where dl is the distance the pulse has moved in time dt, we can write (11) as

$$\sigma A\,dt=\rho A\,dl\,d\dot{v}_L \tag{12}$$

or

$$\sigma=\rho\frac{dl}{dt}dv_L$$

But dl/dt is just the speed of the pulse c_L, so that

$$\sigma = \rho c_L(\Delta v_L) \tag{13}$$

In the same manner it can be shown that for the transverse pulse

$$\tau = \rho c_s(\Delta v_s) \tag{14}$$

where τ is the shear stress, c_s the velocity of propagation of the transverse disturbance, and Δv_s the change in the particle velocity in shear.

1.3.2 Reflection and Superposition of Waves

Any elastic wave will be reflected when it reaches a free surface of the material in which it is traveling. The simplest case occurs when the wave strikes the surface normally. In a longitudinal wave, since the stress normal to the surface at the surface must be zero, the reflected pulse must be opposite in sense to the incident pulse (compression reflected as tension and vice versa). To see this, consider the displacement due to the incident pulse to be $u_I = f(x-ct)$ moving in the positive x direction. After impingement on a free surface, a reflected wave moves in the negative x direction. Let the displacement for the reflected wave be of the form $u_R = g(x+ct)$. At the free boundary, the net stress must be zero:

$$\sigma_{\text{NET}} = \sigma_I + \sigma_R = 0 \qquad \text{at } x = l \tag{15}$$

Since the stress is given by $\sigma = E\varepsilon = E(\partial u/\partial x)$

$$\sigma_{\text{NET}} = E[f'(l-ct) + g'(l+ct)] = 0$$

or (16)

$$f'(l-ct) = -g'(l+ct)$$

Hence, the shape of the reflected pulse is the same as the shape of the incident pulse but is opposite in sign.

The net particle velocity can also be found by superposition.

$$v_{\text{NET}} = v_I + v_R = \frac{\partial u_I}{\partial t} + \frac{\partial u_R}{\partial t} \tag{17}$$

$$= c(-f' + g') \qquad \text{at } x = l \tag{18}$$

$$= 2cg'$$

Hence the particle velocity and also the displacement in a region where the incident and reflected pulses overlap are twice that for either pulse. At a fixed boundary, we require the displacement and particle velocity to vanish. Hence

$$v_{\text{NET}} = -cf'(l-ct) + cg'(l+ct)$$

or

$$f'(l-ct) = g'(l+ct) \tag{19}$$

The net stress

$$\sigma_{\text{NET}} = E\left(\frac{\partial u_I}{\partial x} + \frac{\partial u_R}{\partial x}\right) = E[f'(l-ct)+g'(l+ct)] \tag{20}$$
$$= 2Ef'(l-ct)$$

is doubled at a fixed boundary while the net displacements and particle velocities are zero.

A convenient technique for visualization of the behavior of stress pulses at an interface relies on the linearity of the wave equation which permits superposition of solutions. Figure 2 shows the procedure for a simple square wave shape. Imagine a phantom pulse outside the material of the same shape as the incident pulse but with a stress of opposite sense and traveling toward the surface in such a manner that both the incident (real) pulse and the phantom pulse strike the surface at the same time. Then allow the incident pulse to pass out of the material and the phantom pulse to pass into the material without any distortion.

A transverse disturbance, upon striking a free surface normally, will reflect as a transverse wave.

Recall that for a tensile pulse the displacement and particle velocity are in a direction opposite to that of the pulse while for a compressive pulse the displacement and particle velocity are in the direction of pulse propagation.

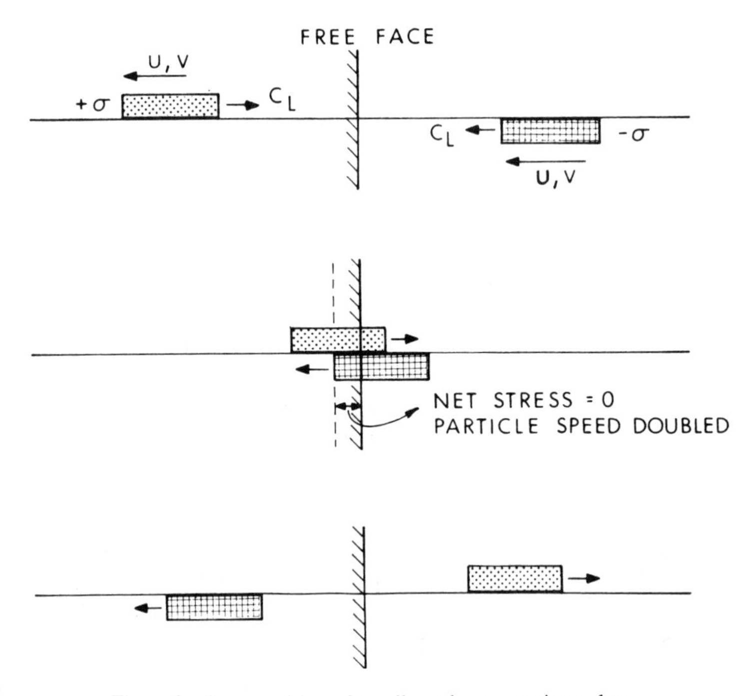

Figure 2 Superposition of tensile and compressive pulses.

Hence, in Figure 2, where the real and phantom pulses overlap, the net stress becomes zero while the net displacement and particle velocities are doubled.

If we consider in a similar fashion the reflection of a stress pulse at a *fixed* boundary (Figure 3) we find that in the portion of the bar where the two pulses overlap the total stress is doubled and the particle speed is zero. Hence, an elastic wave reflected from a fixed-end bar is entirely unchanged in shape or intensity.

When an elastic wave strikes a free surface obliquely the situation is much more complicated. This complication arises from the fact that the energy of the incident wave is partitioned into two reflected waves instead of only one. When either a longitudinal or transverse pulse strikes a surface obliquely both a longitudinal and transverse wave are reflected. One of the most lucid treatments of wave interactions at boundaries is given by Achenbach (1975).

As an example of the ideas just presented, consider the impact of a bar with a rigid surface (or equivalently, the colinear impact of two identical bars) with velocity v_0. The sequence of events is depicted in Figure 4:

1 After impact, a compressive wave of intensity $\rho v_0 c_L$ moves into the bar. At $0 \leq t \leq l/c_L$ the particles engulfed by the wave are brought to rest.

2 At $t = l/c_L$, the bar is stationary but in compression. All of the kinetic

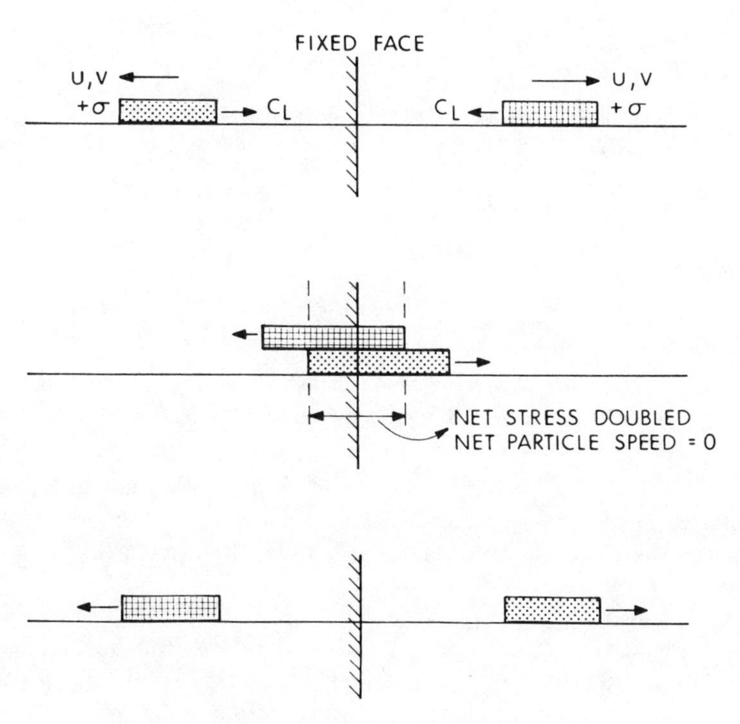

Figure 3 Superposition of tensile pulses.

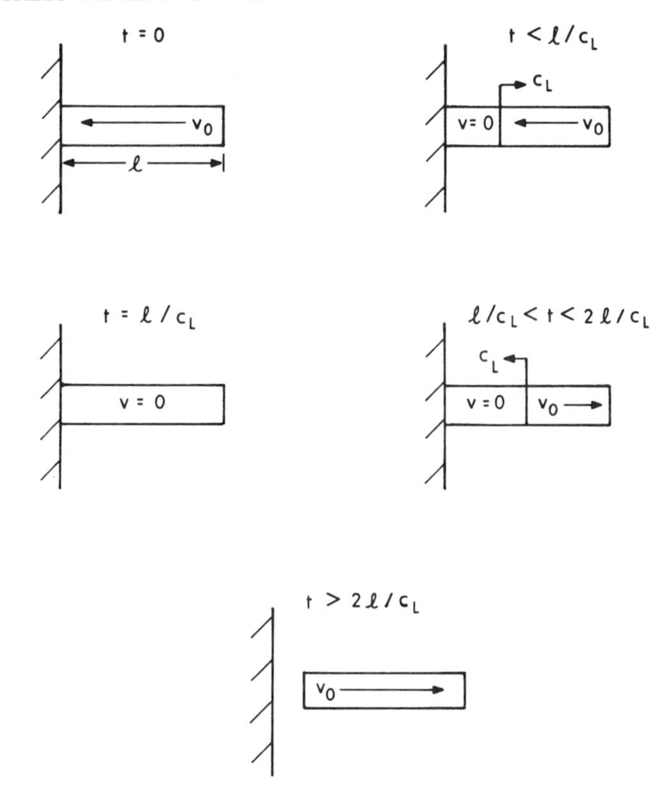

Figure 4 Particle-velocity distribution in impacted bar.

energy has been converted to strain energy:

$$\mathrm{KE}=\tfrac{1}{2}(\text{mass})v_0^2=\tfrac{1}{2}A_0 l\rho v_0^2 \tag{21}$$

$$\mathrm{IE}=(\text{volume})\frac{\sigma^2}{2E}=\frac{A_0 l}{2E}\left(\rho c_L v_0^2\right)=\tfrac{1}{2}A_0 l\rho v_0^2 \tag{22}$$

3 At $t=l/c_L$, the compressive wave is reflected from the rear free surface as a tensile wave. The tensile wave acts as an unloading wave, canceling the effects of the incident compressive wave.

4 At $t=2l/c_L$, the bar is completely stress-free. The reflected tensile wave has conferred a speed v_0 on the particles whose direction is opposite to that at impact. Therefore, at $2l/c_L$, the bar moves away at equal but opposite speed.

Formal mathematical solutions to the above problem can be obtained, for example, from Goldsmith (1960), and Achenbach (1975) but the above technique is quite useful for developing an appreciation of the relationships thus far derived.

Analysis of the transmission of waves in a uniform circular cylinder of finite length is made difficult because of the introduction of the end conditions on the bar. The one-dimensional solutions give results close to reality (except near the ends) so long as the length of the bar is much greater than its diameter. A direct treatment of the problem was achieved by Skalak (1957) who analyzed the impact of two isotropic semiinfinite bars using a superposition technique. Skalak's result, which includes transverse strain effects, is compared to the one-dimensional solution in Figure 5, the principal differences being in a finite decay of the stress pulse and the presence of oscillations due to high frequency components near the head of the pulse. Skalak's solution is also restricted to regions away from the impact end and late times.

An appreciation of the utility of the approximate equations can be gained by examining the results in Figure 6, where the force in a long bar ($L/D=50$) is plotted as a function of axial position at various times. These results were obtained with the EPIC-2 finite element computer code. The two-dimensional axially symmetric equations of motion were solved for the case of a bar of circular cross section ($r=0.254$ cm, $L=2.54$ cm) striking a rigid barrier at 6.1 m/s (20 ft/s). Note that some time is required before the stress state reaches that predicted by the one-dimensional equations.

1.3.3 Stress in Bars with Discontinuous Cross Sections and Different Materials

Consider a bar with discontinuous cross section made of different materials (Figure 7). Assume that a disturbance at the left end of bar 1 has caused an elastic compressive pulse with intensity σ_I to propagate to the right. At the

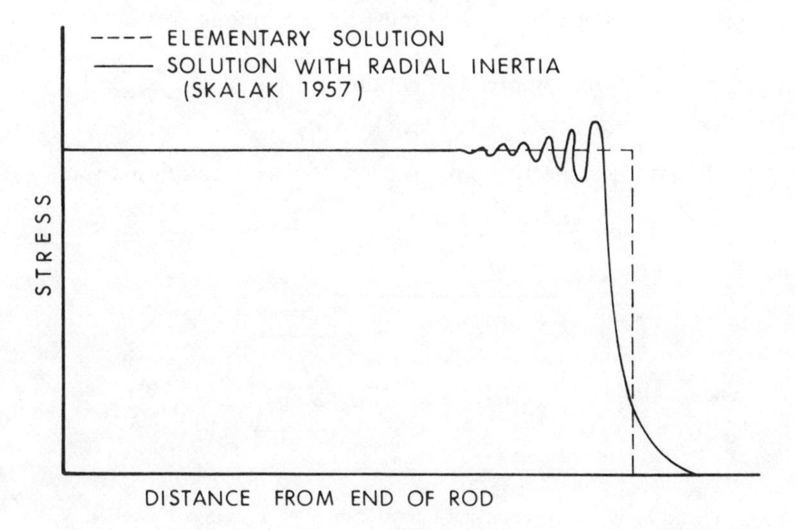

Figure 5 Comparison of exact and elementary solutions for semiinfinite bar.

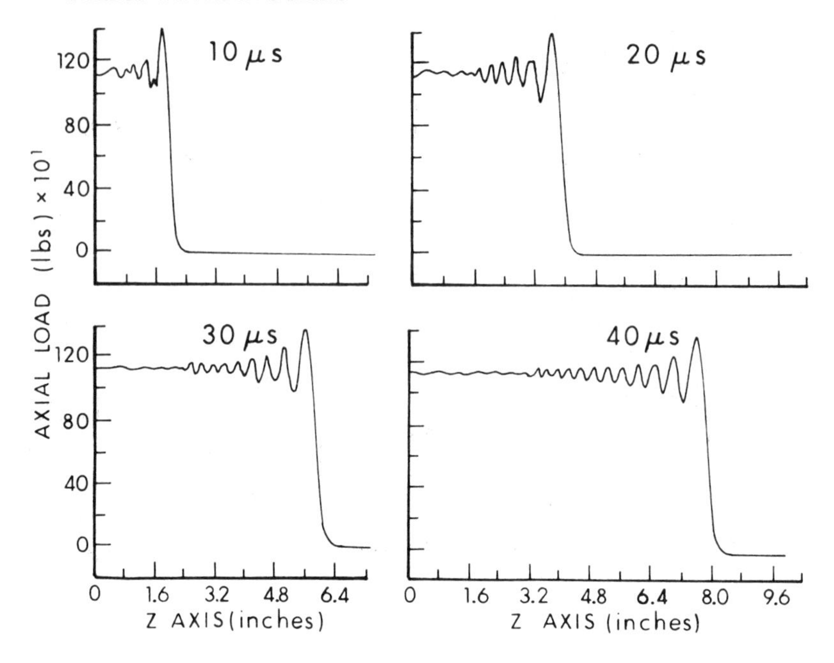

Figure 6 EPIC-2 results at various times for L/D=50, bar-impacting rigid barrier.

interface with bar 2, the wave will be partly transmitted and partly reflected. Call the transmitted wave amplitude σ_T, the reflected wave amplitude σ_R. Two conditions must be satisfied at the interface:

1 The forces in both bars at the interface must be equal.

2 Particle velocities at the interface must be continuous.

Taking σ_R and σ_T to be compressive, condition 1 gives

$$A_1(\sigma_I + \sigma_R) = A_2\sigma_T \tag{23}$$

where A_1, A_2 are the respective cross-sectional areas. Condition 2 gives

$$v_I - v_R = v_T \tag{24}$$

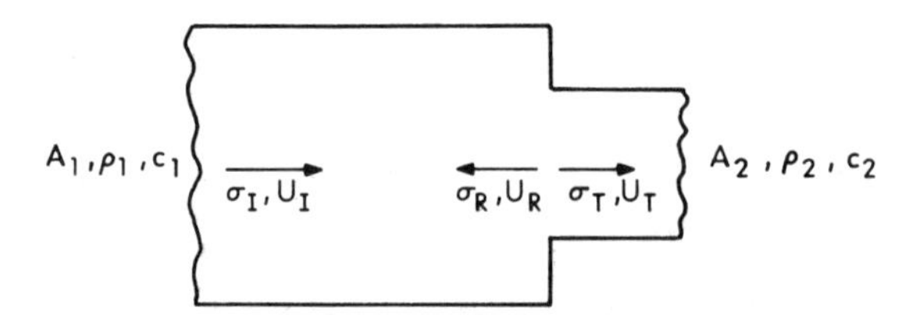

Figure 7 Wave reflection and transmission at changes in cross section.

or, using $\sigma = \rho c v$

$$\frac{\sigma_I}{\rho_1 c_1} - \frac{\sigma_R}{\rho_1 c_1} = \frac{\sigma_T}{\rho_2 c_2} \tag{25}$$

Solving for σ_R, σ_T in terms of σ_I gives

$$\sigma_T = \frac{2A_1 \rho_2 c_2}{A_1 \rho_1 c_1 + A_2 \rho_2 c_2} \sigma_I \tag{26}$$

$$\sigma_R = \frac{A_2 \rho_2 c_2 - A_1 \rho_1 c_1}{A_1 \rho_1 c_1 + A_2 \rho_2 c_2} \sigma_I \tag{27}$$

Consider several implications of the above expressions:

1 If the materials in both bars are identical then $\rho_1 = \rho_2$, $c_1 = c_2$ and

$$\sigma_T = \frac{2A_1 \sigma_I}{(A_1 + A_2)} \tag{28}$$

$$\sigma_R = \frac{(A_2 - A_1)\sigma_I}{(A_1 + A_2)} \tag{29}$$

If $A_2 > A_1$, then σ_T and σ_R will be of the same type. If $A_2 < A_1$, σ_T and σ_R will be of opposite sign.

2 If $A_2/A_1 \to 0$, the rod is effectively free and $\sigma_R \to -\sigma_I$. If $A_2/A_1 \to \infty$, the rod is fixed and $\sigma_R \to \sigma_I$; $\sigma_T \to 0$.

3 For no wave reflection to occur from the discontinuity in the bar,

$$\sigma_R = 0 \therefore A_2 \rho_2 c_2 = A_1 \rho_1 c_1 \qquad \text{and} \qquad \sigma_T = \sigma_I \sqrt{E_2 \rho_2 / E_1 \rho_1} \tag{30}$$

4 In (26) the coefficient of σ_I can never be negative. This means that tension will be transmitted as tension and compression as compression. For a situation wherein $\rho_2 c_2 \gg \rho_1 c_1$ that is, medium 2 is much more rigid than medium 1, the stress of the transmitted pulse is approximately twice the stress of the incident wave.

5 In (27), the coefficient of σ_I can be positive or negative depending on whether $\rho_1 c_1 < \rho_2 c_2$ or $\rho_1 c_1 > \rho_2 c_2$. If the coefficient is negative, an incident compressional stress is reflected as a tensile stress and vice versa. If the coefficient is positive, the incident compressive stress is reflected as a compressive stress. These results are in complete agreement with the laws of conservation of momentum and kinetic energy.

Numerous examples of the application of equations (26)–(27) are given in the book by Johnson (1972) and Rinehart (1960, 1975).

1.4 SHOCK WAVES

Historically, work on shock waves has been done with plate geometries. Plate impact situations generate a state of *uniaxial strain* but three-dimensional stress, whereas in bar experiments a state of *uniaxial stress* was assumed in the one-dimensional approximations. The reason for the change in geometry was the necessity to obtain higher stress amplitudes and higher strain rates. Recall that in the bar impact experiments, plastic flow near the impact end introduced three-dimensional effects (radial inertia, heating) so that one-dimensional theory applied only at points far away from the point of application of the load. With increasing striking velocity a three-dimensional theory is required for complete analysis of experimental results. That, unfortunately, is beyond current capability. Plate geometry offers the opportunity to study materials behavior at higher loads and shorter times while offering again the simplicity of a one-dimensional analysis, this time for uniaxial strain. However, just as bar theories neglected lateral inertia, plate impact analyses neglect effects of thermomechanical coupling, which can be significant at strains exceeding 30% [Lee (1971)]. Much of the initial work assumed hydrodynamic behavior of the material. However, an elastic precursor can produce significant volumetric strain. An elastic unloading wave can significantly change the local state of the material before the arrival of a plastic wave so that finite elastic and plastic effects may need to be accounted for.

The conventional uniaxial stress-strain curve, as depicted by the idealized models of Figure 8, does not adequately represent the state of stress and strain to which a material is subjected under shock loading. Therefore the quantities associated with such curves (elastic modulus, yield strength, ultimate strength, and elongation) are not by themselves appropriate to describe the relative behavior of materials. The following is a brief, simplified summary of shock-wave propagation and relies primarily on the paper by Rolsten (1974) and the report by Oscarson and Graff (1968).

If we visualize a situation where deformation is restricted to *one dimension*, as in the case of plane waves propagating through a material where dimensions and constraints are such that the lateral strains are zero, the characteristic

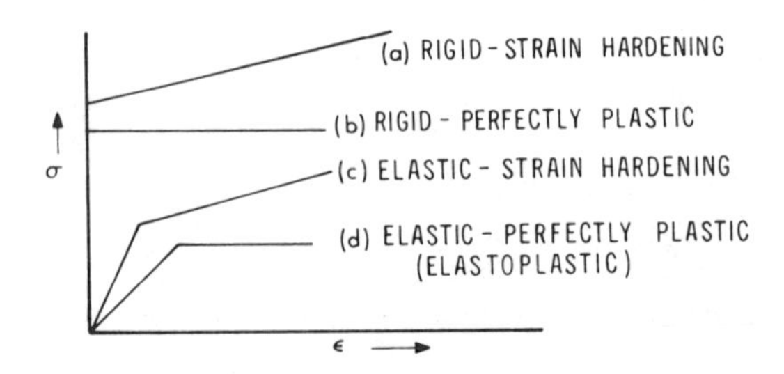

Figure 8 Characteristic uniaxial stress-strain curves.

stress-strain curve takes on the form shown in Figure 9. This situation is commonly referred to as uniaxial strain.

To understand why these changes occur, consider the stress-strain relationship for one-dimensional deformation. In the general case (strains <30%) the three principal strains can be divided into an elastic and a plastic part:

$$\begin{aligned} \varepsilon_1 &= \varepsilon_1^e + \varepsilon_1^p \\ \varepsilon_2 &= \varepsilon_2^e + \varepsilon_2^p \\ \varepsilon_3 &= \varepsilon_3^e + \varepsilon_3^p \end{aligned} \tag{31}$$

Where the superscripts e and p refer to elastic and plastic, respectively, and the subscripts are the three principal directions.

In one-dimensional deformation

$$\varepsilon_2 = \varepsilon_3 = 0 \;\therefore \tag{32}$$

$$\begin{aligned} \varepsilon_2^p &= -\varepsilon_2^e \\ \varepsilon_3^p &= -\varepsilon_3^e \end{aligned} \tag{33}$$

The plastic portion of the strain is taken to be incompressible, so that

$$\varepsilon_1^p + \varepsilon_2^p + \varepsilon_3^p = 0 \tag{34}$$

which gives

$$\varepsilon_1^p = -\varepsilon_2^p - \varepsilon_3^p = -2\varepsilon_2^p \tag{35}$$

since $\varepsilon_2^p = \varepsilon_3^p$ due to symmetry. From (33) we have that

$$\varepsilon_1^p = 2\varepsilon_2^e \tag{36}$$

so that the total strain ε_1 may be written as

$$\varepsilon_1 = \varepsilon_1^e + \varepsilon_1^p = \varepsilon_1^e + 2\varepsilon_2^e \tag{37}$$

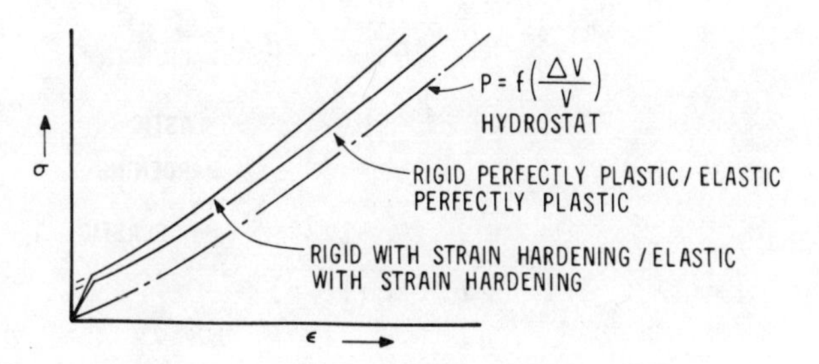

Figure 9 Stress-strain curve for uniaxial strain.

The elastic strain in terms of the stresses and elastic constants is given by

$$\varepsilon_1^e = \frac{\sigma_1}{E} - \frac{\nu}{E}(\sigma_2 + \sigma_3) = \frac{\sigma_1}{E} - \frac{2\nu}{E}\sigma_2 \qquad (\text{since } \sigma_2 = \sigma_3)$$

$$\varepsilon_2^e = \frac{\sigma_2}{E} - \frac{\nu}{E}(\sigma_1 + \sigma_3) = \frac{(1-\nu)}{E}\sigma_2 - \frac{\nu}{E}\sigma_1$$

$$\varepsilon_3^e = \frac{\sigma_3}{E} - \frac{\nu}{E}(\sigma_1 + \sigma_2) = \frac{1-\nu}{E}\sigma_3 - \frac{\nu}{E}\sigma_1 \tag{38}$$

Combining (38) with (37) gives

$$\varepsilon_1 = \frac{\sigma_1(1-2\nu)}{E} + \frac{2\sigma_2(1-2\nu)}{E}$$

The plasticity condition for either the von Mises or Tresca conditions for this case is

$$\sigma_1 - \sigma_2 = Y_0 \tag{39}$$

Using (39) as the definition for σ_2 and inserting into (38) gives $(E/1-2\nu)\varepsilon_1 = 3\sigma_1 - 2Y_0$ or

$$\sigma_1 = \frac{E}{3(1-2\nu)}\varepsilon_1 + \frac{2}{3}Y_0 = K\varepsilon_1 + \frac{2Y_0}{3} \tag{40}$$

where $K = E/3(1-2\nu)$ is called the bulk modulus.

The most important difference between uniaxial stress and uniaxial strain is the bulk compressibility term—the stress now continues to increase regardless of the yield strength or strain hardening. For ballistic impact or other high-rate phenomena where the material does not have time to deform laterally, a condition of uniaxial strain will initially occur. Later on, as lateral deformation takes place, a condition approaching uniaxial stress may occur and the stress will decrease.

For the special case of elastic one-dimensional strain

$$\varepsilon_1 = \varepsilon_1^e$$

$$\varepsilon_2 = \varepsilon_2^e = \varepsilon_3 = \varepsilon_3^e = 0$$

$$\varepsilon_1^p = \varepsilon_2^p = \varepsilon_3^p = 0$$

$$\varepsilon_2^e = 0 = \frac{1-\nu}{E}\sigma_2 - \frac{\nu}{E}\sigma_1$$

$$\sigma_2 = \left(\frac{\nu}{1-\nu}\right)\sigma_1$$

and

$$\varepsilon_1 = \frac{\sigma_1}{E} - \frac{2\nu^2\sigma_1}{E(1-\nu)}$$

or

$$\sigma_1 = \frac{1-\nu}{(1-2\nu)(1+\nu)}E\varepsilon_1 \tag{41}$$

Equation (41) shows that the slope of the elastic line in one-dimensional strain is $(1-\nu)E/[(1-2\nu)(1+\nu)]$. When very high pressures are considered, the pressure-compressibility curve of Figure 9 [$P=f(\Delta V/V)$ where $\Delta V/V$ is the volumetric strain], also known as the Hugoniot curve, is the only one that is considered to describe the behavior of a material. At lower pressures, such as those generated by conventional impacts, considerable deviation from the Hugoniot occurs. For example, the uniaxial strain curve corresponding to the uniaxial stress condition for an elastic, perfectly plastic material (Figure 10) is shown in Figure 11. Note there:

1 The increase in modulus by a factor $(1-\nu)/[(1-2\nu)(1+\nu)]$.
2 The Hugoniot elastic limit σ_{HEL}, the maximum stress for one-dimensional elastic wave propagation (uniaxial strain).
3 The constant deviation from the Hugoniot of the stress σ_1 by $2Y_0/3$, where Y_0 is the static yield strength. If the yield strength changes in a strain-

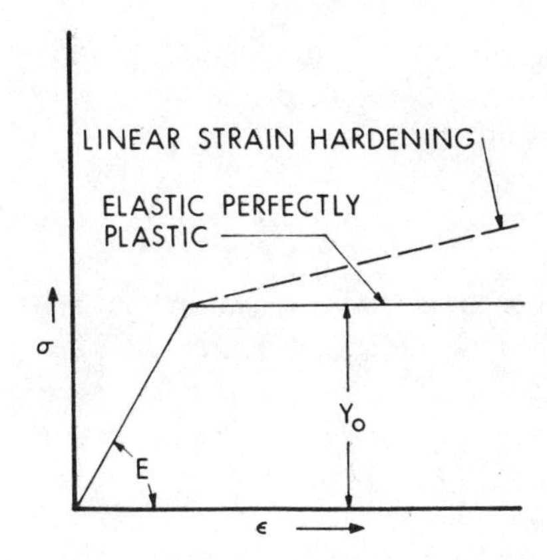

Figure 10 Uniaxial stress state for elastic, perfectly plastic material.

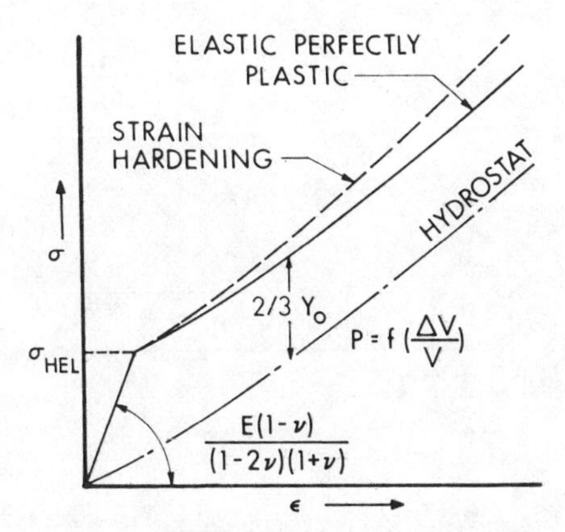

Figure 11 Uniaxial strain curve for elastic, perfectly plastic material.

hardening material, so will the difference between the σ_1 and P curves. A typical loading cycle in uniaxial strain for an elastic, perfectly plastic material is shown in Figure 12. Note that the reverse yielding occurs at point C. If reverse loading occurs, as in stress-wave reflection from a free surface, the line segment CD extends to the negative (tension) region below the strain axis but again different by $2Y_0/3$ from the hydrostat (Hugoniot), assuming tensile and compressive yield strengths are equal.

If the magnitude of the applied stress pulse is above σ_{HEL}, two waves will propagate through the medium, the elastic wave moving with speed

$$c_E^2 = \frac{E(1-\nu)}{\rho_0(1-2\nu)(1-\nu)} \tag{42}$$

followed by a plastic wave moving with speed

$$c_p^2 = \frac{\sigma_B - \sigma_{\text{HEL}}}{\rho_{\text{HEL}}(\varepsilon_B - \varepsilon_A)} \tag{43}$$

In a later discussion of plastic wave propagation it will be shown that the speed of the wave is a function of the slope of the stress-strain curve at a given value of strain and that it can be expressed as

$$c_p(\sigma) = \sqrt{\frac{1}{\rho_0}\frac{d\sigma}{d\varepsilon}} \tag{44}$$

If the applied stress is of finite duration, an elastic unloading wave is generated after removal of the load (Figure 13). The unloading wave travels

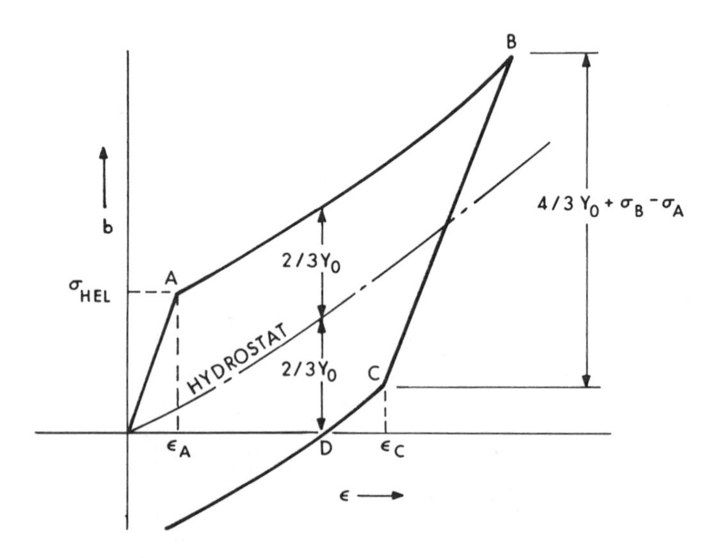

Figure 12 Loading cycle in uniaxial strain.

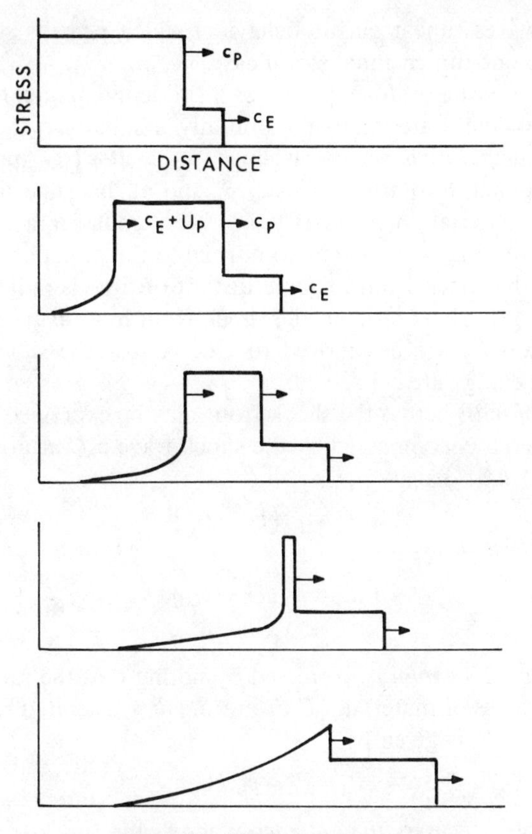

Figure 13 Plastic-wave attenuation.

faster than the compressive wave so that for a short duration pulse the compressive amplitude may be attenuated by unloading from the rear. The point at which this unloading occurs is called the catch-up distance and is usually defined in terms of the incident pulse thickness.

If we have a situation where $c_p(\sigma) > c_E$, conditions have been created for the formation of a steep plastic front. The more rapidly traveling stress components overtake the slower ones. The continuous plastic wave front breaks down and a single discontinuous shock front is formed traveling at a shock velocity U. Across the shock front, there is a discontinuity in stress, density, velocity, and internal energy. Shock waves will form under conditions of extremely high impulsive stress and will propagate in a material in a manner similar to the fluid dynamics situation. It becomes reasonable to consider the solid as behaving like a compressible fluid described by an equation of state (in general a relationship between pressure, density, and internal energy) in the form $P = f(V)$ where $V = 1/\rho$, the specific volume. Shock-wave propagation has inherent simplifying features (simplified equation of state) analogous to the case of elastic waves (linear elastic behavior) which permits some simplified solutions for the one-dimensional strain case.

Consider the case of a uniform pressure P_1 suddenly applied to one face of a plate of compressible material that is initially at pressure P_0. This pulse is propagated by means of a wave traveling at velocity U_s. Application of P_1 compresses the material to a new density ρ_1 and at the same time accelerates the compressed material to a velocity U_p. Now consider a segment of the material (with unit cross-sectional area) normal to the direction of wave travel. The position of the shock front at some instant of time is indicated in Figure 14 by the line AA. A short dt later, the shock front has advanced to BB while the matter initially at AA has moved to CC. Across the shock front, mass, momentum, and energy are conserved.

Conservation of mass across the shock front may be expressed by noting that the mass of material encompassed by the shock wave $\rho_0 U_s\, dt$ now occupies the volume $(U_s - U_p)\, dt$ at density ρ_1

$$\rho_0 U_s = \rho_1 (U_s - U_p)$$

or

$$V_1 U_s = V_0 (U_s - U_p) \tag{45}$$

where $V = 1/\rho$.

Conservation of momentum is expressed by noting that the rate of change of momentum of a mass of material $\rho_0 U_s\, dt$ in time dt accelerated to a velocity U_p by a net force $P_1 - P_0$ is given by

$$P_1 - P_0 = \rho_0 U_s U_p \tag{46}$$

The expression for *conservation of energy* across the shock front is obtained by equating the work done by the shock wave with the sum of the increase of both kinetic and internal energy of the system. Thus

$$P_1 U_p = \tfrac{1}{2}\rho_0 U_s U_p^2 + \rho_0 U_s (E_1' - E_0') \tag{47}$$

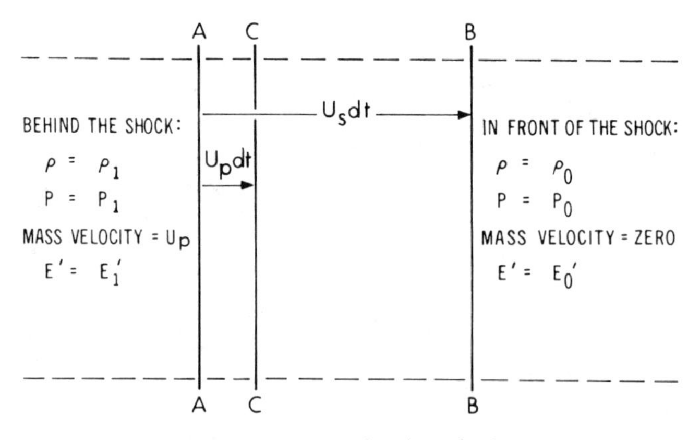

Figure 14 Progress of a plane shock wave.

Equations (45)–(47) contain a total of 8 parameters (ρ_0, ρ_1, P_0, P_1, U_s, U_p, E_1', E_0'). If it is assumed that ρ_0, P_0, E_0' are known, three equations with five unknowns remain. By eliminating U_s and U_p through (46) and (47), the resulting equation is known as the Rankine-Hugoniot relation

$$E_1' - E_0' = \tfrac{1}{2}(V_0 - V_1)(P_1 + P_0) = \frac{1}{2}\left(\frac{1}{\rho_0} - \frac{1}{\rho_1}\right)(P_1 + P_0) \tag{48}$$

Equations (45) (46), and either (47) or (48) are the "jump conditions" that must be satisfied by material parameters on the two sides of a shock front. Hence the states (E_1', ρ_1, P_1) that can be achieved from an initial state (E_0', ρ_0, P_0) have been identified.

Elimination of the particle velocity from the mass and momentum equations results in an expression for the shock velocity of the form

$$U_s^2 = \frac{1}{\rho_0^2} \frac{P_1 - P_0}{V_0 - V_1} \tag{49}$$

with $V_0 = 1/\rho_0$; $V_1 = 1/\rho_1$.

Changes in pressure, density, and internal energy across a shock front can be calculated by measurement of just two parameters, the shock velocity U_s and the particle velocity U_p. The locus of pressure-density states that are attainable by shock loading from a single initial state is called the Rankine-Hugoniot curve of a material, or simply the Hugoniot. The term is frequently applied to those curves representing any two of the five variables—pressure, density, internal energy, shock velocity, and particle velocity. The assumptions made in the development of the Hugoniot were one-dimensional motion, thermodynamic equilibrium ahead of and immediately behind the shock front, and neglect of material rigidity. The first can generally be met experimentally, the second will hold if thermodynamic equilibrium is achieved within a few tenths of a microsecond after passage of a shock, and the third is justified for extremely high pressures (some two orders of magnitude above the yield strength of the material).

The Hugoniot curve for a solid is often fit to experimental data in the straight line form

$$U_s = a + bU_p \tag{50}$$

The physical reason for the good agreement with the linear relationship is not well understood. The constant a represents the wave velocity in an extended medium. The constant b is related to the Gruneisen parameter at low pressure through

$$b = \frac{1 + \Gamma}{2}$$

where $\Gamma = 3\alpha K/\rho_0 C_v$ with

α = coefficient of linear expansion

K = bulk modulus

C_v = specific heat at constant volume

ρ_0 = initial density

When Hugoniot data for a particular material are not available, first approximations may be generated from the bulk sound speed (a) and the value of Γ at zero pressure.

A much more comprehensive treatment of shock waves is to be found in Duvall (1972; 1971; 1961), and Duvall & Fowles (1963), Murri et al. (1974), Rinehart (1975), and Seigel (1977).

Having these relationships and a constitutive equation for the material as well as the conditions of impact, it is possible to infer the stress history within the interior of the material. The inverse problem, in which the material properties are to be found from the plate impact test, is equally important. The procedure for determining a material constitutive equation assuming that both impact conditions and experimental data on the resulting motion of the target free surface are known is outlined in Chapter 4. For a more complete treatment, see Karnes (1968), Oscarson & Graff (1968), and Rolsten (1974).

REFERENCES

Achenbach, J. D. (1975), *Wave Propagation in Elastic Solids*, American Elsevier, New York.

Duvall, G. E. (1961), in P. G. Shewmon and V. F. Zackay, (Eds.), *Response of Metals to High Velocity Deformation*, Interscience, New York.

Duvall, G. E. (1971), in P. Caldirola and H. Knoepfel, (Eds.), *Physics of High Energy Density*, Academic Press, New York.

Duvall, G. E. (1972), in P. C. Chou and A. K. Hopkins, (Eds.), *Dynamic Response of Materials to Intense Impulsive Loading*, U.S. Government Printing Office, Washington, D.C.

Duvall, G. E., and G. R. Fowles (1963), in R. S. Bradley, (Ed.), *High Pressure Physics and Chemistry*, Vol. 2, Academic Press, New York.

Eringen, A. C., and E. S. Suhubi, (1975), *Elastodynamics*, Vols. I and II, Academic Press, New York.

Ewing, M., Jardetzky, W. S., and Press, F. (1957), *Elastic Waves in Layered Media*, McGraw-Hill, New York.

Goldsmith, W. (1960), *Impact*, Edward Arnold, London.

Hopkins, H. G. (1964), in H. Kolsky and W. Prager, (Eds.), *Stress Waves in Anelastic Solids*, Springer-Verlag, Berlin.

Johnson, W. (1972), *Impact Strength of Materials*, Crane, Russak, New York.

Karnes, C. H. (1968), in U. S. Lindholm, (Ed.), *Mechanical Behavior of Materials Under Dynamic Loads*, Springer-Verlag, New York.

Kolsky, H. (1963), *Stress Waves in Solids*, Dover, New York.

Lee, E. H. (1971), in J. J. Burke and V. Weiss, (Eds.), *Shock Waves and the Mechanical Properties of Solids*, Syracuse University Press, Syracuse, N.Y.

Murri, W. J., et al. (1974), in R. H. Wentorf, Jr., (Ed.), *Advances in High Pressure Research*, Vol. 4, Academic Press, New York.

Oscarson, J. H., and Graff, K. F. (1968), Battelle Memorial Institute, BAT-197A-4-3 (AD 669440).

Rinehart, J. S. (1960), Air Force Special Weapons Center, AFSWC-TR-60-7 (AD 236719).

Rinehart, J. S. (1975), *Stress Transients in Solids*, HyperDynamics, Santa Fe, N.M.

Rolsten, R. F. (1974), *Trans. N.Y. Acad. Sci.*, **36**, 416.

Seigel, A. E. (1977), in I. L. Spain and J. Paauwe, (Eds.), *High Pressure Technology*, Vol. II. *Applications and Processes*, Marcel Dekker, New York.

Skalak, R. (1957), *J. Appl. Mech., Trans. ASME*, **24**, 59.

Wasley, R. J. (1973), *Stress Wave Propagation in Solids: An Introduction*, Marcel Dekker, New York.

Whitham, G. B. (1974), *Linear and Nonlinear Waves*, Wiley, New York.

Wood, D. (1963) in P. G. Shewmon and V. F. Zackay, (Eds.), *Response of Metals to High Velocity Deformation*, Interscience, New York.

2

LIMITATIONS OF ELEMENTARY WAVE THEORY

Jonas A. Zukas

The elementary analysis of one-dimensional wave motion presented in Chapter 1 is quite adequate for interpreting many aspects of the elastic behavior of solids under impact loading. In this chapter, the limitations of the elementary approach are exposed through considerations of selected problems of practical interest.

2.1 LONGITUDINAL WAVES IN FINITE LENGTH BARS

As noted in Chapter 1, the elementary theory for wave propagation in circular bars is valid only if the ratio of the bar diameter to the transmitted wavelength is small. These approximate theories can be misleading if a stress pulse contains a wide band of various frequencies. A refinement to the elementary theory was developed by both Love and Rayleigh [Johnson (1972)]. If radial inertia is taken into account, then the equation for one-dimensional wave propagation in rods becomes

$$\frac{E}{\rho}\frac{\partial^2 u}{\partial x^2}=\frac{\partial^2 u}{\partial t^2}-\frac{\nu^2 r^2}{2}\frac{\partial^4 u}{\partial x^2 \partial t^2} \tag{1}$$

where ν is Poisson's ratio, r the radius of the bar, and the other quantities as in (9) of Chapter 1. A solution may be obtained for the relationship between the bar and the phase velocities

$$\frac{c_p}{c_L} = 1 - \nu^2\pi^2\left(\frac{r}{\lambda}\right)^2 \tag{2}$$

where λ represents wavelength and c_L is the bar velocity $\sqrt{E/\rho}$. Equation (2) indicates that in a long bar the speed at which waves are propagated depends on their frequency. Each frequency has its own phase velocity c_p so that a pulse containing a mixture of frequencies will be dispersed.

Conway and Jakubowski (1969) used Love's theory to analyze experimental results of the axial impact of short-length steel bars (length-to-diameter ratios of 4 to 24). Axial strains and durations of impact were calculated using Love's equation for the coaxial impact of identical bars. Axial strains at various locations from the impact end were also computed assuming step- and ramp-pressure loadings at one end. Conway and Jakubowski found that the duration of impact was slightly larger than that predicted by classical theory ($2l/c_L$) and that the discrepancy between experimental results and the classical theory increased as the bars were shortened. The strain-time curves showed finite rise and decay times as well as oscillations that are characteristic of radial inertia effects. Comparison between experimental and analytical results indicated that Love's equation is satisfactory provided the end loading is correctly simulated.

Because analytical solutions can be obtained for only a few idealized situations, many bar impact studies have been performed using numerical methods. Bertholf (1966, 1967) studied the behavior of short (length-to-diameter ratios of 1 and 2) aluminum bars subjected to continuous and step loading. The solution for an applied harmonic displacement closely approached the Pochhammer-Chree solution at distances removed from the point of application of the load. Solutions for a step change in stress agreed with experimental data near the end of the bars (both free- and lubricated-end conditions were considered) and exhibited a region that agreed with the classical solution. Ramamurti and Ramanamurti (1977) developed a finite element method based on triangular elements to study impact of solid and hollow finite-length bars. Analytical results compared favorably with experimental data for aluminum, mild steel, and araldite bars.

The differences between classical and numerical solutions for short bars can be seen in Figures 1–3. The computations were performed with the EPIC-2 finite element program which uses triangular elements and integrates the equations of motion directly, bypassing the stiffness matrix. The material model in EPIC-2 is similar to that in its three-dimensional counterpart, EPIC-3, which is described in Chapter 11. The figures depict the force in a steel bar, four diameters long, impacting a rigid surface at 10 m/s. Superimposed is the classical one-dimensional solution [(9) of Chapter 1]. The effects of radial inertia are easily seen in the oscillations in the force curve. It should be evident that while the classical solution provides some insight into the impact behavior of bars, more refined treatments are required when the conditions

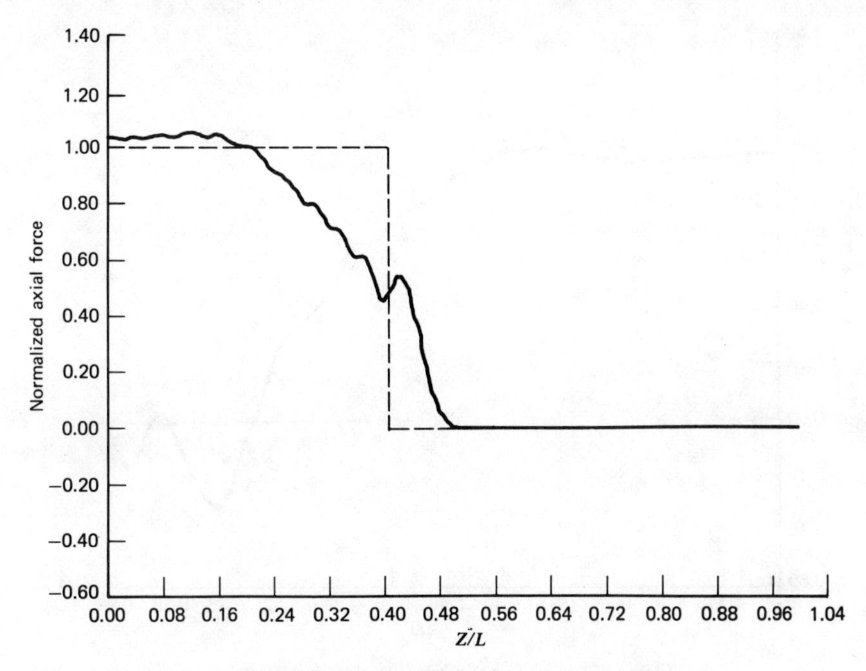

Figure 1 Axial force in bar, 4 μs after impact.

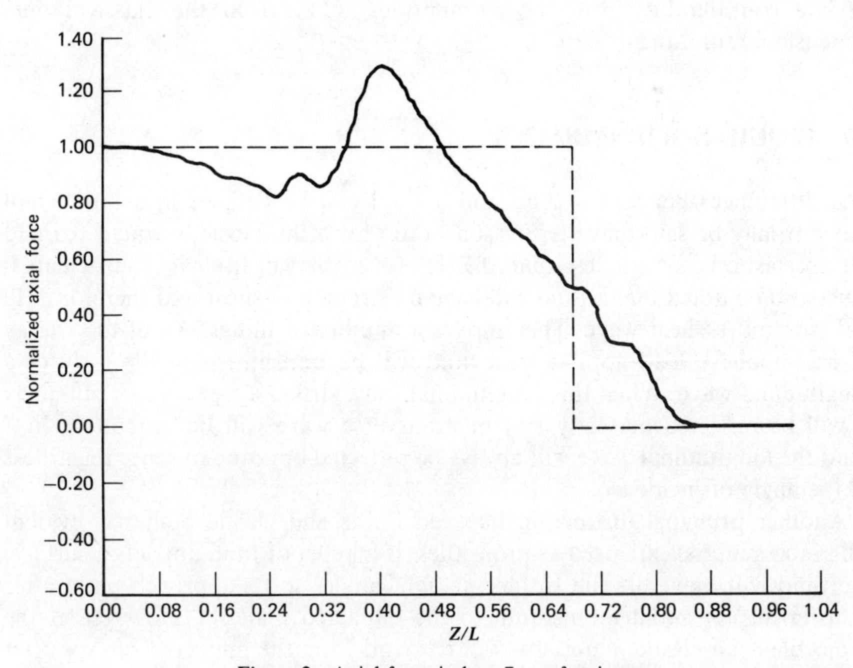

Figure 2 Axial force in bar, 7 μs after impact.

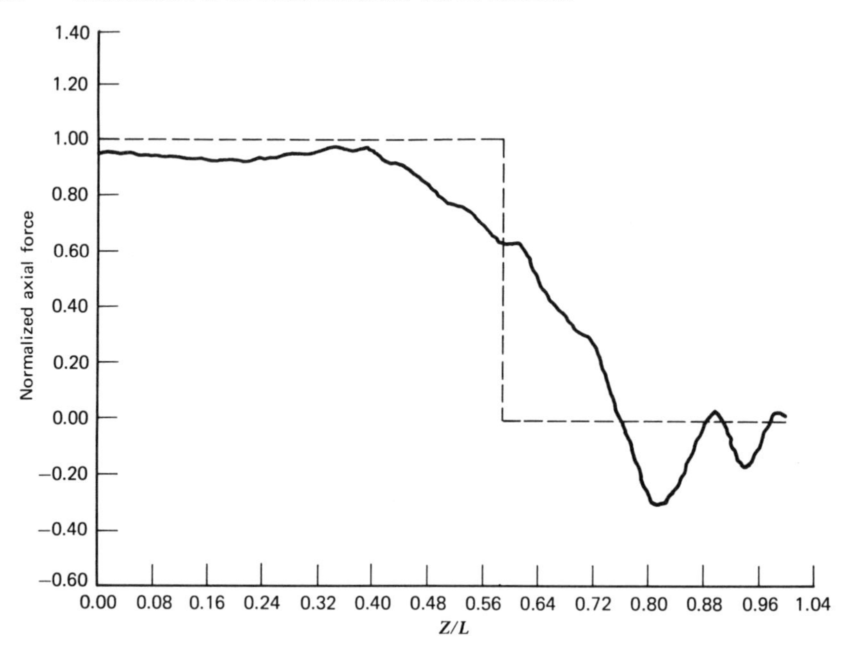

Figure 3 Axial force in bar, 14 μs after impact.

deviate considerably from the assumptions inherent in the classical one-dimensional solutions.

2.2 LIQUID-SOLID IMPACTS

The differences between a solid and a fluid can be defined in a number of ways. It may be said that the Poisson's ratio for a fluid exactly equals 0.5 and for an elastic body it is less than 0.5. However, that in itself is insufficient. It must also be noted that a fluid will have no strength in shear and therefore will not transmit a shear wave. This implies a number of things. All of the energy of an impulsive load applied to a fluid will be transmitted in the form of a longitudinal wave. When this longitudinal wave strikes a free surface obliquely it will be reflected perfectly and no transverse wave will be generated. In a fluid the longitudinal wave will always be reflected opposite in sense regardless of the angle of incidence.

Another principal distinction between fluids and elastic bodies is evident when the materials are used as projectiles. If a pellet of fluid impacts a surface, the fluid will move off the surface at right angles to its approach path, after transferring its initial momentum to the impacted material. However, if the projectile is an elastic material it will rebound from the impacted surface after transferring to the impacted material more than its initial momentum.

Fluid-solid impact situations are of interest because they can easily give rise to stress waves that can cause failure in brittle materials or plastic flow in ductile ones. Solid-liquid impacts can cause erosion on fast-moving turbine blades. Forward regions of high speed aircraft such as radomes and plastic observation windows can be damaged on passage through a rain cloud. Many other applications and descriptions of current efforts to understand erosion phenomena are to be found in the recent books by Springer (1976), Preece (1979), and Adler (1979*a*). Particularly relevant to the discussion that follows are the review articles by Adler (1979*b*) and Brunton and Rochester (1979), which deal with the mechanics of liquid impact.

The type and extent of damage that occurs in liquid-solid impact depends primarily on the size, density, and velocity of the liquid and on the strength of the solid. At a velocity of 750 m/s a 2 mm diameter water drop is capable of fracturing and eroding tungsten carbide and plastically deforming martensitic steels. At velocities near 200 m/s, a single impact may produce no visible damage but repeated impacts will bring about erosion of the solid. The resulting weight loss in a solid does not occur at a constant rate but can be divided into several stages [Brunton and Rochester (1979)]. There is an initial incubation period during which there is no detectable weight loss, although some surface deformation may be noted. This is followed by a period of rapidly increasing material removal as cracks and voids coalesce. Finally, the erosion rate levels off and the weight loss continues at an approximately constant rate. A finer subdivision of this behavior into six stages is given by Adler (1979*b*).

Brunton and Rochester (1979) also summarize the various investigations to relate impact and material variables to erosion rate. No general relationships exist between impact variables (velocity, drop size and shape, impact angle) and erosion damage which would be valid for a wide range of materials. However, empirical relationships are available that show fair agreement with experimental data.

The relationship between erosion rate and impact velocity typically takes the form

$$E \propto (v - v_c)^n \tag{3}$$

where E represents erosion rate, v the impact velocity, v_c the impact velocity below which there is no mass loss, and n a constant dependent on the materials involved. This relationship fits erosion rate data in the second stage described above for many materials. But as with all empirical relationships, it cannot be safely extrapolated to conditions (impact velocities, materials) other than those for which measurements were made.

A similar expression is available to relate the effect of impact angle on erosion. It has been found that for smooth hard surfaces, the normal component of impact velocity is mainly responsible for damage. The tangential component has little effect. Hence the erosion rate expression becomes

$$E \propto (v \cos\theta - v_c)^n \tag{4}$$

where θ is the angle between the normal to the surface and the velocity vector. For roughened surfaces or surfaces that undergo considerable deformation during an impact, the tangential component contributes significantly to damage and the dependence on θ is more complicated. An equation of the type

$$E \propto (v \cos\theta - v_c)^n \sec\theta \tag{5}$$

has been proposed where $\sec\theta$ was introduced to account for the effect of the tangential velocity component on erosion [Brunton and Rochester (1979)].

There have been a number of investigations on effects of drop size. It has generally been found that erosion decreases with drop size when drop diameters are less than 1 mm and is independent of drop size for larger use. Most investigators seem to have adopted a 2 mm drop as a standard size.

Attempts to correlate erosion resistance with a single material property have been unsuccessful to date. Empirical relationships exist that relate erosion resistance to hardness or strain energy to fracture. Neither however provide a satisfactory measure of erosion resistance for the range of engineering materials of interest.

A number of elementary analyses of the interface pressure magnitude have been put forward over the years [Adler (1979*a*, *b*), Brunton and Rochester (1979), Johnson (1972)]. These tend toward the conclusion that the peak pressure generated in the contact area can be found from the water hammer equation

$$P = \rho c_w v \tag{6}$$

where v is the striking velocity, ρ the density and c_w the sound speed in water, which is close to 1500 m/s. Equation (6) holds if the surface is perfectly rigid. Otherwise, for a compressible solid

$$P = v \frac{(\rho c)_w \cdot (\rho c)_s}{(\rho c)_w + (\rho c)_s} \tag{7}$$

where the subscripts s and w represent solid and water, respectively. It is well established however that the peak impact pressure can exceed the water hammer pressure by a considerable amount. This initial compression is progressively relieved until the maximum pressure decays to the stagnation pressure for incompressible flow, that is, $\rho v^2/2$.

Adler (1979*b*) reviews the various approximations that have been made to the general physical problem of fluid-solid impact. From a mathematical standpoint, the situation is difficult because the governing equations and boundary conditions are coupled. Existing analyses can be grouped in four categories:

1. One-dimensional approximations of the interface pressure magnitude between drop and solid.
2. Evaluation of interfacial pressure and flow of a drop on a rigid surface.

3 Evaluation of the stresses and deformations in a solid acted on by a pressure distribution assumed representative for the drop-solid impact at some point in the process.

4 Computation of flow patterns, pressure distributions, and transient stress states developed during collision of a liquid drop and a deformable surface without a priori assumptions regarding the interfacial pressure distribution.

Numerical work in this area is reviewed by Adler (1979*b*) and Brunton and Rochester (1979). Finite difference and finite element schemes have provided insight into the transient nature of water-drop collision and the stress distributions in the target. The removal of material cannot be handled as yet because of the inadequacy of the material failure models contained in current continuum mechanics codes. From the diversity of computational results, it is also evident that a systematic study of the effects of grid size, computational element, artificial viscosity, and the various equation of state formulations for water on parameters of interest must be made before further progress can be expected.

A number of studies have been made of the collapse of a water drop as it strikes a rigid plane. The overall impact sequence is shown in Figures 4–6 for a 1 mm spherical water droplet striking a rigid barrier at 335 m/s. When the water drop impacts the barrier, the contact zone expands while the pressure builds to a maximum. Note that the peak pressure exceeds the water hammer by a factor of 1.2 and occurs at the periphery of the contact area. Once the pressure buildup phase is completed, outflow jetting begins, the pressure drops rapidly, and the drop collapses onto the surface. A stagnation region forms at the center of the drop during the jetting phase. Jetting of the drop occurs at velocities some 4–5 times greater than the impact velocity (Figure 6). The calculations were performed with the EPIC-2 finite element computer program using 496 nodes and 900 elements.

2.3 FRACTURE WITH STRESS WAVES

Problems of material failure near a free surface remote from the area of application of the impulsive load have been studied extensively. Many examples are to be found in Rinehart (1960*a*, *b*; 1975).

For materials strong in compression but weak in tension (rocks, concrete), scabbing or spalling at free surfaces are phenomena to be expected due to reflection of incident-compressive impulses generated by explosives or high speed impact. The problem of spallation is discussed by Oscarson and Graff (1968), Mescall and Papirno (1974), Rinehart (1975), and Kinslow (1976) in some detail. The following simple treatment [Rinehart (1960*a*)] will serve to illustrate the basic principles.

The impulse generated by explosives will be treated here as a triangular (sawtoothed) pulse. Other possible shapes and their treatment are discussed at

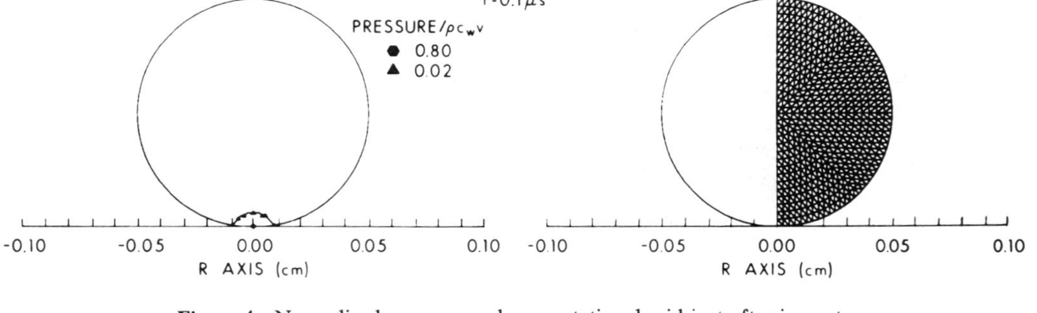

Figure 4 Normalized pressure and computational grid just after impact.

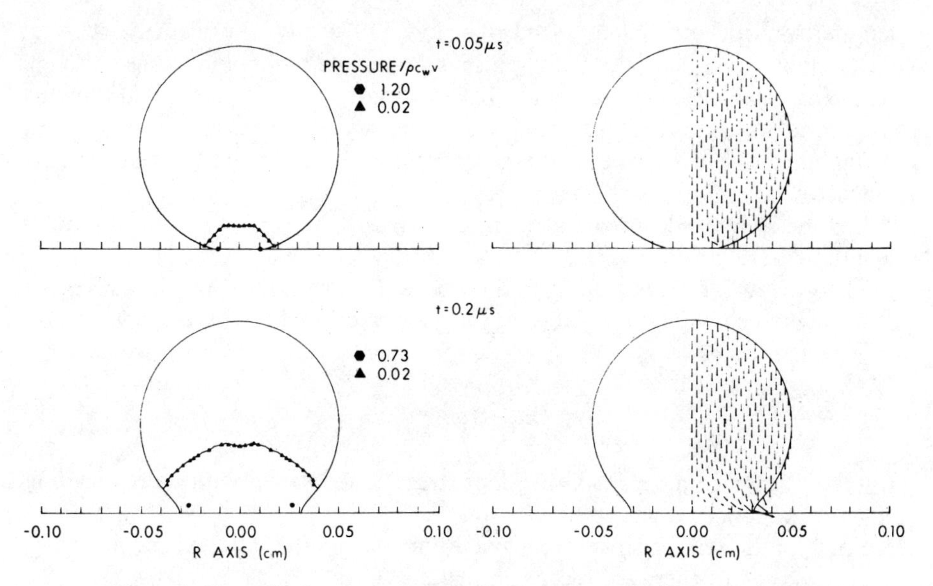

Figure 5 Normalized pressure and velocity vectors at early times after impact.

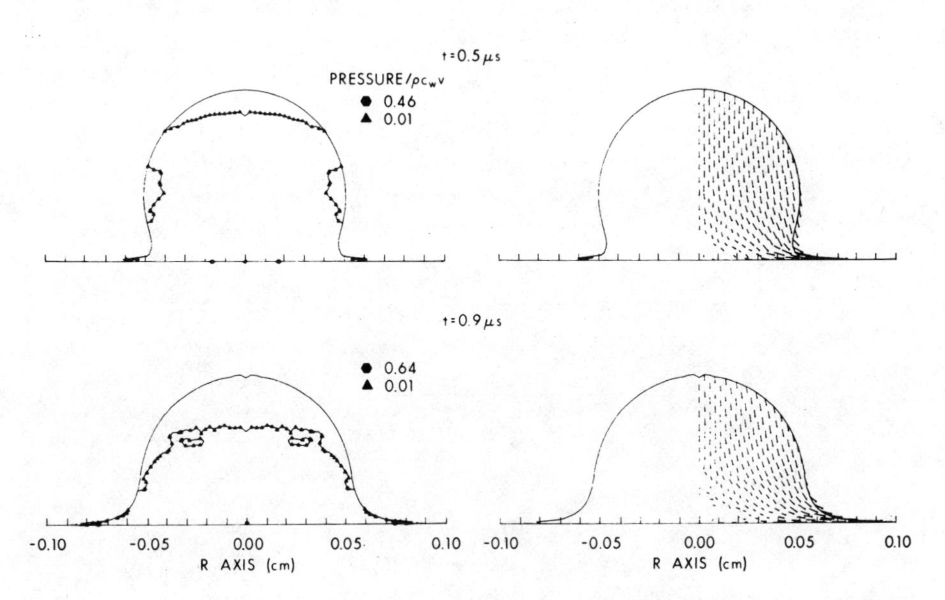

Figure 6 Jetting of water drop.

length by Rinehart (1960*a*) and Kinslow (1976). In the elementary one-dimensional treatment, the stress pulse is propagated without change in shape or intensity. It is further assumed that if a material undergoes fracture when the tensile stress reaches some critical value, say σ_F, then fracture will occur instantaneously in a plane of the geometry being considered where σ_F is first reached.

Let the magnitude of an incident-compressive pulse be σ_m and its pulse length be λ (Figure 7).

The wave will be reflected from the free surface with a net-maximum tensile stress σ_T, which will always occur at the leading edge of the reflected wave and can be defined by

$$\sigma_T = \sigma_m - \sigma_I \tag{8}$$

where σ_I is the compressional-incident stress at the same point as the leading edge of the reflected wave.

Equation (8) will be true at any given time during the reflection of the wave.

Now if $\sigma_m > \sigma_F$ there will occur an instant when

$$\sigma_T = \sigma_F \tag{9}$$

and fracture will form. At this instant

$$\sigma_F = \sigma_m - \sigma_I \tag{10}$$

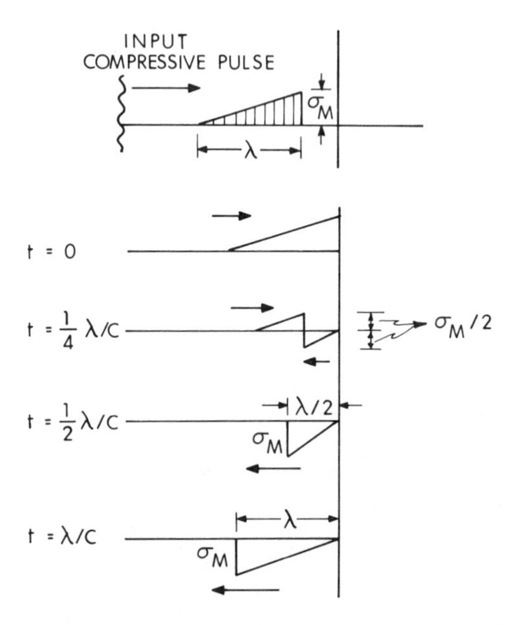

Figure 7 Net stress for reflection of sawtooth pulse at various times [Rinehart (1960a)].

where σ_I can now be defined by geometry as

$$\frac{\sigma_I}{\lambda-2t_1}=\frac{\sigma_m}{\lambda} \tag{11}$$

or

$$\sigma_I=\sigma_m\left(\frac{\lambda-2t_1}{\lambda}\right) \tag{12}$$

Substituting (10) into (8) to eliminate σ_I gives

$$\sigma_F=\sigma_m-\sigma_m\left(\frac{\lambda-2t_1}{\lambda}\right) \tag{13}$$

Now solve for the spall thickness t_1 to get

$$t_1=\frac{\sigma_F}{\sigma_m}\frac{\lambda}{2} \tag{14}$$

If $\sigma_m=\sigma_F$, fracture occurs at a distance $\lambda/2$ from the rear face if it occurs at all. If $\sigma_m<\sigma_F$, no fracture will occur. If $\sigma_m\gg\sigma_F$, multiple fractures can occur.

Assume the latter is the case. We now have a new wave and a new free surface where $\lambda_2=\lambda-2t_1$, and $\sigma_{m_2}=\sigma_I$. Repeating the above process it can be shown that

$$t_2=\frac{\sigma_F}{\sigma_{m2}}\frac{\lambda_2}{2} \tag{15}$$

Further repetition will show that

$$t_n=\frac{\sigma_F}{\sigma_{mn}}\frac{\lambda_n}{2}$$

where the subscript denotes the number of the spall.

To find the velocity imparted to the first spall relate the wave to time rather than distance by $\tau=\lambda/c$ and apply the principle that impulse equals momentum.

The impulse imparted to the spall is the area under that part of the wave trapped by the spall. This impulse can be equated to the momentum of the spall,

$$\frac{\sigma_m+\sigma_I}{2}\frac{2t_1}{c}A=\rho t_1 v_{t_1}A \tag{16}$$

where A is the cross-sectional area involved.

Substituting for σ_I and solving for v_{t_1} it follows that

$$v_{t_1}=\frac{2\sigma_m-\sigma_F}{\rho c} \tag{17}$$

Similarly, the velocity of the second spall is found to be

$$v_{t_2}=\frac{2\sigma_m-3\sigma_F}{\rho c} \tag{18}$$

In general, the velocity of any spall can be found to be

$$v_{t_n} = \frac{2\sigma_m - (2n-1)\sigma_F}{\rho c} \tag{19}$$

It is apparent that the maximum level of stress is reduced by σ_F every time a spall forms. Therefore, the number of spalls that a wave will produce will be

$$n = \frac{\sigma_m}{\sigma_F} \tag{20}$$

Such analyses are useful in obtaining a qualitative appreciation for the damaging effects of stress waves on solids. But experimental evidence indicates that material failure is a time-dependent process. Thus the subject deserves somewhat closer examination.

Recall first the differences between static and dynamic fracture. In the static case, one fracture or crack forms and propagates through the material to separate it into two parts. Fragmentation occurs by repeated branching of the original crack. Dynamic fracture, on the other hand, consists of four basic stages [Shockey et al. (1973)]

1 Rapid nucleation of microfractures at a large number of locations in the material.
2 Growth of the fracture nuclei in a rather symmetric manner.
3 Coalescence of adjacent microfractures.
4 Spallation or fragmentation by formation of one or more continuous fracture surfaces through the material.

Historically the greatest effort has gone into the study of spall fracture. Oscarson and Graff (1968) define spall as material fracture due to the interaction of two or more rarefaction waves. Several types of spall failure have been identified. *Ductile spall* refers to cracking detectable only by microscopic examination. *Phase-transformation spall* occurs in materials that exhibit a pressure-induced phase change. *Ultimate spall* is characterized by sublimation or disintegration of part of the impacted material. Both ultimate and phase-transformation spall occur at relatively high stress levels. Ductile spall occurs at low pressures and involves the same basic mechanisms (crack nucleation and propagation) that are associated with lower strain-rate failures. It is the most extensively studied of the three types. Three levels of ductile spall have been recognized and classified according to the level of damage.

The incipient spall threshold is defined as that combination of stress amplitude *and* pulse duration below which no damage is detected in an impacted specimen, which, after having been sectioned, polished, and etched, is viewed at about $100x$ magnification.

If the combination of stress and pulse duration exceeds the level required for incipient spall, the degree of damage is correspondingly greater. A criterion for quantitative damage assessment has been proposed by Herrmann of Sandia [Oscarson and Graff (1968)], who proposed that damaged specimens, including the spall plane, be cut into tensile members and pulled apart in a tension machine. The residual strength of the tensile member thus provides a quantitative measure of the degree of damage (Figure 8). As the severity of loading increases, the level of damage increases and the residual strength decreases. At some point, the decrease in residual strength stops as loading severity increases. The tensile stress and pulse duration corresponding to this point define the intermediate spall threshold.

If the material is loaded well beyond the intermediate spall threshold, an intact piece will separate from it. The combination of stress and pulse duration at which this occurs is called the complete spall threshold. This threshold is of great practical importance and thus has been extensively studied [e.g., Rinehart (1960a, b; 1975), Johnson (1972)].

The analysis presented at the beginning of this section was based on the so-called "critical-normal fracture stress criterion" which states that a material will spall instantaneously and completely when a unique critical value of normal tensile stress is attained. Sufficient evidence has been accumulated to indicate that a realistic spall criterion must include at least one or more additional parameters to account for the behavior observed in spall experiments. Among these are:

1 Tensile pulse duration at the spall plane.
2 Effects of precompression at the spall plane.
3 Temperature.
4 Strain and stress rates at the spall plane.
5 Spatial stress gradients.
6 Yield strength.
7 Metallurgical characteristics such as grain orientation.

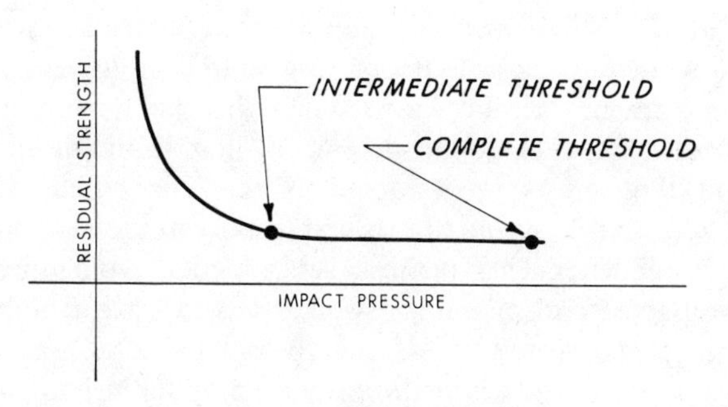

Figure 8 Spall thresholds.

These are considered in some detail in the report by Oscarson and Graff (1968) and experimental evidence cited in support of each. By far, time (pulse duration) appears to exert the greatest influence. The effect is clearly demonstrated by the plate impact experiments on 6061-T6 aluminum of Blincow and Keller. They observed that the flyer velocity required to cause insipient spall decreased with increasing flyer thickness in an approximately linear manner. Since impact velocity is directly related to stress (see Chapters 1 and 4) and flyer thickness can be similarly related to pulse duration, a definite time dependence of spall is implied.

Four types of spall models exist that incorporate some of these effects [Oscarson and Graff (1968)]. Cumulative models account for the effects of stress amplitude and duration in attempting to predict spall for an arbitrary pulse shape from previous experiments with rectangular pulses. Energy models are based on the criterion that the mechanical energy supplied to the spall plane up to the time of fracture is invariant. Stress-rate models use the criterion that the tensile strength of a material is dependent on its tensile unloading characteristics (stress rate or stress gradient). All such models incorporate a number of empirical factors and require a fair degree of material characterization. They have been used for special applications in one- and two-dimensional computer codes with reasonable success. They are generally not available in production codes for reasons mentioned in Chapter 10.

A model that attempts to relate failure on the continuum level with the breaking of atomic bonds has also been developed. Since it omits plastic flow effects, it has not been successful in predicting spall in materials exhibiting a degree of ductility.

Micromechanical aspects of material failure under intense impulsive loading are discussed further in Chapter 9.

2.4 DYNAMIC PLASTIC BUCKLING OF LONG RODS

Consider a long rod of finite length that suddenly experiences an impact load at one end. Assume that the load is sufficiently intense to compress the material beyond its elastic limit. As a result of the impact, an elastic-compressive wave travels toward the opposite end of the rod. This wave is followed by a slower but more intense plastic wave. When the elastic wave that leads the plastic wave front reaches the opposite end of the rod, it will be reflected as a tensile wave if the end is unrestrained. When this reflected tensile wave interacts with the oncoming compressive-plastic wave, the amplitude of the plastic wave is reduced. This process is known as unloading. If the end opposite the impacted end of the rod were fixed to a rigid body, the elastic-compressive wave would reflect off the fixed end as a compressive-plastic wave. The fixed-end situation is well represented by the driving of a spike into a hard media.

If the load at the impact end is removed, an elastic unloading (tensile) wave is generated. Since this moves at a higher velocity than the compressive-plastic wave, it overtakes the plastic wave and reduces its amplitude.

In general, flexural waves will also be present when rods impact rigid surfaces. The flexural waves propagate toward the free end of the rod at velocities that depend the wavelength. The elastic flexure-wave velocity is less than the longitudinal wave velocity and, according to one-dimensional theory, will never exceed 0.5764 $\sqrt{E/\rho}$ regardless of wavelength.

Plastic buckling will occur when the impact velocity is sufficiently large (>60 m/s for most structural materials). The plastic buckling that occurs will be confined within the axial plastic compression wave. The lateral motion that results can be due to several factors, among them:

1. Nonorthogonal impact.
2. Nonuniformity of the rod (i.e., presence of bends).
3. Nonisotropic material properties.
4. Prestresses existing within the rod at impact.

Examples of rods undergoing large plastic deformation are shown in Figures 9–14. The experiments involved the impact of 6061-T6 aluminum rods with a

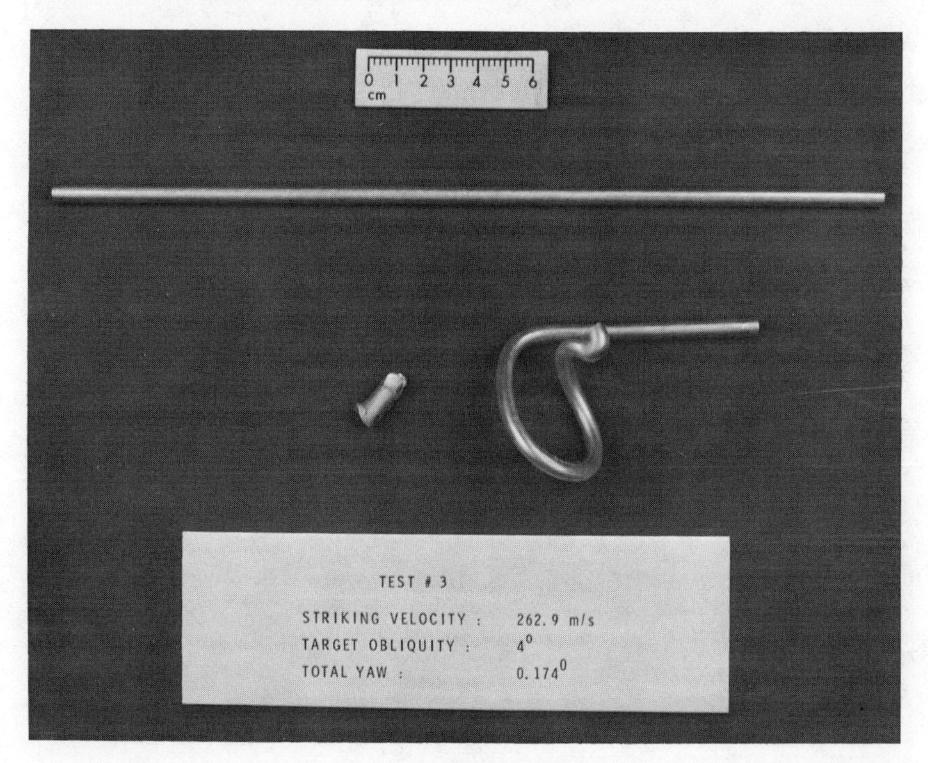

Figure 9 Initial and deformed shapes for test 3.

length-to-diameter (L/D) ratio of 48 ($D=0.635$ cm, $L=30.48$ cm) into a 2.54 cm thick rolled homogeneous armor plate at various velocities and obliquities, as noted in the figures. The experimental setup is shown in Figure 15. Framing camera records were obtained at (approximately) 120-μs intervals to 2–3 ms for each event. The results are similar to those obtained by Valentine and Whitehouse (1972) who had a much more extensive experimental program. Plastic buckling experiments have also been conducted by Weirauch et al. (1970) for copper, steel, and tungsten carbide rods striking metallic targets at velocities of 130–1000 m/s. Normal impact of mild steel rods into mild steel plate has been studied by Grabarek & Ricchiazzi (1968). They used rods with L/D of 10 at velocities of 365–1830 m/s so that for the most part penetration and gross plastic flow, rather than buckling, resulted.

Almost all analyses of dynamic plastic buckling acknowledge their debt to the work of Abrahamson & Goodier (1966). Because of the complexity of the subject, simplifying assumptions are introduced which usually limit the utility of the various models to rather low (<200 m/s) impact velocities. As an example of the analysis typically employed for determination of deformed mode shape (or alternatively half wavelength) of plastically buckled rods, Valentine's approach is presented next at some length.

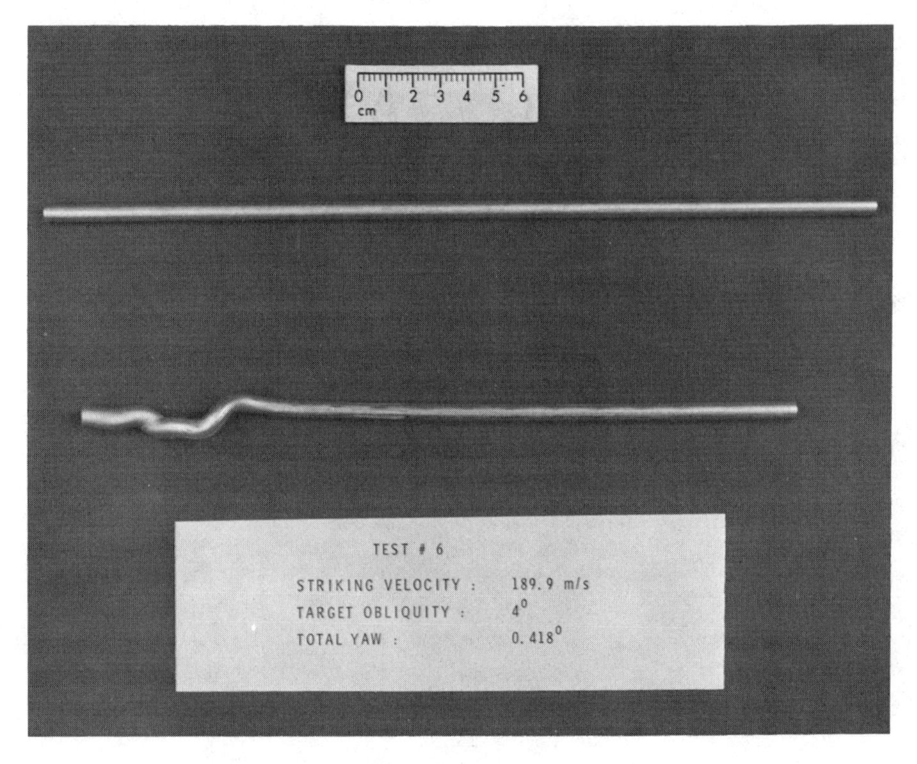

Figure 10 Initial and final rod shapes for test 6.

Valentine employs the principle of virtual work to develop the equations of motion for a rod striking a rigid surface. The development includes the effects of axial, lateral, and rotary inertia for a material whose stress-strain relation is nonlinear. An axial force P and a bending moment M is assumed to act on the rod (Figure 16). The axial displacement is denoted as u, the lateral displacement as y, and the angle of rotation of the differential element as ψ. The cross-sectional area of the rod is denoted by A and its moment of inertia by I. The differential work produced by the bending moment M and the axial force P may be written as

$$dW = M\,d\psi + P\,du \tag{21}$$

The work produced by the inertia forces is

$$dW_I = \rho A\left(\frac{\partial^2 u}{\partial t^2}u + \frac{\partial^2 y}{\partial t^2}y\right) + \rho I\frac{\partial^2 \psi}{\partial t^2}\psi \tag{22}$$

The work produced by the external forces assumed to act on the impacted end of the rod is

$$W_F = -F_y y(0) - F_x u(0) \tag{23}$$

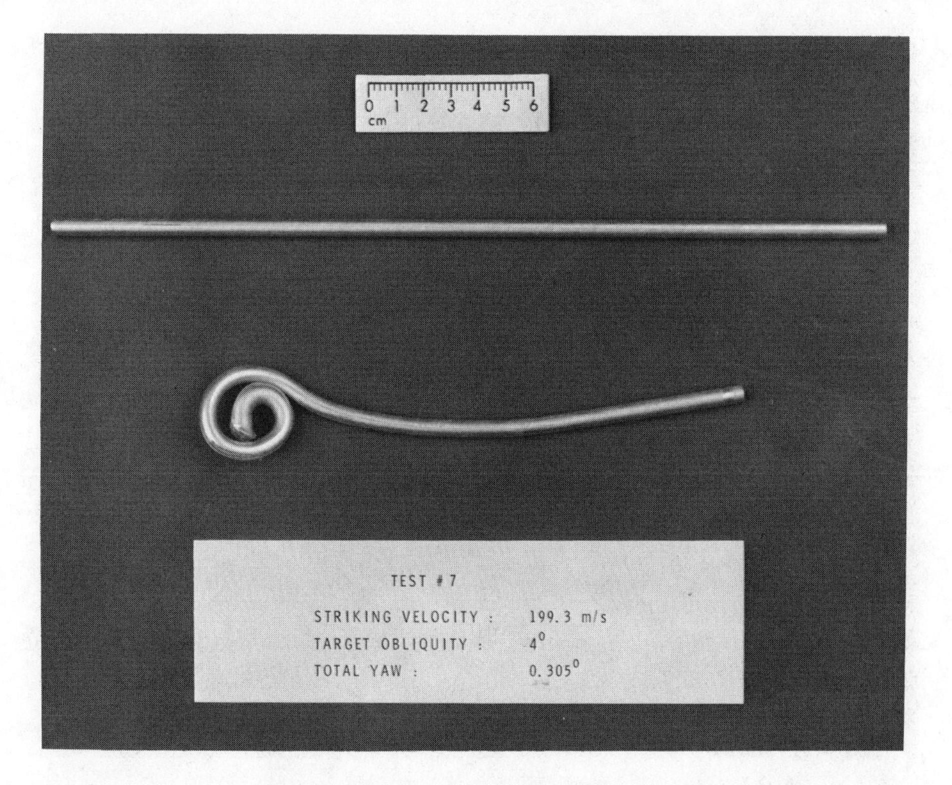

Figure 11 Initial and final rod shapes for test 7.

Summing the work of the individual components, integrating over the length of the rod and applying the principle of virtual work ($\delta W=0$) results in the equations of motion and the boundary conditions.

In the interior of the rod, the following equations of motion hold:

$$\frac{\partial^2 M}{\partial x^2}-\rho I\frac{\partial^4 I}{\partial x^2 \partial t^2}+\rho A\frac{\partial^2 y}{\partial t^2}=0 \tag{24}$$

$$\rho A\frac{\partial^2 u}{\partial t^2}-\frac{\partial P}{\partial x}=0 \tag{25}$$

At the boundary $x=0$, the impacted end, the boundary conditions are:

$$M=0$$

$$P=F_x=0$$

$$\frac{\partial M}{\partial x}-\rho I\frac{\partial^3 y}{\partial t^2 \partial x}+F_y=0 \tag{26}$$

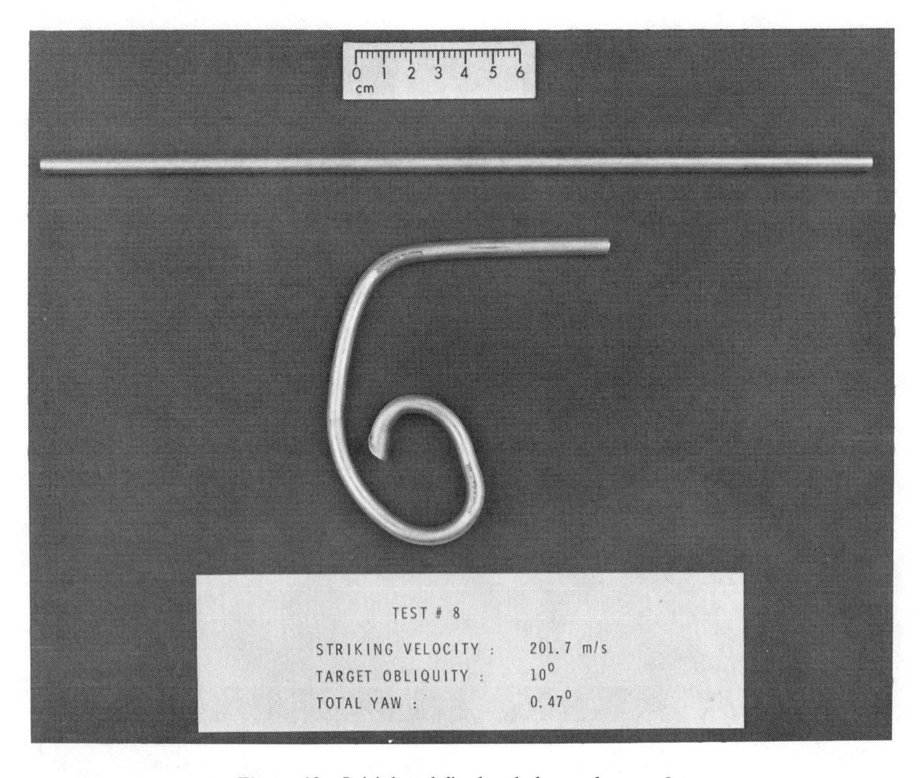

Figure 12 Initial and final rod shapes for test 8.

At the boundary $x=L$, the free end, the boundary conditions are:

$$M=0; \qquad P=0$$

$$\frac{\partial M}{\partial x}-\rho I\frac{\partial^3 y}{\partial t^2\,\partial x}=0 \tag{27}$$

The boundary conditions are those of a free rod and allow deflections and rotations at both the impacted end and the free end. The moment vanishes at both boundaries. At the impacted end, the axial force in the rod is balanced by the axial forcing function and the rotational forces are balanced by the lateral forcing function. At the free end, both the axial and rotational forces vanish.

Note that in equations (24)–(27) the bending moment M and the axial load P appear. These must be expressed in terms of displacements before a solution to the equations can be obtained.

The axial force in the rod may be related to an axial deflection in a straightforward manner. The stress-strain curve for 6061-T6 aluminum is approximated by Valentine as a bilinear relation defined as:

$$\sigma=E\varepsilon \qquad \text{for } \varepsilon<\varepsilon_0$$

$$\sigma=E\varepsilon_0+\beta(\varepsilon-\varepsilon_0) \qquad \text{for } \varepsilon\geqslant\varepsilon_0 \tag{28}$$

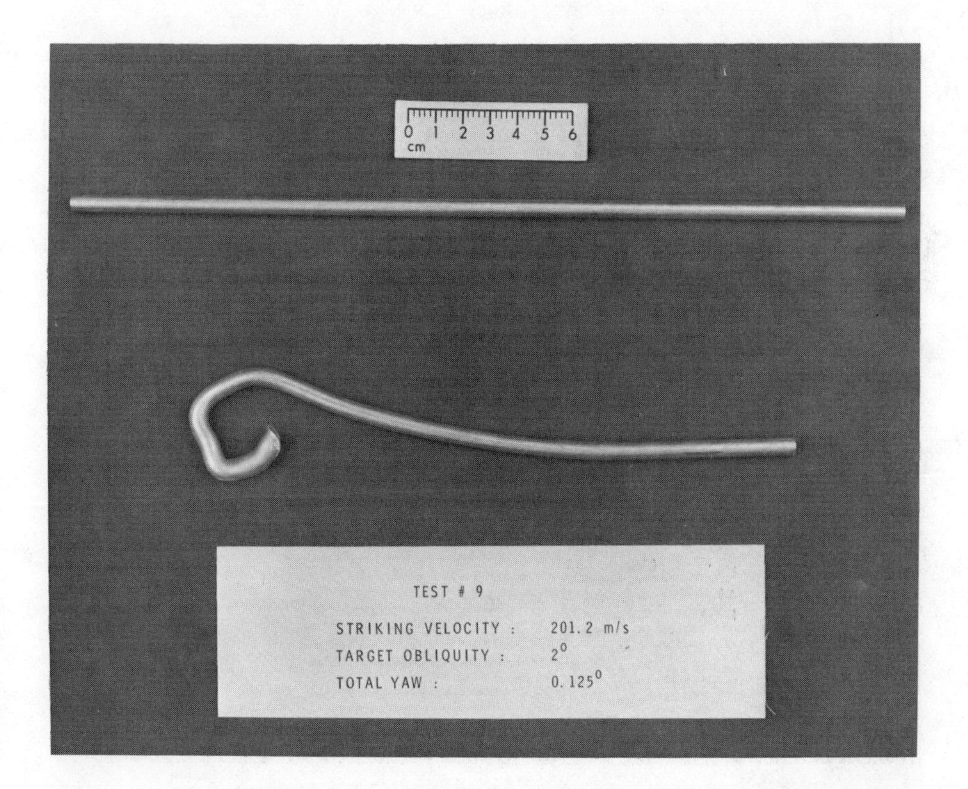

Figure 13 Initial and final rod shapes for test 9.

Where ε_0 is the yield strain and β the slope of the stress-strain curve in the plastic region.

Taking
$$\varepsilon = \frac{\partial u}{\partial x} \text{ and } \sigma = P/A$$

$$P = AE\frac{\partial u}{\partial x} \text{ for } \varepsilon < \varepsilon_0$$

$$P = A\left[E\varepsilon_0 + \beta\left(\frac{\partial u}{\partial x} - \varepsilon_0\right)\right] \text{ for } \varepsilon \geq \varepsilon_0 \tag{29}$$

From (9) it follows that

$$\frac{\partial P}{\partial x} = AE\frac{\partial^2 u}{\partial x^2} \qquad \text{for } \varepsilon < \varepsilon_0$$

$$\frac{\partial P}{\partial x} = A\beta\frac{\partial^2 u}{\partial x^2} \qquad \text{for } \varepsilon \geq \varepsilon_0 \tag{30}$$

This provides the needed connection between axial load and displacement. Note that since the equations (30) differ only by the slope of the stress-strain curve, the latter of (30) may be taken as the general relationship between force

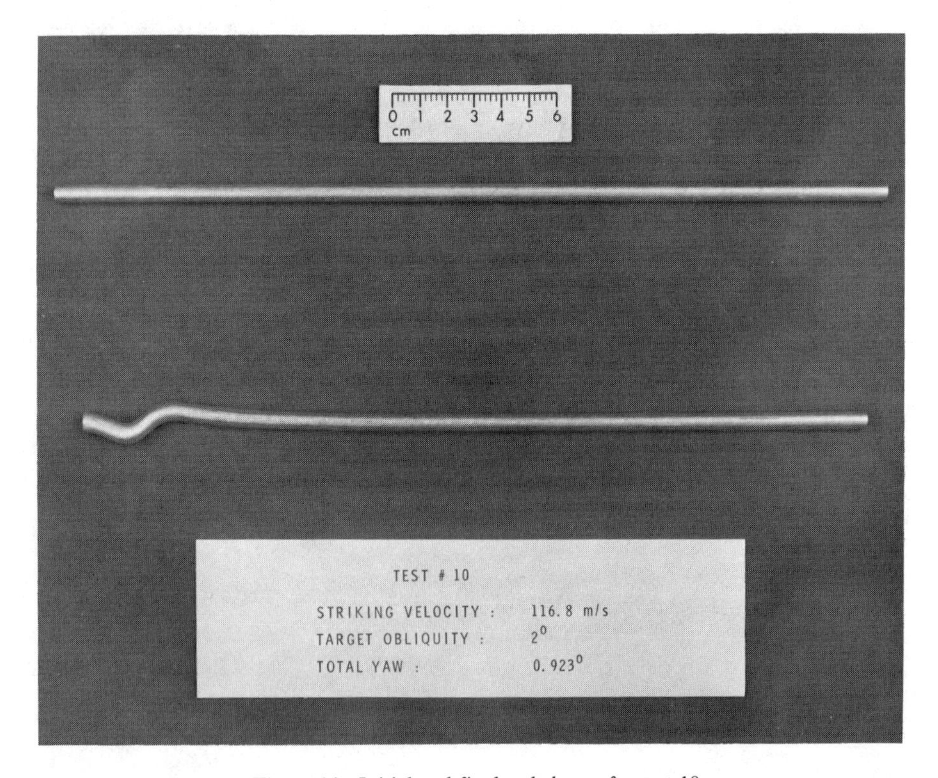

Figure 14 Initial and final rod shapes for test 10.

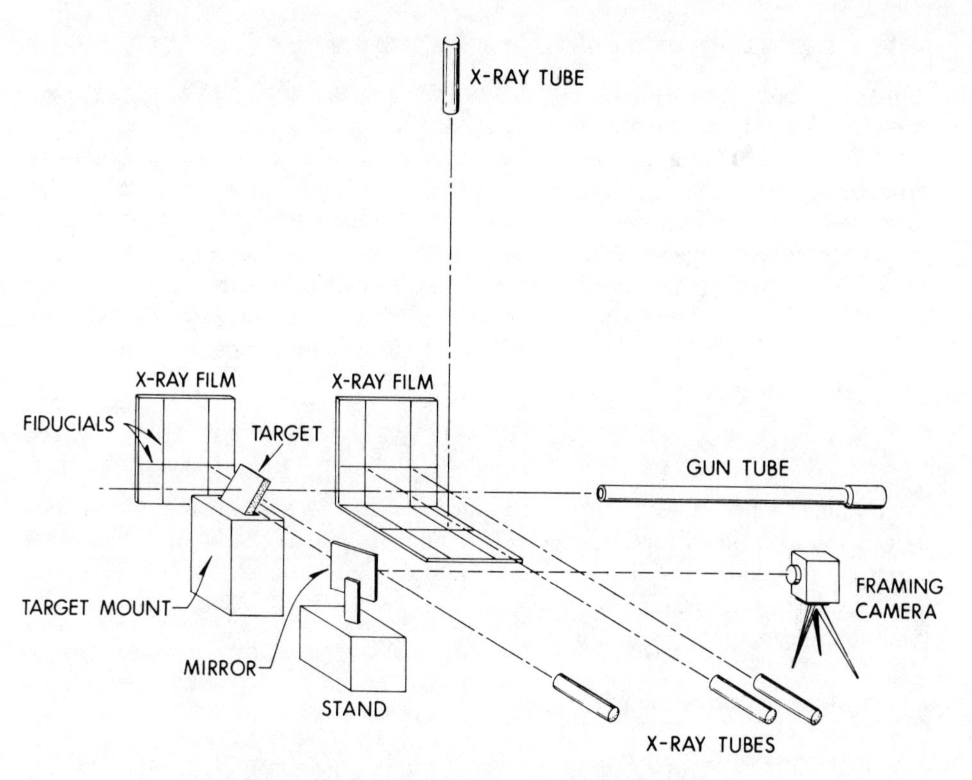

Figure 15 Experimental setup for rod-impact tests.

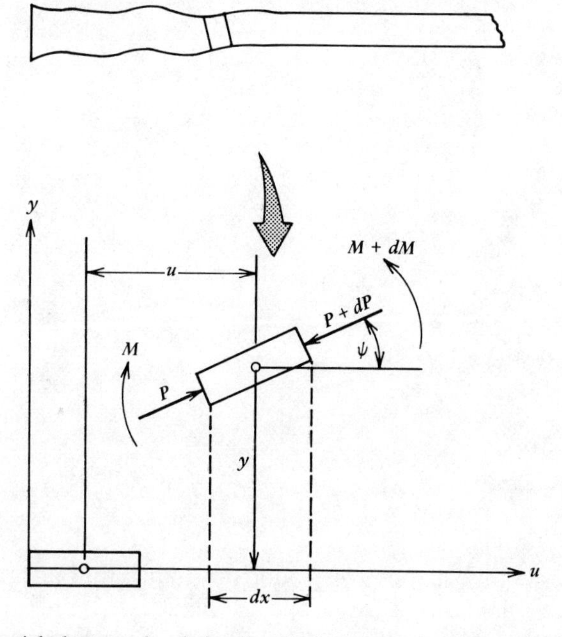

Figure 16 Differential element of rod showing displacements, forces, and moments [Valentine and Whitehouse (1972)].

and displacement provided that β is interpreted as the slope of the stress-strain relation for both elastic and plastic deformation.

A bilinear stress-strain relation is again employed to relate the bending moment M to a lateral deflection. The assumption is also made that the plastic buckling occurs within the axial plastic compression waves.

The necessary relationship is found by assuming that the rod has failed plastically due to axial compression and by superimposing a bending moment on this state [Valentine and Whitehouse (1972)]. By using the definition of bending moment from elementary beam theory and imposing static equilibrium, the relationship

$$M=\frac{\partial^2 y}{\partial x^2}(\beta I_1+EI_2) \tag{31}$$

is obtained where $I=I_1+I_2$ is the moment of inertia about the gravity axis. Equation (31) holds for both the elastic and the plastic moment. For elastic motion, β becomes the elastic modulus E and I_1+I_2 equal I.

Substituting (30) and (31) into (24)–(26) results in the final equations of motion and boundary conditions:

1 In the interior of the rod

$$(EI_2+\beta I_1)\frac{\partial^4 y}{\partial x^4}-\rho I\frac{\partial^4 y}{\partial x^2\partial t^2}+\rho A\frac{\partial^2 y}{\partial t^2}=0 \tag{32}$$

$$\beta\frac{\partial^2 u}{\partial x^2}-\rho\frac{\partial^2 u}{\partial t^2}=0$$

2 At the boundary $x=0$

$$\begin{gathered}\frac{\partial u}{\partial x}=-\frac{F_x}{A\beta}\\ \frac{\partial^2 y}{\partial x^2}=0\\ (EI_2+\beta I_1)\frac{\partial^3 y}{\partial x^3}-\rho I\frac{\partial^3 y}{\partial t^2\partial x}-F_y=0\end{gathered} \tag{33}$$

3 At the boundary $x=L$

$$\frac{\partial u}{\partial x}=0;\qquad \frac{\partial^2 y}{\partial x^2}=0$$

$$(EI_2+\beta I_1)\frac{\partial^3 y}{\partial x^3}-\rho I\frac{\partial^3 y}{\partial t^2\partial x}=0 \tag{34}$$

The axial and lateral equations of motion are coupled through the stress-strain relationship that involves β, the slope of the stress-strain curve in both the elastic and plastic regions. Equations (32)–(34) must be solved as a system, together with the stress-strain relation.

Valentine and Whitehouse (1972) cast (32)–(34) in finite difference form and obtained solutions for a large number of impact situations which were compared with experimental results. Generally good agreement for deformed shape was obtained for the lower velocities (<160 m/s) studied. Agreement between computed and measured deformation profiles became progressively poorer as striking velocities increased much above 160 m/s. This divergence may be attributed to three principal factors:

1 Transverse shear deformation was neglected in deriving (32)–(34). Transverse shear effects play a significant role in flexural wave propagation for wavelengths comparable to the cross-sectional dimensions of the rod.
2 Small strain-small rotation assumptions are made in derivation of the governing equations, for example (31) and the definition of strain as $\varepsilon = \partial u/\partial x$. As impact velocity increases, deformations become progressively larger and the above equations should not be expected to hold.
3 Material parameters (yield strength and the plastic slope) are taken from a *static* stress-strain curve. Dynamic data for 6061-T6 from the free-flight bar impacts of Bell (1974) indicate that the dynamic yield stress is somewhat higher than the static value and the plastic slope considerably lower. For 6061-T6, differences between static and dynamic stress-strain curves are

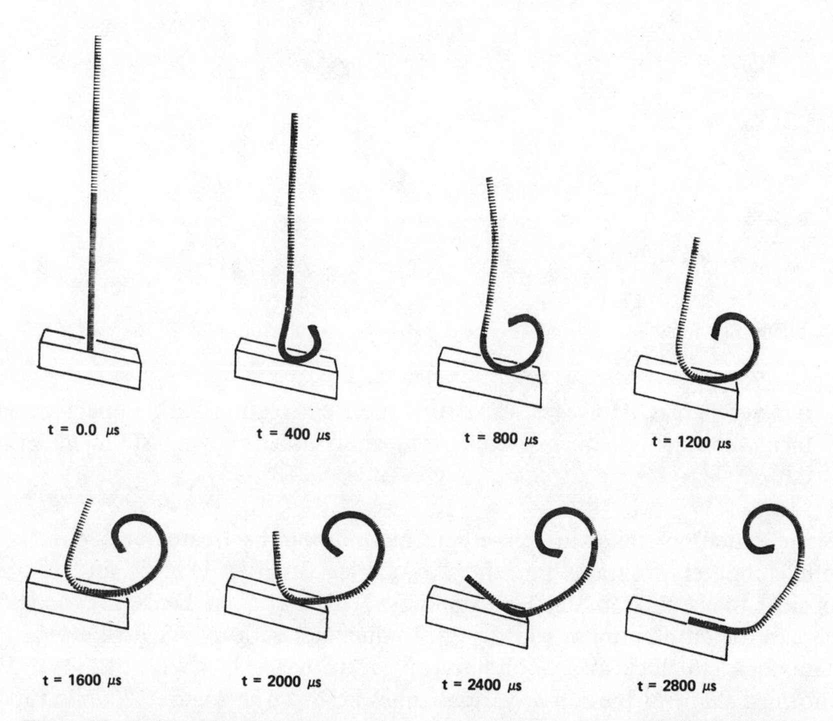

Figure 17 Deformed rod shapes at various stages of oblique impact [Hallquist (1979)].

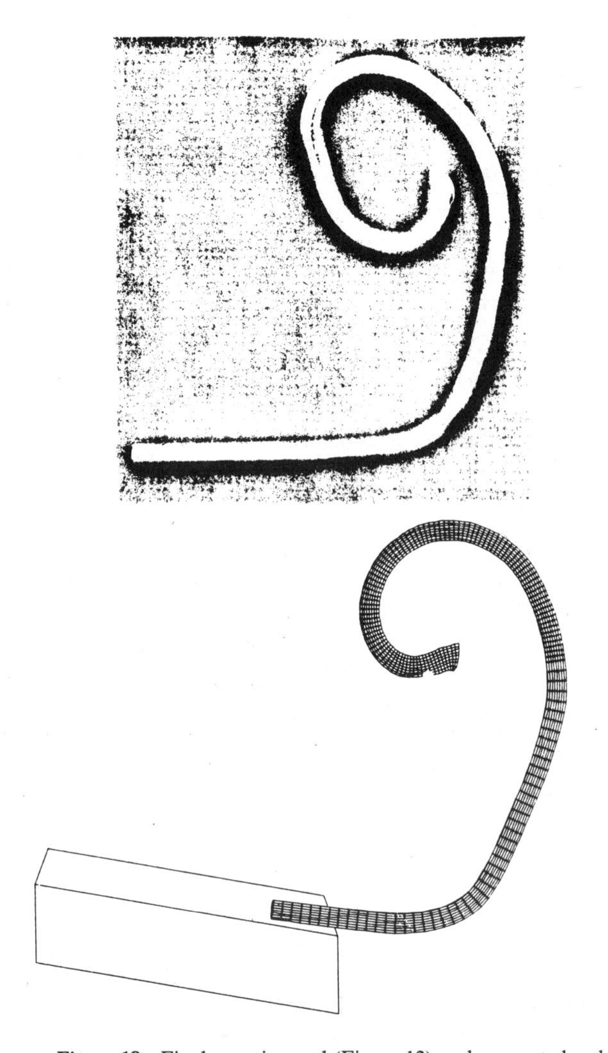

Figure 18 Final experimental (Figure 12) and computed rod profiles [Hallquist (1979)].

not very great. However, as a rule good correlation with impact experiments should not be expected from analyses employing static materials data.

Impact situations involving large deformations can be treated very effectively with computer programs currently available (Chapter 11). As an example, consider the results obtained by Hallquist (1979) with the DYNA3D code for the case of an aluminum rod, length-to-diameter ratio of 48, striking a steel plate, 2.54 cm thick at an obliquity of 10° with a velocity of 201 m/s. The deformed shape of the rod at various times is shown in Figure 17 while Figure 18 shows a comparison between the computed and experimental profiles at approximately 3 ms. The agreement is quite good and serves to demonstrate the utility of such computations in large deformation situations. The calculations were done on a CRAY machine at the Lawrence Livermore National Laboratory and required approximately one and one-half hours of CPU time.

REFERENCES

Abrahamson, R. G., and Goodier, J. N. (1966), *J. Appl. Mech., Trans. ASME*, **33**, 241.

Adler, W. F. (Ed.) (1979a), *Erosion: Prevention and Useful Applications*, ASTM-STP-664.

Adler, W. F. (1979b), in C. M. Preece, (Ed.), *Treatise on Materials Science and Technology, vol. 16, Erosion*, Academic Press, New York.

Bell, J. F. (1974), Ballistic Research Laboratory, BRL-CR-184.

Bertholf, L. D. (1966), Longitudinal Elastic Wave Propagation in Finite Cylindrical Bars, Ph.D. dissertation, Washington State U., Pullman, Wash.

Bertholf, L. D. (1967), *J. Appl. Mech., Trans ASME*, **34**, 725.

Brunton, J. H., and Rochester, M. C. (1979), in C. M. Preece (Ed.), *Treatise on Materials Science and Technology, vol. 16, Erosion*, Academic Press, New York.

Conway, H. D., and Jakubowski, M. (1969), *J. Appl. Mech., Trans. ASME*, **36**, 809.

Grabarek, C., and Ricchiazzi, A. (1968), *Proc. Army Symp. on Solid Mech.*, AMMRC.

Hallquist, J. O. (1979), Lawrence Livermore Laboratory, UCID-17268, Rev. 1.

Johnson, W. (1972), *Impact Strength of Materials*, Crane, Russak, New York.

Kinslow, R. (1976), Army Mobility Equipment R & D Command, Report 2179.

Mescall, J., and Papirno, R. (1974), *Exp. Mech.*, **14**, 257.

Oscarson, J. H., and Graff, K. F. (1968), Battelle Memorial Institute, BAT-197A-4-3 (AD 669440).

Preece, C. M. (Ed.) (1979), *Treatise on Materials Science and Technology, vol. 16, Erosion*, Academic Press, New York.

Ramamurti, V., and Ramanamurti, P. V. (1977), *J. Sound Vib.*, **53**, 529.

Rinehart, J. S. (1960a), Air Force Special Weapons Center, AFSWC-TR-60-7 (AD 236719).

Rinehart, J. S. (1960b) in N. Davids (Ed.), *Stress Wave Propagation in Materials*, Interscience, New York.

Rinehart, J. S. (1975), *Stress Transients in Solids*, HyperDynamics, Santa Fe, N.M.

Shockey, D. A., Seaman, L., and Curran, D. R. (1973), in R. W. Rohde, et al. (Eds.), *Metallurgical Effects at High Strain Rates*, Plenum, New York.

Springer, G. S. (1976), *Erosion by Liquid Impact*, Wiley, New York.

Valentine, M. B., and Whitehouse, G. D. (1972), Air Force Armament Lab, AFATL-TR-72-124 (AD 754241).

Weirauch, G. et al. (1970), Deutsch-Französisches Forschungsinstitut Saint-Louis, Report 13/70.

Wright, T. W. (1973), Ballistic Research Laboratory, BRL-MR-2296.

3

DAMAGE IN COMPOSITE MATERIALS DUE TO LOW VELOCITY IMPACT

Longin B. Greszczuk

Although advanced composites have been accepted as engineering materials and the design/analysis techniques for the response of composite materials and structures to static loads are well established, no comparable techniques are available for design of advanced composites against foreign-object impact. Only recently have concentrated efforts been initiated to understand the impact response of composites through rigorous analyses and meaningful tests, as those described in a recent ASTM (1975) publication. Although the response of composite materials to particle or foreign-body impact could be studied using empirical or semiempirical approaches, this appears undesirable because of the large and costly efforts that would be required to cover the various combinations of constituent materials, layups, stacking sequences, and constructions. For example, at present time at least 38 different graphite and carbon fibers are available commercially. The Young's moduli of the various fibers range from 6×10^6 to 80×10^6 psi, while their tensile strength ranges from 110 to $\sim$450 ksi. Thus even for graphite-epoxy composites the determination of which of the available fibers is optimum for impact response would constitute a monumental task. When designing for impact response, it appears desirable to have a criterion for determining how the various properties of the target and the impact parameters influence target damage. The analytical approach described here is oriented toward that goal.

3.1 THEORY DEVELOPMENT

The approach in studying the response of isotropic and composite materials to low velocity impact is shown in Figure 1. The three major steps of the approach are: (1) determination of impactor-induced surface pressure and its distribution, (2) determination of internal stresses in the composite target caused by the surface pressure, and (3) determination of failure modes in the target caused by the internal stresses. For the most general case the target is assumed to be a multilayer, generally orthotropic solid, whereas the impactor is assumed to be a body of revolution. Also it is assumed that (1) the target and the impactor are linear elastic, (2) impact duration is long compared to stress-wave transit times in the impactor (or target or finite thickness), and (3) the impact is normal to the target surface.

3.1.1 Pressure Distribution

The magnitude and distribution of surface pressure in the target caused by impact can be obtained by analytically combining the dynamic solution to the problem of impact of solids with the static solution for the pressure between two bodies in contact, similar to the method described by Timoshenko (1934) for impact of spheres.

Spherical Impactor, Isotropic Target. Denoting the mass and the velocity of the impactor as m_1 and v_1 respectively and the target mass and its velocity as m_2 and v_2, the rates of change of velocity during impact (as the two bodies come in contact) are

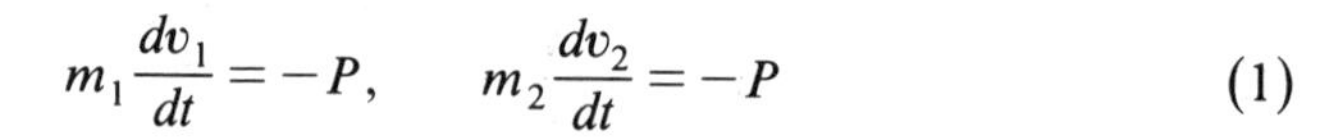

$$m_1 \frac{dv_1}{dt} = -P, \qquad m_2 \frac{dv_2}{dt} = -P \tag{1}$$

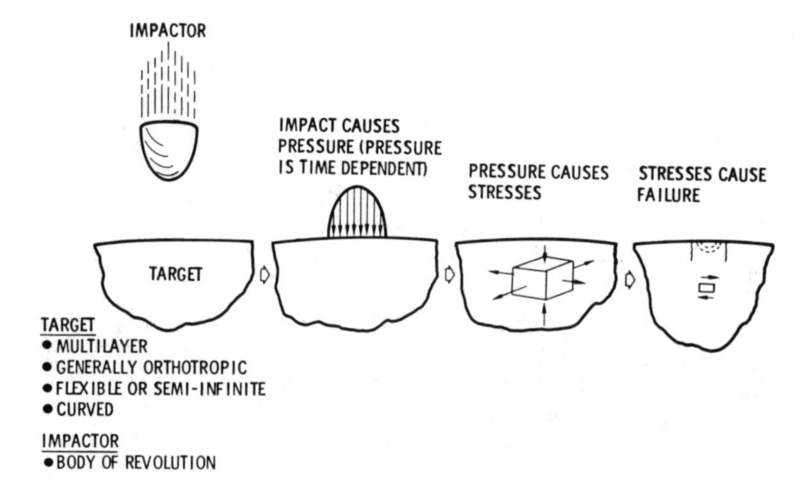

Figure 1 Essential features of the approach.

If we denote by α the distance that the impactor and the target approach one another because of local compression at the point of contact, the velocity of this approach is

$$\dot{\alpha}=v_1+v_2 \tag{2}$$

Results obtained by Rayleigh (1906) show that if the contact duration between the impactor and the target is very long in comparison with their natural periods, vibrations of the system can be neglected. It can therefore be assumed that the Hertz law

$$P=n\alpha^{3/2} \tag{3}$$

that was established for statical conditions applies also during impact. The term n is defined as

$$n=\frac{4\sqrt{R_1}}{3\pi(k_1+k_2)} \tag{4}$$

where R_1 is the radius of a spherical impactor or indenter,

$$k_1=\frac{1-\nu_1^2}{\pi E_1} \tag{5}$$

$$k_2=\frac{1-\nu_2^2}{\pi E_2} \tag{6}$$

where E and ν are the Young's modulus and Poisson's ratio respectively, and the subscripts 1 and 2 refer to the impactor and the target. Differentiating (2), combining it with (1), and substitution of (3) into resultant equation yields

$$\ddot{\alpha}=nM\alpha^{3/2} \tag{7}$$

where

$$M=\frac{1}{m_1}+\frac{1}{m_2} \tag{8}$$

Proceeding now as in Timoshenko (1934), if both sides of (7) are multiplied by $\dot{\alpha}$ and the resultant equation is integrated the following results

$$(\dot{\alpha}^2-v^2)=-\tfrac{4}{5}Mn\alpha^{5/2} \tag{9}$$

where v is the approach velocity of the two bodies at $t=0$, that is, at the beginning of impact. Maximum deformation, α_1, occurs when $\dot{\alpha}=0$ and is

$$\alpha_1=\left(\frac{5v^2}{4Mn}\right)^{2/5} \tag{10}$$

An alternate way of arriving at relationship given by (10) is to start with the energy balance of the system. Assuming that the target is semiinfinite and stationary and the impactor is moving at velocity v_1, the energy balance becomes

$$\tfrac{1}{2}m_1v_1^2=\int_0^{\alpha_1}P\,d\alpha \tag{11}$$

Substitution of (3) into (11) followed by evaluation of the resulting integral gives

$$\tfrac{1}{2}m_1v_1^2=\tfrac{2}{5}n\alpha_1^{5/2} \tag{12}$$

which when solved for α_1 is identical to the result given by (10). It is noted that in this case, $v_1\equiv v$ and $M\equiv 1/m_1$. Substitution of (10) into (3) gives the following final relationship

$$P=n^{2/5}\left(\frac{5v^2}{4M}\right)^{3/5} \tag{13}$$

For the case of the Hertzian contact problem involving a sphere pressed onto a flat surface by a force P, the relationship between P and a, the radius of the area of contact is [Hertz (1881)]

$$a=\left[\frac{3\pi P}{4}(k_1+k_2)R_1\right]^{1/3} \tag{14}$$

Combining (14) and (13) the maximum radius of the area of contact between a flat target and a spherical impactor then becomes

$$a=(R_1)^{1/2}\left(\frac{5v^2}{4Mn}\right)^{1/5} \tag{15}$$

It has been shown [Hertz (1881), Timoshenko (1934), Belajef (1924)] that the pressure distribution over the area of contact is

$$q_{x,y}=q_0\left[1-\frac{x^2}{a^2}-\frac{y^2}{a^2}\right]^{1/2} \tag{16}$$

where q_0 is the surface pressure at the center of area of contact, at $x=y=0$. At the boundary of the surface

$$\frac{x^2}{a^2}+\frac{y^2}{a^2}=1 \tag{17}$$

and therefore

$$q_{x,y}=0 \tag{18}$$

By summing the pressures acting on the area of contact and equating the result to P, we obtain

$$q_0=\frac{3P}{2\pi a^2} \tag{19}$$

Combining (13), (15), (16), and (19) and introducing polar coordinates, the following equation is obtained for the magnitude and distribution of the surface pressure

$$q_r=\left(\frac{3n}{2\pi R_1}\right)\left(\frac{5v^2}{4nM}\right)^{1/5}\left[1-\left(\frac{r}{a}\right)^2\right]^{1/2} \tag{20}$$

and

$$q_0 = \left(\frac{3n}{2\pi R_1}\right)\left(\frac{5v^2}{4nM}\right)^{1/5} \tag{21}$$

Equations (13), (15), and (20) are the final equations that give the impact force, radius of the area of the contact, and magnitude and distribution of the surface pressure in terms of the impact velocity, geometry of the impactor, as well as the elastic properties and masses of the impactor and the target. Equations developed herein can also be used to determine the variation of the above noted parameters (P, a, q_r, and q_0) with time, as discussed in more detail in Section 3.1.2.

General Case of Impact Between Two Nonisotropic Bodies of Revolution. The same approach as described in the preceding section can be used for investigating the more general case of impact between two arbitrary bodies of revolution made of transversely isotropic and orthotropic materials, including targets made of laminated composites. Theory extension to arbitrary bodies of revolution requires that appropriate equations corresponding to the solution of the contact problem between such bodies be used. The solution to the more general contact problem can be found in the works of Hertz (1881, 1895), Whittemore and Petrenko (1921), and is also given by Timoshenko (1934), Belajef (1945), and in numerous other references.

If a solid (or impactor designated by subscript 1) having, at the point of contact, principal radii of curvature R_{1m} and R_{1M} is pressed by a force P into a target having principal radii of curvature R_{2m} and R_{2M}, the area of contact will be elliptical with major and minor axes of ellipse being [Whittemore and Petrenko (1921)]

$$a = m\left[\frac{3\pi}{2}P(k_1' + k_2')C_R\right]^{1/3} \tag{22}$$

$$b = r\left[\frac{3\pi}{2}P(k_1' + k_2')C_R\right]^{1/3} \tag{23}$$

where C_R is a term that takes into account the curvature effect

$$C_R^{-1} = \frac{1}{R_{1m}} + \frac{1}{R_{2m}} + \frac{1}{R_{1M}} + \frac{1}{R_{2M}} \tag{24}$$

k_1' and k_2' are still to be defined parameters that take into account the elastic properties of the impactor and the target; and m, r, and s are parameters that are a function of R_{1m}, R_{1M}, R_{2m}, and R_{2M} and are given in Table 1 as a function of θ where

$$\theta = \arc\cos\left\{C_R\left[\left(\frac{1}{R_{1m}} - \frac{1}{R_{1M}}\right)^2 + \left(\frac{1}{R_{2m}} - \frac{1}{R_{2M}}\right)^2 + 2\left(\frac{1}{R_{1m}} - \frac{1}{R_{1M}}\right)\left(\frac{1}{R_{2m}} - \frac{1}{R_{2M}}\right)\cos 2\phi\right]^{1/2}\right\} \tag{25}$$

Table 1 Values Of Parameters m, r, and s[a]

θ	0°	10°	20°	30°	40°	50°	60°	70°	80°	90°
m	∞	6.612	3.778	2.731	2.136	1.754	1.486	1.284	1.128	1.00
r	0	0.319	0.408	0.493	0.567	0.641	0.717	0.802	0.893	1.00
s	–	0.851	1.220	1.453	1.637	1.772	1.875	1.994	1.985	2.00

[a] From Whittemore and Petrenko (1921).

and ϕ is the angle between normal planes containing curvatures $1/R_{1m}$ and $1/R_{2m}$.

The relationship between the contact force P and the combined deformation of both solids at the point of contact can be expressed in terms of similar parameters and is

$$\alpha = s\left[\frac{9\pi^2 P^2 (k_1' + k_2')^2}{256 C_R}\right]^{1/3} \tag{26}$$

or expressing the above relationship in a form similar to (3)

$$P = n'\alpha^{3/2} \tag{27}$$

where now

$$n' = \left(\frac{16}{3\pi(k_1' + k_2')}\right)\left(\frac{C_R}{s^3}\right)^{1/2} \tag{28}$$

Substitution of n' for n in (10), and combination of result with (27) gives

$$P = n'\left(\frac{5v^2}{4Mn'}\right)^{3/5} \tag{29}$$

In the case of contact problem involving solids of revolution the pressure distribution has been shown to be of the form [Belajef (1945)]

$$q_{x,y} = q_0\left[1 - \frac{x^2}{a^2} - \frac{y^2}{b^2}\right] \tag{30}$$

where x and y are the coordinate axes in the directions of the axes of the ellipse, a and b respectively. Integrating the surface pressure over the elliptical area of contact and equating the result to P we now obtain

$$q_0 = \frac{3P}{2\pi ab} \tag{31}$$

Combining (29), (22), (23), (26), and (31), the following expressions are obtained for the major and minor axes of the area of contact, maximum deformation at the impact site, and maximum surface pressure

$$\frac{a}{m}=\frac{b}{r}=\left[\left(\frac{3\pi}{2}\right)(k_1'+k_2')(C_R)(n')^{2/5}\left(\frac{5v^2}{4M}\right)^{3/5}\right]^{1/3} \tag{32}$$

$$\alpha_1=\left(\frac{5v^2}{4Mn'}\right)^{2/5} \tag{33}$$

$$q_0=\frac{1}{\pi^{4/3}}\left\{\left(\frac{\left(\frac{3}{2\pi}\right)^{1/3}(n')^{2/15}}{mr[(k_1'+k_2')C_R]^{2/3}}\right)\left(\frac{5v^2}{4M}\right)^{1/5}\right\} \tag{34}$$

The terms k_1' and k_2' appearing in the (32), (33), and (34) take into account the elastic properties of the impactor and the target. For the case of impact between isotropic solids k_1' and k_2' are as defined by (5) and (6). If the target is made of transversely isotropic material the following expression for k_2' can be derived from the results given by Conway (1956)

$$k_2'=\frac{\sqrt{A_{22}}\left\{\left(\sqrt{A_{11}A_{22}}+G_{zr}\right)^2-\left(A_{12}+G_{zr}\right)^2\right\}^{1/2}}{2\pi\sqrt{G_{zr}}\left(A_{11}A_{22}-A_{12}^2\right)} \tag{35}$$

where

$$\begin{aligned} A_{11}&=E_z(1-\nu_r)\beta \\ A_{22}&=\frac{E_r\beta(1-\nu_{zr}^2\delta)}{1+\nu_r} \\ A_{12}&=E_r\nu_{zr}\beta \\ \beta&=\frac{1}{1-\nu_r-2\nu_{zr}^2\delta} \\ \delta&=\frac{E_r}{E_z} \end{aligned} \tag{36}$$

and E, G, and ν are the Young's modulus, shear modulus, and Poisson's ratio of the target, while subscripts r and z denote the radial and thickness directions, respectively, z being in the direction of impact. For a planar isotropic material, the properties in the r plane are independent of the orientation.

If the impactor is also made of transversely isotropic material, the expression for k_1' will be similar to k_2' except that the various elastic constants appearing in (35) and (36) will be those of the impactor material.

Although no closed-form solution has been found for k_2' for generally orthotropic solids and its derivation appears to be extremely complex, an approximate numerical solution for k_2' of generally orthotropic solids shows k_2' to be relatively insensitive to the inplane fiber orientation [Greszczuk and Chao (1975)].

For a target made of orthotropic material with inplane anisotropy ratio of $E_L/E_T=14.3$ and impacted by a rigid spherical impactor, the area of contact was slightly elliptical. The ratio of the major to minor axis of the ellipse was only 1.07. Experimental studies by Moon (1972) have also shown that when an isotropic sphere is pressed into an orthotropic target made of unidirectional composite the area of contact is only slightly elliptical. As noted by Greszczuk (1975), the properties that influence k_2' the most are those associated with the thickness direction, that is, the direction of impact, z. Because of the weak dependence of k_2' on the inplane properties of the target, (35) can, as a first approximation, be used for generally orthotropic materials if appropriate, say average, inplane properties (E_r, ν_r) are used. The parameter k_2' for a given generally orthotropic material can also be obtained experimentally from a static indentation test. For a spherical indenter $(R_{1m}=R_{1M}=R_1)$ made of isotropic material and a flat target $(R_{2m}=R_{2M}=\infty)$, $C_R=R_1/2$, and $s=8$, and (27) when solved for k_2' yields

$$k_2' = \frac{4}{3\pi P}\left(\alpha^3 R_1\right)^{1/2} - \frac{1-\nu_1^2}{\pi E_1} \tag{37}$$

Thus by conducting the static indentation test and measuring load, P, versus deformation, α, the parameter k_2' can be determined from (37). The properties entering in the second term on the right side of (37) are the elastic properties of the indenter. If the indenter is rigid $(E_1 \gg E_2)$ compared to the target, the second term may be neglected. Figure 2 shows some typical results on load-deformation behavior of a plexiglass target indented by a steel sphere. The theoretical result predicted from (3) is also shown therein. The deviation of experimental data from theoretical prediction above $P \approx 100$ lb is because of inelastic deformation of the target. The value of k_2' in the inelastic range can also be determined from (37).

Impact Response of Flexible Target. The impact behavior of plates and beams made of isotropic materials has been studied rigorously and extensively by Goldsmith (1960). The response of anisotropic laminated plate to impact has been investigated analytically by Sun and Chattopadhyay (1975) and by Moon (1972). Greszczuk and Chao (1975) investigated both analytically and experimentally the response of composite plates to low velocity impact. Experimental investigation of the impact response of stiffened and unstiffened composite plates has been conducted by Starnes et al. (1978) and Williams et al. (1979).

For a flexible, plate-type target, the surface pressure, area of contact, and impact duration will be functions of the parameters entering in (32)–(34) as well as plate bending stiffness and boundary conditions. For a given impact velocity the magnitude of dynamic force P will decrease as the target flexibility increases (or target thickness decreases). Increase in target flexibility will also increase contact duration and decrease the area of contact. An approximate solution (subject to the assumptions made at the beginning of this chapter) for

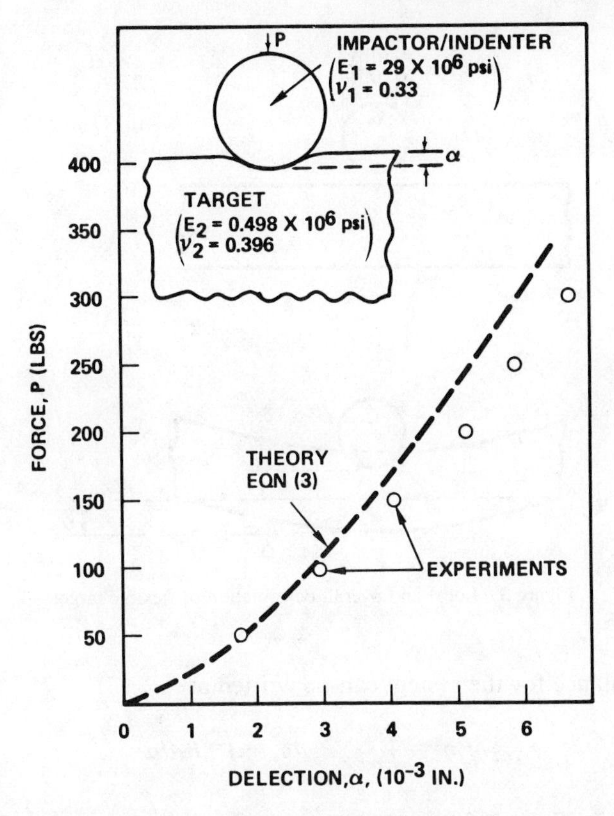

Figure 2 Test-theory comparison of force versus deformation for isotropic target made of plexiglass.

the impact response of flexible composite plates can be obtained by considering the deformations shown in Figure 3. At the point of contact, the plate undergoes the Hertzian contact deformation, α, as well as plate bending deformation, δ_p. The Hertzian force-deformation relationship for the contact problem was already given as

$$P_c = n'\alpha^{3/2} \tag{37}$$

whereas the force-deflection relationship for a plate subjected to a concentrated load will be of the form

$$P_p = K_p \delta_p \tag{38}$$

where subscripts c and p refer to the contact problem and plate respectively and K_p is the spring constant for the plate. K_p will be a function of the elastic constants of the plate material as well as plate boundary conditions. Assuming the plate to be stationary and the approach velocity of impactor being $v = v_1$,

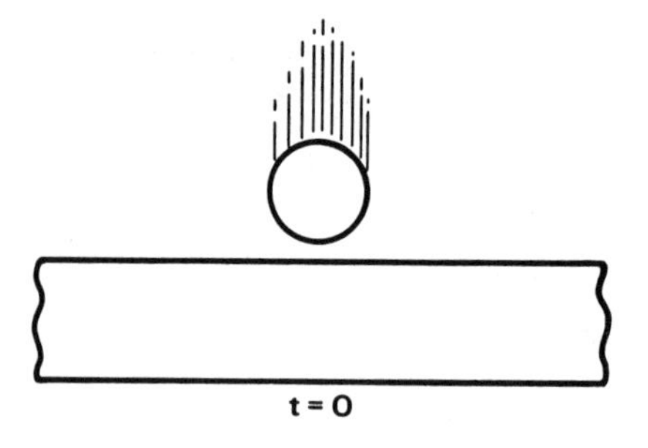

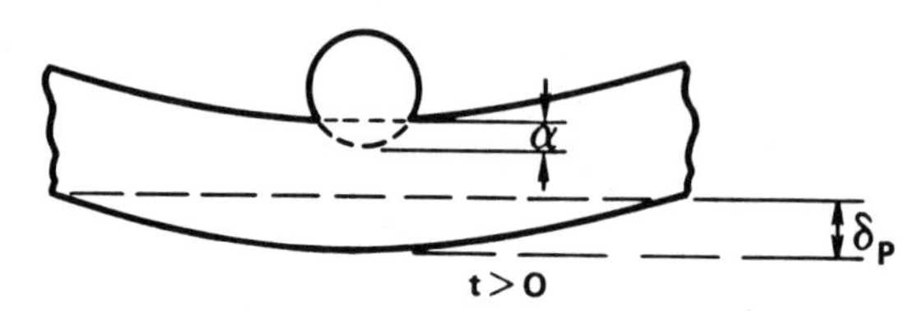

Figure 3 Local and overall deformation of flexible target.

the energy balance for the system can be written as

$$\tfrac{1}{2}m_1v^2 = \int_0^{\delta_{\text{MAX}}} P_p\, d\delta_p + \int_0^{\alpha_1} P_c\, d\alpha \tag{39}$$

Substitution of (27) and (38) into (39) and evaluation of the integrals yields

$$\tfrac{1}{2}m_1v^2 = \tfrac{1}{2}K_p\delta_p^2 + \tfrac{2}{5}n'\alpha_1^{5/2} \tag{40}$$

Finally combining (27), (28), and (40) and noting that $P_c = P_p = P$ we obtain

$$\tfrac{1}{2}m_1v^2 = \frac{1}{2}\left(\frac{P^2}{K_p}\right) + \frac{2}{5}\left(\frac{P^{5/3}}{n'^{2/3}}\right) \tag{41}$$

For a circular isotropic or pseudoisotropic composite plate of radius R and thickness h clamped along the outer boundary [Roark (1954)]

$$K_p = \frac{P}{\delta} = \frac{4\pi E_r h^3}{3(1-\nu_r^2)R^2} \tag{42}$$

whereas for a plate with simply supported edges

$$K_p' = \frac{4\pi E_r h^3}{3(1-\nu_r)(3+\nu_r)R^2} \tag{43}$$

Considering the case of a circular plate with simply supported boundaries, substitution of (28) and (43) into (41) yields

$$\tfrac{1}{2}m_1 v^2 = P^2\left[\frac{3(1-\nu_r)(3+\nu_r)R^2}{8\pi E_r h^3}\right.$$

$$\left. + P^{5/2}\left\{\frac{2s}{5}\left[\frac{3\pi(k_1'+k_2')}{16\sqrt{C_R}}\right]^{2/3}\right\}\right] \tag{44}$$

The first term on the right side of (44) gives the plate bending effect and the second term gives the Hertzian contact effect. If $t \gg 1$ it can be shown that (44) reduces to (29).

Equation (44) can be solved for P to obtain the relationship between the impact force, impact velocity, and properties of the impactor and a flexible target. If the target is made of composite material having layup other than pseudoisotropic, appropriate expressions have to be used for K_p.

Typical results on influence of plate target thickness, h, impact velocity, v, and plate boundary conditions on the dynamic force, P, are shown in Figures 4 and 5. These results are based on the following properties of a spherical

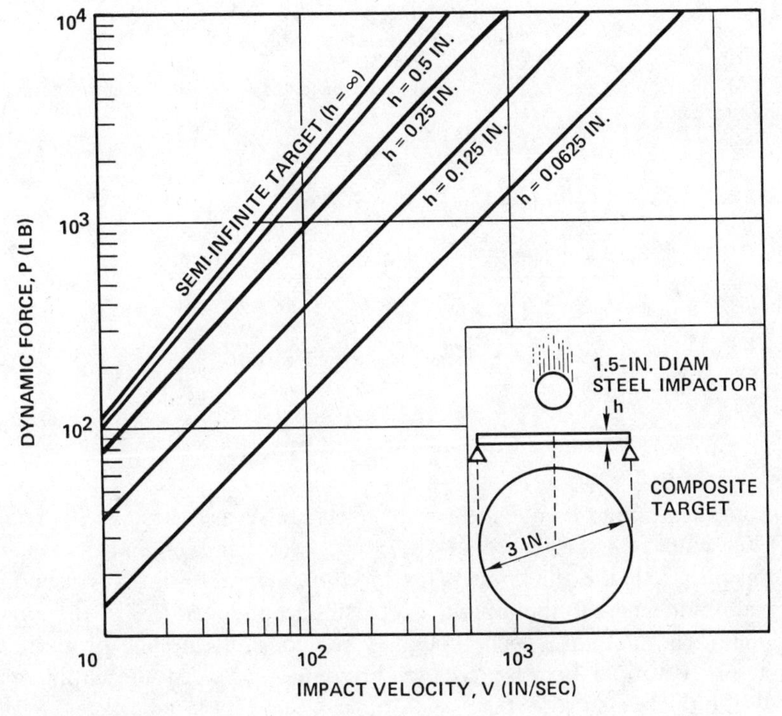

Figure 4 Dynamic force versus impact velocity and thickness of circular composite target with simply supported boundaries.

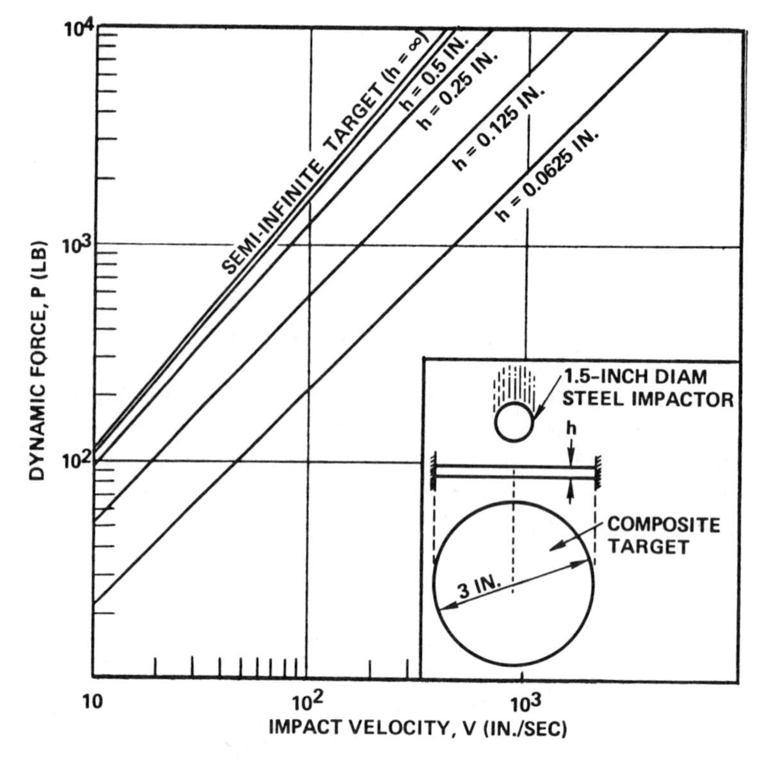

Figure 5 Dynamic force versus impact velocity and thickness of circular composite target with clamped boundaries.

impactor and a composite plate target:

Steel Impactor	Composite Target
$E_1 = 29 \times 10^6$ psi	$E_r = 7.42 \times 10^6$ psi
$\nu_1 = 0.33$	$E_z = 1.72 \times 10^6$ psi
$\rho = 0.288$ lb/in.^{3}	$G_r = 2.83 \times 10^6$ psi
	$G_{zr} = 0.60 \times 10^6$ psi
	$\nu_r = 0.31$
	$\nu_{zr} = 0.06$

The properties of the composite target given above are those for a Thornel 300/5208 with fibers oriented at (0, +60, −60) and laminates dispersed uniformly through the thickness. This type of fiber pattern gives pseudoisotropic laminate whereby the inplane properties are independent of the angular coordinate. The laminate, even though it has constant inplane properties, is transversely isotropic because of the differences between the inplane and through-the-thickness properties.

A comparison of impact force calculated from (49) and test data obtained on pseudoisotropic composite plate is shown in Figure 6.

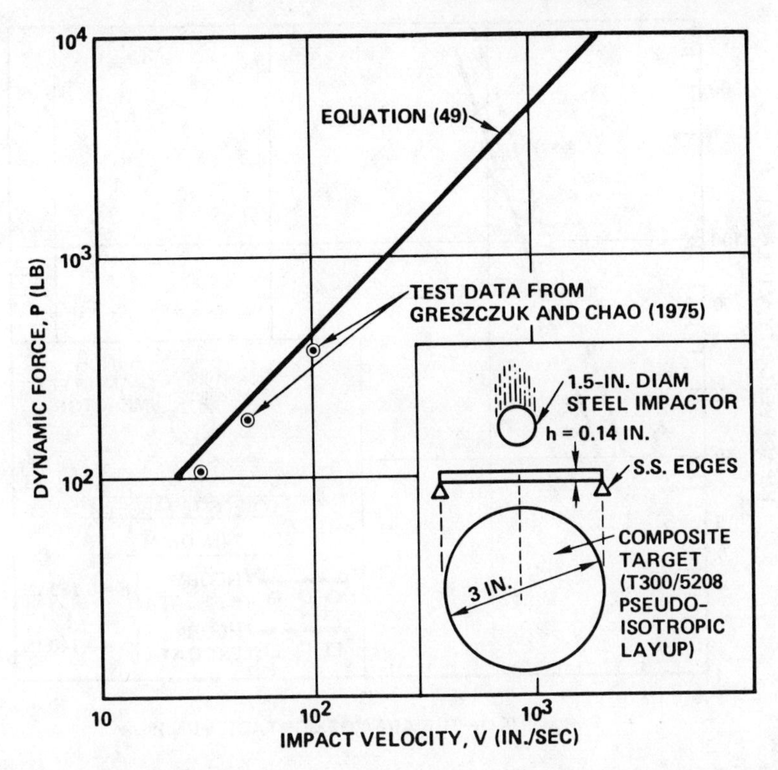

Figure 6 Test-theory comparison of dynamic force versus impact velocity for flexible composite target impacted by a steel ball.

As noted at the beginning of this section, target flexibility will also have an effect on the area of contact. The latter can be calculated approximately by first calculating force P from (41). The radii of the area of contact can then be determined directly from (22) and (23) if the calculated value of P is substituted therein. Figure 7 shows the variation of the radius of the area of contact as a function of impact velocity for Thornel 300/5208 pseudoisotropic composites plates of two different thicknesses. The solid and dashed lines are the analytical results calculated by the technique described above while the circles and squares denote test data.

3.1.2 Impact Duration

The maximum pressure q_0 occurs at a time $0.5t_0$ where t_0 is the impact duration. The latter can be determined using an approach similar to that described by Timoshenko (1934). From the problem of impact of two bodies [(9)] $(\dot{\alpha}^2 - v^2) = -\frac{4}{5}Mn\alpha^{5/2}$ or solving for $\dot{\alpha}$ and generalizing the above equation by replacing n with n'

$$\dot{\alpha} = \left(v^2 - \tfrac{4}{5}Mn'\alpha^{5/2}\right)^{1/2} \tag{45}$$

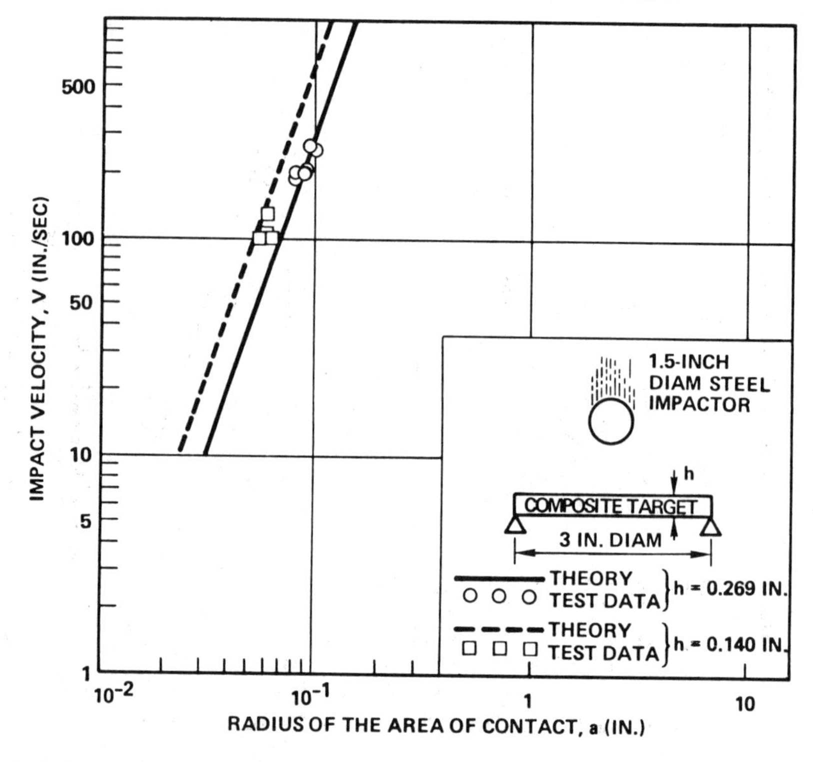

Figure 7 Test-theory comparison of the radius of area of contact in thornel 300/5208 pseudoisotropic composite targets impacted by a 1.5-in. diameter steel impactor.

Substituting $\dot{\alpha}=d\alpha/dt$ in (45) and solving for dt

$$dt=\frac{d\alpha}{\left(v^2-\frac{4}{5}Mn'\alpha^{5/2}\right)^{1/2}} \tag{46}$$

Combining (46) with (33) and integrating gives

$$t=\frac{2\alpha_1}{v}\int_0^x\frac{dx}{\left(1-x^{5/2}\right)^{1/2}} \tag{47}$$

where $x=(\alpha/\alpha_1)$. The total impact duration, t_0, is obtained by intergrating between the limits $x=0$ and $x=1$ and is

$$t_0=2.94\frac{\alpha_1}{v}=2.94\left(\frac{5}{4Mn'v^{1/2}}\right)^{2/5} \tag{48}$$

To verify the accuracy of (48), measurements were made of the contact duration between aluminum and composite plates impacted by steel spheres [Greszczuk (1980)]. The experimental setup that was employed is shown in Figure 8. Basically when the impactor was released the electric circuit was

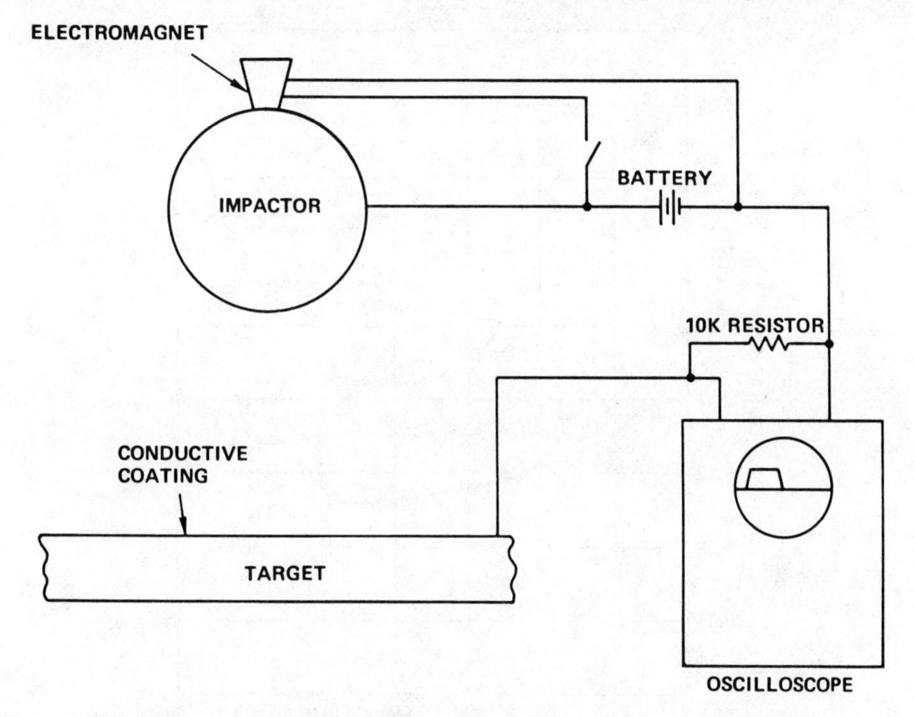

Figure 8 Schematic of test setup for measuring contact duration during impact.

closed from the time the impactor first contacted the target to the time it bounced off. The latter time duration was measured using an oscilloscope. Comparison of measured contact durations with values calculated from (48) is shown in Figure 9. Figure 10 shows experimental results on the variation of contact duration with target thickness.

The variation of surface pressure q_0, radii of the area of contact, a and b, and surface pressure distribution, q_r, with time can be determined by first numerically integrating (47) and determining α/α_1 as a function of time t/t_0. The resultant plot is shown in Figure 11. The curve can be approximated fairly accurately by an equation

$$\alpha = \alpha_1 \sin \frac{\pi t}{t_0} \tag{50}$$

or substituting (48) for t_0

$$\alpha = \alpha_1 \sin \frac{\pi t v}{2.94 \alpha_1} \tag{51}$$

Substitution of (27) into (22), (23), and (31) followed by substitution of (51) into the resultant equations yields the following expressions for a, b, and q_0 as

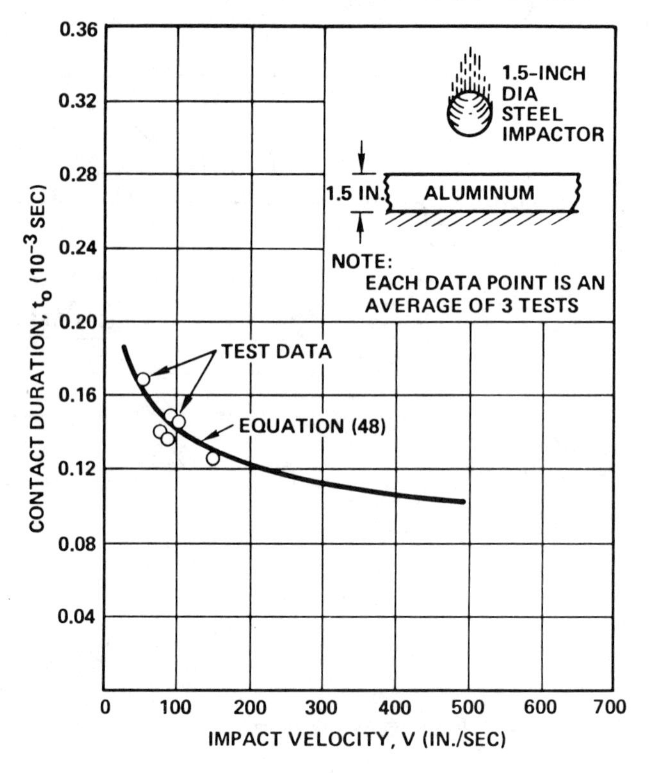

Figure 9 Test-theory comparison of contact duration in an aluminum target impacted by a steel sphere.

a function of time, t

$$\frac{a(t)}{m} = \frac{b(t)}{r} = \left(\frac{4C_R}{s} \alpha_1 \sin \frac{\pi t v}{2.94\alpha_1} \right)^{1/2} \tag{52}$$

$$q_0(t) = \frac{3n's}{8\pi C_R m r} \left(\alpha_1 \sin \frac{\pi t v}{2.94\alpha_1} \right)^{1/2} \tag{53}$$

whereas the distribution of surface pressure follows from (30).

3.1.3 Internal Stresses Caused by Impact Pressure

Knowing the surface pressure, its distribution, as well as the area of contact, all as a function of impact velocity and time, the time-dependent internal triaxial stresses in isotropic, multilayered orthotropic, or anisotropic targets can be determined using various finite-element computer codes or in some instances from closed form solutions.

Semiinfinite Isotropic Solid. In the case of semiinfinite isotropic solid subjected to surface pressure q_r distributed according to (16), the internal triaxial

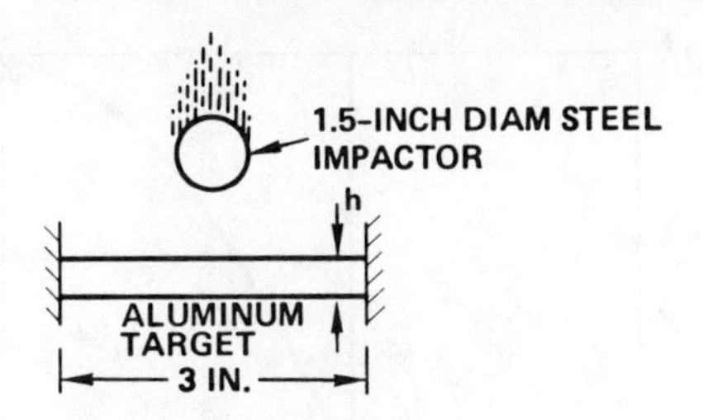

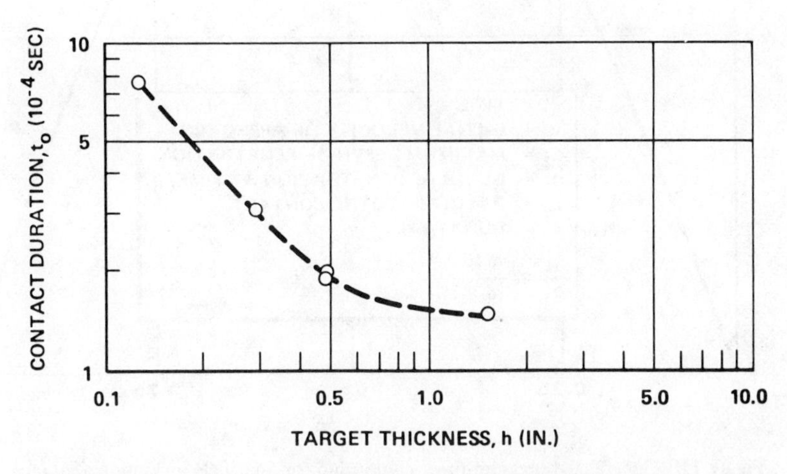

Figure 10 Influence of target thickness on contact duration ($v=100$ in./s).

stresses can be determined from the equations given by Love (1929), Huber (1956), Belajef (1945), Sneddon (1951), or from the curves in a paper by Morton and Close (1922). Figure 12 shows internal triaxial stress contours for σ_r, σ_z, and σ_θ. The stresses have been normalized with respect to maximum surface pressure q_0 (at $r=0$) while the dimensions have been normalized with respect to the radius of the area of contact, a. In addition to showing the results from a closed form solution, Figure 12 also shows the stresses determined from SAAS III finite element code developed by Crose and Jones (1968). The maximum tensile, compressive and shear stresses (σ_t, σ_c, and σ_s respectively) that occur in targets made of isotropic materials are related to the surface pressure by the following simple equations [Timoshenko (1934)],

$$\sigma_t = \left(\frac{1-\nu_2}{3}\right) q_0(t) \tag{54}$$

$$\sigma_c = q_0(t) \tag{55}$$

$$\sigma_s = \left[(1+\nu)(s\cot^{-1}s-1)+\frac{3}{2(1+s^2)}\right] q_0(t) \tag{56}$$

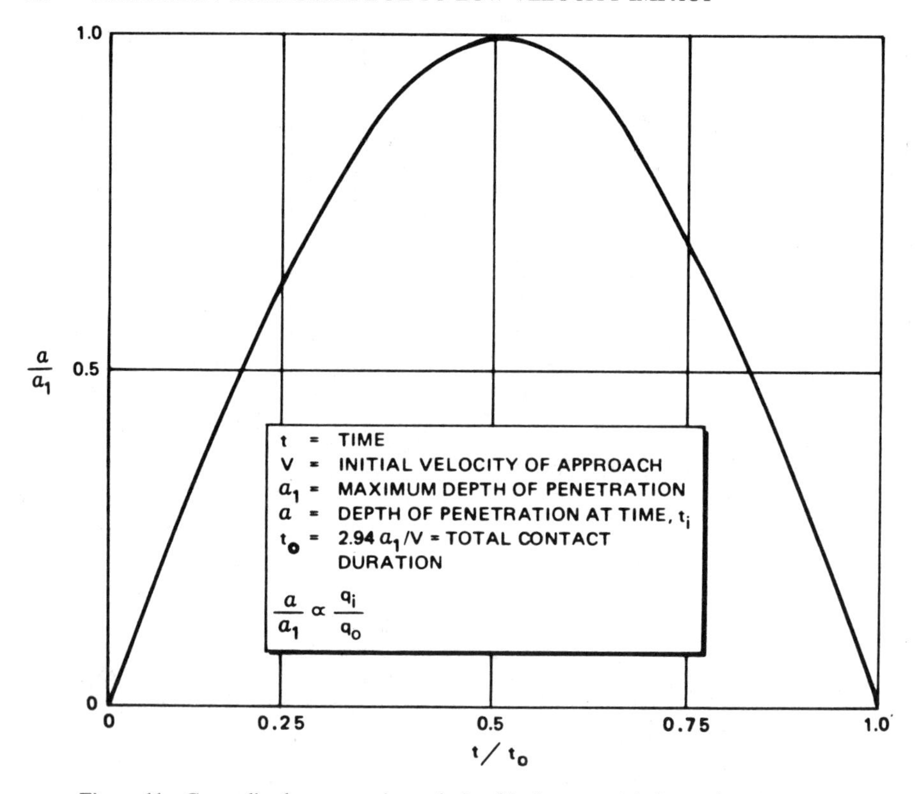

Figure 11 Generalized pressure-time relationship for a particle impacting a target.

with the maximum σ_s occurring at $s \approx 2/(1+\nu)\pi$ where $s = z/a$ and where $q_0(t)$ is the maximum surface pressure at a given time, t. The locations of maximum stresses are noted in Figure 13. Thus for brittle materials the failure due to impact would be expected to be governed by tensile strength and would initiate at the periphery of the area of contact. For materials with low shear-strength impact would produce subsurface shear failure.

For the case of flexible isotropic targets subjected to impact, calculation of the three-dimensional internal stress state caused by impact-induced surface pressure can best be carried out by finite element codes similar to those described in the next section.

Semiinfinite and Flexible Composite Targets. The internal triaxial stresses in semiinfinite or flexible composite targets subjected to impact-induced surface pressure can be determined using finite-element computer codes such as SAAS III [Crose and Jones (1968)] or the more general ASAAS code [Crose (1971)]. The SAAS III is applicable to transversely isotropic materials whereas the modified version of ASAAS [Greszczuk and Chao (1975)] is applicable to multilayer, generally orthotropic solids having orthogonal symmetry. Other

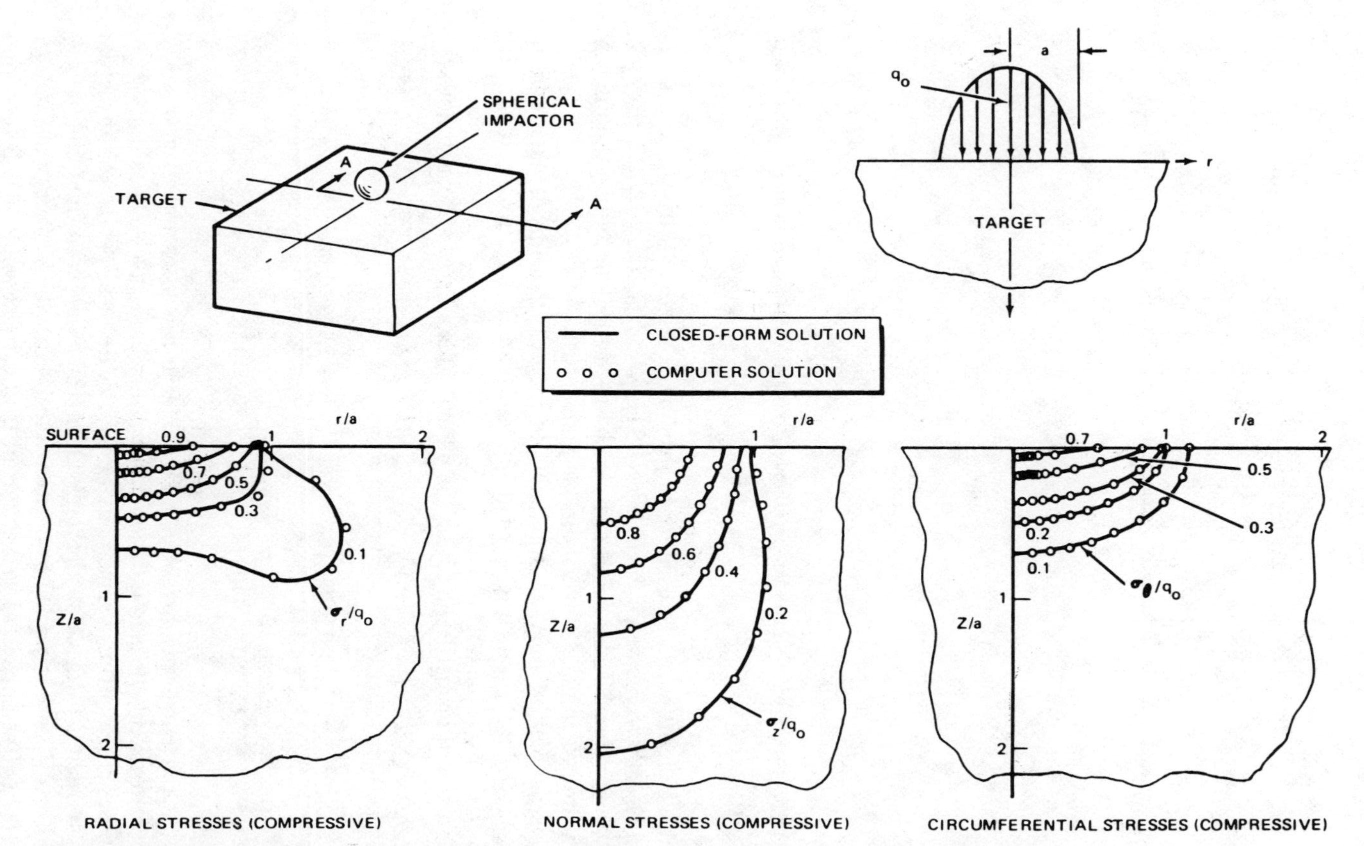

Figure 12 Comparison of closed form and computer solutions for internal triaxial stresses in a solid subjected to surface pressure caused by impact.

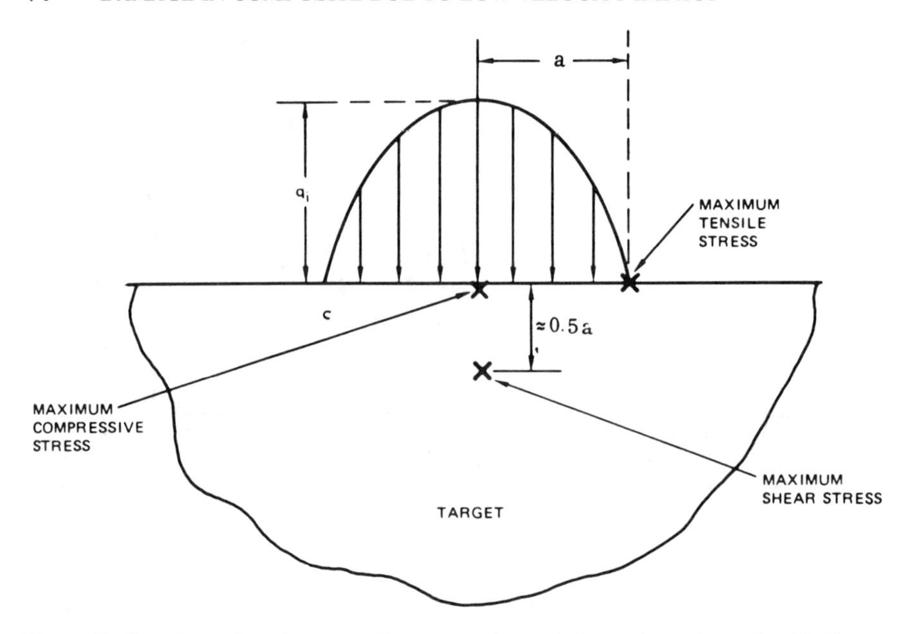

Figure 13 Locations of maximum tensile, compressive, and shear stresses in an elastichalf space under surface loading.

available codes that can be used for determining the internal triaxial stresses in generally orthotropic solids include: NASTRAN, ANSYS, MARC, SAP IV, N1SA, and others. These codes can also be used for determining the internal triaxial stresses in generally orthotropic plates.

The sensitivity of internal stress state to target orthotropy is illustrated in Figure 14 which shows the normal, σ_z, and radial, σ_r, stress contours for transversely isotropic targets with $E_r/E_z=0.35$ and $E_r/E_z=2.86$.

The material properties used for the numerical examples were as follows:

Case I ($E_r/E_z=0.35$)	Case II ($E_r/E_z=2.86$)
$E_r=2.9\times10^6$ psi	$E_r=8.3\times10^6$ psi
$E_z=8.3\times10^6$ psi	$E_z=2.9\times10^6$ psi
$v_r=0.144$	$v_r=0.260$
$v_{zr}=0.260$	$v_{zr}=0.091$
$G_{rz}=0.86\times10^6$ psi	$G_{rz}=0.86\times10^6$ psi

Thus for a given surface pressure, q_0, increase in E_z causes any given σ_z-stress contour to extend deeper into the target whereas a given stress contour σ_r diminishes in extent. Conversely, as E_r is increased (E_z decreases) any given σ_r-stress extends further into the target. If the materials corresponding to Case I and Case II were impacted at a given velocity, v_1, the surface pressures, q_0, generated by the impact, as well as the areas of contact would also be different for the two materials. Denoting the material in Case I by subscript I and

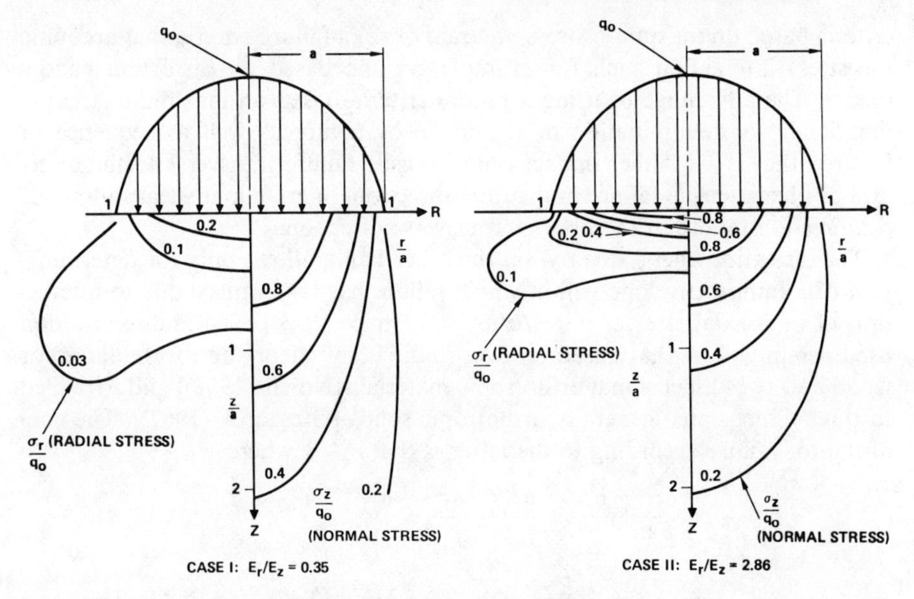

Figure 14 Internal stress distribution in semiinfinite, multilayer, transversely isotropic material resulting from surface pressure caused by foreign-object impact.

material in Case II by subscript II for a given impact velocity the ratios of the maximum surface pressures will be approximately $(q_0)_{\mathrm{I}}/(q_0)_{\mathrm{II}} \propto [(E_z)_{\mathrm{I}}/(E_z)_{\mathrm{II}}]^{4/5} \simeq 2.32$ while the ratios of the radii of the area of contact will be approximately $a_{\mathrm{I}}/a_{\mathrm{II}} \propto [(E_z)_{\mathrm{II}}/(E_z)_{\mathrm{I}}]^{1/5} \simeq 0.81$.

Thus to make a true comparison of the impact-induced stress state in the two materials requires that the results in Figure 14 be expressed in terms of $(q_0)_{\mathrm{I}}$ and a_{I} or $(q_0)_{\mathrm{II}}$ and a_{II}.

Results similar to those shown in Figure 14 can be readily obtained for generally orthotropic materials using finite element code such as ASAAS. Once the internal stress state caused by impact is known, the extent of material damage can be determined by applying appropriate failure criteria to the multiaxial stress state within the body.

3.1.4 Failure Criteria

The final step in the theory development is to establish failure modes from internal triaxial stresses caused by impact-induced surface pressure, as well as the time sequence for occurrence of the various failure modes. This can be done by applying appropriate failure criteria to the impact-induced triaxial stress state at each point in the target. As no experimentally verified dynamic failure criteria exists now for composites, therefore to get some insight into the type and extent of impact damage, applicability of static failure criteria will be assumed. In conjunction with the latter, one has a choice to use: (1) a failure criteria based on maximum stress or strain or (2) a failure criteria that accounts for stress interaction such for example, as one based on distortion energy theory. The advantage of using a failure criteria based on maximum stress is that it allows determination of the mode of failure as well as sequence of failures, that is—whether impact causes tensile failure transverse to the direction of fibers, tensile failure in the fiber direction, interlaminar shear failure, or compression failure in the fiber or transverse direction.

The distortion energy theory, on the other hand, allows only for determination of a failure envelope within which failure has taken place due to interaction of multiaxial stresses [Greszczuk and Chao (1975)]. The failure criterion used herein will be that based on distortion energy theory and formulated for wood and two-dimensional orthotropic materials [Norris (1962)] and extended to three-dimensional generally orthotropic solids [Greszczuk (1967)]. The condition for failure according to the latter is that $\sigma_e \geqslant 1$ where

$$\begin{aligned}\sigma_e = &\frac{\sigma_{11}^2}{F_{11}^2} + \frac{\sigma_{22}^2}{F_{22}^2} + \frac{\sigma_{33}^2}{F_{33}^2} + \frac{\sigma_{12}^2}{F_{12}^2} + \frac{\sigma_{13}^2}{F_{13}^2} + \frac{\sigma_{23}^2}{F_{23}^2} \\ &- \left(\frac{\sigma_{11}}{F_{11}}\right)\left(\frac{\sigma_{22}}{F_{22}}\right)\left[\frac{(1+2\nu_{21}-\nu_{23})E_1+(1+2\nu_{12}-\nu_{13})E_2}{\{(2+\nu_{12}+\nu_{13})(2+\nu_{21}+\nu_{23})E_1E_2\}^{1/2}}\right] \\ &- \left(\frac{\sigma_{11}}{F_{11}}\right)\left(\frac{\sigma_{33}}{F_{33}}\right)\left[\frac{(1+2\nu_{31}-\nu_{32})E_1+(1+2\nu_{13}-\nu_{12})E_3}{\{(2+\nu_{12}+\nu_{13})(2+\nu_{31}+\nu_{32})E_1E_2\}^{1/2}}\right] \\ &- \left(\frac{\sigma_{22}}{F_{22}}\right)\left(\frac{\sigma_{33}}{F_{33}}\right)\left[\frac{(1+2\nu_{23}-\nu_{21})E_3+(1+2\nu_{32}-\nu_{31})E_2}{\{(2+\nu_{21}+\nu_{23})(2+\nu_{31}+\nu_{32})E_2E_3\}^{1/2}}\right]\end{aligned} \tag{57}$$

where σ_e is an equivalent stress as defined above, σ's are the impact-induced direct and shear stresses, F's are the allowable strength properties of material associated with the three orthogonal directions, and E's and ν's are the Young's moduli and Poisson's ratios associated with directions 1, 2, and 3. To take into account differences in tensile and compressive strength in any given direction, allowable compressive strength, F_c is used if σ is negative and allowable tensile strength is used if σ is positive. Typical σ_e stress contours for a semiinfinite target made of Thornel 300/5208 pseudoisotropic laminate subjected to a 48 in./s impact by a 1.5-in. diameter steel spherical impactor are shown in Figure 15. Failure occurs at points where $\sigma_e \geqslant 1$. Thus for the example shown, all damage would have been below the impacted surface as none of the stress contours, $\sigma_e \geqslant 1$, extend to the surface.

This approach may also be used for calculating the impact damage in targets made of isotropic materials. Moreover, if for a given impact velocity, v, the impact damage initiates at $t<t_0$, the extent of damage at t_0 can also be established by this approach provided that degradation of target properties is taken into account. The latter can be handled by iterative technique.

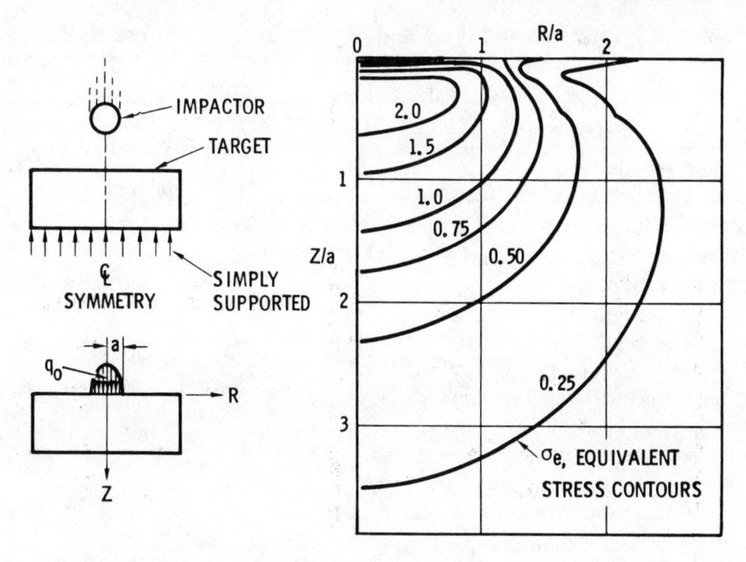

Figure 15 Equivalent stress contours for thornel 300/epoxy subjected to $q_0 = 100$ ksi.

3.2 THEORY APPLICATION

Using the approach described in the preceding section, a number of numerical examples have been worked out on the impact response of targets made of composite materials. The examples have been chosen so as to illustrate how the impact response and failure modes are influenced by the fiber and matrix properties, fiber orientation, stacking sequence, and target thickness.

3.2.1 Influence of Fiber and Matrix Properties

To establish how fiber properties influence impact response of graphite-epoxy composites, three types of composites were selected for analytical studies: Celion GY70/epoxy (made with ultra-high-modulus graphite fibers); Modmor II/epoxy (made with high-modulus, moderate-strength graphite fibers); and Thornel 300/epoxy (made with high-strength, moderate-modulus graphite fibers).

To minimize the number of variables, the target was assumed to be flat, the fiber layup was pseudoisotropic, and the target thickness in the direction of impact large compared to radius of the impactor, so that only local stresses and deformations had to be considered. The impactor was assumed to be a 1.5-in. diameter steel sphere. Properties of the composite targets used in the impact analysis are given in Table 2. Properties of the steel impactor were taken as: $E_1 = 29 \times 10^6$ psi, $\nu = 0.33$, and $\rho = 0.288$ lb/in.^{3} Pertinent results on the influence of fiber properties on the damage zone in graphite-epoxy composites are shown in Table 3. Using volume of the damage zone as a criterion to rank

Table 2 Properties of the Candidate Composite Materials

Property	Units	GY70/E	HMS/E	T300/E	GL/E
		Material			
Young's modulus, E_r	10^6 psi	14.8	11.10	8.0	4.42
Young's modulus, E_z	10^6 psi	1.01	1.4	1.77	2.9
Shear modulus, G_z	10^6 psi	0.60	0.85	0.85	0.86
Poisson's ratio, ν_r	—	0.318	0.305	0.30	0.32
Poisson's ratio, ν_z	—	0.005	0.009	0.0202	0.091
Tensile strength, σ_{rt}	10^3 psi	33.4	49.7	76.2	91.5
Tensile strength, σ_{zt}	10^3 psi	3.0	7.6	6.8	11.0
Compressive strength, σ_{rc}	10^3 psi	33.4	47.4	79.1	75.7
Compressive strength, σ_{zc}	10^3 psi	27.0	32.6	33.0	29.0
Shear strength, σ_{zr}	10^3 psi	8.6	13.7	17.6	9.0

materials with respect to impact resistance, Table 3 shows Thornel 300/epoxy to be more impact resistant than the other two materials.

To establish if response to impact loading of composites made with Thornel 300 fibers could be further improved, studies were conducted on the extent of the damage zone as influenced by matrix properties. Here again the target was assumed to be semiinfinite, fiber layup pseudoisotropic, and the impactor a 1.5-in. diameter steel sphere. Candidate resins selected included: high-modulus resin ($E_r = 1.0 \times 10^6$ psi); moderate-modulus resin ($E_r = 0.5 \times 10^6$ psi); and low-modulus resin ($E_r = 0.25 \times 10^6$ psi). The various elastic and strength properties of unidirectional composites and pseudoisotropic laminates made with Thornel 300 fibers and various resins were calculated using available analytical

Table 3 Damage Zone for Constant Impact Velocity ($v = 100$ in./s)

Material	$q_0 \times 10^{-3}$ (psi)	a (in.)	Y_Z (in.)	Y_R (in.)	$\bar{Y}^a$ (in.)
GY70/epoxy	98	0.088	0.186	0.195	0.190
Modmor II/ epoxy	124	0.083	0.171	0.161	0.166
T300/epoxy	134	0.081	0.180	0.141	0.159

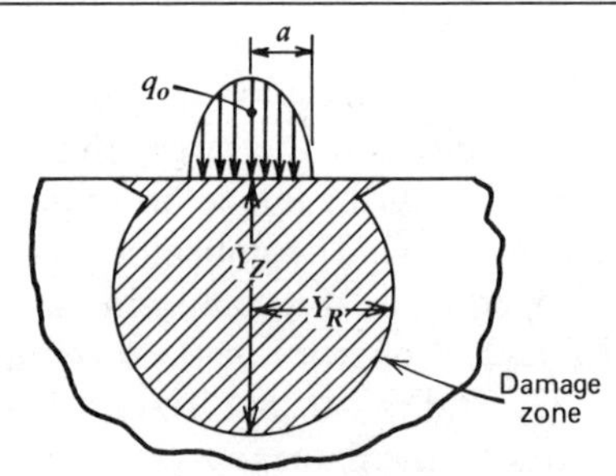

[a]Equivalent radius of damage zone defined as $\bar{Y} = \sqrt{Y_Z Y_R}$.

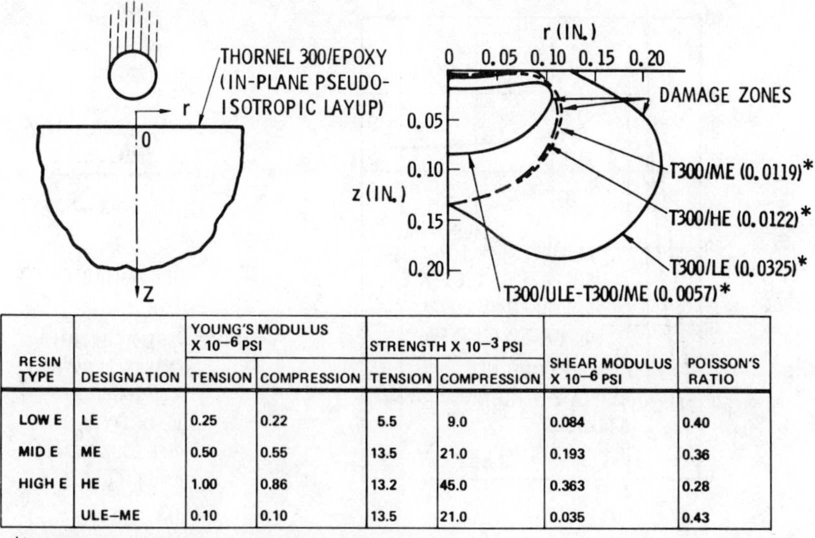

RESIN TYPE	DESIGNATION	YOUNG'S MODULUS X 10^{-6} PSI		STRENGTH X 10^{-3} PSI		SHEAR MODULUS X 10^{-6} PSI	POISSON'S RATIO
		TENSION	COMPRESSION	TENSION	COMPRESSION		
LOW E	LE	0.25	0.22	5.5	9.0	0.084	0.40
MID E	ME	0.50	0.55	13.5	21.0	0.193	0.36
HIGH E	HE	1.00	0.86	13.2	45.0	0.363	0.28
	ULE–ME	0.10	0.10	13.5	21.0	0.035	0.43

* CROSS-SECTIONAL AREA OF DAMAGE ZONE ($IN.^2$)

Figure 16 Damage resulting from a 100 in./s impact by a 1.5-in. diameter steel sphere into semiinfinite composite targets made of Thornel 300 fibers and different resins.

techniques [see for example Greszczuk (1975)]. Knowing the various properties of the target materials, the damage zones resulting from impact were determined. Figure 16 shows the damage zones in composites made with the three resins noted, as well as a hypothetical resin (designated as ULE-ME) having ultra-low modulus and moderate strength. Matrix properties are shown to have a significant influence on the extent of impact damage. To minimize the impact damage requires that the matrix have high strength and low modulus.

3.2.2 Influence of the Target Thickness

As discussed in Sections 3.1.1 and 3.1.2 target thickness and/or flexibility influences the magnitude and distribution of surface pressure, area of contact, and contact duration. Target thickness also affects the location where failure due to impact initiates, as shown in Figure 17. Critical surface pressure, q_0^*, at which failure initiates is plotted there as a function of H, the plate thickness normalized with respect to the radius of the area of contact, a. The results are for targets made of Thornel 300/5208 plates having pseudoisotropic layup and impacted by a 1.5-in. diameter steel sphere. As expected, for thin plates the impact damage initiates on the bottom surface and is governed by plate bending stresses. As the thickness of the target increases, plate bending stresses become small and impact damage is from local contact stresses. Typical results on the growth of damage zone with time in thin and semiinfinite targets are shown in Figures 18 and 19.

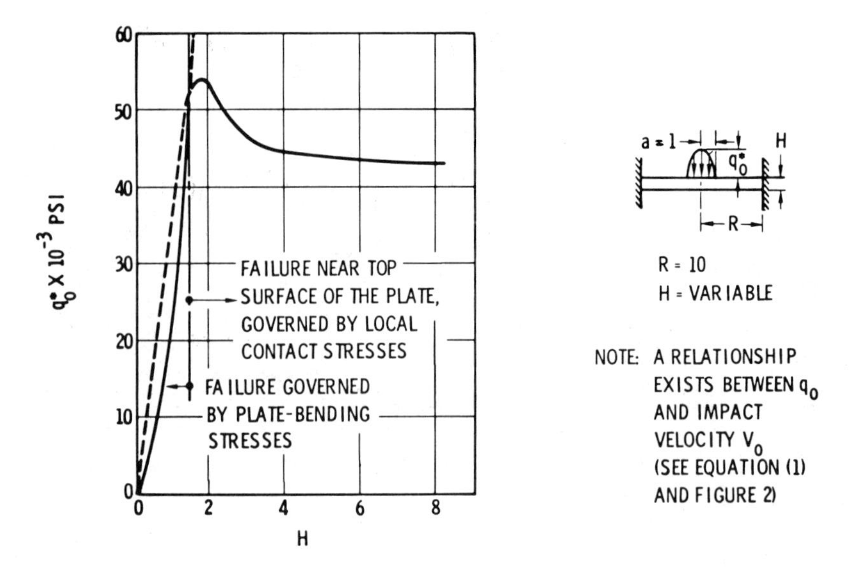

Figure 17 Influence of plate thickness on maximum surface, pressure q_0^* at which failure initiates at a point in a plate.

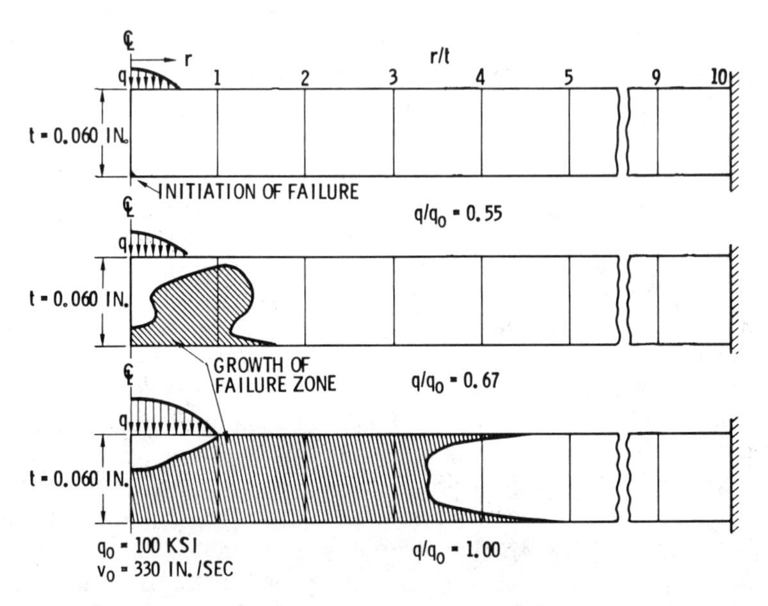

Figure 18 Increase in damage zone in a Thornel 300/epoxy composite plate as a function of increasing surface pressure and/or increasing impact velocity (pseudoisotropic layup).

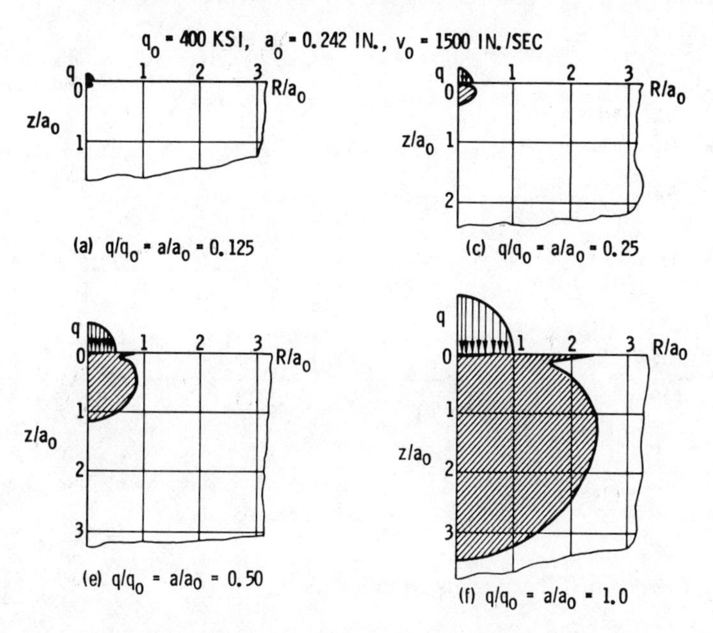

Figure 19 Increase in damage zone (shaded area) in a Thornel 300/epoxy pseudoisotropic, semiinfinite composite target with increasing surface pressure (or impact velocity).

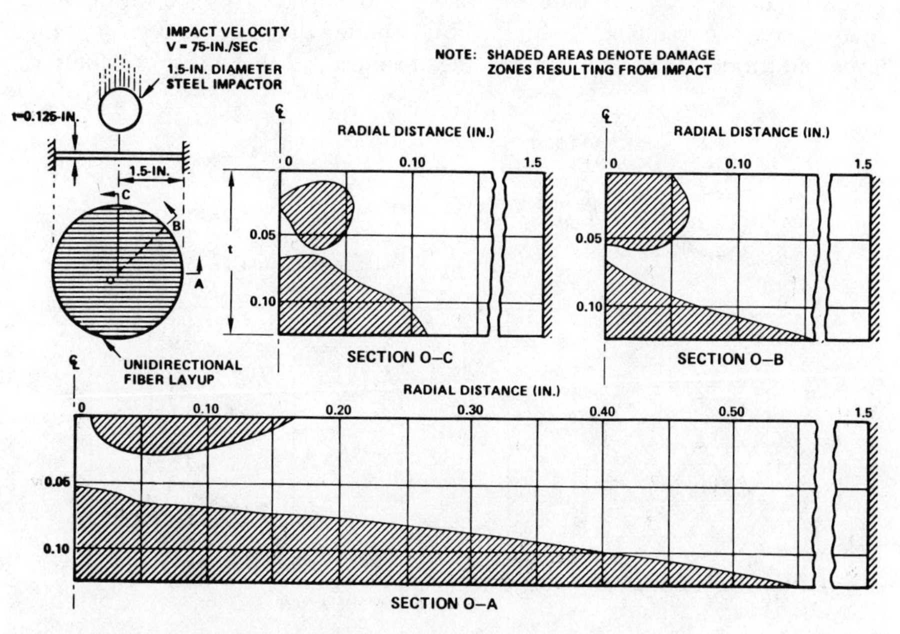

Figure 20 Foreign-object, impact-induced damage in a unidirectional Thornel 300/epoxy composite plate.

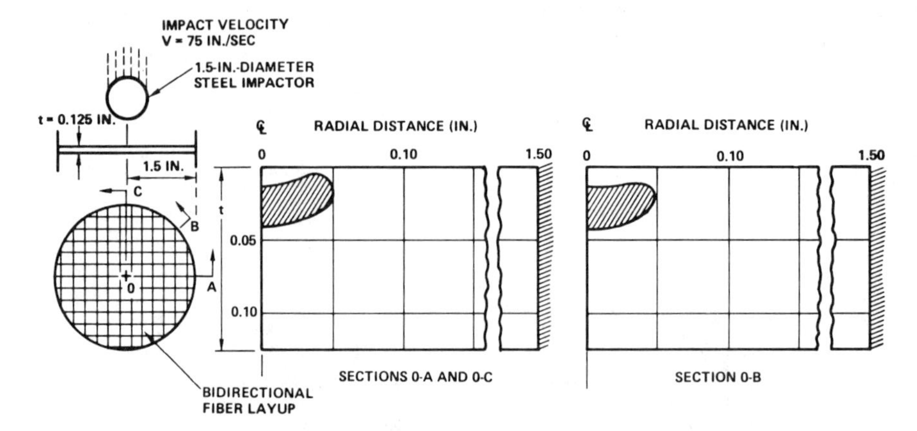

Figure 21 Foreign-object, impact-induced damage in a 1 : 1 bidirectional Thornel 300/epoxy composite plate.

3.2.3 Influence of Fiber Orientation

Another variable with significant influence on the type and extent of impact damage is fiber orientation. Figures 20, 21, 22, and 23 show the damage zones in unidirectional, bidirectional, and tridirectional (pseudoisotropic) circular plates clamped along the outer periphery and impacted by 1.5-in. diameter steel sphere. The results were obtained by assuming the plates to consist of thin layers uniformly dispersed through the thickness so that the plate material

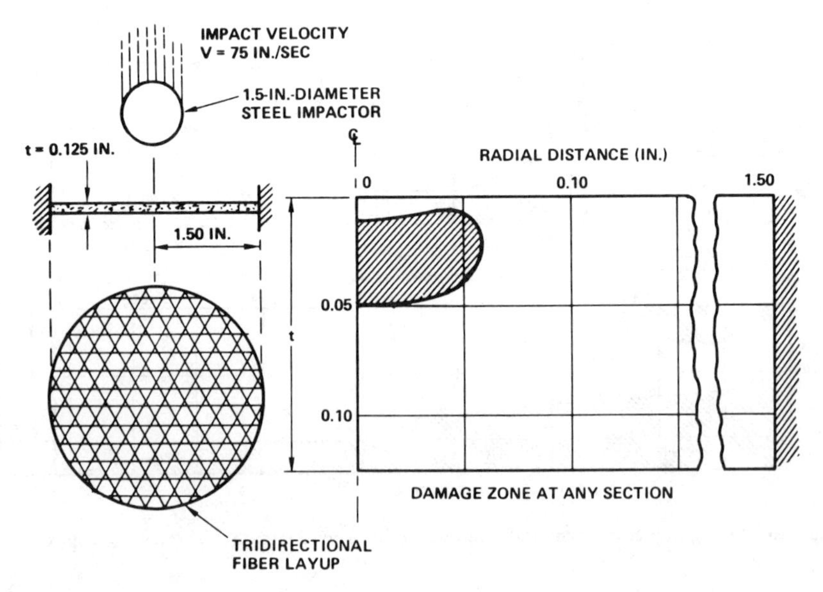

Figure 22 Foreign-Object, impact-induced damage in tridirectional Thornel 300/epoxy composite (inplane-pseudoisotropic material).

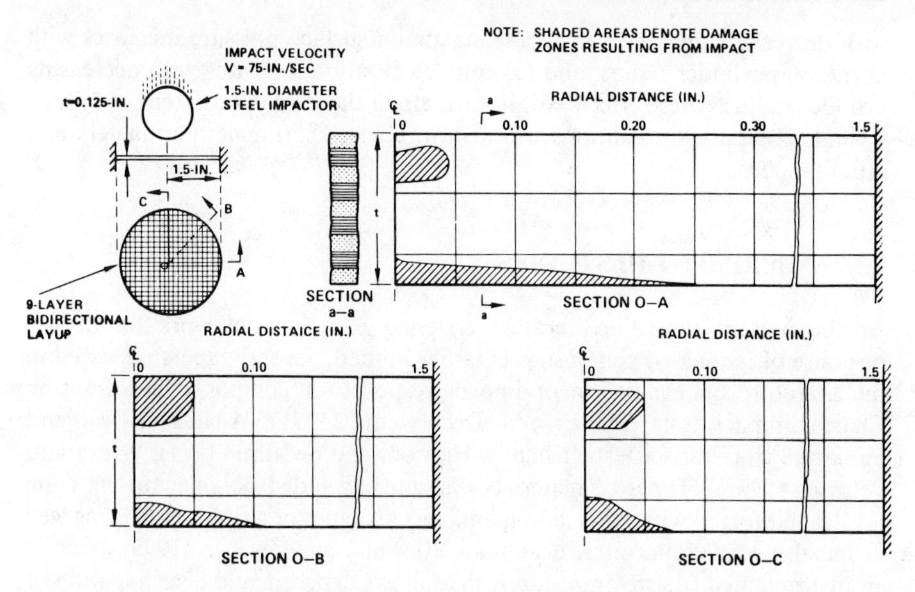

Figure 23 Foreign-object, impact-induced damage in a bidirectional Thornel 300/epoxy plate having a nine-layer dispersed construction.

could be considered as an equivalent homogeneous orthotropic material. The results in Figure 23 were calculated assuming that the bidirectional composite plate consists of nine discrete layers having fibers oriented at 0° and 90°. Any one given layer was considered as homogeneous orthotropic, whereas the composite was treated as multilayer heterogeneous orthotropic. Comparison of results in Figures 21 and 23 shows that to obtain an accurate determination of the damage zone requires that the composite be analyzed as a multilayer, heterogeneous orthotropic rather than a homogeneous orthotropic material. From the results presented above, it is apparent that the impact-induced damage zone is minimized if the layers are dispersed through the thickness and the fibers are placed in a bidirectional layup. This conclusion as well as the results on the influence of fiber and matrix properties have been confirmed experimentally as discussed in Section 3.3.

3.2.4 Influence of Target Curvature

As discussed in Section 3.1.1, wherein pertinent equations are given, target curvature affects both the magnitude and distribution of surface pressure caused by impact as well as the shape of the area of contact. Some general observations on influence of target curvature on impact parameters for a case of a cylindrical target impacted by a solid sphere [noted by Greszczuk and Chao (1975] are: (1) area of contact is elliptical and approaches a circle as the radius of the cylinder increases, (2) the area of contact decreases with decreasing cylinder radius, (3) maximum load, resulting from impact, decreases with decreasing cylinder radius, (4) maximum surface pressure increases with decreasing cylinder radius, and (5) contact duration increases with decreasing cylinder radius. These effects will in turn affect the mode and extent of failure. Cylinder boundary conditions will also influence the impact parameters and failure modes.

3.3 EXPERIMENTAL STUDIES

At the present time the literature covering experimental work on impact response of laminated composites is rather limited. Several papers appeared in the literature on evaluation of impact response of composite beams using Charpy impact tests [Novak and Decrescente (1971), Adams (1976)] and miniature izod impact tests [Chamis, Hanson, and Serafini (1971), Winsa and Petrasek (1971)]. Types of materials evaluated included polymer matrix composites reinforced with glass, boron and various types of graphite fibers, as well as metal matrix composites. Beaumont, Riewald, and Zweben (1973) describe an instrumented Charpy impact test that allows experimental determination of energy required to initiate damage, energy absorbed during damage propagation, and energy to produce catastrophic failure. The latter reference [*ASTM* (1973)] also contains a paper by Opplinger and Slepetz on impact damage tolerance of graphite/epoxy sandwich panels and several papers comparing ballistic impact response of metals and composites. Recent experimental work on response of composite plates to impact is described in publications of Greszczuk and Chao (1975), Starnes et al. (1978), and Williams et al. (1979). A paper by Figge et al. (1973) contains experimental results on response of composite cylinders to impact loading. Some of the more significant results on the impact response of composite plates are described in Section 3.3.1.

3.3.1 Experimental Setup and Instrumentation

To experimentally establish the modes of failure and the threshold energy levels required to cause impact damage in graphite-epoxy composites, ball-drop impact tests were conducted on various types of composite plates [Greszczuk and Chao (1975)]. The test variables were: reinforcing fibers, matrix material, fiber orientation, stacking sequence, plate thickness, and impact velocity. The test setup is shown in Figure 24. Several of the composite plates were instrumented with strain gages to measure strain versus time. Typical strain-time traces for 0.139-in. thick Thornel 300/5208 composite plates with $[(0,-60,60)_4]_S$ layup and subjected to 50- and 100-in./s impacts by a 1.5-in. diameter steel sphere are shown in Figure 25. The two curves shown in Figure 25 are outputs from two biaxial gages located on the bottom surface of the plates (opposite the impact site). In addition to monitoring strains, observations were also made of the failure modes caused by impact. These are discussed in sections that follow.

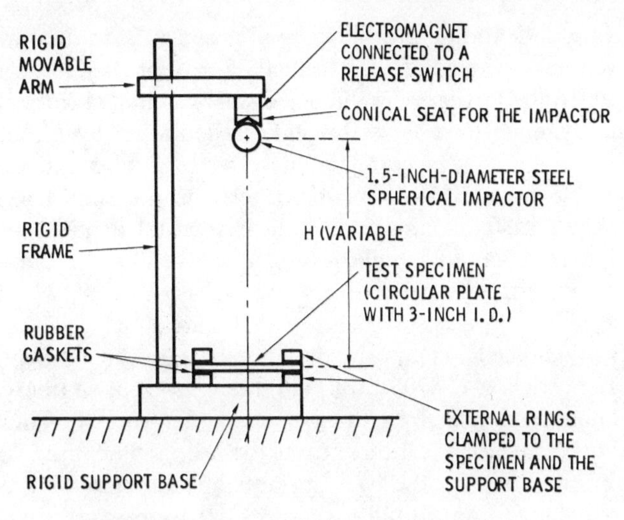

Figure 24 Test setup for low-velocity impact.

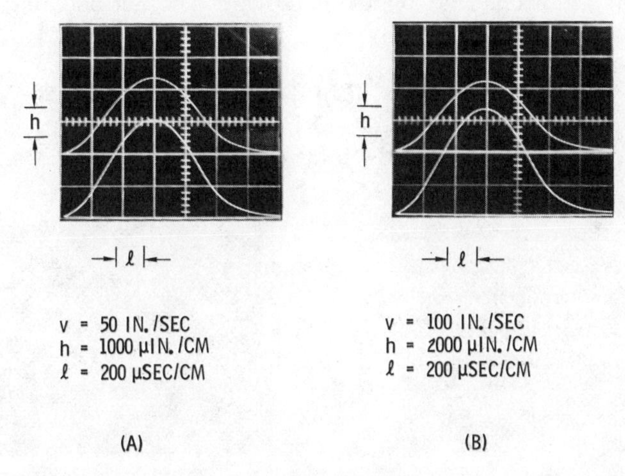

Figure 25 Impact-induced strains in pseudoisotropic Thornel 300/5208 composite plates.

3.3.2 Experimentally Observed Failure Modes

One of the assumptions in the impact theory described in Section 3.1 was that at low impact velocities the impact response can be predicted using quasidynamic, non-stress wave approach. This assumption was found to be valid in so far as determination of impact force, area of contact, and contact duration is concerned, as shown previously by the test-theory correlation of these parameters. It now remains to examine the failure modes caused by low-velocity impact. Typical failure modes produced in ATJ-S graphite by low-velocity impact and static indentation are shown in Figure 26. Both the static indentation and impact tests were conducted using 0.25-in. diameter steel spheres. The velocities at which the targets were impacted were calculated from (13) so as to give the same dynamic load P as the static indentation load. As seen from Figure 26 the failure modes correlate quite well. Similar results but for a Thornel 300/5208 pseudoisotropic composite plate are shown in Figure 27. Both the backface surface damage as well as internal damage due to static indentation and impact are similar. Additional results on impact-induced failure modes in composite plates are shown in Figures 28, 29, and 30. All plates were tested using test setup shown in Figure 24. Figure 28 shows the backface impact damage in composite plates made with three different types of reinforcing fibers: Celion GY70 (ultra-high-modulus graphite fibers), Modmor II (high-modulus, moderate-strength fibers), and Thornel 300 (high-strength, moderate-modulus fibers). In all three cases the resin system used was Narmco 5208 epoxy. Prior to impact, the back surfaces of the plates were coated with chalk dust to enhance the visibility of cracks that formed on impact. The pertinent information for the plates and impact parameters are given in Figure 28. By comparing the impact damage in plates made from the three different materials, it is obvious that plates made with Thornel 300 fibers were far

FRACTURE IN GRAPHITE SUBJECTED TO STATIC INDENTATION AND IMPACT USING A 1/4-INCH-DIAMETER STEEL BALL

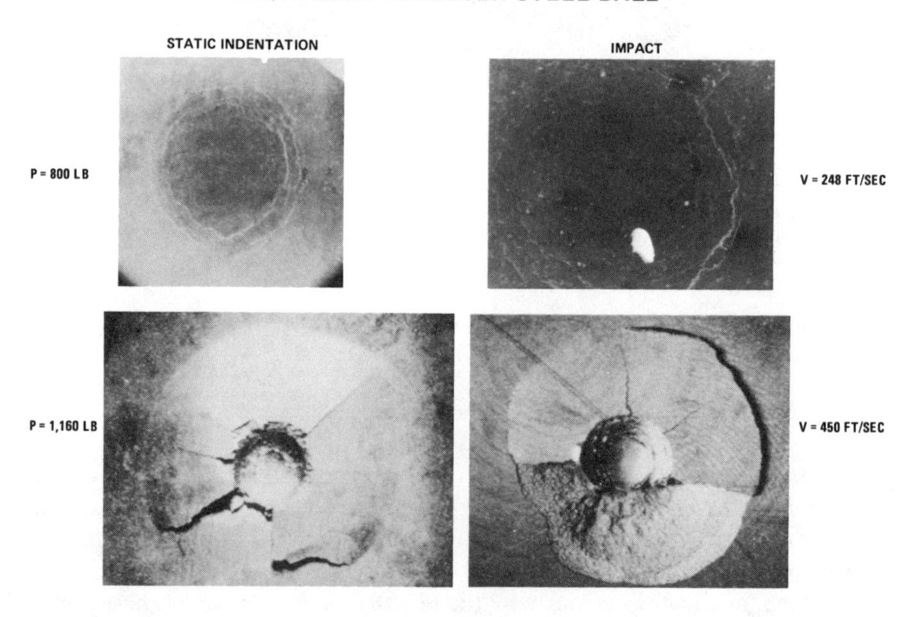

Figure 26 Fracture in graphite subjected to static indentation and impact using a $\frac{1}{4}$-in. diameter steel ball.

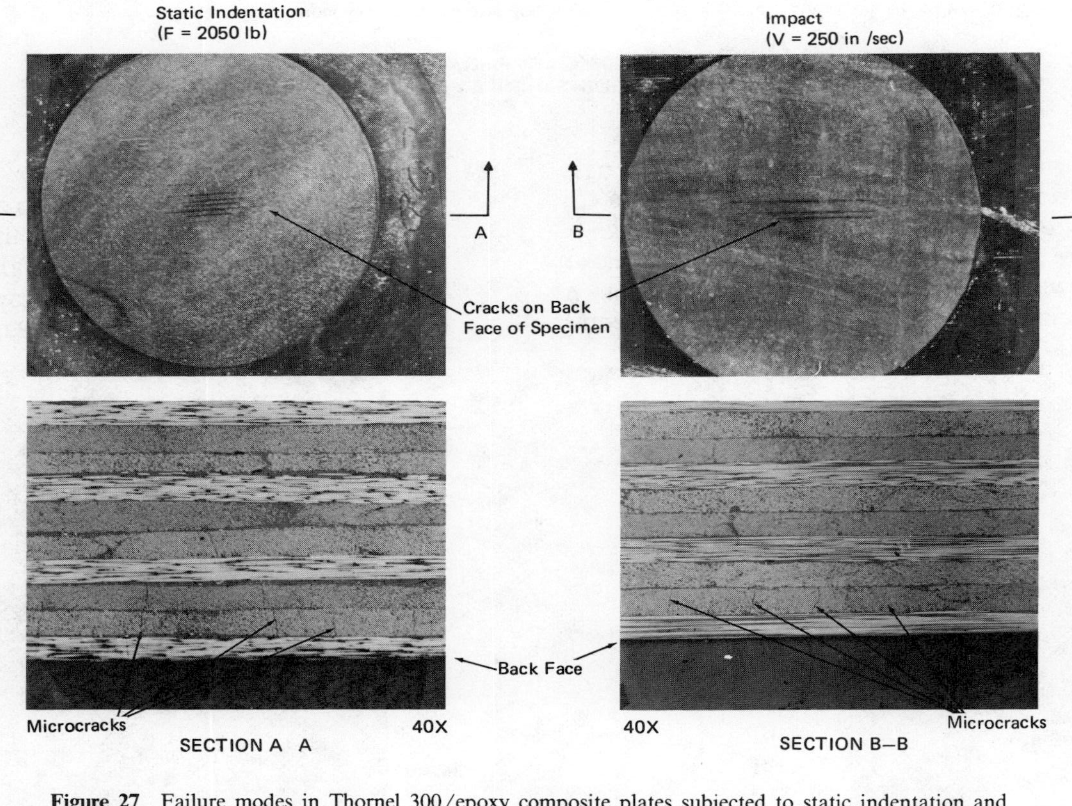

Figure 27 Failure modes in Thornel 300/epoxy composite plates subjected to static indentation and impact.

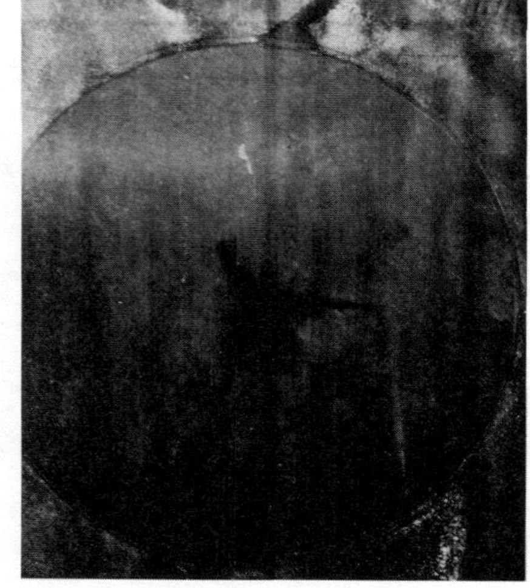

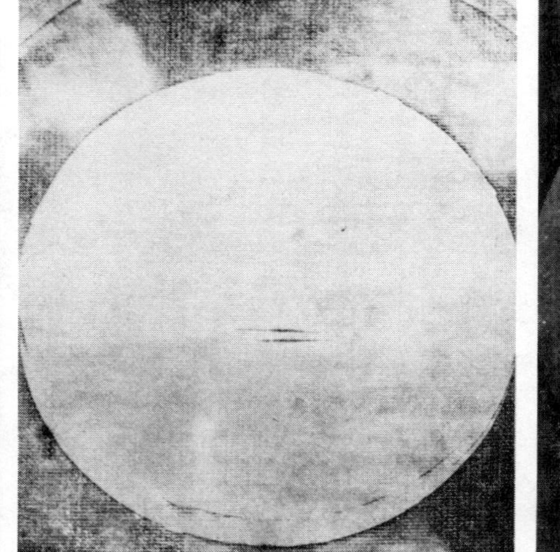

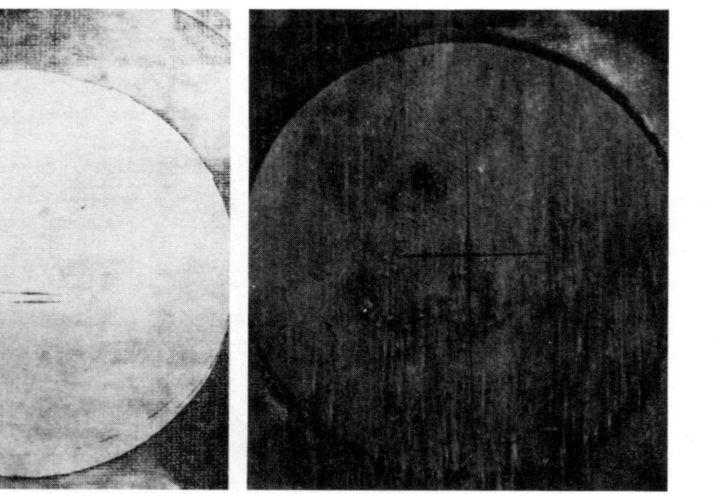

MATERIAL:	CELION GY70/5208	MODMOR II/5208	THORNEL 300/5208
IMPACT VELOCITY:	V = 100 IN./SEC	V = 100 IN./SEC	V = 100 IN./SEC
PLATE THICKNESS:	t = 0.132 IN.	t = 0.167 IN.	t = 0.139 IN.
FIBER VOLUME:	k_f = 61.7%	k_f = 55.9%	k_f = 58.0%
VOID CONTENT:	k_v = 0.12%	k_v = 0.81%	k_v = 0.93%
	SPEC 1.1a	SPEC 1.1c	SPEC 1.1b

NOTE: RADIUS OF PLATE WAS 1.5 INCHES; IMPACTOR WAS A 1.5-INCH-DIAMETER STEEL SPHERE; MATERIAL LAYUP WAS $[(0,-60, 60)_4]_S$

Figure 28 Impact damage in pseudoisotropic composite plates reinforced with three different types of fibers.

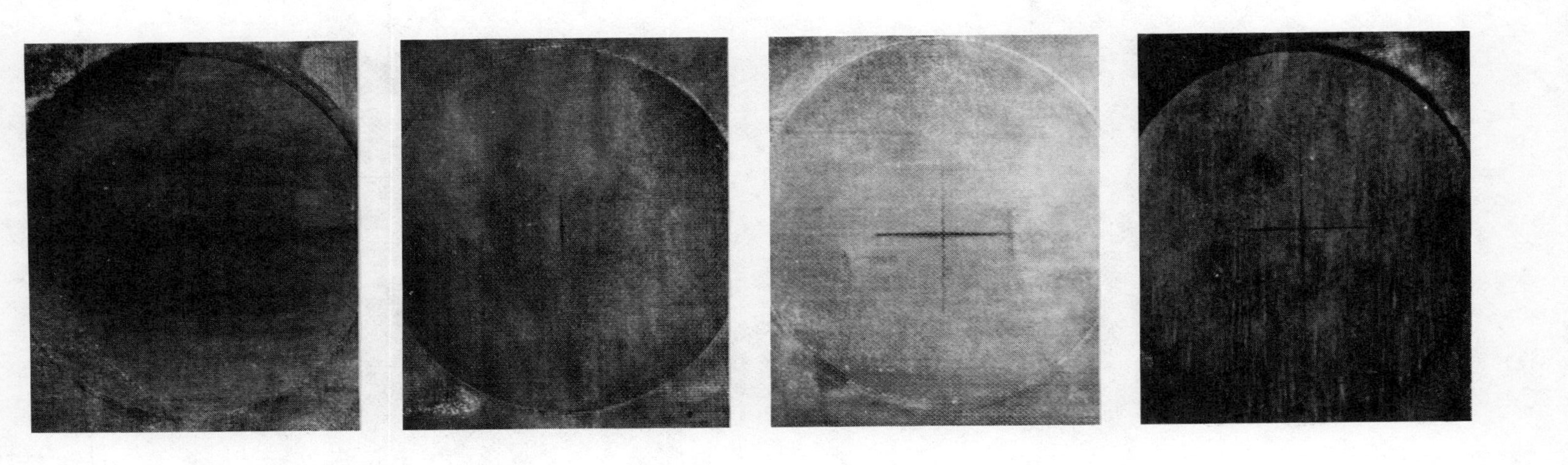

UNIDIRECTIONAL LAYUP	1:1 BIDIRECTIONAL LAYUP	2:1 BIDIRECTIONAL LAYUP	TRIDIRECTIONAL LAYUP
V = 100 IN./SEC	V = 100 IN.SEC	V = 100 IN./SEC	V = 100 IN./SEC
t = 0.139 IN.	t = 0.143 IN.	t = 0.141 IN.	t = 0.139 IN.
k_f = 58.5%	k_f = 56.2%	k_f = 58.1%	k_f = 58.0%
k_v = 0.85%	k_v = 0.92%	k_v = 0.70%	k_v = 0.93%
SPEC 1.3a	SPEC 1.3b	SPEC 1.3c	SPEC 1.1b

Figure 29 Influence of fiber layup on impact damage in composite plates made of Thornel 300/5208.

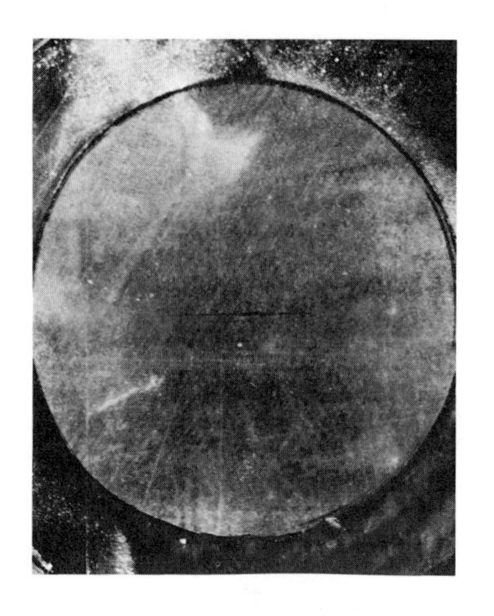

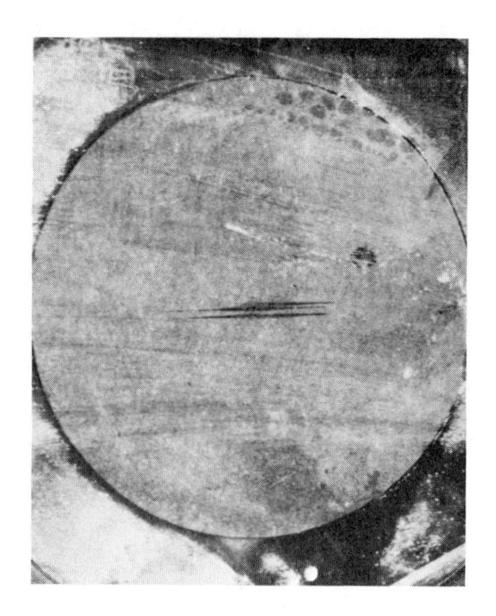

V = 75 IN./SEC

V = 100 IN./SEC

t = 0.066 IN., k_f = 63.2%, k_v = 1.03% (NO DAMAGE AT V = 50 IN./SEC)
(A)

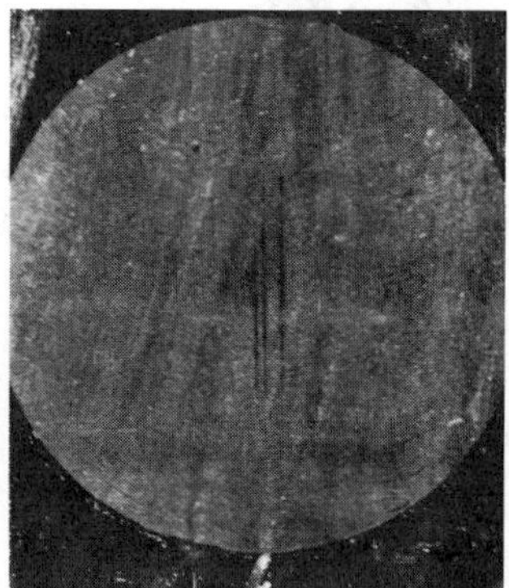

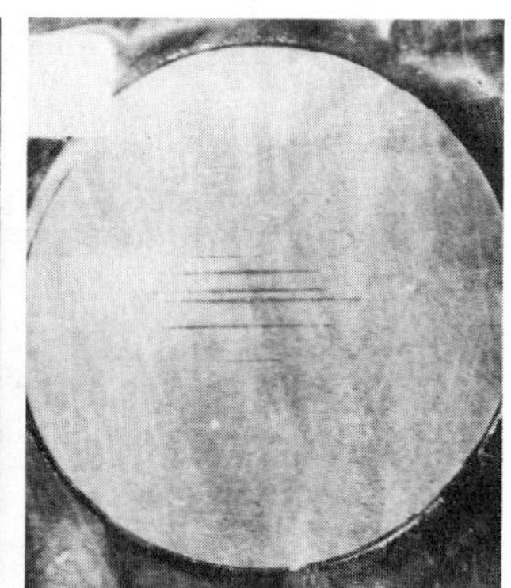

V = 200 IN./SEC

V = 250 IN./SEC

V = 266 IN./SEC

t = 0.269 IN., k_f = 58.5%, k_v = 1.07% (NO DAMAGE AT V = 192 IN./SEC)
(B)

Figure 30 Influence of plate thickness and impact velocity on impact damage in pseudoisotropic Thornel 300/5208 composite plates.

superior in resisting damage than those made with GY70 or Modmor II fibers. It is noted here that the Modmor II plates were ≈20% thicker than those of Thornel 300 or GY70. Thus if the Modmor II plates were of the same thickness as the other plates, more damage than shown in Figure 28 would be expected.

Figure 29 shows typical backface impact damage in composite plates made with Thornel 300/5208. Four different fiber layups were employed: unidirectional, 1:1 bidirectional, 2:1 bidirectional, and tridirectional (or pseudoisotropic). Rating for impact damage resistance was found to be as follows (starting with the best): (1) 1:1 bidirectional, (2) tridirectional, (3) 2:1 bidirectional, and (4) unidirectional.

In the case of the unidirectional specimen, the crack formed on impact progressed through the width of the specimen and only the support ring stopped further crack growth. In the case of plates made with multidirectional fibers, the reinforcement normal to the crack direction acted as the crack stopper and thus limited crack growth.

The results on the plate thickness effects and damage versus increasing impact velocity shown in Figure 30 are self-explanatory. As expected, increasing the thickness of the plates causes increased resistance to damage. From the experimental results obtained by Greszczuk and Chao (1975), the estimated

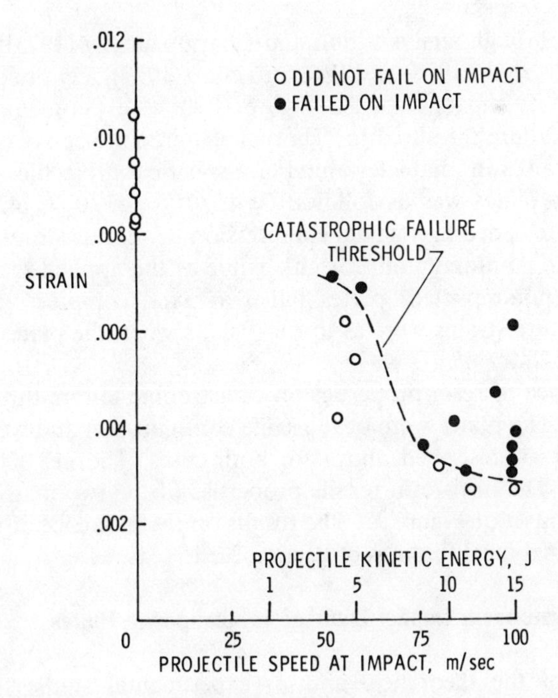

Figure 31 Effect of projectile speed on catastropic failure strain [from Stranes, et al. (1978)].

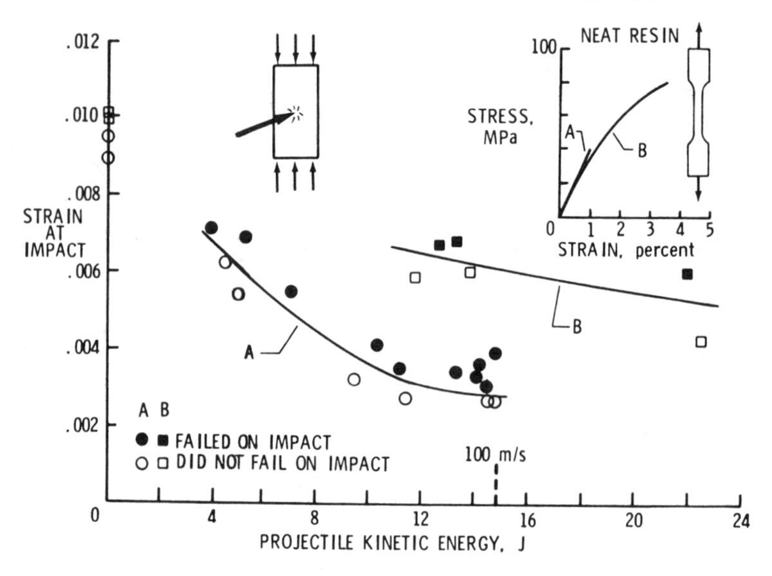

Figure 32 Damage tolerance improvement using alternate matrix. Projectile used was 1.27-cm diameter aluminum sphere [from Williams, et al. (1979)].

threshold impact velocities for initiation of damage in 0.066-, 0.139-, and 0.269-in. thick composite plates were estimated to be approximately 75, 100, and 200 in./s respectively.

As shown through analysis [Sun and Chattopadhyay (1975)] and by experiments [Starnes et al. (1978) and Williams et al. (1979)] any preload on the plate will influence its impact response. Figure 31 shows experimental results on the catastrophic failure threshold for Thornel 300 graphite-epoxy composite plates impacted by a 0.5-in. diameter aluminum spherical projectile. The fiber layup in the 48 ply plates was as follows: $(\pm 45/0_2/\pm 45/0_2/\pm 45/0/90)_{2S}$. The plates were preloaded in uniaxial compression to various strain levels and then impacted with an aluminum projectile while at the applied load. Whereas the undamaged (nonimpacted) plates failed at axial compressive strains above 0.008, the failure strains were as low as 0.0031 when the plates were impacted at 100 m/s (328 ft/s).

The influence of resin properties on catastrophic failure threshold is shown in Figure 32. The plates were of the same configuration and were tested in the same manner as described above. In both cases, Thornel 300 was the reinforcing fiber. The neat-resin tensile properties of the two matrix materials are shown in the inset of Figure 32. The results on the influence of resin properties confirm the analytical findings of Section 3.2.1.

3.3.3 Conclusions on Impact Damage in Composite Plates

The results of the theoretical and/or experimental studies show that: (1) Resistance to damage increases as the fiber strength increases and the fiber modulus decreases. (2) Resistance to damage increases as the Young's modulus of the matrix decreases and the strength of the matrix increases. (3) Bidirectional layup is more efficient in resisting damage than tridirectional or unidirectional layup. (4) Construction using complete dispersion of layers (having different fiber orientations) through the thickness is more resistant to damage than that in which the layers were not dispersed. (5) Impact can cause extensive internal damage (matrix cracking and delaminations) with very little or no visible damage on the outer surfaces. (6) In thick targets impact causes local, subsurface damage, whereas in thin targets the damage initiates on the backface. (7) Preload on the target affects the magnitude of the impact velocity that produced catastrophic failure and (8) Target curvature affects the impact parameters and failure modes.

REFERENCES

Adams, D. F. (1977), in *Composite Materials: Testing and Design (Fourth Conference)* ASTM STP **617**, 409.

ASTM (1975), *Foreign Object Impact Damage in Composites*, ASTM STP **568**, 183.

Belajef, W. M. (1924), *Memoirs on Theory of Structures*, Izdatelstwo Puti, St. Petersburg.

Belajef, W. M. (1957), *Works in the Theory of Elasticity and Plasticity*, Gosudarstvennoje Izdatelstwo Tekh-Teor, St. Petersburg.

Beaumont, P. W. R., Riewald, P. G., and Zweben, C. (1975), in *Foreign Object Impact Damage in Composites*, ASTM STP **568**, 134.

Chamis, C. C., Hanson, M. P., and Serafini, T. T. (1972), in *Composite Materials: Testing and Design (Second Conference)*, ASTM STP **497**, 324.

Conway, H. D. (1956), *Z. Angew. Math. Phys.*, **7**, 460.

Crose, J. G. (1971), The Aerospace Corporation, TR-0172 (S2816-15)-1.

Crose, J. G., and Jones, R. M. (1968), The Aerospace Corporation, TR-0200(54950)-1.

Figge, J. E., Henshaw, J., Roy, P. A., and Oster, E. F. (1974), in *Composite Materials: Testing and Design (Third Conference)*, ASTM STP **546**.

Goldsmith, W. (1960), *Impact*, Edward Arnold, London.

Greszczuk, L. B. (1967), McDonnell Douglas Astronautics Company Report, DAC60869.

Greszczuk, L. B. (1975a), in *Foreign Object Impact Damage in Composites*, ASTM STP **568**, 183.

Greszczuk, L. B. (1975b), *AGARD Conf. Proc.* No. 163.

Greszczuk, L. B. (1980), McDonnell Douglas Astronautics Company Contract Report.

Greszczuk, L. B., and Chao, H. (1975), U.S. Army Air Mobility R and D Center, USAAMRDL-TR-75-15.

Hertz, H. (1881), *J. Reine Ang. Math.*, **92**, 156.

Hertz, H. (1895), *Gesammelte Werke*, **1** Leipzig, 155.

Huber, M. T. (1956), *Pisma*, **2**, Warsaw.

Love, A. E. H. (1929), *Phil. Trans. Roy. Soc. Lond.* A, **228**.

Moon, F. C. (1972), National Aeronautics and Space Administration, NASA CR-121110.

Morton, W. B., and Close, L. J. (1922), *Phil. Mag., S. G.*, **43**, No. 254, 320.

Norris, C. B. (1950), Forrest Products Laboratory, FPL 1816.

Novak, R. C., and DeCrescente, M. A. (1972), in *Composite Materials: Testing and Design (Second Conference)*, ASTM STP **497**, 311.

Opplinger, D. W., and Slepetz, J. M. (1975), in *Foreign Object Impact Damage in Composites*, ASTM STP **568**, 30.

Lord Rayleigh (1906), *Phil. Mag.*, **11**, 283.

Roark, R. J. (1954), *Formulas for Stress and Strain*, McGraw-Hill, New York.

Sneddon, I. (1951), *Fourier Transforms*, McGraw-Hill, New York.

Starnes, J. H. Jr., Rhodes, M. D., and Williams, J. G. (1978), National Aeronautics and Space Administration, NASA TM 78796.

Sun, C. T., and Chattopadhyay, S. (1975), *J. Appl. Mech. Trans. ASME.*, **42**, No. 3, 693.

Timoshenko, S. (1934), *Theory of Elasticity*, McGraw-Hill, New York.

Whittemore, H. L., and Petrenko, S. N. (1921), *U.S. Bur. Standards Tech.* Paper 201.

Williams, J. G., Anderson, M. S., Rhodes, M. D., and Starnes, J. H. (1979), National Aeronautics and Space Administration, NASA TM 80077.

Winsa, E. A., and Petrasek, D. W. (1972), in *Composite Materials: Testing and Design (Second Conference),* ASTM STP **497**, 350.

4

ELASTIC – PLASTIC STRESS WAVES

Theodore Nicholas

A material subjected to stresses through some external dynamic loading or testing apparatus will experience deformations that can be determined from the equilibrium equations or equations of motion and the material's constitutive relations. If the intensity or rate of loading is high enough, the inertia forces in the material have to be considered. Disturbances are then propagated through the body as stress waves. For loadings that cause stresses below the yield limit in the material, elastic stress waves are generated. If the intensity of loading causes stresses beyond the yield limit, then inelastic stress waves are generated. This chapter primarily deals with inelastic stress waves in metallic materials, generally referred to as plastic waves. The analysis of plastic wave propagation phenomena requires a mathematical description of the behavior of the material as a function of strain rate, or a constitutive relation. However, the constitutive relation is often the quantity that is being sought in plastic wave propagation studies. It is this dilemma that makes it difficult to properly interpret experimental data obtained under conditions of high rates of loading where wave propagation phenomena must be considered. It has also motivated the search for theories of elastic-plastic and shock-wave propagation which are both realistic in their physical basis and are mathematically tractable. The study of plastic wave propagation thus serves two functions: it attempts to explain the response of materials to intense dynamic loading, and it serves as a basis for determining dynamic material properties.

The subject of plastic waves has been treated extensively in the literature and reviewed from various points of view. Wave problems in the theory of plasticity are treated extensively in the book by Nowacki (1978). A general discussion of plastic waves is provided by Clifton (1974). The mathematical basis for plastic wave propagation is treated in detail by Cristescu (1967). Plastic waves are also reviewed by Craggs (1961), while a survey of viscoelastic waves is provided by Hunter (1960). The earlier work in plastic waves is reviewed from a nonmathematical viewpoint by Hopkins (1963). A review of theoretical treatments and experimental methods in dynamic plasticity is presented by Campbell (1973), and discussed in terms of the rate-controlling mechanisms. Plastic wave theory, especially for uniaxial strain conditions, is reviewed by Herrmann (1969). Chou and Hopkins (1973) also treat high-amplitude stress waves, primarily shock waves, which are also treated in Chapter 1. A number of pertinent papers are to be found in the symposium proceedings of Kolsky and Prager (1964).

This chapter deals primarily with waves of uniaxial stress in long rods, but other configurations are discussed and the relationships with uniaxial strain waves are considered. Although the strain rates and stress levels obtained in long rod experiments are considerably lower than those obtained in uniaxial-strain or flat-plate experiments, the methodology developed for deducing the constitutive equations and predicting the wave propagation phenomena is common to both. The added complexities of thermodynamic considerations and combined deviatoric and hydrostatic stress and strain states in uniaxial-strain investigations make it more difficult to solve even the simplest wave propagation problems. This will become quite apparent when the difficulties in the simpler uniaxial-stress case are outlined.

4.1 WAVES OF UNIAXIAL STRESS IN LONG RODS

4.1.1 Analysis by Rate-Independent Theory

A major portion of the experimental work on plastic wave propagation has dealt with longitudinal waves in wires, rods, or bars which can be readily instrumented along their length to detect strains or particle velocities which, in turn, can be used to deduce the nature of wave propagation along the rod. The earliest work on the propagation of stress waves in impacted bars is that of St. Venant (1868) who related the stress in a bar to the velocity of an impacting mass. Hopkinson (1901) performed the first experiments in plastic wave propagation using elastic wave theory to explain the propagation of stress pulses in annealed iron wires. The foundation of the rate-independent (RI) theory of plastic wave propagation was provided by Donnell (1930) who studied the effect of a nonlinear stress-strain law on the propagation of inelastic stress in a bar. He considered a material with a bilinear stress-strain relation as shown in Figure 1 and predicted that two distinct wave fronts would propagate through the material. The velocity of each wave front has its own characteristic speed dependent on the respective moduli of the elastic and plastic regions, E and E_1, resulting in the wave profile shown in Figure 1. The analysis of Donnell thus extended the concept of elastic waves propagation in

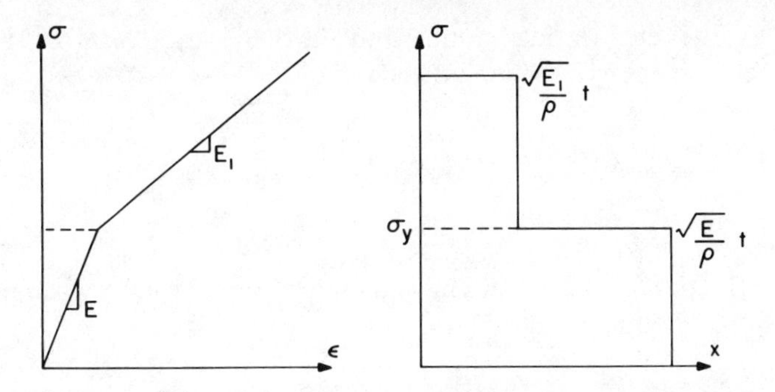

Figure 1 Stress-strain relation and wave profile for bilinear material.

a bar by considering the material to be bilinear elastic and independent of strain rate. It was not until World War II that the first plausible one-dimensional, finite-amplitude plastic wave propagation theory was developed independently by Karman (1942) in the United States, Taylor (1942) in England, and Rakhmatulin (1945a) in Russia. The theory assumed that the behavior of a material could be described by a single-valued relation between stress and strain in uniaxial stress and that the stress-strain curve was concave toward the stress axis. The theory further implied that the stress-strain relation was the one obtained in a conventional quasistatic tensile test. The theory was totally uniaxial in nature, neglecting any three-dimensional effects such as might arise because of lateral inertia. Only axial displacements and stresses were considered. Using a lagrangian coordinate system with the x axis parallel to the bar axis as shown in Figure 2, consider an element of rod of original length dx. The unbalanced force in the x direction is

$$dF = A\frac{\partial \sigma}{\partial x}dx \tag{1}$$

and acts on an element whose mass is $\rho A\,dx$, where ρ and A are the mass density and cross-sectional area, respectively, and σ is the axial engineering stress. It is assumed that the material behavior can be described by a single-valued relation of the form

$$\sigma = \sigma(\varepsilon) \tag{2}$$

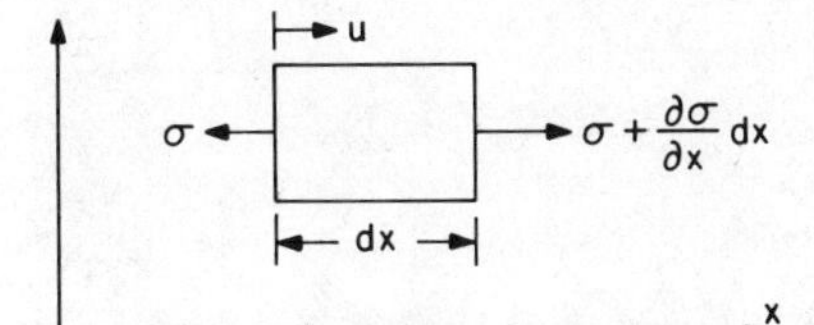

Figure 2 Nomenclature for element of rod.

where ε is the engineering strain, and further, strain can be related to displacements in the x direction, u, then

$$\varepsilon=\frac{\partial u}{\partial x} \tag{3}$$

$$v=\frac{\partial u}{\partial t} \tag{4}$$

where v is the particle velocity. The equation of motion for longitudinal stress in a bar or rod is then

$$\rho\frac{\partial^2 u}{\partial t^2}=\frac{d\sigma}{d\varepsilon}\frac{\partial \varepsilon}{\partial x} \tag{5}$$

The first problem solved was that of a tensile impact at the end of a semiinfinite bar, imparting to it a constant velocity at $t=0$, $x=0$. For a bar originally occupying the region $-\infty$ to 0, the boundary conditions are

$$\begin{aligned} &u=v_1 \quad @x=0 \qquad (v_1=\text{constant})\\ &u=0 \quad @x=-\infty \end{aligned} \tag{6}$$

Three solutions to (5) can be found which are valid in different regions:

$$\text{(1)} \qquad u=v_1[t+\frac{x}{c_1}] \qquad \text{or} \qquad \varepsilon=\frac{\partial u}{\partial x}=\frac{v_1}{c_1}=\text{constant} \tag{7}$$

$$\text{(2)} \qquad \frac{E}{\rho}=\frac{x^2}{t^2} \tag{8}$$

This solution is obtained by letting $\xi=x/t$, thus $\varepsilon=f(\xi)$ since $E=E(\varepsilon)$. We can then write

$$u=\int_{-\infty}^{x}\frac{\partial u}{\partial x}dx=\int_{-\infty}^{x}f(\xi)\,dx=t\int_{-\infty}^{x}f(\xi)\,d\xi \tag{9}$$

Since $\xi=x/t$, we have

$$\frac{\partial \xi}{\partial x}=\frac{1}{t}, \qquad \frac{\partial \xi}{\partial t}=-\frac{x}{t^2}=-\frac{\xi}{t} \tag{10}$$

and from

$$\frac{\partial u}{\partial x}=\frac{\partial u}{\partial \xi}\frac{\partial \xi}{\partial x}=\frac{1}{t}\frac{\partial u}{\partial \xi} \tag{11}$$

and

$$\frac{\partial u}{\partial t}=\frac{\partial u}{\partial \xi}\frac{\partial \xi}{\partial t}=-\frac{\xi}{t}\frac{\partial u}{\partial \xi} \tag{12}$$

we get

$$\frac{\partial^2 u}{\partial t^2}=f'(\xi)\frac{\xi^2}{t}, \qquad \frac{\partial^2 u}{\partial x^2}=f'(\xi)\frac{1}{t} \tag{13}$$

Substitution into (5) yields

$$f'(\xi)[\rho\xi^2 - E] = 0 \tag{14}$$

But $f'(\xi)=0$ corresponds to ε=constant which is the first solution (7); the remainder yields the second solution (8).

$$(3) \qquad \varepsilon = 0 \tag{15}$$

The three solutions are pieced together to give the total solution [Karman and Duwez (1950)] as follows:

$$\text{for } |x| < c_1 t, \qquad \varepsilon = \text{constant} = \frac{v_1}{c_1} = \varepsilon_1 \tag{16a}$$

$$\text{for } c_1 t < |x| < c_0 t, \qquad E(\varepsilon) = \frac{x^2}{t^2} \tag{16b}$$

$$\text{for } |x| > c_0 t, \qquad \varepsilon = 0 \tag{16c}$$

The solution for strain as a function of $\xi = x/t$ is presented in Figure 3 which shows the two wave fronts traveling at different velocities. The magnitude of the elastic wave front is the yield strain, ε_y.

The three main features of the constant-velocity boundary value problem using the rate-independent (RI) theory are (1) the plastic wave front of constant amplitude, (2) the relation between ε_1 and v_1, and (3) the strain distribution between elastic and plastic wave fronts as given by the second solution.

Introducing the definition of engineering strain (3) into (5) leads to the wave equation

$$\frac{\partial^2 u}{\partial t^2} - c^2(\varepsilon)\frac{\partial^2 u}{\partial x^2} = 0 \tag{17}$$

where

$$c^2(\varepsilon) = \frac{1}{\rho}\frac{d\sigma}{d\varepsilon} \tag{18}$$

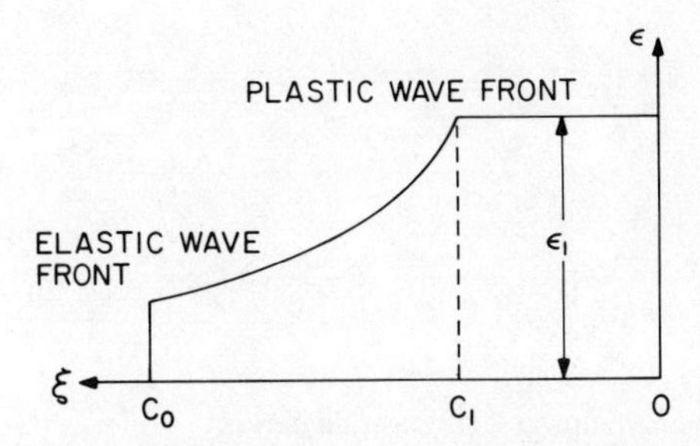

Figure 3 Strain distribution in rod produced by constant velocity impact at end.

The primary feature of the RI theory is the prediction, (18), that each level of stress or strain propagates with its own characteristic velocity dependent on the slope of the tangent to the stress-strain curve at that point, that is, the tangent modulus. Because of the assumption of a concave-down stress-strain curve, the higher the stress or strain, the lower the plastic wave speed for that level of stress or strain. Thus, initially sharp pulses tend to disperse as they propagate down the bar.

Removal of the restriction of a concave-down stress-strain curve was first considered by White and Griffis (1948) who deduced that a concave-up stress-strain curve must result in the formation of shock fronts. Consider the concave-up stress-strain curve depicted in Figure 4. Since the slope governing the wave speed is an increasing function of strain, higher strain increments will travel faster than lower strain increments and must eventually overtake them. The physical impossibility of having a large strain increment without a lower strain increment requires that a shock front be formed. If point A in Figure 4 represents the maximum stress or strain imparted to a specimen, then a shock wave will eventually form of that magnitude and travel at a velocity determined by the slope of the line OA. The concave-up stress-strain curve is a regular feature of waves of uniaxial strain at high pressures. Lee (1966) points out, however, that for compression impact in bars or rods, when lateral inertia and thermodynamic considerations are neglected, shock waves eventually must develop as impact velocities increase because the maximum engineering strain is unity while the maximum stress (force divided by original cross-sectional area) can become infinite.

Shock waves can also form if the impact velocity exceeds the wave velocity in the material. Ting (1978) discusses both supersonic impacts, where the impact velocity exceeds the elastic wave velocity, and transonic impacts, where the impact velocity is less than the elastic velocity but greater than the velocity of the slowest wave in the RI theory. The true wave speed, c^*, is shown to exceed the plastic wave speed $c=\sqrt{d\sigma/\rho\, d\varepsilon}$, and is calculated from

$$c^* = v + (1+\varepsilon)c \tag{19}$$

where v is the particle velocity given by

$$v = -\int_0^{\varepsilon} c\, d\varepsilon \tag{20}$$

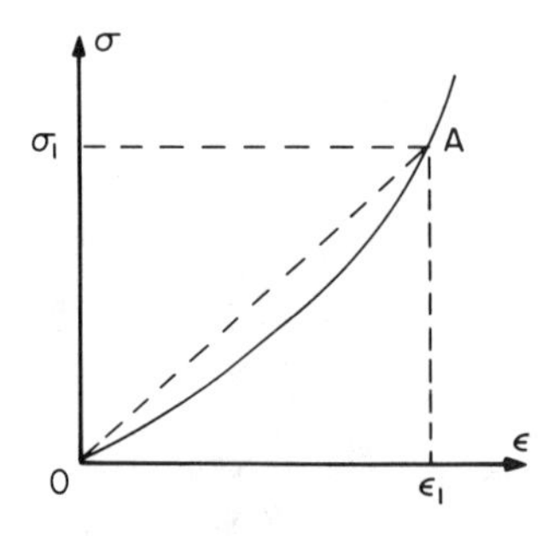

Figure 4 Concave-upward, stress-strain curve.

Up to this point only loading conditions have been considered in semiinfinite bars with nondecreasing boundary conditions. If unloading occurs due to removal of the stress or velocity boundary condition or due to reflections from a free end of a bar, then the interactions of the loading and unloading waves have to be considered. One of the first analyses of this type of problem was presented by Lee (1952) where the plastic unloading boundary was determined for an impacting cylinder. In the numerical problem solved, a slight difference in the permanent strain was detected based on unloading considerations compared to the predictions due to the first passage of the wave in a semiinfinite bar. The difficulties in the mathematical analysis arise due to the different constitutive relations for loading or unloading and the associated differences in wave velocities. Consider a semiinfinite bar of a rate-independent material subjected to an impact of duration T. The stress distribution some time t_1 after impact is as shown by curve 1 in Figure 5. The maximum stress propagates at a velocity c_1 given by the Karman solution which is slower than the elastic wave velocity c_0 (see also Figure 3). An arbitrary stress level will propagate at its own characteristic velocity, c, and will have traveled a distance ct_1. Since $c_1 < c_0$, the distance between the elastic and plastic wave fronts will gradually increase. Meanwhile, an elastic unloading wave traveling at c_0 remains a fixed distance c_0T behind the elastic wave front. As the plastic wave front spreads out, it eventually gets cut off by the elastic unloading wave that overtakes it. The shape of the wave at some later time is depicted by curve 2 in Figure 5. The plastic wave front will be overtaken by the elastic unloading wave in a time

$$t = \frac{c_0 T}{c_0 - c_1} \tag{21}$$

at a distance along the bar

$$l_1 = \frac{c_0 c_1 T}{c_0 - c_1} \tag{22}$$

Eventually, the entire plastic wave will be diminished until only the elastic portion of length c_0T remains.

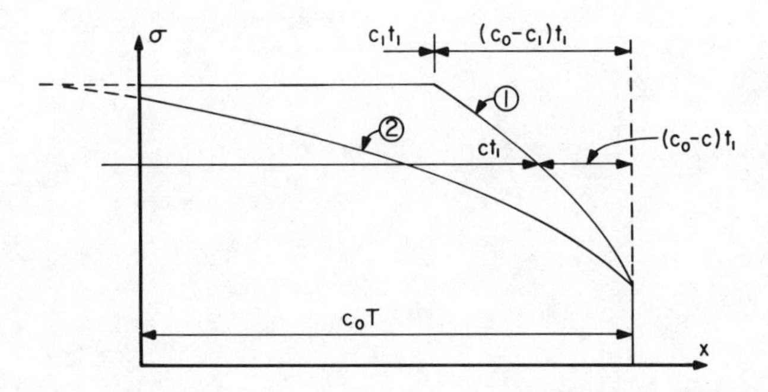

Figure 5 Wave profiles in rod of elastic-plastic material showing unloading effect.

Another feature of the RI theory is the concept of a critical velocity in tensile impact beyond which instantaneous failure will occur at the impacted end. The critical velocity concept can be demonstrated by making use of the expression for the constant velocity at the impact end in the form

$$v_1 = \int_0^{\varepsilon_1} c(\varepsilon)\, d\varepsilon \tag{23}$$

where v_1 is the velocity at $x=0$ corresponding to a maximum strain ε_1 and $c(\varepsilon)$ is as given by (18). Equation (23) is easily derived from the solution to the problem of a constant velocity impact at the end of a semiinfinite bar by evaluating $u(0,t)/t$ and imposing a change in variables. For the stress-strain curve shown in Figure 6, at a strain ε_m the wave speed for a strain increment becomes zero because the tangent to the curve is horizontal. The critical velocity for impact is defined as

$$v_c = \int_0^{\varepsilon_m} c(\varepsilon)\, d\varepsilon \tag{24}$$

For velocities above v_c, the integral (23) cannot exist. What this means physically is that the energy of impact cannot propagate away from the impact point because of the zero velocity of additional strain and thus instantaneous failure occurs.

4.1.2 Method of Characteristics

A powerful mathematical tool for the solution to many problems in wave propagation is the method of characteristics. The method, described in Hopkins (1968) and Karpp and Chou (1973), for example, can be applied to systems of equations in the space and time variables x and t to determine the values of the dependent variables such as σ, ε, and v numerically. The method is applicable to equations of the form

$$a_{ij}\frac{\partial u_j}{\partial x} + b_{ij}\frac{\partial u_j}{\partial t} = R_i \tag{25}$$

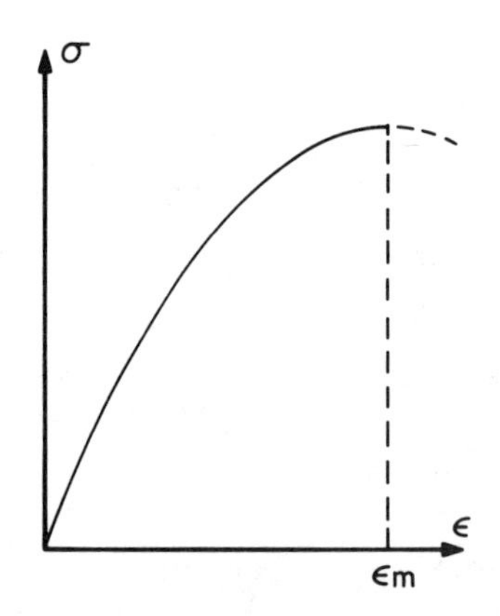

Figure 6 Stress-strain curve illustrating critical velocity concept.

where repeated subscripts indicate summation. These are a set of quasilinear equations, that is, linear in the first derivatives, in the variables u_j, where the coefficients a_{ij}, b_{ij}, and R_i may be functions of u_j, x, and t. Consider, as an example, the problem of waves propagating in a long rod governed by the RI theory of plastic wave propagation. The equations can be written in terms of the strain, ε, and particle velocity, v, as

$$\frac{\partial v}{\partial t} - c^2 \frac{\partial \varepsilon}{\partial x} = 0 \tag{26}$$

$$\frac{\partial v}{\partial x} - \frac{\partial \varepsilon}{\partial t} = 0 \tag{27}$$

where

$$c = c(\varepsilon) = \left(\frac{1}{\rho} \frac{d\sigma}{d\varepsilon} \right)^{1/2} \tag{28}$$

Note that the derivative of a function $f(x,t)$ in the s direction in the x-t plane is

$$\frac{df}{ds} = \frac{\partial f}{\partial x}\frac{dx}{ds} + \frac{\partial f}{\partial t}\frac{dt}{ds} \tag{29}$$

where the s direction is defined by

$$\frac{dx/ds}{dt/ds} = \frac{dx}{dt} \tag{30}$$

Taking a linear combination of (26) and (27) multiplied by α_1 and α_2, respectively,

$$\alpha_1 \frac{\partial v}{\partial t} + \alpha_2 \frac{\partial v}{\partial x} - \alpha_1 c^2 \frac{\partial \varepsilon}{\partial x} - \alpha_2 \frac{\partial \varepsilon}{\partial t} = 0 \tag{31}$$

The first two terms represent dv/ds where the direction s_1 is given by $\alpha_2/\alpha_1 = dx/dt$. The last two terms represent $d\varepsilon/ds_2$ where the direction s_2 is given by $\alpha_1 c^2/\alpha_2 = dx/dt$. We require that the s_1 and s_2 directions be identical, defined by τ, thus

$$\frac{dx}{dt} = \tau = \frac{\alpha_2}{\alpha_1} = \frac{\alpha_1 c^2}{\alpha_2} \tag{32}$$

which results in two equations for α_1 and α_2:

$$\tau\alpha_1 - \alpha_2 = 0$$

$$c^2\alpha_1 - \tau\alpha_2 = 0 \tag{33}$$

For a solution, the necessary and sufficient condition is that the determinant of the coefficients of α_1 and α_2 be zero, resulting in

$$\tau^2 - c^2 = 0 \tag{34}$$

which, from (32), provides

$$\frac{dx}{dt}=\tau=\pm c \tag{35}$$

as the characteristic directions in the x-t plane. From (32) we have the following:

$$\text{on } \frac{dx}{dt}=c, \qquad \frac{\alpha_2}{\alpha_1}=c \tag{36}$$

$$\text{on } \frac{dx}{dt}=-c, \qquad \frac{\alpha_2}{\alpha_1}=-c \tag{37}$$

Substituting these values into (31), we obtain

$$\text{along } \frac{dx}{dt}=\pm c, \qquad \left(\frac{\partial v}{\partial t}\pm c\frac{\partial v}{\partial x}\right)-c\left(c\frac{\partial \varepsilon}{\partial x}\pm\frac{\partial \varepsilon}{\partial t}\right)=0 \tag{38}$$

Finally, noting that the magnitude of a directional derivative is obtained from

$$A_1\frac{\partial u}{\partial x}+A_2\frac{\partial u}{\partial t}=\sqrt{A_1^2+A_2^2}\,\frac{du}{ds} \tag{39}$$

we obtain the expressions along the characteristic directions:

$$\text{along } \frac{dx}{dt}=\pm c, \qquad dv\mp c\,d\varepsilon=0 \tag{40}$$

Introducing the integral

$$\phi=\int_0^{\varepsilon} c\,d\varepsilon \tag{41}$$

we can write, after integration of (40),

$$\text{along } \frac{dx}{dt}=\pm c, \qquad v\mp\phi=\text{constant} \tag{42}$$

For the problem of an instantaneously applied stress or velocity at the end of a semiinfinite bar, the characteristics and solution are as depicted in Figure 7. The solution is readily seen to be that derived previously and given by (16). The straight line characteristics have slopes ranging from c_0, corresponding to elastic behavior, to c_1, which corresponds to the maximum strain. Along each straight line characteristic, ε is constant because the slope $c=c(\varepsilon)$ depends on ε. Furthermore, since $\sigma=\sigma(\varepsilon)$ and $\phi=\phi(\varepsilon)$ from (41), then with the use of the condition (42) it is readily seen that σ, ε, v, and ϕ are all constant along the straight line characteristics $dx=c\,dt$.

For the case of a stress or velocity applied gradually over a time period τ and remaining constant thereafter as the boundary condition, the characteristics and solution are as depicted in Figure 8. In this case, there is a region of elastic behavior behind the leading wave front followed by a plastic region. In both cases, a uniform region is found behind $dx=c_1\,dt$ which corresponds to the slowest wave velocity for the highest stress or strain amplitude. These simple

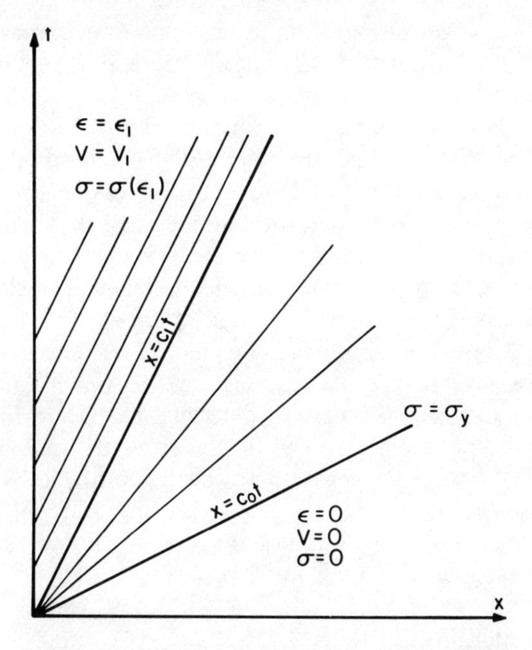

Figure 7 Simple-wave solution for instantaneously applied stress or velocity.

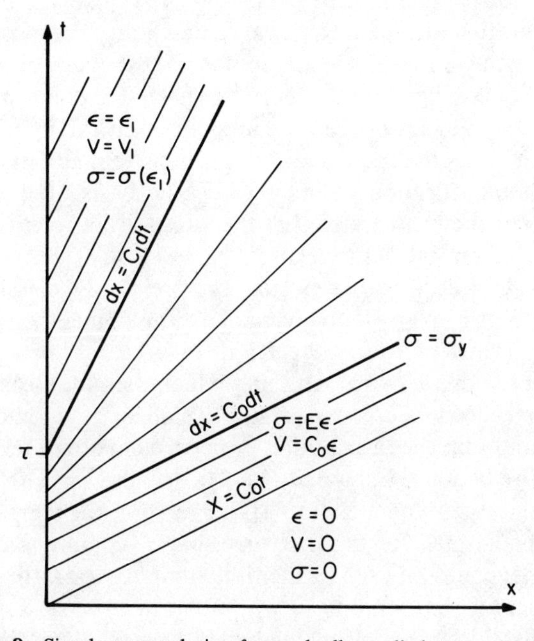

Figure 8 Simple-wave solution for gradually applied stress or velocity.

wave solutions for waves propagating in only one direction are dependent on $\sigma=\sigma(\varepsilon)$ being a single-valued monotonically increasing function and $c=c(\varepsilon)$ being a decreasing function of strain which is equivalent to the condition of a concave-down stress-strain curve or the absence of shock waves.

When unloading is considered, the constitutive equation of the material can be either the single-valued function $\sigma=\sigma(\varepsilon)$ for increasing stresses or an elastic relation for decreasing stresses. It becomes necessary, then, to distinguish between loading and unloading regions in the characteristic plane to apply the right relations along the proper characteristic directions. For elastic behavior, it is to be recalled, the characteristics are straight lines of (constant) slope c_0.

Tuschak and Schultz (1971) provide one method for determining the unloading boundary for several types of problems in one-dimensional elastic-plastic stress waves in a rod. The problem of determining the moving elastic-plastic boundaries due to unloading waves of biaxial stress in combined longitudinal and torsional plastic wave propagation has been considered by Clifton (1968). In general, however, the problem of both loading and unloading waves is complicated because the unloading boundary is not known and can only be obtained through a solution of the entire problem.

4.1.3 Experimental Investigations

The earliest experimental verification of the RI theory was undertaken by Duwez and Clark (1947) who measured the distribution of permanent strain in copper wires that were impacted by falling weights. They achieved good correlation between the permanent strain and the velocity of impact using the static stress-strain curve in the theoretical calculations. They also demonstrated the existence of the critical velocity in tension above which instantaneous rupture occurs as predicted by the theory [see (24)]. Careful examination of the data revealed discrepancies in the measured strain at the impacted end which were explained by imperfect reflections of plastic strain and inaccuracies in the measurement of the duration of impact. On the basis of their experimental evidence, however, they concluded that the stress-strain relation under impact conditions differs from the static relation.

Karman and Duwez (1950) reported on previously classified work performed during World War II involving the development of the theory and experiments performed to verify the RI theory. Using impact loading via a falling weight on copper wires and permanent strain measurements, they verified the existence of the constant strain plateau as a function of impact velocity, but found that the distribution of permanent strain deviated considerably from the predictions. Campbell (1952) used the same (RI) plastic wave theory to calculate the dynamic stress-strain relation for copper under tensile impact and demonstrated for the first time that a separate stress-strain curve, not the static one, could be used in the finite-amplitude plastic wave theory to explain experimental observations.

The most serious doubts about the validity of the RI theory were cast by the results of experiments involving the measurement of the velocity of propagation of small increments of stress in bars already prestressed into the plastic region. Since the velocity of propagation of a given stress level depends on the tangent modulus [see (18)], incremental waves in a plastically prestressed material should be expected to propagate with a velocity that is less than the elastic bar velocity. The first experiments of this type were performed by Bell (1951) who observed that an incremental deformation wave always traveled with the elastic wave velocity in prestressed steel bars. Increments of both loading and unloading were used, but no sharp plastic wave front was observed in any case; the velocity was always c_0—the elastic wave velocity. In an elaborate and one of the most referenced series of experiments, Sternglass and Stuart (1953) measured the velocity of propagation of incremental pulses in prestressed cold-rolled copper strips. They observed that the wave front always propagates at c_0 and, furthermore, that the velocity of propagation of any part of the incremental pulse was always much greater than that predicted by the RI theory using the tangent modulus. They noted dispersion in the stress pulses and found that the maximum strain was not predicted from the theory using the static stress-strain curve. Realizing that these experiments had involved the application of a dynamic stress superimposed on a static prestress, Alter and Curtis (1956) performed experiments on lead bars by applying a dynamic prestress and incremental stress pulse in the form of two closely spaced impacts. However, even in this situation the incremental wave traveled with the elastic wave velocity and it was concluded that a rate-dependent theory would be necessary to predict the experimental observations. It must be pointed out, however, that these latter experiments used lead as a test material which is generally believed to display rate-sensitive material behavior.

Other investigations involving incremental waves in prestressed bars included the work of Bell and Stein (1962) who used a dynamic prestress and concluded, based on wave velocity measurements, that only the incipient portion of the incremental stress propagates with the elastic velocity. Bianchi (1964) used prestressed annealed copper and concluded that a strain rate-dependent form of constitutive equation gives a good description of the overall wave propagation phenomena. However, for constant-velocity end conditions and long times, his results were those predicted by the RI theory, that is, the RI theory was adequate for asymptotic solutions for long time and for predicting permanent deformation.

In the incremental pulse tests in prestressed bars cited above, it should be noted that the analysis is purely one-dimensional, that is, radial motion and corresponding lateral inertia effects are neglected. The actual wave propagation phenomenon is much more complicated. Errors involved in the simple theory were estimated by DeVault (1965) who performed a simple analysis based on an approximate wave equation that includes corrections for lateral inertia. The analysis showed that the errors are such that a rate-independent material might show what appears to be a strain-rate effect. However, Hunter and Johnson (1964) investigated geometric dispersion of an incremental wave in a prestressed rod of a rate-independent material and found that velocities higher than the plastic wave speed were obtained, but they were no higher than one-half the elastic velocity. They concluded, therefore, that strain-rate effects must be an important factor in explaining the experimental results of Sternglass and Stuart. Lee (1970), in an analysis of combined stress, suggested the possibility that the elastic incremental-wave velocity might be attributed to the influence of combined stresses, but he carefully pointed out that quantitatively this explanation was not realistic and that rate of strain effects were more probably the cause.

To avoid the possible complications of lateral inertia and three-dimensional effects, the use of torsional waves in thin-walled tubes was adopted by several investigators as an alternate approach to the study of plastic wave propagation. Among the first torsional incremental wave propagation experiments were those reported by Convery and Pugh (1968) that confirmed the results of Bell, Sternglass, and Stuart, which showed incremental waves travel at the elastic velocity. Convery and Pugh could not achieve good rise times in their incremental pulses, however, making it difficult to analyze the data in detail. Yew and Richardson (1969) performed similar experiments in torsion on annealed and cold-worked copper. Using slow continuous loading to avoid creep, they observed that low-amplitude incremental pulses travel at close to the elastic wave velocity, in contradiction to the rate-independent theory, although large amplitude waves propagated with velocities predicted by the RI theory, as shown in Figure 9. They concluded that, since there were no three-dimensional effects to obscure the data in torsion, the material must be rate-dependent. In the same series of experiments, however, measurements of wave velocities in unstressed bars were in very good agreement with the predictions of the RI theory as shown in Figure 10. Because of the large rise time in the generated

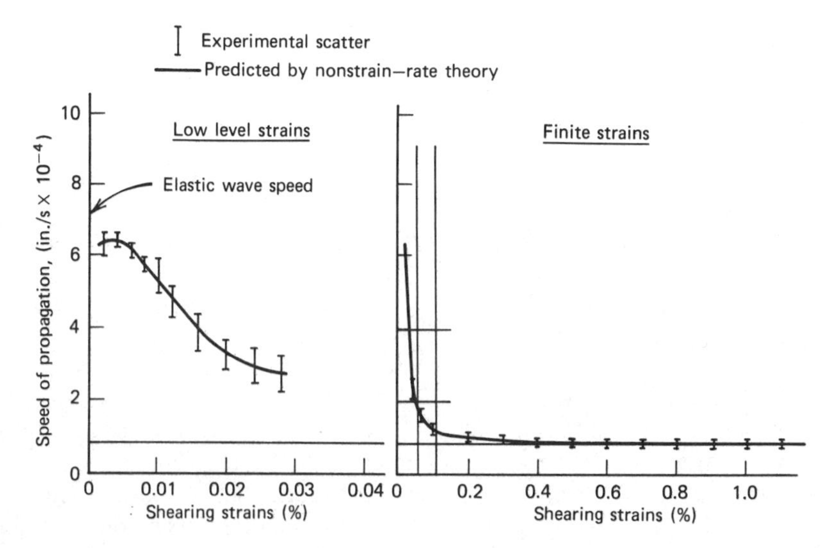

Figure 9 Speed of propagation of shearing strain in prestressed copper specimen [Yew and Richardson (1969)].

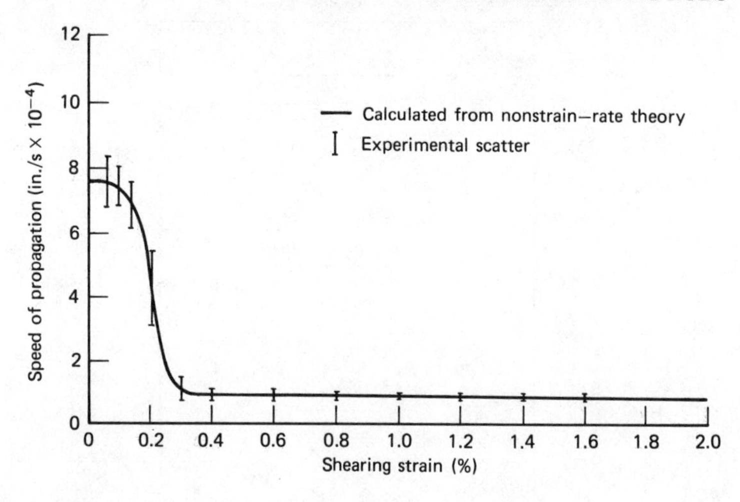

Figure 10 Speed of propagation of shearing strain in stress-free copper specimen [Yew and Richardson (1969)].

torque of approximately 80 μs, thereby invalidating the assumption of an idealized step function in time of the applied velocity, they concluded that the fact that the data shown in Figure 10 are in close agreement with the RI theory is not conclusive evidence of rate insensitivity of the test material.

Campbell and Dowling (1970) performed a series of experiments using mild steel, copper, and aluminum prestressed in torsion. They noted that the propagation speed of a stress wave set up by a shear stress increment is essentially that of elastic waves for a range of values of prestrain as shown in Figure 11 for copper. Note the predictions of the RI theory based on the quasistatic stress-strain curve shown by the broken line. They concluded that a rate-dependent constitutive law that predicts instantaneous plastic strain increments must be used to describe those materials. Nicholas and Campbell (1972) performed incremental wave experiments on a high-strength aluminum alloy in torsion. For this material, which was found to be effectively insensitive to strain rate in constant-strain-rate torsional Hopkinson bar tests, a wave front traveling at the elastic shear-wave velocity was found in all cases, although this overstress was sustained for a very short time period and its peak magnitude was only some 4% of the static flow stress. Klepaczko (1972) performed incremental wave experiments in annealed aluminum, annealed copper, and deep drawn steel in tension when the specimens were preloaded at a constant strain rate. The precursor was again found to propagate at c_0 in apparent verification of a rate-dependent theory. Klepaczko pointed out, however, that these types of experiments must be performed under constant-strain-rate preloading, since either constant-stress or constant-strain preloading will lead to creep or stress relaxation, respectively, for a material that is strain-rate dependent. He further emphasized both the improved accuracy in the use of

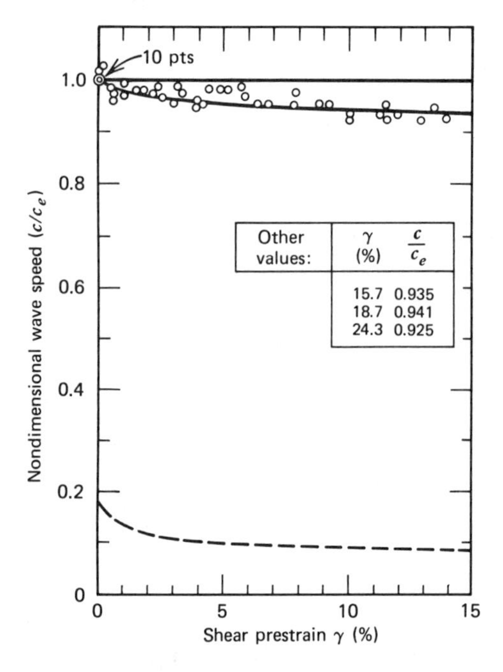

Figure 11 Ratio of incremental-wave speed to elastic-wave speed in copper. Broken line indicates solution from RI theory [Campbell and Dowling (1970].

incremental wave-speed measurements over those in unstressed rods as well as the importance of strain-rate history in the material response.

4.1.4 Analysis by Strain-Rate Dependent Theory

From the time of the earliest experiments attempting to verify the RI theory of Karman, Taylor, and Rakhmatulin it was clear that there were aspects of wave propagation phenomena that were not readily explainable by the RI theory and that there were certain aspects such as the elastic velocity of incremental waves that apparently required a rate-dependent constitutive equation in the formulation of the problem. The most commonly used form of rate-dependent theory, which will subsequently be referred to as the RD theory, was proposed by Malvern (1951a, b). He proposed a stress-strain-strain rate law to model dynamic behavior of materials of the form

$$\sigma = f(\varepsilon) + \ln(1 + b\dot{\varepsilon}_p) \tag{43}$$

where $f(\varepsilon)$ is the stress from a quasistatic stress-strain curve, $\dot{\varepsilon}_p$ is the plastic strain rate, and dots denote time derivatives. This expression can be rewritten in the form

$$\dot{\varepsilon}_p = \frac{1}{b}\left[\exp\left(\frac{\sigma - f(\varepsilon)}{a}\right) - 1\right] \tag{44}$$

which shows that the plastic strain rate is a function of the overstress, $\sigma - f(\varepsilon)$, or the difference between the instantaneous stress and the value that would occur in a quasistatic test at the same value of strain. The more general form of the overstress function would be

$$E\dot{\varepsilon}_p = F[\sigma - f(\varepsilon)] \tag{45}$$

where F may be an arbitrary function. The Malvern law, which was also proposed independently by Sokolovsky (1948), decomposes the total strain rate into an elastic and a plastic component and assumes that the elastic strain rate is related to the stress rate through Hooke's law. It further allows for a general form of the plastic strain rate function

$$E\dot{\varepsilon}_p = g(\sigma, \varepsilon) \tag{46}$$

or

$$E\dot{\varepsilon} = \dot{\sigma} + g(\sigma, \varepsilon) \tag{47}$$

In numerical studies of plastic-wave propagation phenomena in rods, the linear form of the overstress function

$$g(\sigma, \varepsilon) = k[\sigma - f(\varepsilon)] \tag{48}$$

has been used extensively in the literature, primarily because of its computational simplicity, particularly in the early days of dynamic plasticity when computer technology was far less sophisticated than today.

The above formulation of Malvern, which decomposes the strain rate into an elastic and a plastic portion, implies that a material is brought to a state of incipient plastic flow after a given amount of elastic strain, independent of the elastic strain rate, but that the plastic flow requires time in which to become appreciable so that the additional strain beyond the static yield strain is mainly elastic. This explains the propagation of stress increments at the elastic wave velocity in a prestressed bar since time is required for plastic flow to occur. Malvern's first calculations did not, however, predict a region of uniform strain near the impacted end of a bar subjected to constant velocity impact as predicted with the RI theory and verified in subsequent experiments. Figure 12 shows the results of the RI and RD theories, where the broken line represents the solution neglecting strain-rate effects (RI theory) of Karman, Taylor, and Rakhmatulin. In addition to the lack of a plateau, the RD theory also shows an increased strain near the impacted end which was in disagreement with the early experimental findings. Although gradual application of the impact or alternate forms of the plastic-strain-rate function might be expected to improve the agreement between experiment and theory, Malvern considered it likely that any form of the RD theory would fail to predict the uniform strain plateau.

Plass and Wang (1959) extended the work of Malvern, studying both a linear and an exponential law for dynamic overstress based on results obtained experimentally for copper and pearlitic steel. They found that the exponential

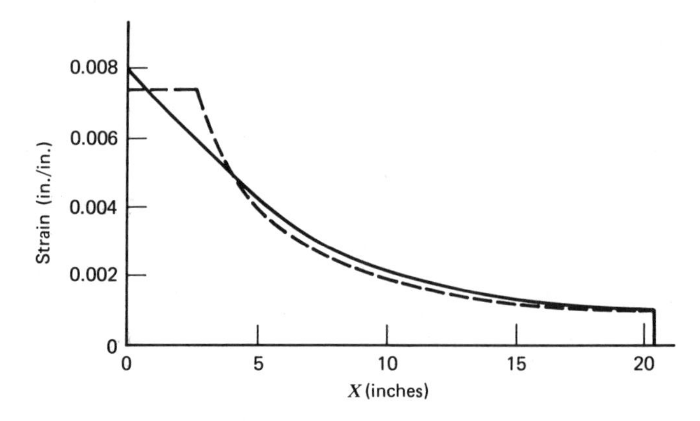

Figure 12 Strain distribution in a wire produced by constant velocity at end. Comparison of results of RI theory (broken line) and RD theory (solid line) at 102.4 μs [Malvern (1951a)].

law gave better prediction for larger plastic strain, while for smaller strains the two laws were not discernible. Both the rate-dependent and rate-independent theories were generalized by Lubliner (1964) in a theory of plastic wave propagation formulated on the basis of a general quasilinear constitutive equation. He showed that both theories were special cases of a generalized theory and showed conditions under which one or the other could be valid.

The solution to the equations of the RD theory can be obtained using the method of characteristics, discussed in Section 4.1.2. The equations governing the propagation of waves in a long bar or rod for the RD theory are

$$\frac{\partial \sigma}{\partial x} = \rho \frac{\partial v}{\partial t} \tag{49}$$

$$\frac{\partial \varepsilon}{\partial t} = \frac{\partial v}{\partial x} \tag{50}$$

$$E\frac{\partial \varepsilon}{\partial t} = \frac{\partial \sigma}{\partial t} + g(\sigma, \varepsilon) \tag{51}$$

where the most general form of the plastic-strain-rate function, $g(\sigma, \varepsilon)$, has been used. This set of quasilinear differential equations may be integrated numerically along the characteristic directions where the following differential relations hold true:

$$\text{along } \frac{dx}{dt} = c_0 \qquad d\sigma - \rho c_0 \, dv = -g(\sigma, \varepsilon)\, dt \tag{52a}$$

$$\text{along } \frac{dx}{dt} = -c_0 \qquad d\sigma + \rho c_0 \, dv = -g(\sigma, \varepsilon)\, dt \tag{52b}$$

$$\text{along } \frac{dx}{dt} = 0 \qquad E\, d\varepsilon = d\sigma + g(\sigma, \varepsilon)\, dt \tag{52c}$$

It is to be noted that the characteristic directions are fixed straight lines in the x-t or characteristic plane which greatly simplifies the numerical integration. The solution is obtained by a numerical forward integration procedure using the initial conditions along the x-axis and the boundary conditions that are prescribed along the t-axis. The region ahead of the leading elastic wave front, $x=c_0t$, is at rest as shown in Figure 13 for assumed zero initial conditions. Consider two points, A and B, on either side of the elastic wave front along a characteristic line $dx=-c_0\,dt$. Applying the relation (52b), noting that $\sigma_A=\epsilon_A=v_A=g_A=0$, and approximating the term $g\,dt$ by $(g_A+g_B)\Delta t/2$, we have

$$\sigma_B+\rho c_0 v_B=-\tfrac{1}{2}g_B\Delta t \tag{53}$$

But Δt may be made arbitrarily small, such that

$$\sigma_B=-\rho c_0 v_B \tag{54}$$

is valid for all points "B" along the leading wave front. The solution is obtained by numerically integrating along the leading wave front from the origin and then integrating point by point in the interior region. To solve for the values of the variables at point F in the interior, as shown in Figure 13, we make use of the previously calculated values at points C, D, and E in the differential relations (52). For example, along $dx/dt=c_0$ we have

$$\sigma_D-\sigma_C-\rho c_0(v_D-v_C)=-\tfrac{1}{2}(g_D+g_C)\Delta t \tag{55}$$

where Δt is as shown. Applying the appropriate relations in addition along paths EF, CE, DF, and CF, we can derive the following expressions for the

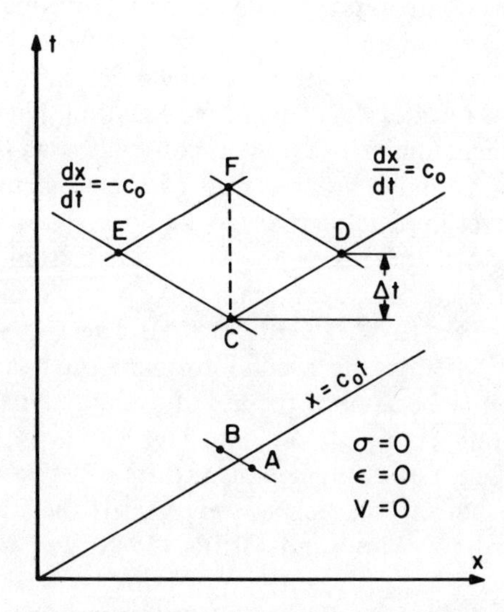

Figure 13 Characteristic plane showing leading wave front and interior mesh points.

values at point F:

$$\sigma_F = \sigma_E + \sigma_D - \sigma_C - g_F \frac{\Delta t}{2} + g_C \frac{\Delta t}{2} \tag{56}$$

$$v_F = v_E + v_C - v_D + \frac{\Delta t}{2\rho c_0}(g_E - g_D) \tag{57}$$

$$E\varepsilon_F = E\varepsilon_C + \sigma_F - \sigma_C + g_F \Delta t + g_C \Delta t \tag{58}$$

Note that in the above expressions the term $g_F(\sigma_F, \varepsilon_F)$ appears in the solution for σ_F, v_F, and ε_F. The solution can only be obtained by an iterative procedure by assuming values of g and then calculating them. It is apparent that caution must be exercised in choosing the size of the mesh to insure numerical stability and to justify the approximation of $g\,dt$ by an average value of g multiplied by Δt. Examples of the numerical procedures employed in plastic wave propagation problems involving use of the Malvern form of the RD constitutive equation can be found in Wood and Phillips (1967) and Nicholas (1973a). A thorough treatment of the mathematics of plastic wave propagation in bars using various forms of rate-dependent equations is presented in Cristescu (1967).

4.1.5 Experimental Verification of Theories

The apparent inability of either the RI or RD theory to predict all of the details of plastic wave propagation, coupled with the uncertainties regarding the accuracy of the one-dimensional assumptions, fueled a major controversy over the necessity of including strain-rate effects in a constitutive model for impacts in rods or bars. An additional factor, which must have played an important part in the controversy, has to be the relative mathematical simplicity of the RI theory for certain boundary value problems. Nonetheless, a large number of experimental and analytical investigations were conducted to better define the nature and form of the constitutive equation that is valid in dynamic plasticity. An excellent summary is presented by Hopkins (1966).

The existing theories that have been utilized for predicting the propagation of uniaxial stress waves in metallic rods can be broken down into three general categories: (1) the RI theory using the static stress-strain curve, (2) the RI theory using a single dynamic, stress-strain curve, and (3) the RD theory in the form originally proposed by Malvern (1951b). There is some philosophical disagreement as to whether the second formulation can truly be called a rate-independent theory because of the use of a single dynamic curve which is different than the static stress-strain curve. The RI theory was formulated on the basis of the existence of a single-valued relation between stress and strain, it was not specified that this relation was necessarily the static one. This point has been emphasized by White and Griffis (1948) and by Bell (1965) who argues that nonlinear wave propagation phenomena are different than quasi-static loading and thus there is no reason to expect that the governing stress-strain relation should be the same.

Malvern (1965) reported the results of longitudinal wave propagation studies in annealed aluminum bars using two types of measuring devices. In addition to strain gages as in most of the other investigations reported, he employed an electromagnetic velocity transducer [Ripperger and Yeakley (1963), Efron and Malvern (1969)], to measure particle velocity at various sections along the bar. The velocity measurements were found to be consistent with a RI theory based on a single dynamic curve which was only slightly higher than the static curve. The strain gage measurements indicated lower propagation speeds than those found with the velocity transducer and thus cast doubt on the validity of strain gage measurements, at least during that time period. Malvern concluded, however, that the RD theory was not invalidated since the two theories predict virtually the same thing for a material with a very slight rate dependence but the RI theory is easier to apply and therefore preferable in this situation.

One of the most extensive series of investigations involving measurements of displacements in plastic wave propagation using a diffraction grating technique was conducted by Bell (1968). The optical technique of Bell (1956) allows extremely precise measurements of surface strains to be made over very short gage lengths. Using this technique, which allowed strains as high as 3% to be measured accurately, Bell (1959, 1960a) found from constant-velocity impact tests on annealed aluminum bars that the velocity of propagation of each level of strain was constant and, furthermore, in very good agreement with the propagation velocities determined from the RI theory using the slopes on the static stress-strain curve. He also found that strain levels for various impact velocities were those predicted from the RI theory and static curve at positions more than 2 diameters from the impacted end. Deviations within 1 diameter of the impact end were attributed to large radial accelerations associated with rapidly increasing strain and a corresponding large dilatation which was observed experimentally. Further experiments by Bell involving constant-velocity boundary conditions achieved through the symmetrical free flight impact of identical bars showed that the RI theory is in close agreement with experimental observations in several completely annealed metal polycrystal and single crystal rods [Bell (1965)]. Comparison of experimental and theoretical strain-time profiles using the RI theory is illustrated in Figure 14. The maximum strain is well predicted as is the velocity at the lower strain levels. The lower velocity at the higher strain levels, indicated by the rounding off of the strain-time profile, is attributed to the delay in traversing the first diameter because of three-dimensional effects. Sperrazza (1962) verified some of Bell's conclusions in experiments on pure lead bars and found that a single dynamic curve gave consistent results using the RI theory of plastic wave propagation. It can be noted here again that lead is generally considered to be a highly strain-rate-sensitive material.

The discrepancies between the RI and RD theories appear to be most pronounced at early times after impact and from observations made near the impacted end of rods. Bell (1960a, b) suggested that the observed strain-rate effect previously observed in high-purity copper and aluminum might be the result of the mechanics of the development of the plastic wave front near the

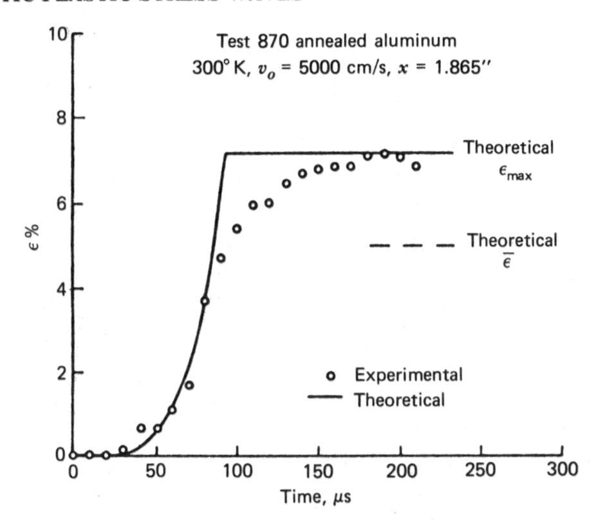

Figure 14 Comparison of experimental and theoretical strain-time profiles from RI theory [Bell (1968)].

impact end of the rod. It had already been pointed out by Alter and Curtis (1956) and Malvern (1951b), and demonstrated mathematically by Rubin (1954), that strain-rate-dependent solutions asymptotically approach those solutions obtained from the RI theory as time increases. Thus, one is not surprised by results such as those of Bodner and Clifton (1967), and Clifton and Bodner (1966) who investigated elastic-plastic pulse propagation at distances far from the impacted end in annealed, commercially pure aluminum bars. They found that the general features of the RI theory were verified using a single dynamic stress-strain curve that did not differ appreciably from the quasistatic curve, but which also exhibited a Bauschinger effect.

Cristescu and Bell (1970) looked at several different boundary conditions to find which one best simulated experimental data from symmetric free flight impact experiments. They reasoned that up to 1 diameter from the impacted end there is a boundary or wave initiation effect that is not one-dimensional. The best correlation was obtained with a boundary condition which consists of an instantaneous jump in stress to 80% of the maximum stress and then a further increase over 40 μs. Using the RI theory, they found that one can quite accurately describe many of the main features of both the loading and unloading process in a bar of finite length. The main discrepancies were attributed to the fact that the problem can no longer be considered one-dimensional near the two ends of the bar. Although the use of torsional waves to study plastic wave propagation in unstressed rods or tubes, as first demonstrated by Baker and Yew (1966), eliminates the complications of three-dimensional effects, the inability to achieve high-amplitude torsional impacts with very rapid rise times and corresponding high stresses and strain rates near the point of wave initiation has been a major drawback to this approach for studying dynamic plasticity effects in metals.

4.1.6 Remarks on the Strain Plateau

The existence of a strain plateau near the impacted end of a long rod, which is predicted by the RI theory and has been verified in numerous experimental investigations [Cristescu (1968a)], has been the cause of a large part of the controversy in verifying a particular theory. As noted previously, the failure of the calculations of the RD theory to predict this plateau became a major obstacle to the acceptance of the RD theory, at least in the form proposed by Malvern. A numerical study, by Wood and Phillips (1967), however, showed that the strain plateau near the impact end can always be expected to appear using the RD theory of Malvern. They pointed out, however, that the plateau could be missed if the calculations were not carried out for a sufficient duration in time. Figure 15 shows the time history of strain at various positions along the bar for an assumed linear form of the overstress function

$$E\dot{\varepsilon}_p = k\left[\sigma - f(\varepsilon)\right] \tag{59}$$

At all stations the strain asymptotically approaches the maximum value as determined by the static stress-strain curve and the applied stress at the boundary (see Figure 16). The stress-strain histories at various positions along the bar for this same problem are shown in Figure 16. The dynamic stress-strain curves show that the nearer a location is to the impact end, the higher will be the dynamic stress with the value approaching the static value as distance is increased. Strain-rate effects are seen to be most important near the impact end and at earlier times. This points out the general problem of using residual strain measurements to verify a particular wave theory because calculations must consider all wave reflections and interactions for sufficient time to predict residual strains accurately. Assumptions such as that of a semiinfinite bar may thus be invalidated. The work of Wood and Phillips, and that of Malvern (1965) also demonstrated the sensitivity of calculations to the assumed boundary conditions. In both works, the strain plateau near the end of

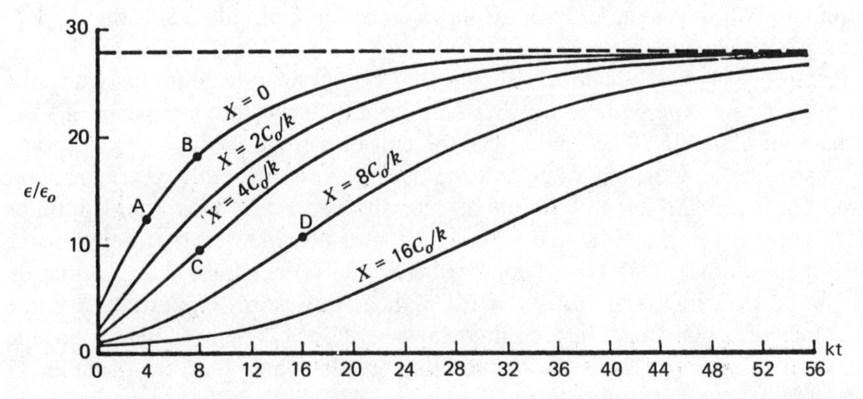

Figure 15 Strain-time history for various stations along bar subjected to constant-stress boundary condition at end [Wood and Phillips (1967)].

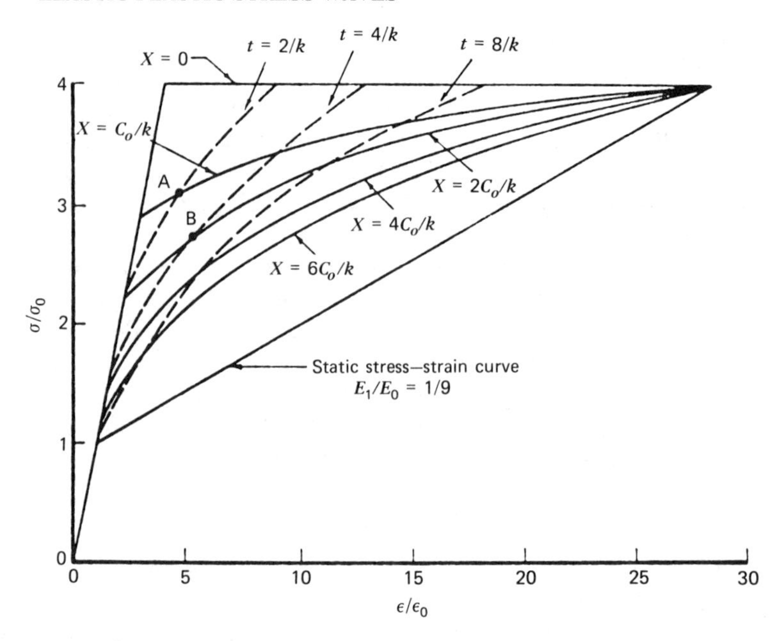

Figure 16 Stress-strain history for selected stations along bar subjected to constant-stress boundary condition at end [Wood and Phillips (1967)].

an impacted bar was predicted, provided that the boundary condition at the impact end was that of constant stress. Figure 17 shows the results of Malvern's calculation assuming a constant stress boundary condition and demonstrates clearly the existence of a plateau of constant strain. The plateau is at the same strain level as predicted by the RI theory for the same end condition, while the calculated end velocity was reported to become almost instantaneously about 90% of the value predicted by the RI theory, and then rise to almost exactly the predicted value in about 12 μs. Contrast these results with those of previous calculations that were made assuming a constant-velocity boundary condition and did not display a constant strain plateau using the RD theory as shown in Figure 12.

Rubin (1954) had previously shown that the strain-rate solutions using the semilinear rate-type equation of Malvern approach the rate-independent solutions asymptotically. The strain near the end of a semiinfinite rod was shown to approach the Karman value as time became very large for either constant stress or constant-velocity boundary conditions. Both Wood and Phillips (1967) and Efron and Malvern (1969) have also obtained asymptotic plateaus with the semilinear rate-type theory. Suliciu et al. (1972) studied the formation of the plateau in detail from a mathematical viewpoint. Conditions for the existence of a plateau were presented for a semilinear or quasilinear rate-type constitutive equation, to be discussed later. It was shown that for a semilinear model only an asymptotic plateau can exist. With the quasilinear model of the

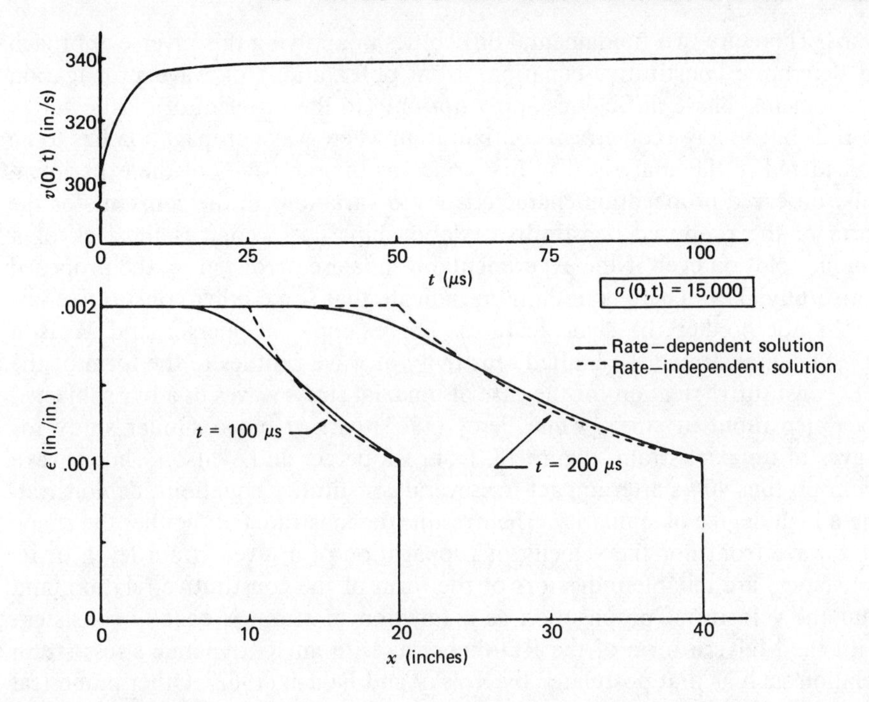

Figure 17 Solution for constant-stress impact at end of bar. Comparison of results of RI theory and RD theory [Malvern (1965)].

form

$$\frac{\partial \varepsilon}{\partial t} = \phi \frac{\partial \varepsilon}{\partial t} + \psi \tag{60}$$

absolute plateaus can exist. Bianchi (1964), in comparing experimental results on propagation of incremental pulses in prestressed copper bars with numerical calculations, had noted that the semilinear rate-type theory, and possibly even more a nonlinear form of theory, gives a good description of the whole phenomenon for any value of time, distance from impact end, and static preload. He noted, however, that the asymptotic behavior for constant-velocity boundary conditions was given by the solution using the RI theory that was consistent with the observations of Rubin (1954).

4.1.7 Sensitivity of Equations to Experimental Observations

The entire concept of trying to determine a material constitutive equation from wave propagation studies whose analysis depends on the same constitutive equation has to be questioned as to the viability of the approach [Ripperger and Simon (1970)]. As noted previously, the RI and RD theories predict many of the same features. Even if one form of a theory could be generally accepted, the methodology for determining the empirical constants is not straightforward. There are two fundamental difficulties in applying this inverse approach to determine constitutive equations from observations of wave propagation phenomena. These difficulties apply not only to the problem of plastic waves in rods but to any geometrical configuration where wave propagation has to be considered in the analysis. The first concerns the sensitivity of the experimentally observed propagation characteristics to variations in the constants or the form of the proposed constitutive relationship. The second is the lack of a unique solution even if the experimental results are predicted by the proposed constitutive law. There is nothing to indicate that some other constitutive law might not predict the same features. For example, Ripperger and Watson (1968) demonstrated the limited sensitivity of wave profiles to the form of the RD constitutive relation for the case of uniaxial stress waves in a bar subjected to a step input in stress, while Percy (1963) carried out a similar study for waves of uniaxial strain. Figure 18, from Ripperger and Watson, shows wave front profiles 40 μs after impact for several constitutive equations, demonstrating a high degree of similarity. Their results demonstrate that neither the shape of a wave front nor the velocity of propagation of a given strain level, or its constancy, are reliable indicators of the form of the constitutive relation, and that the velocity of propagation as a function of strain is equally consistent with the Malvern form of the RD theory as with an RI dynamic stress-strain relation such as that postulated by Kolsky and Douch (1962). Other numerical results by Wood and Phillips (1967) and Lawson and Nicholas (1972) demonstrate the similarities between the RI and RD theories regarding the shape of stress profiles propagating down a bar. Nicholas (1973b) carried out calcula-

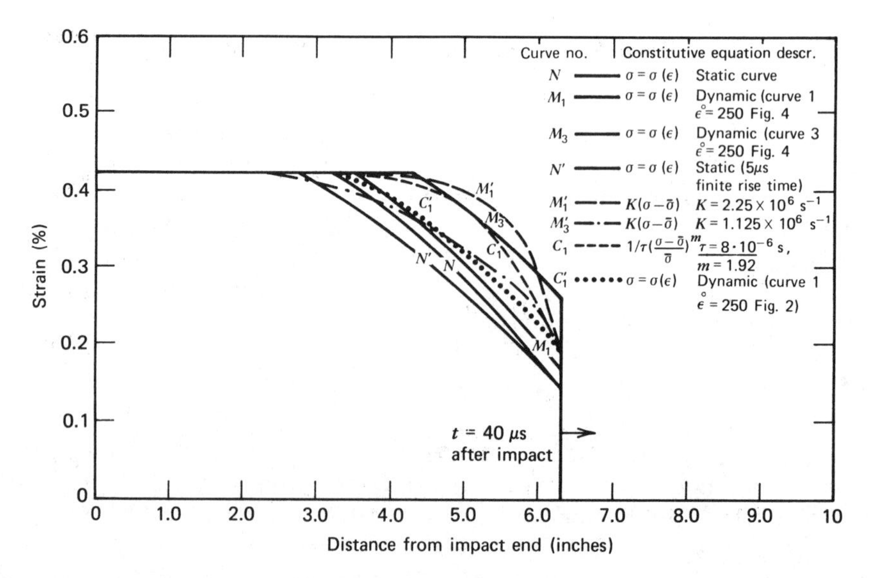

Figure 18 Calculated wave profiles at 40 μs for several constitutive equations [Ripperger and Watson (1968)].

tions using the RI theory with both a static and a single dynamic stress-strain curve as well as with the RD theory for a constant-velocity boundary condition applied gradually over 10 μs. Figure 19, from that study, shows calculated strain-time profiles for the three cases, where nondimensional variables $\varepsilon^* = E\varepsilon/\sigma_y$ and $x^* = 10^{-6}x/ct$ are used. The quantity $x^* = 1$ represents the distance traveled at the elastic wave speed in 1 μs. The RD theory used an exponential form of the overstress

$$\dot{\varepsilon}_p^* = K \exp\left[\left(\frac{\sigma^* - f(\varepsilon^*)}{a}\right) - 1\right] \tag{61}$$

where $\sigma^* = \sigma/\sigma_y$. The equation was chosen as being representative of a large body of data in the literature which shows an approximately linear dependence of stress with log strain rate above a transition strain rate, and constant stress or rate-independence below that rate. The constants a and K can be chosen to provide the appropriate degree of rate dependence and the transition strain rate. In addition to the high degree of similarity in the strain-time profiles shown in Figure 19, calculation of the wave velocities for the three constitutive models shown in Figure 20 demonstrate the lack of sensitivity to the model. The results from the RD theory and the RI theory using a single dynamic curve are essentially indistinguishable, while the RI theory with the static curve

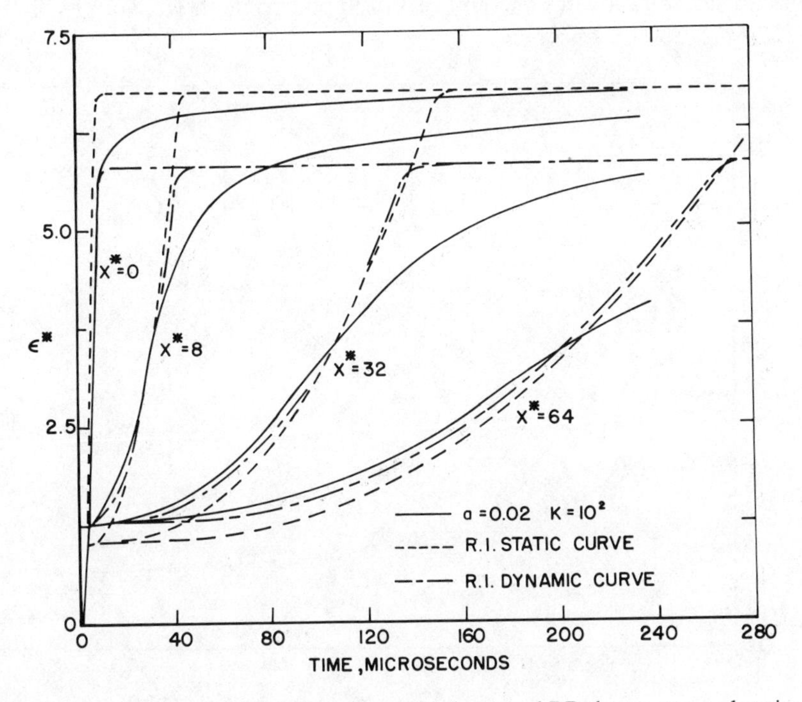

Figure 19 Comparison of strain-time profiles of RI theory and RD theory at several stations for constant-velocity impact at end.

is only slightly different. Comparing the results in Figure 20 with those of Yew and Richardson in Figure 10, it becomes apparent that the conclusion that the experimental data were in agreement with predictions of the RI theory might equally be made for the RD theory. From the calculations using the RD theory to obtain the wave profiles of Figure 19, the plastic strain-rate history is shown for various locations along the bar in Figure 21. Although there is a reasonable amount of variation in strain rate, a value of $\dot{\varepsilon}_p^* = 4 \times 10^3$ can be chosen somewhat arbitrarily as representing the average dynamic rate encountered in this simulated experiment. Using this value, three sets of constants for the RD theory (61), were chosen representing various degrees of rate sensitivity and transition strain rate. They were all chosen, however, to provide the same value of stress at $\dot{\varepsilon}_p^* = 4 \times 10^3$. The curves of stress as a function of strain rate for the three sets of constants are shown in Figure 22. Also shown for comparison are two curves for the linear overstress model, also in nondimensional form. In the latter case, changing the constant K serves to move the curve horizontally along the log strain-rate axis. For the exponential form of the overstress function, strain-time profiles were computed for the three sets of constants shown in Figure 22. These profiles, shown at different locations along the bar in Figure 23, demonstrate a very high degree of similarity. It can be concluded that it may not be as important to have the right form for a plastic wave theory, either RI or RD, as to provide the proper value of the dynamic stresses at the average strain rates encountered in an experiment. Conversely, the test

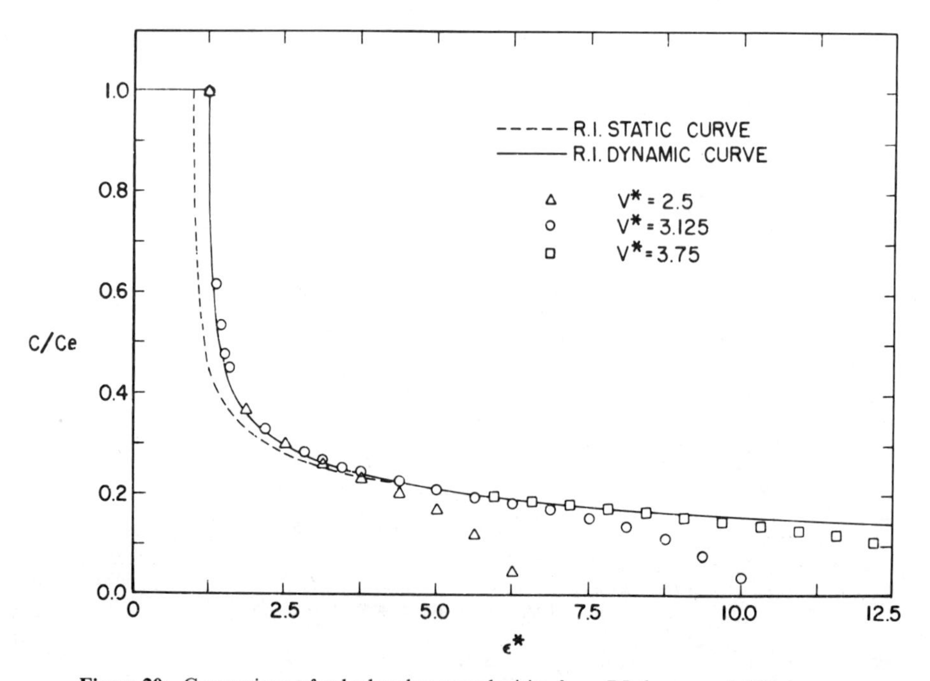

Figure 20 Comparison of calculated wave velocities from RI theory and RD theory.

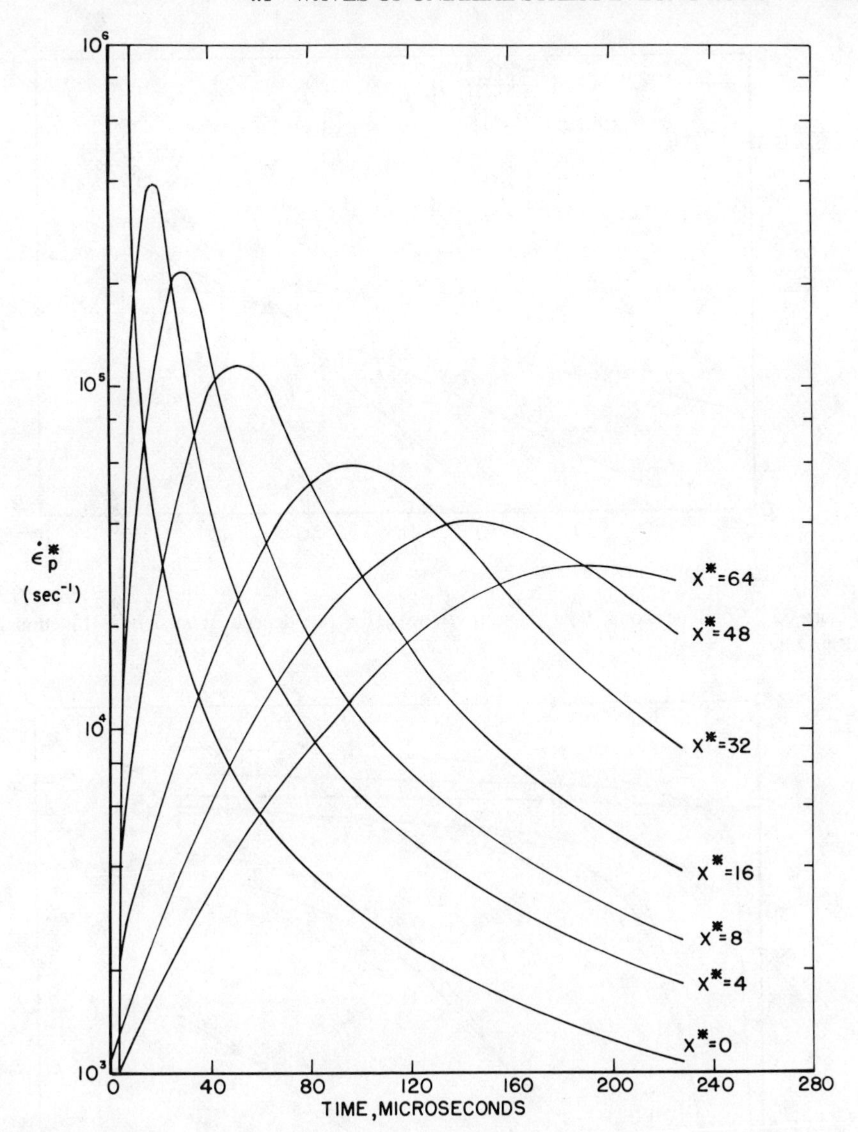

Figure 21 Plastic strain-rate histories at several stations along bar subjected to constant-velocity impact at end.

of the validity of a given theory is to demonstrate it in a series of experiments covering a wide range of strain rates. This is not always possible, however, because the material response plays an important role in determining the strain rates encountered in a plastic wave propagation experiment.

In another study, Herrmann and Lawrence (1978) demonstrated the sensitivity of the stress wave propagation and pulse attenuation to the material model

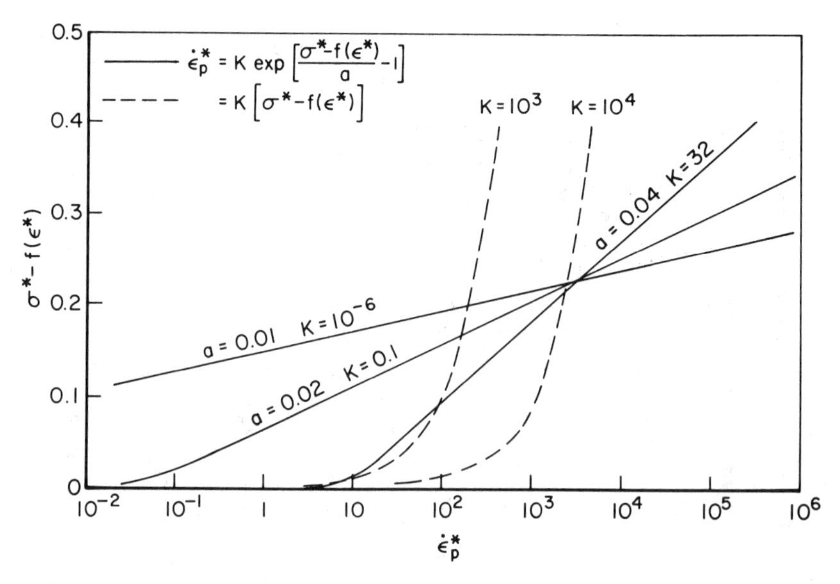

Figure 22 Nondimensional flow stress (overstress) for rate-dependent models as function of strain rate.

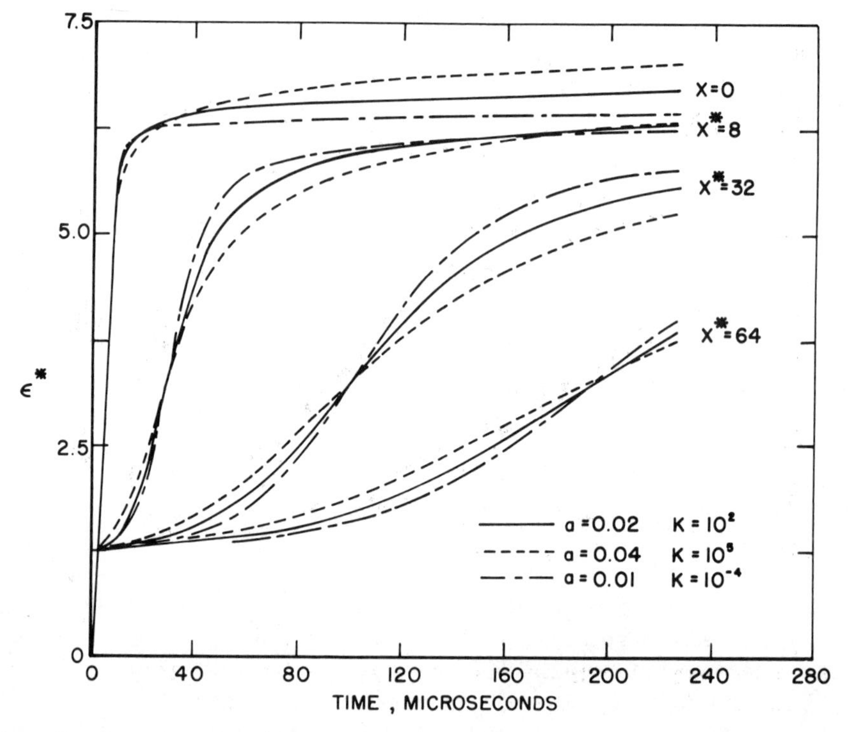

Figure 23 Comparison of strain-time profiles at several stations for three sets of constants in RD theory. Boundary condition is constant velocity.

for a wide class of materials. These and other results should serve to demonstrate the great care that must be exercised in interpreting the results of plastic-wave propagation experiments in terms of an assumed theory and, particularly, the lack of uniqueness when a given theory is found to model experimental results.

4.1.8 Other Constitutive Models

Most of the works cited have dealt with the semilinear form of the RD theory as proposed by Malvern (1951a, b), in which the plastic strain rate is a function of stress and strain. One result of this formulation is the prediction that the velocity of an incremental wave is the elastic wave speed as observed in numerous experiments. The failure of this formulation to predict all of the aspects of plastic waves in rods leads to the consideration of a somewhat more general quasilinear rate-type constitutive equation [Cristescu (1963), Lubliner (1964)], of the form

$$\dot{\varepsilon}_p = \phi(\sigma, \varepsilon)\dot{\sigma} + \psi(\sigma, \varepsilon) \tag{62}$$

where the functions ϕ and ψ govern the instantaneous and noninstantaneous response, respectively. The governing equations for the propagation of waves in a long rod using this model are,

$$\frac{\partial \sigma}{\partial x} = \rho \frac{\partial v}{\partial t} \tag{63}$$

$$\frac{\partial \varepsilon}{\partial t} = \frac{\partial v}{\partial x} \tag{64}$$

$$\frac{\partial \varepsilon}{\partial t} = \frac{1}{E}\frac{\partial \sigma}{\partial t} + \phi(\sigma, \varepsilon)\frac{\partial \sigma}{\partial t} + \psi(\sigma, \varepsilon) \tag{65}$$

This set of quasilinear partial differential equations can be solved using the method of characteristics from which the following relations are obtained:

$$\text{along } \frac{dx}{dt} = c(\sigma, \varepsilon), \qquad d\sigma = -\rho c\, dv + \rho c^2 \psi\, dt \tag{66a}$$

$$\text{along } \frac{dx}{dt} = -c(\sigma, \varepsilon), \qquad d\sigma = \rho c\, dv + \rho c^2 \psi\, dt \tag{66b}$$

$$\text{along } \frac{dx}{dt} = 0, \qquad d\varepsilon_e = \frac{1}{E} d\sigma$$

$$d\varepsilon_p = \phi\, d\sigma + \psi\, dt \tag{66c}$$

where subscripts e and p denote the elastic and plastic components of strain and where

$$c = c_0(1 + E\phi)^{-1/2} \tag{67}$$

Note that in this case the characteristics are neither constant nor straight lines but depend on the solution itself. The numerical integration of these equations, therefore, becomes a rather tedious task.

One result from the use of this more general form of constitutive relation (62) is that the incremental wave speed given by (67) is not necessarily the elastic wave speed c_0. Most experimental observations, however, have indicated that incremental waves travel at or near the elastic velocity, c_0, indicating that ϕ does not play an important role in describing the material behavior. Campbell and Dowling (1970) found a small decrease in incremental wave speed with increasing prestrain for copper (see Figure 11). However, because of the small magnitude of this decrease, they concluded that ϕ may be neglected. Yew and Richardson (1969) found an even greater decrease in the incremental wave speed in copper (see Figure 9), but the wave speeds were still substantially above those predicted by the RI theory. Campbell (1973) points out that there appears to be no theoretical basis for assuming that the instantaneous response of metallic materials is other than elastic, although part of the inelastic deformation may occur sufficiently rapidly to be considered as instantaneous.

By assuming there is no instantaneous plastic response ($\phi=0$), which is equivalent to assuming that incremental waves travel at the elastic velocity, c_0, the more general form of the constitutive equation reduces to

$$\dot{\varepsilon}_p=\psi(\sigma,\varepsilon)=\frac{1}{E}g(\sigma,\varepsilon) \tag{68}$$

which is the form originally proposed by Malvern (1961a, b) (46). Banerjee and Malvern (1975) performed numerical computations using the semilinear and quasilinear rate-type equations as well as a model of Perzyna (1963) which is an exponential overstress version of the Malvern equation. They considered the problem of torsional waves in a prestressed rod and performed an analysis of the experimental results of Yew and Richardson (1969). Yew and Richardson had concluded that high strain levels propagate at rates predicted by the RI theory while small strain levels propagate at c_0 as predicted by the RD theory (see Figure 9). Using the experimental strain-time profile at one station as the input function, Banerjee and Malvern's computations led them to conclude that reasonable agreement with experiments could be obtained for all three rate-type models investigated at intermediate strain levels, but not at high and low strains. Results of their computations are shown in Figure 24. From their comparisons of strain-time profiles, the quasilinear model was considered somewhat better than the others in predicting wave profiles and velocities, but all three rate-type models were considered to give better results than the RI theory.

A procedure for determining the functions in the more general quasilinear form of constitutive relation (62), as well as a detailed mathematical discussion of other forms of rate-dependent constitutive equations, is presented by Cristescu (1972). He concludes that several experimentally observed features of

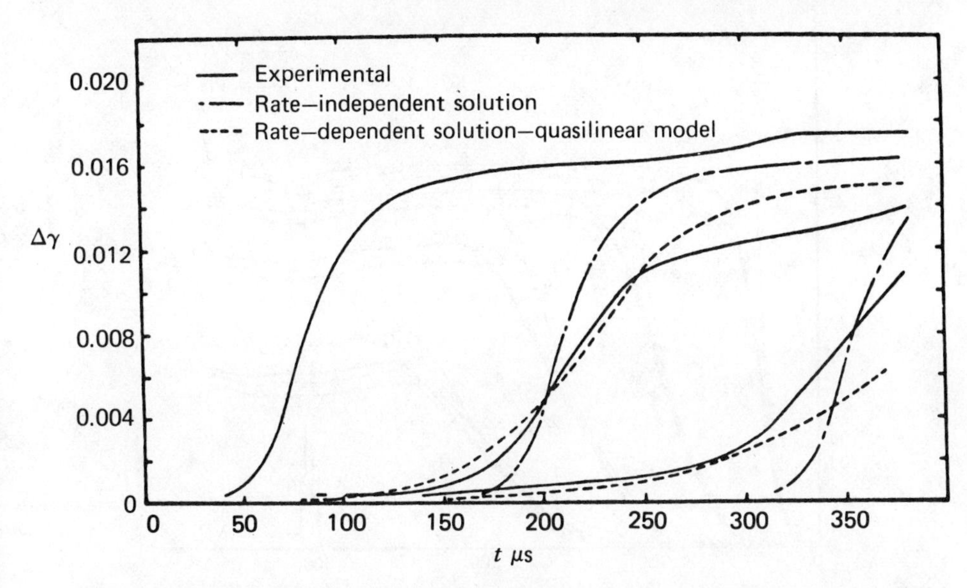

Figure 24 Incremental strain-time plots for quasilinear model at 0.38, 1.50, and 2.75 in. from impact end. First curve (0.38 in.) is taken as input for calculations [Banerjee and Malvern (1975)].

plastic wave propagation in rods can be described with a general rate-type theory as depicted in Figure 25 which shows the results of calculated wave profiles using various forms of the general rate theory. However, since many of the apparent rate effects are significant primarily very close to the impacted end, Cristescu concludes that the validation of the theory is highly questionable because of the complex three-dimensional stress state near the impact end which cannot be described by a one-dimensional wave theory. Karnes and Bertholf (1970) performed a detailed three-dimensional analysis of the wave profiles near the impacted end of a bar using a finite-difference numerical scheme. Using a rate-independent plasticity theory with a uniaxial stress quasistatic stress-strain curve and several dynamic curves, they calculated wave profiles as shown in Figure 26. The values of Y_0 are the proportional limit in Kbars for a single dynamic stress-strain curve for use in the RI theory. The calculations at a distance $\frac{1}{4}$ diameter from the impacted end show substantial differences in the longitudinal surface strain histories, but these differences quickly diminish at a distance only $\frac{1}{2}$ diameter from the end. The interpretation of these results, however, as well as those of Cristescu in Figure 25 and Banerjee and Malvern in Figure 24 should be considered in terms of the previous discussion on the sensitivity of wave profiles to the form of the constitutive relation.

One of the primary features of the constitutive models considered for dynamic plasticity has been mathematical tractability. The RI theory is convenient to use and is obtained in closed-form solution for the case of constant

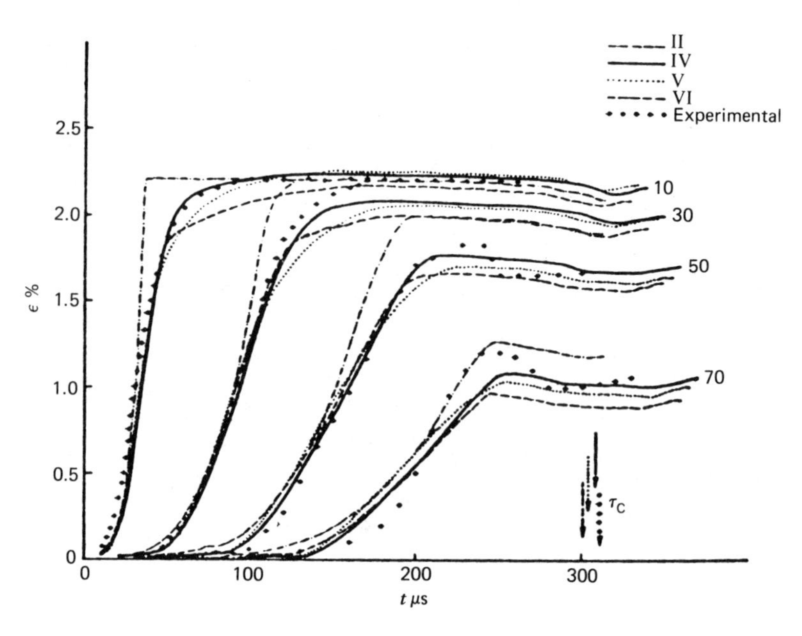

Figure 25 Comparison of strain-time profiles for various forms of constitutive equation (1 D means 1 diameter from impact end, etc.) [Cristescu (1972)].

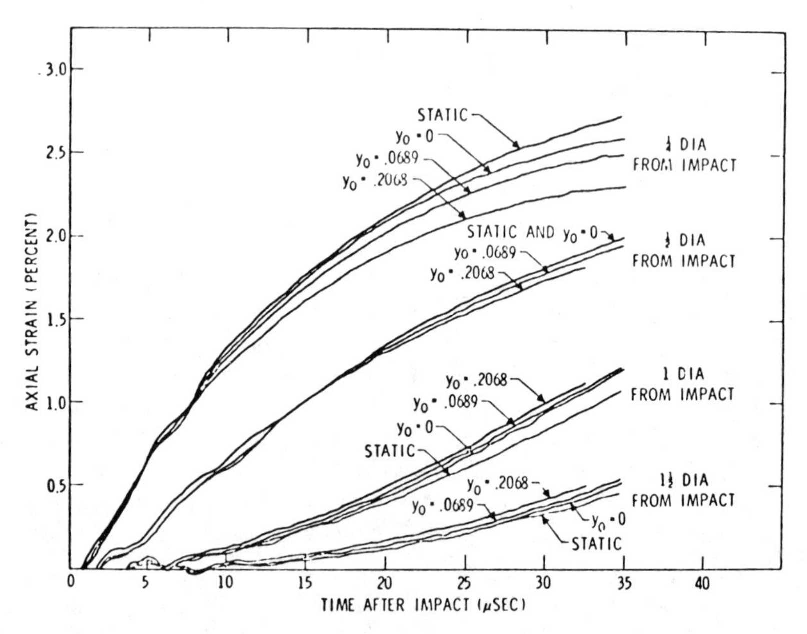

Figure 26 Computed longitudinal surface-strain histories at various positions from impact end [Karnes and Bertholf (1970)].

velocity applied to one end of a semiinfinite bar. The RD theory of Malvern is readily solved using the method of characteristics that involves relatively simple, numerical integration schemes. The more complex forms of the RD theory, though perhaps more physically meaningful and versatile, present formidable obstacles when applied to boundary value problems because the characteristic directions are not only not straight lines in the characteristic plane but also depend on the solution itself. An alternate approach to the method of characteristics for solving plastic wave propagation problems is a direct finite difference approach. Karnes and Bertholf (1970) employed such a technique to solve the two-dimensional problem of axisymmetric impact at the end of two identical bars. In this technique, the exact equations of motion are solved in an approximate manner through the use of a finite difference scheme. Although this approach is commonly used in shock wave studies, little attention has been paid to the technique for the simple one-dimensional (or two-dimensional) problem of elastic-plastic stress waves in long rods. The advantage of such a technique is that other forms of constitutive models could be investigated with little additional difficulty.

An example of a model that shows great potential for describing strain-rate effects is an incremental elastic-viscoplastic, strain-hardening, constitutive equation which Bodner and Partom (1974) suggested as appropriate for dynamic structural problems. The model, which is discussed in Chapter 8, is based on dislocation dynamics concepts and involves the use of a state variable that characterizes a materials resistance to plastic flow which, in turn, depends on plastic work. This model, which has been used to characterize both constant strain-rate and variable strain-rate experiments in uniaxial tension, has also been applied recently to the solution of plastic wave propagation problems by incorporating it into a finite-difference wave propagation code [Bodner (1980)]. The results for constant velocity impact in compression at the end of a semiinfinite rod of a rate-dependent material are presented in Figures 27 and 28. The material constants for a highly rate-dependent commercially pure titanium are taken from Bodner and Partom (1975) and provide an increase in flow stress of nearly 40% from a constant strain rate of 3×10^{-3} to 3 s^{-1}. In the problems modeled, strain rates of the order of 10^3 s^{-1} are achieved. The problem is treated as one of uniaxial stress in the x direction with inertia considered only in the x direction. However, transverse plastic strains are considered that couple with the equation of motion through the generalized three-dimensional constitutive model. Several features of interest are observed in the figures. In Figure 27, the plastic strain level tends toward a plateau near the impact end, $x=0$, even though the material exhibits a very strong strain-rate sensitivity. The corresponding stresses show pronounced stress relaxation, especially near the impact end, and the lack of a very sharp elastic front as predicted by the classical one-dimensional theories. Another feature of this plot is that the very front of the stress appears to propagate at the dilatational wave velocity, denoted by c_0, while the main front propagates at the elastic bar velocity, denoted by c_e. In Figure 28, stress and total axial strain are shown at

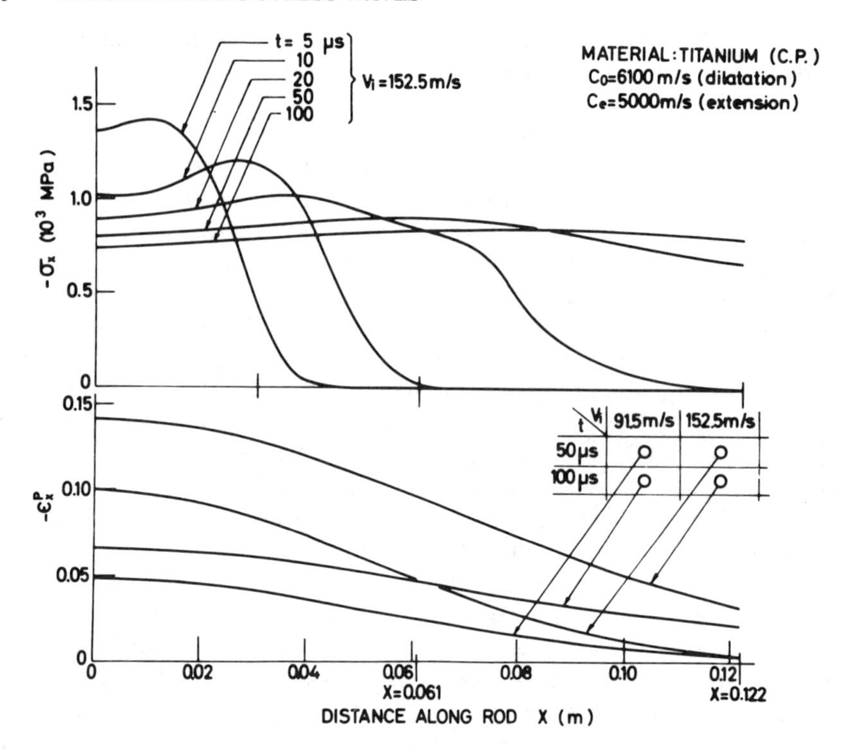

Figure 27 Stress and plastic-strain distribution in a semiinfinite bar subjected to constant-velocity impact at end [Bodner (1980)].

the impact end and at two additional locations along the bar. Except for the steep rise at $x=0$ for a very short time, the strain rates are nearly constant and decay slowly. For the time periods indicated, equal strain levels do not appear to propagate at constant velocity as predicted by the RI theory. Also, a pronounced stress peak at the impact end is observed which is consistent with observations reported by Bell (1968) and noted by Cristescu (1967). These results demonstrate the capability of an alternate form of constitutive equation to model many of the features that have been previously attributed to being characteristic of either the RI or RD theories. Furthermore, they illustrate the application of finite-difference computer codes to simple-wave propagation problems which heretofore have been treated mainly through solutions by the method of characteristics for a limited class of constitutive models.

Another constitutive model, which has shown capability for modeling both constant and variable strain rate experiments, discussed in Chapter 8, is the nonlinear hereditary-type stress-strain relation of Rabotnov and Suvorova (1978). They outline the method of solution for longitudinal wave propagation problems. It would appear that solutions of this type, as well as solutions involving incorporation of various constitutive models used to explain constant strain-rate data into finite difference formulations, could provide additional

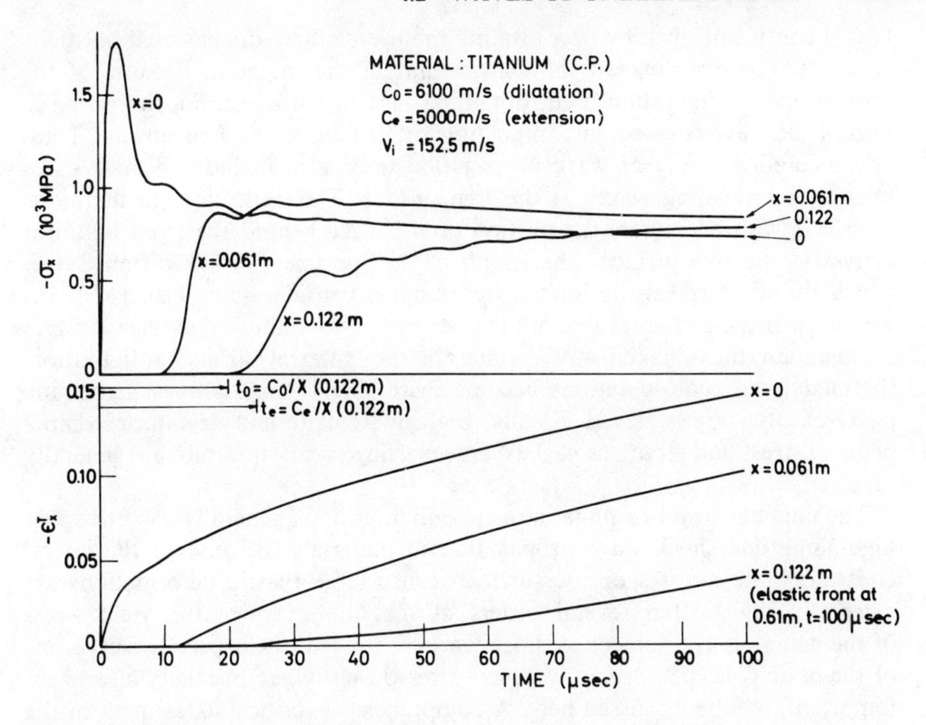

Figure 28 Stress and total-strain histories in a semiinfinite bar subjected to constant-velocity impact at end [Bodner (1980)].

insight into the nature of plastic wave propagation and an understanding of dynamic material behavior.

4.2 WAVES OF UNIAXIAL STRAIN

4.2.1 Introduction

Because of the numerous problems associated with determining dynamic properties of materials from studies of longitudinal wave propagation in bars or rods, other configurations have been sought. The most common technique involves the study of plane waves using the flat plate configuration to generate high amplitude waves of uniaxial strain [Karnes (1968)]. By impacting a flat plate whose diameter is large compared to its thickness in a direction perpendicular to its surface, a plane wave can be generated. Before the arrival of release waves from the edges of the plate, the center portion is in a confined state of one-dimensional or uniaxial strain. In this configuration, since there is only one component of displacement or strain in the direction of the wave propagation, no assumptions regarding the neglect of radial or lateral inertia have to be made. The stress state, however, is three-dimensional because of the lateral constraint, thereby necessitating the use of three-dimensional plasticity concepts if stresses exceed the elastic limit of the material. Because of the geometrical configuration used, obtaining measurements during the propagation of the wave is essentially impossible other than at the free surface. Thus, any theoretical study of wave propagation must also include an analysis of release or unloading waves at the free surface. This reflection, furthermore, interferes with and alters the portion of the wave behind the front before it arrives at the free surface. The length of the specimen, or the distance over which the wave travels, is limited by practical considerations because of the geometry of the configuration and the necessity of avoiding edge-release waves to maintain the uniaxial strain state. In the uniaxial strain configuration, thermodynamic considerations become increasingly important as higher impact velocities are achieved. Finally, both hydrostatic and deviatoric components of stress and strain, as well as large elastic and plastic strain are generally involved.

The uniaxial strain or plate impact configuration is generally used to study high-amplitude shock wave propagation in materials. Because of the lateral constraint, the stresses or pressures to cause large plastic deformations are extremely high, often several orders of magnitude above the yield stress of the material. The subject of shock waves is treated in Chapter 1. Only some of the basic concepts, primarily at low stress levels where plasticity effects are important, will be discussed here. A comprehensive, critical assessment of the status of the response of solids to shock compression including treatment of plasticity effects can be found in Davison and Graham (1979). Plasticity effects are also reviewed and critiqued in NMAB (1980). Continuum theories of plasticity applied to uniaxial strain problems are discussed by Clifton (1974) and Davison et al. (1977). Indepth reviews of the topic of viscoplastic behavior in uniaxial strain wave propagation are given by Wilkins (1964), Hopkins (1966), Herrmann (1969), Herrmann and Nunziato (1973), and Herrmann (1976).

The continuum theoretical analysis of elastic-plastic waves of uniaxial strain was first addressed by Wood (1952) who stressed the importance of hydrostatic compressibility in determining the nature of a wave. A systematic investigation of wave propagation treating elastic and plastic waves and the formation of shock waves was presented by Morland (1959) who presented the equations for a material exhibiting linear stress-strain behavior at low stresses and concave-up, stress-strain behavior in the plastic region. This analysis was an extension of that of Wood to allow for variations in the elastic constants with increasing pressure.

4.2.2 Analysis

The equations of motion and continuity are effectively the same as for the uniaxial stress case. Using σ_x, v_x, and ε_x to represent the components of stress, particle velocity, and strain in the direction of propagation, the equations of motion and continuity are

$$\rho_0 \frac{\partial v_x}{\partial t} = \frac{\partial \sigma_x}{\partial x} \tag{69}$$

$$\frac{\partial \varepsilon_x}{\partial t} = \frac{\partial v_x}{\partial x} \tag{70}$$

where ρ_0 is the initial density that can be considered constant if the dilatation is small. If changes in density are to be considered, particularly at high pressures, then conservation of mass requires that

$$\rho(1-\varepsilon_x)=\rho_0 \tag{71}$$

where ε_x is the compressive engineering strain. To complete the formulation of the problem, the constitutive equation relating σ_x to ε_x to describe the material behavior is also needed. Material behavior is generally assumed to be independent of strain rate for computational convenience. The stress-strain relation in the form $\sigma_x = \sigma_x(\varepsilon_x)$ cannot be determined directly but must be determined indirectly from experimental data through the application of plasticity concepts. Denoting by subscript y the transverse direction and noting that y can represent any direction normal to the direction of wave propagation for an isotropic material, the constitutive relation is derived by breaking up the total strains into elastic and plastic parts, assumed additive

$$\begin{aligned} \varepsilon_x &= \varepsilon_x^e + \varepsilon_x^p \\ \varepsilon_y &= \varepsilon_y^e + \varepsilon_y^p \end{aligned} \tag{72}$$

where superscripts e and p denote the elastic and plastic components, respectively. The decomposition of total strain into elastic and plastic components requires that the strains are small. Waves of finite strain, which occur under high pressures, are not considered in this treatment. Three basic assumptions are made regarding the stress-strain behavior of the material: (1) the elastic strains are related to the stresses through the equations of elasticity, (2) plastic flow is incompressible, and (3) strain-rate effects are not present in the constitutive equation. From assumption (1) we have

$$\begin{aligned} \varepsilon_x^e &= \frac{1}{E}\left[\sigma_x - 2\nu\sigma_y\right] \\ \varepsilon_y^e &= \frac{1}{E}\left[(1-\nu)\sigma_y - \nu\sigma_x\right] \end{aligned} \tag{73}$$

The condition of incompressible plastic flow requires

$$\varepsilon_x^p + 2\varepsilon_y^p = 0 \tag{74}$$

while the assumption of uniaxial strain conditions, that is, no lateral motion, requires

$$\varepsilon_y^e + \varepsilon_y^p = 0 \tag{75}$$

Combining (73), (74), and (75), considering only elastic behavior, leads to the uniaxial-strain elastic stress-strain relation

$$\sigma_x = \frac{E(1-\nu)}{(1+\nu)(1-2\nu)}\varepsilon_x = \left(K + \frac{4G}{3}\right)\varepsilon_x \tag{76}$$

where K is defined as the elastic bulk modulus and G is the shear modulus. For behavior beyond the elastic region, the yield condition is applied on the deviatoric components of stress. The relation between the spherical components of stress and strain is assumed to remain elastic for low pressures. It is assumed that the yield criterion for the stress deviator is the same in uniaxial strain as in uniaxial stress, leading to the stress-strain law in the form

$$\sigma_x - \sigma_y = Y(\varepsilon_x^p) \tag{77}$$

which is equivalent to the Mises-Henky yield condition or the maximum shear-stress yield criterion in plasticity. The function $Y(\varepsilon_x^p)$ is the stress-plastic strain relation from a uniaxial tension or compression test under the assumption of isotropic work hardening material behavior. Since a direct relation is to be made with behavior in uniaxial strain under wave propagation conditions, the uniaxial stress behavior should theoretically be obtained under adiabatic conditions.

Consider first the simple case of elastic, perfectly plastic behavior, that is, Y=constant=Y_0. Up to the yield point, the behavior is elastic and given by (76) as shown on the first part of the uniaxial strain line in Figure 29. Above the yield point, the yield criterion (77), is introduced. Combining this with the previous relations that apply to conditions of uniaxial strain, the following

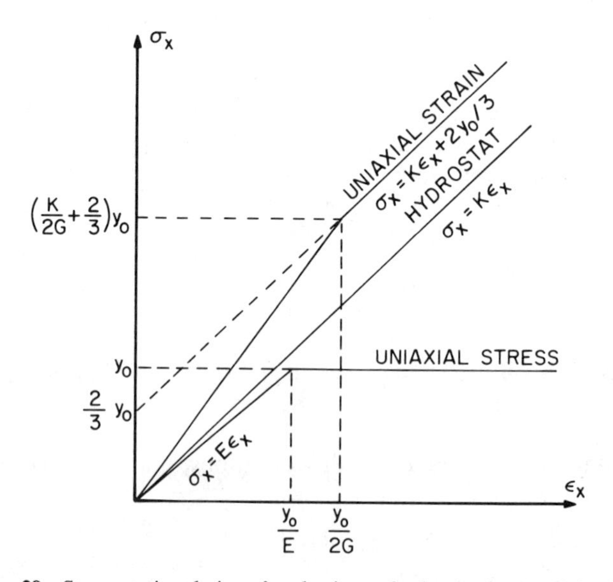

Figure 29 Stress-strain relations for elastic, perfectly plastic material behavior.

relation is easily derived:

$$\sigma_x = K\varepsilon_x + \frac{2Y_0}{3} \tag{78}$$

which is valid only above yield. This expression is also plotted in Figure 29. The yield point in uniaxial strain is easily derived by solving (76) and (78) simultaneously to give the coordinates of the yield point as

$$\sigma_x = \left(\frac{K}{2G} + \frac{2}{3}\right) Y_0$$

$$\varepsilon_x = \frac{Y_0}{2G} \tag{79}$$

This yield stress has been termed the Hugoniot elastic limit (HEL) [Minshall (1965)]. Note that these relations have been derived assuming elastic behavior among the spherical components and elastic, perfectly plastic behavior among the deviatoric components. The hydrostat is shown in Figure 29 as a straight line that assumes a constant bulk modulus, K. It can be seen that the uniaxial strain curve is parallel to the hydrostat and is offset by $2Y_0/3$. Also shown in Figure 29 is the uniaxial-stress stress-strain curve for this condition.

The equations of motion, continuity, and stress-strain relation may be combined in the form

$$\frac{\partial^2 u}{\partial t^2} = \frac{1}{\rho_0} \frac{\partial \sigma_x}{\partial \varepsilon_x} \frac{\partial^2 u}{\partial x^2} \tag{80}$$

which is formally equivalent to the wave equation of Karman (1942) for the rate-independent theory of plastic wave propagation in long thin rods [See (17) and (18)]. As in that case, the velocity of propagation of a strain increment is given by

$$c = \left(\frac{1}{\rho_0} \frac{\partial \sigma_x}{\partial \varepsilon_x}\right)^{1/2} \tag{81}$$

Thus, for purely elastic behavior in both the spherical and deviatoric components of stress, a single elastic wave exists in the uniaxial strain configuration whose velocity is given by

$$c_e = \left[\frac{(K + 4G/3)}{\rho}\right]^{1/2} \tag{82}$$

For spherical elastic but elastic, perfectly plastic deviatoric response, a plastic wave also exists moving at velocity

$$c_p = \left[\frac{K}{\rho}\right]^{1/2} \tag{83}$$

This dual wave structure results from the bilinear stress-strain response shown in Figure 29 and is equivalent to the problem of waves in a bilinear material in uniaxial stress first addressed by Donnell (1930), shown in Figure 1.

The modulus governing plastic wave speeds in (83) is lower than that for elastic waves in (82) because of plastic flow in shear. For low stresses, the bulk modulus, K, can be considered constant. As stresses are increased further, the bulk modulus continues to increase, leading to a concave-up stress-strain curve. As in the uniaxial stress case discussed previously [White and Griffis (1948)] higher stress amplitudes travel at greater velocities than lower ones and eventually overtake them, leading to the formation of plastic shock waves. At sufficiently high pressures, the plastic shock waves can overtake the elastic precursor and form a single shock wave.

The problem of wave propagation in uniaxial strain is further complicated by unloading considerations, as also occurs in the uniaxial stress case. In the uniaxial strain case, however, reversed plasticity is encountered without reversing the applied load, resulting in both elastic and plastic unloading waves. Wood (1952) demonstrated the calculation of the uniaxial strain curve from simple tension and compression test data including unloading effects. His uniaxial stress curve is given in the general form

$$\sigma = Y(\varepsilon^p) \tag{84}$$

and is based on actual data for 24S-T aluminum alloy as shown in Figure 30. The elastic constants are assumed constant for moderate values of hydrostatic stress, equivalent to the assumption of elastic spherical behavior. The calculated uniaxial strain curve is shown in Figure 31, where the points $0, e, 2, 3, 4, 5$ correspond to the equivalent points $0, e', 2', 3', 4', 5'$ in Figure 30. Up to point

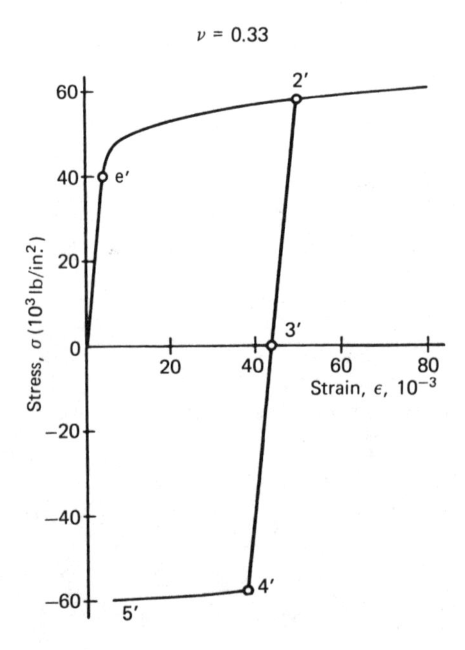

Figure 30 Stress-strain relation for 24S-T aluminum alloy in simple compression and tension [Wood (1952)].

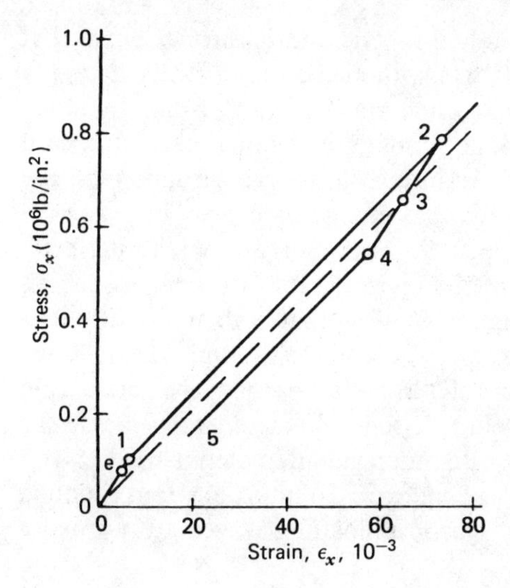

Figure 31 Stress-strain relation for 24S-T aluminum alloy in unaxial strain [Wood (1952)].

2′, no unloading has occurred and the derived equations are valid, the elastic condition being used between 0 and e' and the yield condition (77) between e' and 2′, where e' has been defined as the elastic limit. Beyond the elastic limit, the equations can be rearranged in the form

$$\sigma_x = K\varepsilon_x + \tfrac{2}{3}Y(\varepsilon_x^p) \tag{85}$$

which shows that the uniaxial strain curve is offset from the hydrostat by an amount equal to two-thirds the flow stress in uniaxial stress for the corresponding value of plastic strain. The stress, therefore, is comprised of a hydrostatic stress component and a contribution of the shear strength of the material. At high hydrostatic stresses, the shear strength will eventually become negligible. In addition, the bulk modulus for metals is generally an increasing function of pressure that would appear as a concave-up, stress-strain curve for the hydrostat and, therefore, for the uniaxial strain $\sigma_x(\varepsilon_x)$ curve.

The uniaxial-strain, stress-strain curve can often be approximated by two straight lines of slopes $K+4G/3$ and K intersecting at the HEL which is shown as point 1 in Figure 31. Barker et al. (1964) point out, however, that when an experiment involves relatively small plastic strains, the details of the transition from the elastic to the plastic region become important since the deviation from the bilinear approximation can amount to an appreciable part of the total stress. For this particular example, the calculated wave velocity for strain increments in the plastic region, 1-2, is less than 1% higher than that calculated from the hydrostat, and approximately 18% slower than the elastic velocity. In uniaxial stress in long bars, on the other hand, the plastic wave velocities are generally an order of magnitude slower than the elastic bar velocity.

Beyond point 2′ in Figure 30 unloading occurs. The unloading from point 2′ to 4′ is assumed elastic and parallel to 0-e'. Using the elastic relations in differential form, the curve 2-4 in Figure 31 is constructed. Along this unloading path the stress state becomes hydrostatic and it can be shown that point 3 is equivalent to 3′ in Figure 30. Thus, the hydrostatic stress state 3 is independent of path of loading, being reached via 0-*e*-1-2-3 or directly along the hydrostat $\sigma_x = K\varepsilon_x$. Beyond some point 4′, reverse yielding takes place and the yield condition is again imposed, resulting in the construction of the segment 4-5 in Figure 31. It is thus shown that unloading can produce reverse yielding before the stress σ_x has dropped to zero. For an elastic, perfectly plastic material, using the nomenclature of Figure 31, the stress profile for an impact of short duration is shown in Figure 32. Note that both an elastic and a plastic shock front are produced, traveling at the elastic and plastic wave velocities given by (82) and (83), respectively. In addition, there are both elastic and plastic unloading waves each traveling at their characteristic velocities as shown. These results are based on a rate-independent material model. The existence of both elastic and plastic unloading waves in uniaxial strain is unlike the case in uniaxial stress where only elastic unloading waves are normally encountered.

The yield condition (77), given in terms of plastic strain, can be formulated in terms of plastic work. Fowles (1961) presents a detailed treatment of the incorporation of work hardening into the yield condition,

$$\sigma_x - \sigma_y = Y(W_p) \tag{86}$$

resulting in the relation

$$\sigma_x = \sigma_m + \tfrac{2}{3}Y(W_p) \tag{87}$$

which demonstrates that in the plastic range the stress is larger than the corresponding hydrostatic stress or mean stress, σ_m, for the same strain by an amount equal to $2Y/3$. The difference between the uniaxial-strain and hydro-

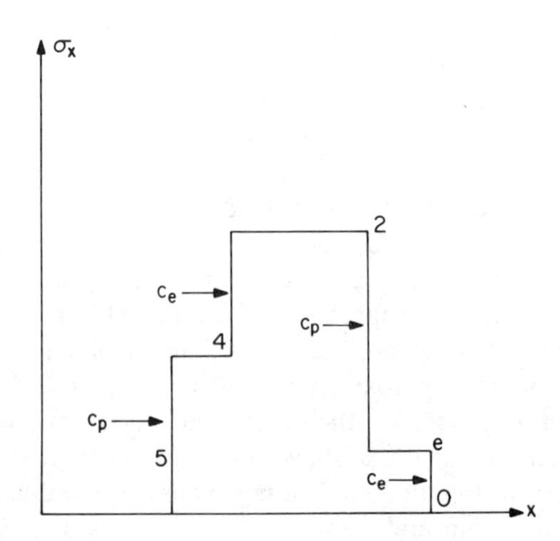

Figure 32 Stress profile for elastic, perfectly plastic material under uniaxial-strain conditions.

static stress-strain curves depends only on the yield stress as a function of strain, which can be obtained from uniaxial stress data provided the comparison is made at the appropriate corresponding strains based on the equivalence of plastic work.

4.2.3 Comparisons With Uniaxial Stress Data

Fowles (1961) obtained some of the earliest experimental results verifying the general predictions of the elastic-plastic model in hardened and annealed 2024 aluminum and found no significant strain-rate effects. Later, Barker et al. (1964) conducted an experimental investigation to measure the dynamic stress-strain relation in aluminum in uniaxial strain from plate impacts. They compared their results to predictions based on uniaxial stress data using the rate-independent, stress-strain assumption and the procedure described in Section 4.2.2, assuming that the yield condition is a function of plastic work as formulated by Morland (1959). To attain a high degree of precision in constructing the uniaxial strain curve from uniaxial stress data, they used values of elastic constants obtained from ultrasonic measurements and performed their calculations to determine offsets from the elastic curves rather than total values as shown in Figure 33. Utilizing expressions derived by Fowles (1961),

$$Y_H = P + \tfrac{2}{3} Y_s \tag{88}$$

$$\varepsilon_h = \tfrac{3}{2}\varepsilon_s - \frac{Y_s}{6K} \tag{89}$$

where the symbols are as shown in Figure 33, and P is the hydrostatic component of the yield stress, they constructed an accurate uniaxial strain

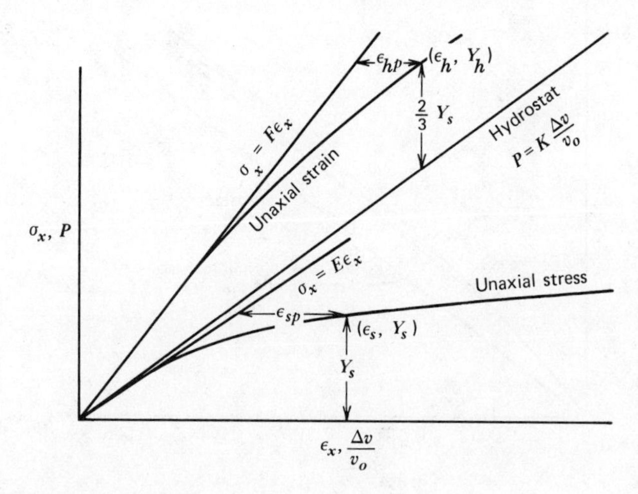

Figure 33 Stress-strain plot defining terms in calculating uniaxial-strain curve from uniaxial-stress curve [Barker at al. (1964)].

curve. Use was also made of a relation among the offsets that they derived,

$$\varepsilon_{hp} = \tfrac{3}{2}\left(\frac{(F-K)}{F}\right)\varepsilon_{sp} \tag{90}$$

where the symbols F and K represent the elastic slopes as shown in Figure 33, which correspond to the slopes of the elastic portion of the uniaxial strain curve and the hydrostat, respectively, as also shown in Figure 29. As shown in Figure 33, ε_{hp} is the offset of the uniaxial strain curve from the line having a slope F, while ε_{sp} is the corresponding offset of the uniaxial stress curve from the line of slope E, Young's modulus, which also corresponds to the total longitudinal plastic strain. The constructed uniaxial strain curve is shown in Figure 34 along with the measured relations from the various plate-impact experiments. It can be seen that the measured values were consistently higher than the predicted curve from quasistatic uniaxial stress data on 6061-T6 aluminum. The tailing off of each experimental curve is attributed to a drop in strain rate near the end of each test, while the position of the static curve from uniaxial stress data indicates a strain-rate effect. The combined effect suggests that a different stress-strain curve exists for each strain rate. To check this hypothesis, Barker et al. conducted an incremental wave-propagation experiment, similar to that used by Sternglass and Stuart (1953) and others to demonstrate strain-rate effects in wave propagation in long rods. In this case, the material was prestressed dynamically with a low-amplitude stress wave into the plastic region and shortly thereafter, in about 2 μs, subjected to a higher amplitude wave. As in the uniaxial stress experiments, a small elastic precursor was observed traveling at approximately the elastic velocity. These data were

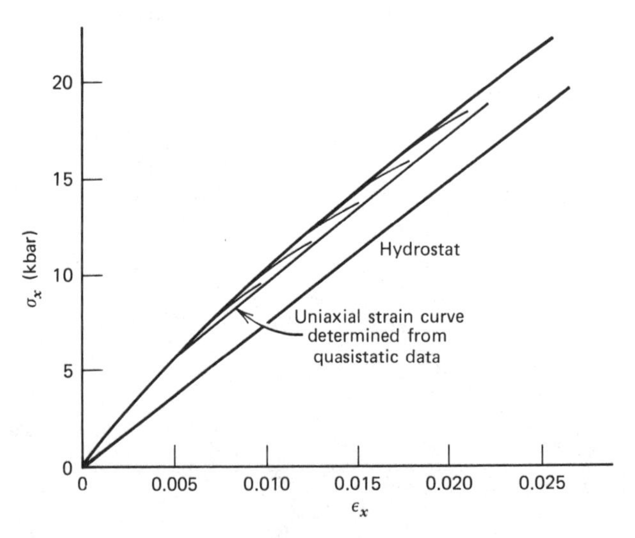

Figure 34 Experimentally measured dynamic stress-strain relation. Relation determined from quasistatic uniaxial-stress data is shown for comparison [Barker et al. (1964)].

interpreted as demonstrating a strain-rate effect in aluminum under shock loading. Barker et al. also measured the stress-strain relation upon unloading and again found deviations from that predicted from the rate-independent, elastic-plastic theory, which they attributed to strain-rate and Bauschinger effects. An earlier investigation to determine the stress-strain relation from plate-impact experiments by Lundergan and Herrmann (1963) suggested the existence of a strain-rate effect in aluminum but failed to positively identify it.

Little work has been done using a constitutive model incorporating strain-rate effects. Jones et al. (1968) utilized uniaxial stress data obtained over a wide range of loading rates to predict wave profiles in uniaxial strain conditions in 1060-0 aluminum. At relatively low pressure levels, they found their best agreement with observed wave profiles using a model that predicts increasing strain-rate sensitivity at higher strain rates. Butcher and Karnes (1966) computed uniaxial strain-wave profiles based on uniaxial stress data for 6061-T6 aluminum which shows a linear variation of stress with log strain rate for rates below $10^3\ s^{-1}$ and a linear variation with strain rate for higher rates. Using this rate-dependent formulation, they found good agreement with experimental data from plate impact experiments using targets of different thicknesses. They observed that the effect of strain rate in uniaxial strain is to cause stress increments to propagate at (nonconstant) velocities different than those predicted by the rate-independent theory, thereby affecting the nature of the plastic wave front although not changing the peak stresses significantly. Butcher and Karnes also noted from their calculations that, contrary to the rate-independent theory, after passage of the wave front and the attainment of nearly constant stress, the plastic strain rate is far from zero and persists at a moderately high level for a considerable time. The constant stress results from a balance of decrease with decreasing strain rate and increase from work hardening, which tend to cancel each other.

4.2.4 Interpretation of Experimental Data

Methods of evaluating material response functions for the plate impact experiment and results of interpretation of experimental data are summarized in Herrmann (1976). Determination of both strain-rate and strain-hardening characteristics are discussed. The large number of variables involved in the plate-impact experiment, however, makes it quite difficult to deduce the nature of dynamic material behavior using the indirect method of wave propagation analysis. As previously noted, the stress-strain relation in uniaxial strain is dominated by the hydrostatic behavior, especially at higher pressures. Higher strain rates are generally achieved at higher pressures experimentally, so that rate effects, even if more extensive at higher rates, are less significant in the overall stress-strain behavior. Furthermore, the plastic wave velocity is governed by the slope of the stress-strain relation that is very insensitive to strain hardening or strain-rate effects because of the large contribution of the hydrostat that does not involve strain-rate effects because of the absence of shear stresses. Note also that strain rate is a variable function, in general, in wave propagation experiments, and depends not only on the experimental conditions such as planarity of impact, but also on the material response itself during wave propagation. Because of the complexity of the mathematics, most of the attention in elastic-plastic wave propagation in uniaxial strain has been confined to use of a rate-independent theory, analogous to the approach used extensively in uniaxial stress wave studies in long rods. The use of a rate-independent theory with a single, dynamic stress-strain curve, is an effective method of incorporating rate effects into a mathematical formulation in both the uniaxial stress and uniaxial strain cases. Incorporation of strain-rate effects in the uniaxial stress case presented numerous problems from the point of view of interpreting experimental results, and is no less difficult in the uniaxial strain case.

Several approaches have been used to identify strength and strain-rate effects in the uniaxial strain configuration. One aspect of wave interactions that is particularly sensitive to the influence of yield strength is the attenuation of a shock wave because of the unloading stresses behind it. The influence of yield strength in such a situation with moderate unloading behind the shock wave was treated by Lee and Liu (1964). They compared an elastic-plastic solution to a hydrodynamic solution and a rigid-plastic solution. The hydrodynamic solution, which neglects yield influences, showed an appreciable difference in attenuation characteristics while the rigid-plastic solution was found to be a poor approximation since the elastic resilience in shear has an appreciable influence on the unloading characteristics.

Another aspect of uniaxial-strain wave propagation which provides a considerable amount of information on strain-rate effects is observation of the amplitude of the elastic precursor at low stress levels where both an elastic and plastic shock front exist (see Figure 32). The amplitude of the elastic precursor is directly related to the dynamic yield strength of the material, commonly referred to as the Hugoniot elastic limit (HEL). If the HEL diminishes as the wave travels through the material, that is the amplitude "*oe*" of the elastic precursor in Figure 32 decreases with distance traveled, some form of stress relaxation or strain-rate effect is indicated. Experimentally, the peak amplitude of the elastic precursor has been observed to decrease with increasing propagation distance in a variety of polycrystalline materials. Such behavior and its relation to dislocation dynamics is reviewed by Jones (1973) and discussed by Clifton (1980). Duvall (1964) showed how the decay of the elastic precursor wave can be related to material relaxation mechanisms. His equations are formally equivalent to the rate-dependent equations of Malvern (1951b) for the uniaxial stress case except for different physical constants. Relaxation of stress waves was also treated by Taylor (1965) in terms of dislocation dynamics.

The interpretation of experimental data is not always straightforward, however, when trying to identify strain-rate effects. Strain-rate effects may result in a portion of the plastic wave propagating with nearly the same velocity as the elastic wave, making it difficult to determine the true elastic limit in precursor decay studies. Additional difficulties in interpreting experimental data are because of the smoothness of the stress-strain curve that transitions from an elastic region to the plastic region gradually for real materials. Finally, especially for rate-dependent materials, high plastic strain rates occur only in the vicinity of the impact face and can decay quite rapidly within a small distance of propagation. The actual conditions that occur at the impact face, which depend a great deal on the planarity of impact, play an important role in determining the material behavior and may obscure strain-rate effects [Butcher and Karnes (1966)]. Although the study of elastic precursor decay is a useful method for determining strain-rate effects in metals, when rate effects are small it may be difficult to obtain meaningful experimental data because of the small specimen thicknesses required.

In the case when the shock front in a plate-impact experiment separates into a elastic and a plastic component with the elastic front traveling at the higher velocity, information on strain-rate effects may also be obtained from a study of the plastic wave front after it evolves into a steady propagating wave after a short distance is traveled. Johnson and Barker (1969) studied steady plastic wave profiles in 6061-T6 aluminum that result from the competing effects of nonlinear material behavior which leads to shock wave formation and rate-dependent material properties which tend to spread out the plastic wave front. This method provides an alternate approach to the study of strain-rate sensitivity in metals. To completely define the wave propagation phenomenology in uniaxial strain as well as to ascertain the role of strain-rate in the material constitutive behavior, a wide variety of experimental techniques should be employed.

4.3 STRESS WAVES IN OTHER GEOMETRIES

4.3.1 Waves in Strings or Wires

The problems of lateral inertia and three-dimensional effects in longitudinal waves in rods can be partially overcome by using small-diameter wires where these effects are reduced. Extremely high strain rates cannot be achieved, however, because increasing impact velocities will cause instantaneous failure in tensile impact or buckling in compression. There is little to be gained over the bar or rod experiments because of this. An alternate approach is to subject wires to transverse impact and monitor the propagation of transverse waves. The mechanics of wave propagation in extensible strings is discussed in detail by Cristescu (1967). The first analysis of the problem of an extensible string subjected to a constant velocity from a transverse impact was presented by Rakhmatulin (1945b). The assumption in the analysis was that of a rate-independent material, or at least one that could be described by a single stress-strain relation.

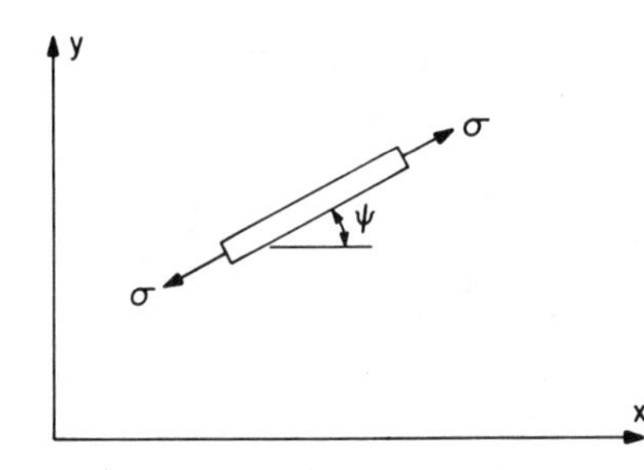

Figure 35 Nomenclature for element of wire.

Consider an element of wire in the *x-y* plane that makes an angle ψ with the *x*-axis as shown in Figure 35 and let u and v be the particle displacements in the x and y directions, respectively. The equations of motion for the wire subjected to an axial stress σ are:

$$\rho\frac{\partial^2 v}{\partial t^2}=\frac{\partial}{\partial x}(\sigma\sin\psi) \tag{91}$$

$$\rho\frac{\partial^2 u}{\partial t^2}=\frac{\partial}{\partial x}(\sigma\cos\psi) \tag{92}$$

where ρ is the mass density in the initial undeformed state. Let λ be the stretch ratio or ratio of current length to initial length. From geometry considerations, it is easily shown that

$$\frac{\partial u}{\partial x}=\lambda\cos\psi-1 \tag{93}$$

$$\frac{\partial v}{\partial x}=\lambda\sin\psi \tag{94}$$

For a solution to the problem, it is convenient to let v_1 and v_2 be particle velocities along and transverse to the wire, respectively. The equations of motion can then be written as:

$$\frac{\partial v_1}{\partial t}-v_2\frac{\partial\psi}{\partial t}-\frac{1}{\rho}\frac{\partial\sigma}{\partial x}=0 \tag{95}$$

$$\frac{\partial v_2}{\partial t}+v_1\frac{\partial\psi}{\partial t}-\frac{\sigma}{\rho}\frac{\partial\psi}{\partial x}=0 \tag{96}$$

$$\frac{\partial v_1}{\partial x}-v_2\frac{\partial\psi}{\partial x}-\frac{\partial\lambda}{\partial t}=0 \tag{97}$$

$$\frac{\partial v_2}{\partial x}+v_1\frac{\partial\psi}{\partial x}-\lambda\frac{\partial\psi}{\partial t}=0 \tag{98}$$

These four equations are seen to involve functions and derivatives with respect to x and t of the five variables v_1, v_2, ψ, σ, and λ. We can eliminate one of the variables through the introduction of a constitutive equation

$$\sigma=\sigma(\lambda) \tag{99}$$

with restrictions

$$\frac{d\sigma}{d\lambda} \geqslant 0 \qquad \frac{d^2\sigma}{d\lambda^2} \leqslant 0 \qquad \text{for } \lambda \geqslant 1 \tag{100}$$

These are equivalent to the assumption of a rate-independent material with a concave-down, stress-strain curve that precludes the formation of shock waves. The problem to be solved is that of an infinite wire subjected to a transverse impact at $x=0$ which imparts a constant transverse velocity v_0 in the y direction. The first solution to this problem was by Rakhmatulin (1945b) who assumed the shape of the deformed wire to obtain a solution. His approach was analogous to that of Karman (1942) for the constant velocity axial impact problem in a long bar for a rate-independent material (see Section 4.1.1). A direct solution may also be obtained by the method of characteristics. Along $dx/dt = \tau$, we obtain simple wave solutions, where the characteristic directions are given by

$$\tau = \pm c = \pm \left[\frac{1}{\rho} \frac{d\sigma}{d\lambda} \right]^{1/2} \tag{101a}$$

$$\tau = \pm \bar{c} = \pm \left(\frac{\sigma}{\rho\lambda} \right)^{1/2} \tag{101b}$$

By symmetry, consider only solutions for $dx/dt = +c, +\bar{c}$.

$$\text{along} \quad \frac{dx}{dt} = c, \qquad v_1 = -\int c\, d\lambda$$

$$v_2 = \text{constant} \tag{102}$$

$$\psi = \text{constant}$$

$$\text{along} \quad \frac{dx}{dt} = \bar{c}, \qquad \lambda = \text{constant}$$

$$v_1 = \alpha \sin\psi + \beta \cos\psi - \lambda \bar{c} \tag{103}$$

$$v_2 = \alpha \cos\psi - \beta \sin\psi$$

where α and β are arbitrary constants. Note that longitudinal wavelets each propagate with their characteristic velocity, c, but that there is only one transverse velocity, $\bar{c}$. The solution for the various regions shown in Figure 36 is summarized below:

$$\text{in (1)} \qquad v_1 = v_2 = \psi = 0, \qquad \lambda = \lambda_0 \tag{104a}$$

$$\text{in (2)} \qquad v_1 = -\int_{\lambda_0}^{\lambda_2} c\, d\lambda, \qquad v_2 = \psi = 0, \qquad \lambda = \lambda_2 \tag{104b}$$

$$\text{in (3)} \qquad v_1 = -\int_{\lambda_0}^{\lambda_2} c\, d\lambda, \qquad \tan\psi = \frac{-v_0}{v_1 + \lambda_2 \bar{c}} \tag{104c}$$

$$v_0^2 = -2\lambda_2 \bar{c} v_1 - v_1^2$$

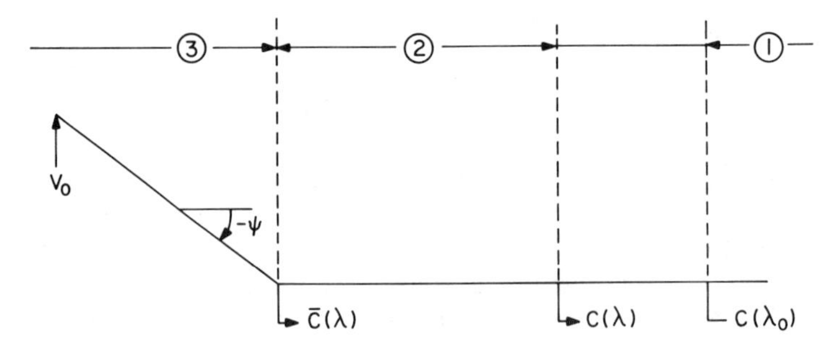

Figure 36 Illustration of solutions in various regions of transversely impacted wire.

Note that if λ_0, v_0, and $\sigma=\sigma(\lambda)$ are known, as well as c and $\bar{c}$ from (101), then the solution is determinate. Alternately, experimental data can be used to construct the curve $\sigma=\sigma(\lambda)$ from measurements of ψ and $\bar{c}$ for various prestress levels. One of the first experimental investigations using this technique to study plastic wave propagation in metals was conducted by Schultz et al. (1967) using aluminum and copper wires with several levels of static prestrain. Although some consistency with the predictions of the rate-independent theory were noted for one aluminum, the behavior of two other aluminums and copper differed from that predicted by the theory at high strain levels, leading to the conclusion that constitutive models incorporating rate dependence should be investigated. As in most of the experiments in prestressed bars or rods, it was again noted that small strain increments propagate elastically from a state of static prestrain. There appears to be no inherent advantage in using the propagation of transverse waves in strings over the longitudinal case because higher strain rates are not achievable.

4.3.2 Wave Propagation in Beams

High-velocity impacts on structures can also cause inelastic stress waves to propagate throughout the structure. Although this is an important problem from an analysis point of view, it is not a very convenient method for deducing material constitutive relations and, hence, will not be treated in detail here. However, the nature of wave propagation possesses similarities to features found in rods and strings.

One of the earliest investigations into transverse impact on beams was conducted by Duwez et al. (1950) for a long beam impacted at the free end with a constant velocity. A moment-curvature relation was assumed based on the (static) stress-strain curve of the material. It was observed that the strain in the beam was not propagated at constant velocity. The concept of rigid plastic behavior and of plastic hinges in beams under impact loading was introduced by Lee and Symonds (1952) to study the large plastic deformations of beams under transverse impact. Parkes (1955) applied an ideal rigid-plastic analysis to the dynamic problem of a cantilever impacted at its free end and achieved satisfactory correlation with experiments for the prediction of permanent deformation at points remote from the impact. Ting (1964) extended the work of Parkes to include a strain-rate sensitivity and found good experimental correlation. Bodner and Symonds (1962) conducted impact experiments on mild steel and aluminum beams to evaluate the rigid-plastic assumptions and found that a strain-rate-dependent rigid-plastic analysis gave excellent agreement with experimental results.

For high strain rates, rotatory inertia and transverse shear deformation can become important in the beam analysis, as well as the effects of strain rate on the material properties. The numerical traveling wave solution for this type of problem was presented by Plass (1955) in which Malvern's (1951a, b) rate-sensitive constitutive model was used. The equations of motion are obtained by applying the principles of conservation of angular and linear momentum to an element of a beam of length dx:

$$\frac{\partial M}{\partial x} - Q = \rho I \frac{\partial \omega}{\partial t} \tag{105}$$

$$\frac{\partial Q}{\partial x} = \rho A \frac{\partial v}{\partial t} \tag{106}$$

where M and Q are the bending moment and shear force, respectively, ρ is the mass density, A the cross-sectional area, I the moment of inertia, ω the angular velocity, and v the transverse velocity in the y direction. Denoting the curvature by k and the shear deformation by γ, the equations of continuity are

$$\frac{\partial k}{\partial t} = \frac{\partial \omega}{\partial x} \tag{107}$$

$$\frac{\partial \gamma}{\partial t} = \frac{\partial v}{\partial x} + \omega \tag{108}$$

For quasistatic loading, or for solutions assuming rate-independent material behavior, the constitutive equations take the form

$$M_s = EIk \tag{109}$$

$$Q_s = GA_s \gamma \tag{110}$$

where the subscript s has been added to note static or rate-independent behavior. E and G are Young's modulus and the shear modulus, respectively, while A_s is the effective cross-sectional area that incorporates the conventional shear correction factor.

For rate-dependent material behavior, the constitutive equations take the form

$$EI \frac{\partial k}{\partial t} = \frac{\partial M}{\partial t} + K_c(M - M_s) \tag{111}$$

$$GA_s \frac{\partial \gamma}{\partial t} = \frac{\partial Q}{\partial t} + K_s(Q - Q_s) \tag{112}$$

where K_c and K_s are constants. The forms of these two equations are analogous to the linear overstress form of Malvern (1951a) for uniaxial stress states (48), and decompose the deformation rates into elastic and plastic components. Here, the elastic components are the force rates and the plastic components are proportional to the difference between the actual force and the force that would occur in a static test. The set of quasilinear differential equations is solvable using the method of characteristics where the characteristic directions are

$$\frac{dx}{dt}=0 \text{ (twice)}, \qquad \pm C_m, \pm C_q \tag{113}$$

where the quantities

$$C_m=\left(\frac{E}{\rho}\right)^{1/2} \tag{114}$$

$$C_q=\left(\frac{GA_s}{\rho A}\right)^{1/2} \tag{115}$$

are the longitudinal and shear wave speed, respectively, in a beam. Kawashima (1978) conducted both a theoretical and experimental investigation of transverse impact at the end of a long beam of high purity aluminum, using the above analysis which considers the effects of rotatory inertia and shear forces and incorporates strain-rate effects in the form proposed by Malvern. The material parameters were obtained from separate compression and torsion tests. It was found that the effect of the shear force is less than the strain-rate effect, but cannot be neglected. Results using the rate-dependent theory were in better agreement than those using a rate-independent formulation.

4.3.3 Biaxial Stress Waves

There has been little work in dynamic plasticity involving subjecting materials to high rates of loading under biaxial stress states and even less in the study of the propagation of elastic-plastic stress waves of combined stress excluding the simple case of uniaxial strain where transverse stresses, but not strains, occur. One of the earliest investigations of biaxial stress waves was that of Gerard and Papirno (1957) who investigated the dynamic stress-strain characteristics of aluminum and mild steel through expansion of thin spherical diaphragms by a sudden pressure pulse. Spherical and cylindrical waves will not be treated in this chapter because their use in studying dynamic material behavior has been minimal because of both experimental difficulties and mathematical complexity. The mathematical concepts are similar to those employed in waves of uniaxial strain. The reader is referred to the works by Craggs (1961) and Nowacki (1978) for further details.

Waves of combined stress involving subjecting statically pretorqued tubes to longitudinal impact and observing the propagation of plastic waves down the tube were studied by Lipkin and Clifton (1970a, b). They performed experiments on annealed 3003-H14 aluminum alloy, making measurements of both longitudinal and shear strain-time profiles at stations along the specimen. Their analysis was based on assumed rate-independent isotropic work-hardening, elastic-plastic material behavior.

Complete agreement between theory and experiment was not obtained because constant state regions predicted by the RI theory were not observed. Furthermore, the permanent shear strain was much less than predicted and the magnitude of the jump in strain rate at an unloading wave did not agree with the calculations. These observations led the authors to conclude that the assumption of strain-rate-independent material behavior was inadequate and that a rate-dependent analysis using the exponential overstress form of the generalized law of Perzyna (1963) might qualitatively explain the results. The case of kinematic hardening used by Lipkin and Clifton was extended to a combination of isotropic and kinematic hardening by Goel and Malvern (1970) who showed that the hardening law can make a significant difference in the results in biaxial plastic wave propagation. Banerjee and Malvern (1974) later analyzed these experimental results using the semilinear rate-type theory with suitably chosen rate-sensitivity parameters. The rate-dependent results gave an overall better representation of the experimental data for wave velocities, even for a relatively rate-insensitive material, although neither the RI or RD theories gave good agreement with observed strain-time profiles. The calculations also demonstrated the mathematical difficulties of reproducing relatively rate-independent material behavior with a rate-dependent analysis. The more general problem of dynamic loading in both tension and torsion was solved theoretically by Cristescu (1968b) for several forms of rate-dependent constitutive equations. The coupling of plastic waves and the concepts of loading and unloading were discussed. A generalization of one-dimensional constitutive equations for rate-sensitive plastic materials to general states of stress is presented by Perzyna (1963).

Although strain-rate effects can be studied in uniaxial-stress wave-propagation experiments with some degree of certainty, uniaxial-strain or plate-impact experiments provide relatively little information on plasticity or strain-rate effects because the stress state is primarily hydrostatic. Abou-Sayed et al. (1976) developed an experimental technique to generate plane waves of combined pressure and shear through the impact of two skewed flat plates. This method, involving combined compression and shear, provides an opportunity to explore varied stress paths to determine the flow characteristics of materials yet is relatively simple to analyze. Abou-Sayed and Clifton (1977a, b) performed experiments on oblique impact of aluminum on aluminum, and aluminum on fused silica targets and compared their data to analytical results from a numerical solution. The results illustrate the advantage of this technique over normal plate-impact experiments for developing constitutive equations in that the longitudinal wave profiles are less sensitive to the dynamic flow characteristics than transverse wave profiles because of the strong in-

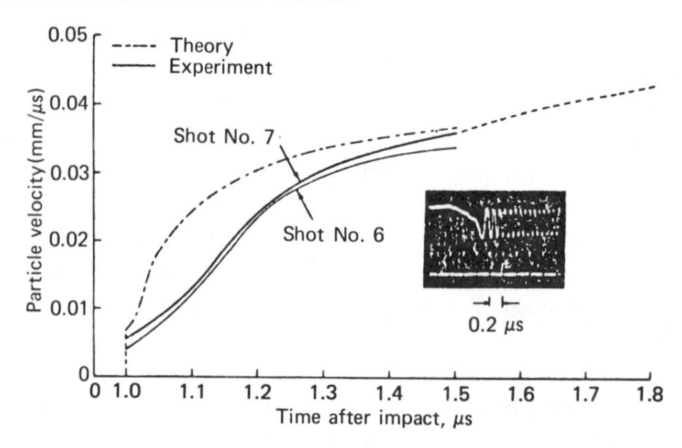

Figure 37 Transverse velocity-time profile at rear surface [Kim and Clifton (1980)].

fluence of the hydrostatic pressure on the longitudinal wave profile. Using highly sophisticated normal and transverse interferometric velocity measuring devices, Kim and Clifton (1980) have performed pressure-shear impact experiments on 6061-T6 aluminum. The results show that the experimental transverse velocity-time profiles following the elastic precursor rise initially more slowly than predicted by analysis as shown in Figure 37. This is attributed primarily to the lack of account for stress path dependence of plastic flow. In a review article entitled "Plastic wave theory-supported by experiments?" Clifton (1980) raises the question of the ability of existing theories to predict the results of plastic wave experiments. Referring to the inadequacy of existing wave theory to predict the decay of the elastic precursor in uniaxial-strain experiments in single crystals and the lack of agreement between theory and experiment in pressure-shear experiments [Kim and Clifton (1980)], he concludes that serious discrepancies still exist between theory and experiment. There is still a definite requirement for the further development of elastic-plastic wave theories and an understanding of dynamic plastic-deformation mechanisms.

REFERENCES

Abou-Sayed, A. S., Clifton, R. J., and Hermann, L. (1976), *Exp. Mech.*, **16**, 127.

Abou-Sayed, A. S., and Clifton, R. J. (1977a), *J. Appl. Mech., Trans. ASME*, **44**, 79.

Abou-Sayed, A. S., and Clifton, R. J. (1977b), *J. Appl. Mech., Trans. ASME*, **44**, 85.

Alter, B. E. K., and Curtis, C. W. (1956), *J. Appl. Phys.*, **27**, 1097.

Baker, W. E., and Yew, C. H. (1966), *J. Appl. Mech., Trans. ASME*, **33**, 917.

Banerjee, A. K., and Malvern, L. E. (1974), *J. Appl. Mech., Trans. ASME*, **41**, 615.

Banerjee, A. K., and Malvern, L. E. (1975), *Int. J. Solids Struct.*, **11**, 347.

Barker, L. M., Lundergan, C. D., and Herrmann, W. (1964), *J. Appl. Phys.*, **35**, 1203.

Bell, J. F. (1951), Tech. Report No. 5, US Navy Contract N6-ONR-243, Johns Hopkins University.

Bell, J. F. (1956), *J. Appl. Phys.*, **27**, 1109.

Bell, J. F. (1959), *J. Appl. Phys.*, **30**, 196.

Bell, J. F. (1960a), *J. Appl. Phys.*, **31**, 277.

Bell, J. F. (1960b), *J. Appl. Phys.*, **31**, 2188.

Bell, J. F. (1965), in N. J. Huffington, Jr. (Ed.), *Behavior of Materials Under Dynamic Loading*, ASME, New York, p. 19.

Bell, J. F. (1968), *The Physics of Large Deformation of Crystalline Solids*, Springer-Verlag, New York.

Bell, J. F., and Stein, A. (1962), *J. Mec.*, **1**, 395.

Bianchi, G. (1964), in H. Kolsky and W. Prager (Eds.), *Stress Waves in Anelastic Solids*, Springer-Verlag, Berlin, p. 101.

Bodner, S. R. (1980), personal communication.

Bodner, S. R., and Clifton, R. J. (1967), *J. Appl. Mech.*, *Trans. ASME*, **34**, 91.

Bodner, S. R., and Partom, Y. (1974), in J. Harding, (Ed.), *Mechanical Properties at High Rates of Strain*, Institute of Physics, London, p. 102.

Bodner, S. R., and Partom, Y. (1975), *J. Appl. Mech.*, *Trans. ASME*, **42**, 385.

Bodner, S. R., and Symonds, P. S. (1962), *J. Appl. Mech.*, *Trans. ASME*, **29**, 719.

Butcher, B. M., and Karnes, C. H. (1966), *J. Appl. Phys.*, **37**, 402.

Campbell, J. D. (1973), *Mater. Sci. Eng.*, **12**, 3.

Campbell, J. D., and Dowling, A. R. (1970), *J. Mech. Phys. Solids*, **18**, 43.

Campbell, W. R. (1952), *Proc. SESA*, **10**, 113.

Chou, P. C., and Hopkins, A. K., Eds., (1973), *Dynamic Response of Materials to Intense Impulsive Loading*, Air Force Materials Laboratory, Wright-Patterson AFB, Ohio.

Clifton, R. J. (1968), *J. Appl. Mech.*, *Trans. ASME*, **35**, 782.

Clifton, R. J. (1974), in S. Nemat-Nasser (Ed.), *Mechanics Today*, Vol. 1, Pergamon, New York, p. 102.

Clifton, R. J. (1980), in J. Harding (Ed.), *Mechanical Properties at High Rates of Strain*, Institute of Physics, London, p. 174.

Clifton, R. J., and Bodner, S. R. (1966), *J. Appl. Mech.*, *Trans. ASME*, **33**, 248.

Convery, E., and Pugh, H. L. D. (1968), *J. Mech. Eng. Sci.*, **10**, 153.

Craggs, J. W. (1961), in I. N. Sneddon and R. Hill (Eds.), *Progress in Solid Mechanics*, Vol. II, North Holland, Amsterdam, p. 141.

Cristescu, N. (1963), *Bull. Acad. Pol. Sci.*, **11**, 129.

Cristescu, N. (1967), *Dynamic Plasticity*, North Holland, Amsterdam.

Cristescu, N. (1968a), *Appl. Mech. Rev.*, **21**, 659.

Cristescu, N. (1968b), in U. S. Lindholm (Ed.), *Mechanical Behavior of Materials under Dynamic Loads*, Springer-Verlag, New York, p. 329.

Cristescu, N. (1972), *Int. J. Solids Struct.*, **8**, 511.

Cristescu, N., and Bell, J. F. (1970) in M. F. Kanninen et al. (Eds.), *Inelastic Behavior of Solids*, McGraw-Hill, New York, p. 397.

Davison, L., and Graham, R. A. (1979), *Phys. Rep.*, **55**, 255.

DeVault, G. P. (1965), *J. Mech. Phys. Solids*, **13**, 55.

Donnell, L. H. (1930), *Trans. ASME*, **52**, 153.

Duvall, G. E. (1964), in H. Kolsky and W. Prager (Eds.), *Stress Waves in Anelastic Solids*, Springer-Verlag, Berlin, p. 20.

Duwez, P. E., and Clark, D. S. (1947), *Proc. ASTM*, **47**, 502.

Duwez, P. E., Clark, D. S., and Bohnenblust, H. F. (1950), *J. Appl. Mech., Trans. ASME*, **17**, 27.

Efron, L., and Malvern, L. E. (1969), *Proc. SESA*, **26**, 255.

Gerard, G., and Papirno, R. (1957), *Trans. ASM*, **49**, 132.

Goel, R. P., and Malvern, L. E. (1970), *J. Appl. Mech., Trans. ASME*, **37**, 1100.

Herrmann, W. (1969), in J. Miklowitz (Ed.), *Wave Propagation in Solids*, ASME, New York, p. 129.

Herrmann, W. (1976), in E. Varley (Ed.), *Propagation of Shock Waves in Solids*, ASME, New York, p. 1.

Herrmann, W., and Lawrence, R. J. (1978), *J. Eng. Mater. Tech., Trans. ASME*, **100**, 84.

Hopkins, H. G. (1963), *Progress in Applied Mechanics*, Macmillan, New York, p. 55.

Hopkins, H. G. (1968), in J. Heyman and F. A. Leckie (Eds.), *Engineering Plasticity*, Cambridge Univ. Press, Cambridge, p. 277.

Hopkinson, J. (1901), Coll. Sci. Paper, Cambridge Univ. Press, **2**, 326.

Hunter, S. C. (1960), in I. N. Sneddon and R. Hill (Eds.), *Progress in Solid Mechanics*, Vol. I, North Holland, Amsterdam, p. 3.

Hunter, S. C., and Johnson, I. A. (1964), in H. Kolsky and W. Prager (Eds.), *Stress Waves in Anelastic Solids*, Springer-Verlag, Berlin, p. 149.

Johnson, J. N., and Barker, L. M. (1969), *J. Appl. Phys.*, **40**, 4321.

Jones, A. H., Maiden, C. J., Green, S. J., and Chin, H. (1968), in U. S. Lindholm (Ed.), *Mechanical Behavior of Materials under Dynamic Loads*, Springer-Verlag, New York, p. 254.

Jones, O. E. (1973), in R. W. Rohde et al. (Eds.), *Metallurgical Effects at High Strain Rates*, Plenum, New York, p. 33.

Karman, T. von (1942), Nat. Def. Res. Counc. Rep. A-29.

Karman, T. von, and Duwez, P. (1950), *J. Appl. Phys.*, **21**, 987.

Karnes, C. H. (1968), in U. S. Lindholm (Ed.), *Mechanical Behavior of Materials under Dynamic Loads*, Springer-Verlag, New York, p. 270.

Karnes, C. H., and Bertholf, L. D. (1970), in M. F. Kanninen et al. (Eds.), *Inelastic Behavior of Solids*, McGraw-Hill, New York, p. 501.

Karpp, R., and Chou, P. C. (1973), in P. C. Chou and A. K. Hopkins (Eds.), *Dynamic Response of Materials to Intense Impulsive Loading*, Air Force Materials Laboratory, Wright-Patterson AFB, Ohio, p. 283.

Kawashima, S. (1978), in K. Kawata and J. Shioiri (Eds.), *High Velocity Deformation of Solids*, Springer-Verlag, Berlin, p. 351.

Kim, K. S., and Clifton, R. J. (1980), *J. Appl. Mech., Trans. ASME*, **47**, 11.

Klepaczko, J. (1973), in A. Sawczuk (Ed.), *Symposium on Foundations of Plasticity*, Noordhoff, Amsterdam, p. 451.

Kolsky, H., and Douch, L. S. (1962), *J. Mech. Phys. Solids*, **10**, 195.

Kolsky, H., and Prager, W., Ed., (1964), *Stress Waves in Anelastic Solids*, Springer-Verlag, Berlin.

Lawson, J. E., and Nicholas, T. (1972), *J. Mech. Phys. Solids*, **20**, 65.

Lee, E. H. (1952), *Quart. Appl. Math.*, **10**, 335.

Lee, E. H. (1966), *Proc. 5th US Natl Cong. Appl. Mech.*, p. 405.

Lee, E. H. (1970), in M. F. Kanninen et al. (Eds.), *Inelastic Behavior of Solids*, McGraw-Hill, New York, p. 423.

Lee, E. H., and Liu, D. T. (1964), in H. Kolsky and W. Prager (Eds.), *Stress Waves in Anelastic Solids*, Springer-Verlag, Berlin, p. 239.

Lee, E. H., and Symonds, P. S. (1952), *J. Appl. Mech., Trans. ASME*, **19**, 308.

Lipkin, J., and Clifton, R. J. (1970a), *J. Appl. Mech., Trans. ASME*, **37**, 1107.

Lipkin, J., and Clifton, R. J. (1970b), *J. Appl. Mech., Trans. ASME*, **37**, 1113.

Lubliner, J. (1964), *J. Mech. Phys. Solids*, **12**, 59.

Lysne, P. C., Boade, R. R., Percival, C. M., and Jones, O. E. (1969), *J. Appl. Phys.*, **40**, 3786.

Malvern, L. E. (1951a), *J. Appl. Mech., Trans. ASME*, **18**, 203.

Malvern, L. E. (1951b), *Quart. Appl. Math.*, **8**, 405.

Malvern, L. E. (1965), in N. J. Huffington, Jr. (Ed.), *Behavior of Materials Under Dynamic Loading*, ASME, New York, p. 81.

Morland, L. W. (1959), *Phil. Trans. Roy. Soc. London A*, **251**, 341.

NMAB (1980), Tech. Rep. NMAB-356, National Materials Advisory Board, National Academy of Sciences, Washington, D.C.

Nicholas, T. (1973a), *J. Appl. Mech., Trans. ASME*, **40**, 277.

Nicholas, T. (1973b), Air Force Materials Laboratory Report, AFML-TR-73-73, Wright-Patterson AFB, Ohio.

Nicholas, T., and Campbell, J. D. (1972), *Exp. Mech.*, **12**, 441.

Nowacki, W. K. (1978), *Stress Waves in Non-Elastic Solids*, Pergamon, Oxford.

Parkes, E. W. (1955), *Proc. Roy. Soc. London A*, **228**, 462.

Percy, J. H. (1963), Proc. Symp. Structural Dynamics under High Impulse Loading, Tech. Rep. ASD-TDR-63-140, Wright-Patterson AFB, Ohio, p. 123.

Perzyna, P. (1963), *Quart. Appl. Math.*, **20**, 321.

Plass, H. J. Jr. (1955), *Proc. 2nd Midwest Conf. on Solid Mech.*, p. 109.

Plass, H. R. Jr., and Wang, N. M. (1959), Proc. 4th Midwest Conf. on Solid Mech., p. 331.

Rabotnov, Yu N., and Suvorova, J. V. (1978), *Int. J. Solids Struct.*, **14**, 173.

Rakhmatulin, K. A. (1945a), *Prikl. Mat. Mekh.*, **9**, 91.

Rakhmatulin, K. A. (1945b), *Prikl. Mat. Mekh.*, **9**, 449.

Ripperger, E. A. (1965), in N. J. Huffington, Jr. (Ed.), *Behavior of Materials Under Dynamic Loading*, ASME, New York, p. 62.

Ripperger, E. A., and Simon, R. (1970), in M. F. Kanninen et al. (Eds.), *Inelastic Behavior of Solids*, McGraw-Hill, New York, p. 543.

Ripperger, E. A., and Watson, H. (1968), in U. S. Lindholm (Ed.), *Mechanical Behavior of Materials under Dynamic Loads*, Springer-Verlag, New York, p. 294.

Ripperger, E. A., and Yeakley, L. M. (1963), *Exp. Mech.*, **3**, 47.

Rubin, R. J. (1954), *J. Appl. Phys.*, **25**, 528.

Saint-Venant, B. de (1868), *Comptes Rendus*, **66**, 650.

Schultz, A. B., Tuschak, P. A., and Vicario, A. A. Jr. (1967), *J. Appl. Mech., Trans. ASME*, **34**, 392.

Sokolovsky, V. V. (1948), *Prikl. Mat. Mekh.*, **12**, 261.

Sperazza, J. (1962), *Proc. 4th U.S. Nat. Cong. Appl. Mech.*, p. 1123.

Sternglass, E. J., and Stuart, D. A. (1953), *J. Appl. Mech., Trans. ASME*, **20**, 427.

Taylor, G. I. (1942), British Ministry of Home Security, Civil Defense Res. Comm. Rep. RC329.

Taylor, J. W. (1965), *J. Appl. Phys.*, **36**, 3146.

Ting, T. C. T. (1964), *J. Appl. Mech., Trans. ASME*, **31**, 38.

Ting, T. C. T. (1978), in K. Kawata and J. Shioiri (Eds.), *High Velocity Deformation of Solids*, Springer-Verlag, Berlin, p. 305.

Tuschak, P. A., and Schultz, A. B. (1971), *J. Appl. Mech., Trans. ASME*, **38**, 888.

White, M. P., and Griffis, Le Van (1948), *J. Appl. Mech., Trans. ASME*, **15**, 256.

Wilkins, M. L. (1964), in B. Alder et al. (Eds.), *Methods of Computational Physics*, Vol. 3, Academic, New York, p. 211.

Wood, D. S. (1952), *J. Appl. Mech., Trans. ASME*, **19**, 521.

Wood, E. R., and Phillips, A. (1967), *J. Mech. Phys. Solids*, **15**, 241.

Yew, C. H., and Richardson, H. A., Jr. (1969), *Exp. Mech.*, **9**, 366.

5

PENETRATION AND PERFORATION OF SOLIDS

Jonas A. Zukas

Situations involving impact—the collision of two or more solids—are currently receiving widespread attention. At one time, impact problems were primarily of concern to the military. Now however, as civilian technology grows more sophisticated, severe demands are being made on the behavior of materials under very short duration loading. Safe and cost-effective design demands a rigorous understanding of the behavior of materials and structures subjected to intense impulsive loading for such diverse applications as:

1 Safe demolition of prestressed concrete structures.

2 The transportation safety of hazardous materials.

3 Crashworthiness of vehicles and protection of their occupants or cargo.

4 Safety of nuclear-reactor containment vessels subjected to missile impact from external sources (tornadoborne debris, aircraft) or internal ones (extreme pressures from reactor excursions, debris, and fragments from failed components).

5 The design of lightweight armor systems, including fabric body armors, for protection of police officers, executives in business, and government and military personnel.

6 The vulnerability of military vehicles, aircraft, and structures to impact and explosive loading.

7 The erosion and fracture of solids subjected to multiple impacts by liquid and solid particles.

8 Protection of spacecraft from meteroid impact.

9 Explosive forming and welding of metals.

The study of impact phenomena involves a variety of classical disciplines. In the low-velocity regime (<250 m/s) many problems fall in the area of structural dynamics. Local indentations or penetrations are strongly coupled to the overall deformation of the structure. Typically, loading and response times are in the millisecond regime. As the striking velocity increases (0.5–2 km/s) the response of the structure becomes secondary to the behavior of the material within a small zone (typically 2–3 projectile diameters) of the impact area. A wave description of the phenomena is appropriate and the influences of velocity, geometry, material constitution, strain rate, localized plastic flow, and failure are manifest at various stages of the impact process. Typically, loading and reaction times are on the order of microseconds. Still further increases in impact velocity (2–3 km/s) result in localized pressures that exceed the strength of the material by an order of magnitude. In effect, the colliding solids can be treated as fluids in the early stages of impact. At ultra-high velocities (>12 km/s) energy deposition occurs at such a high rate that an explosive vaporization of colliding materials results.

Impact phenomena can be characterized in a number of ways: according to the impact angle, the geometric and material characteristics of the target or projectile, or striking velocity. The latter approach is adopted in Table 1, which provides a short classification of impact processes as a function of striking velocity, V_s, and strain rate, $\dot{\varepsilon}$. The impact-velocity ranges should be considered only as reference points. In fact, these transitions are extraordinarily flexible since deformation processes under impact loading depend on a long series of parameters in addition to impact velocity.

A complete description of the dynamics of impacting solids would demand that account be taken of the geometry of the interacting bodies, elastic, plastic, and shock-wave propagation, hydrodynamic flow, finite strains and deforma-

Table 1 Impact Response of Materials

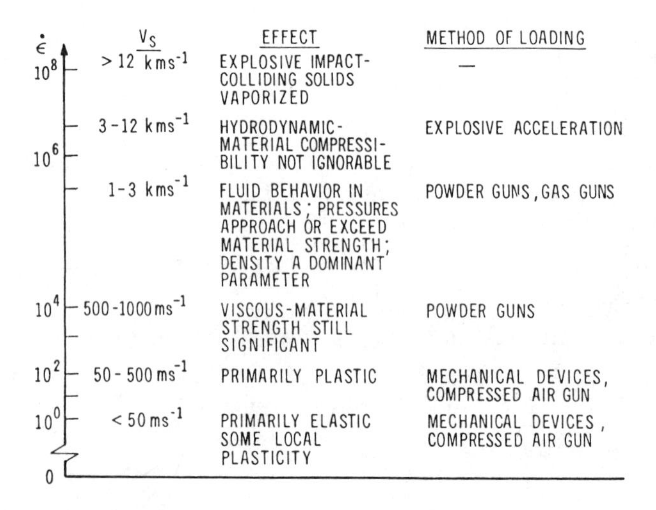

$\dot{\epsilon}$	V_s	EFFECT	METHOD OF LOADING
10^8	> 12 $km s^{-1}$	EXPLOSIVE IMPACT-COLLIDING SOLIDS VAPORIZED	—
10^6	3-12 $km s^{-1}$	HYDRODYNAMIC-MATERIAL COMPRESSIBILITY NOT IGNORABLE	EXPLOSIVE ACCELERATION
	1-3 $km s^{-1}$	FLUID BEHAVIOR IN MATERIALS; PRESSURES APPROACH OR EXCEED MATERIAL STRENGTH; DENSITY A DOMINANT PARAMETER	POWDER GUNS, GAS GUNS
10^4	500-1000 ms^{-1}	VISCOUS-MATERIAL STRENGTH STILL SIGNIFICANT	POWDER GUNS
10^2	50 - 500 ms^{-1}	PRIMARILY PLASTIC	MECHANICAL DEVICES, COMPRESSED AIR GUN
10^0	< 50 ms^{-1}	PRIMARILY ELASTIC SOME LOCAL PLASTICITY	MECHANICAL DEVICES, COMPRESSED AIR GUN
0			

tions, work hardening, thermal and frictional effects, and the initiation and propagation of failure in the colliding materials. An analytical approach would not only be quite formidable, but would also require a degree of material characterization under high strain-rate loading that could not be attained in practice. Hence, much of the work in this field has been experimental in nature.

Penetration may be defined as the entrance of a missile into a target without completing its passage through the body [Backman and Goldsmith (1978)]. This generally results in the embedment of the striker and formation of a crater. If the projectile rebounds from the impacted surface or penetrates along a curved trajectory emerging through the impacted surface with a reduced velocity, the process is termed a ricochet. Perforation, in contrast, is the complete piercing of a target by the projectile. Such processes occur in a time frame of several to several hundred microseconds. Targets and projectiles are usually severely deformed during such encounters.

This chapter concentrates on the phenomena that lead to penetration and perforation of solids in the intermediate velocity regime (0.5–2 km/s). The basic principles for treating lower velocity impacts from a wave propagation viewpoint are presented in Chapters 1 and 2. Other applications and reviews are available in the literature [Goldsmith (1960, 1966, 1967, 1973), Kornhauser (1964), Johnson (1972)]. Hypervelocity impact (3–12 km/s) is considered in Chapter 6 and has been treated extensively in various symposia (1955–63, 1969) and review articles [Hopkins and Kolsky (1960), Herrmann and Jones (1961), Pitek and Hammitt (1966), Vinson (1968), Kinslow (1970), Swift (1972), Johnson (1972), Letho (1972), and Davison and Graham (1979)]. Since the emphasis in this chapter is on materials response—loading and response times in the submillisecond regime—problems in structural dynamics such as the response of beams, plates, and shells to impulsive loading are beyond its scope. Reviews of these problems have been written by (among others) Johnson (1972), Rawlings (1974), Levy and Wilkinson (1976), Baker (1976), Ross et al. (1977), Wright and Baron (1979), and Jones (1975, 1978a, b, 1979).

5.1 PENETRATION AND PERFORATION

Any item capable of being launched can become a projectile. Military projectiles are probably the most familiar, but they form only a small subset of possible missiles. During the demolition of buildings made of prestressed concrete, scabs or spall fragments may be formed as a result of rapid unloading and the reflection of compressive stress waves from a free surface. In hard concrete or steel, these may acquire speeds of the order of 100 m/s and hence can be lethal [Johnson (1979)].

Telephone poles, cars, and assorted tornadoborne debris as well as aircraft can damage reactor containment structures. Similarly, components of failed industrial machinery such as turbine blades and pipes can result in high-speed, irregular missiles that must either be contained or deflected in a direction where they will do no damage. Multiple impacts by water, ice, and sand particles constitute a hazard to high-speed aircraft and reentry vehicles. In particular, the erosion of turbine blades by water droplets condensing from high-pressure steam is well-known [Brunton and Rochester (1979)].

High-speed water jets have many advantages over conventional drilling machines for mining applications [Summers (1979), Huszarik et al. (1979)]. The reduction of coal dust in the air virtually removes the danger of explosions. Noise levels are considerably reduced posing less hazard to machine operators. In addition, the corrosive effects of water under high pressure produce greater yields than conventional mining techniques.

Large and small meteorites have always been a hazard to space vehicles. Considerable work has been done in the past in attempting to understand the basic physics of hypervelocity impact and evaluation of various designs for spacecraft protection. Much of this work is described in Chapter 6. With the increasing interest in the use of space for industrial applications and basic research, questions of protection of manned and unmanned space vehicles from high-velocity meteorite impact have again arisen.

Some commonly occurring projectile geometries are shown in Figure 1. Projectile weights may vary from hundreds of kilograms for failed turbine blade or flywheel components to fractions of a gram for secondary spall particles resulting from a high-velocity impact. They may consist of a single material or combinations of materials (Table 2). Depending on the relative strengths and densities of colliding materials, projectiles may emerge undamaged (Figure 2), plastically deformed (Figure 3), or fractured or shattered (Figure 4) as a result of their encounter with an obstacle (target).

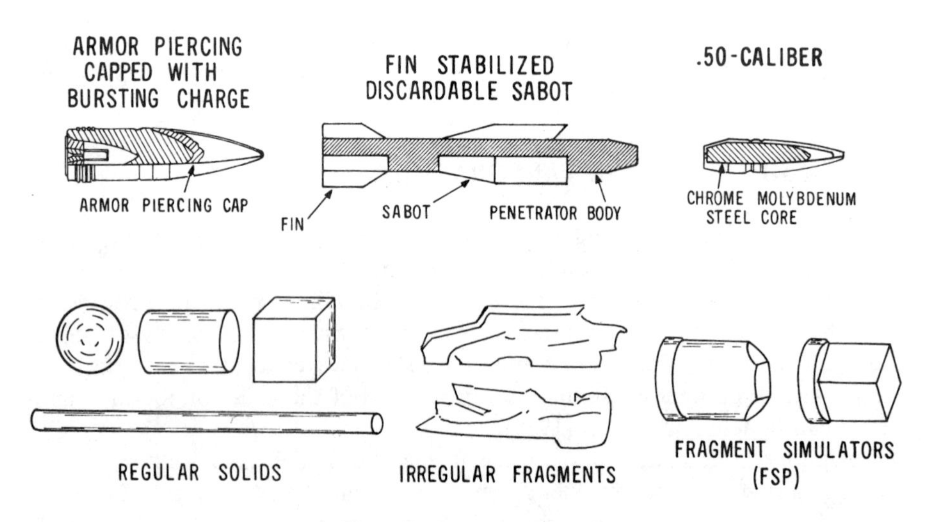

Figure 1 Examples of projectiles.

Table 2 Projectile Characteristics

Geometry			
Basic Shape	*Solid Rod*	*Nose Configuration*	*Cone*
	Sphere		*Ogive*
	Hollow Shell		*Hemisphere*
	Irregular Solid		*Right Circular Cylinder*
Material			
Density	*Lightweight*		
	Wood, Plastics, Ceramics, Aluminum		
	Intermediate		
	Steel, Copper		
	Heavy		
	Lead, Tungsten		
Flight characteristics			
Trajectory	*Straight (stable)*	*Impact Condition*	*Normal*
	Curved (stable)		*Oblique*
	Tumbling (unstable)		
Final Condition			
Shape	*Undeformed*	*Location*	*Rebound*
	Plastically Deformed		*Partial Penetration*
	Fractured		*Perforation*
	Shattered		

As with projectiles, almost any stationary or moving object can become a target. Because impact effects in many materials tend to be highly localized for striking velocities exceeding 500 m/s, only a small portion of the overall structure need be considered in analytical or experimental studies. Even with this simplification, target geometries can still be quite complex (Table 3). Materials typically encountered are similar to those for projectiles—hard and soft metals, concrete, ceramics, rock, sand, clay, snow, plastics, wood, water—to name but a few.

5.1.1 Physical Phenomena in Impacting Solids

Consider the events that occur in projectile and target during impact. For purposes of this discussion, consider the projectile to be in the form of a long rod, generally cylindrical in shape, with conical, ogival, hemispherical, or flat nose. When such a projectile strikes a target, strong compressive waves propagate into both bodies. If the impact velocity is sufficiently high, relief waves will propagate inward from the lateral-free surfaces of the rod and cross at the centerline, creating a region with high-tensile stress. This tensile region can cause fracture in sufficiently brittle materials such as high-strength steels.

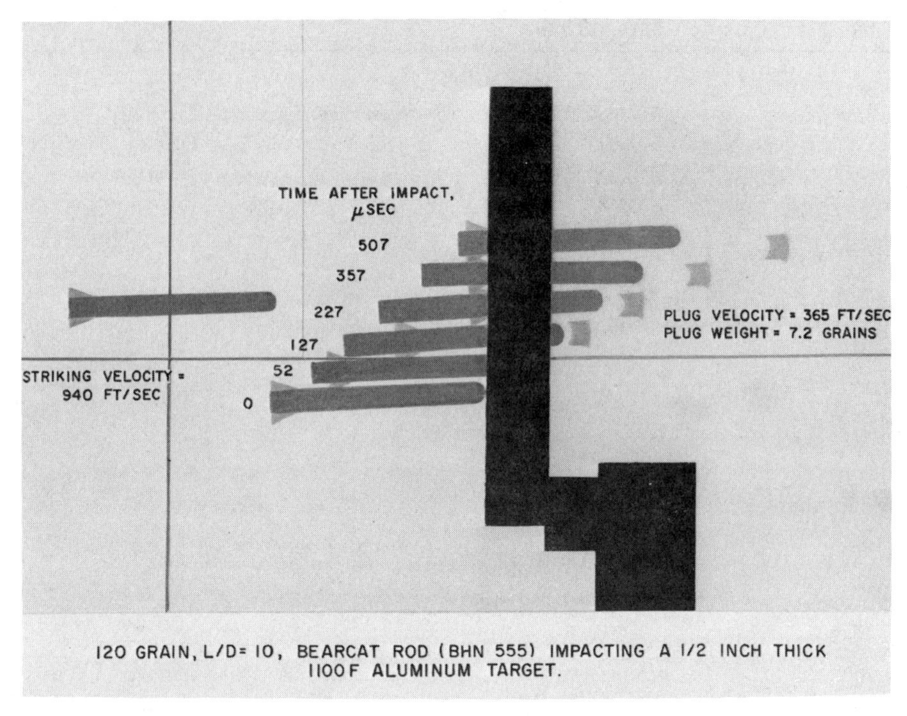

Figure 2 Perforation resulting in plug formation [Courtesy of C. Grabarek and A. Ricchiazzi].

This effect will be enhanced if centerline porosity or other imperfections exist in the projectile material. For normal impacts, the state of stress is clearly two-dimensional. For oblique incidence, there is the additional complication of bending stresses because of the asymmetry of the loading. Under the proper combinations of projectile geometry, material characteristics and impact velocity, the combined bending and tensile stresses can lead to projectile failure and ricochet.

The initial compression wave in the target is followed quickly by a release wave. When the initial compressive wave reaches a free boundary in the target, an additional release wave is generated. If the combination of load intensity (tensile) and duration exceeds a critical value for the target material, failure will be initiated.

The damage produced in colliding solids at typical ordnance velocities and high obliquities is shown in Figure 5. Plates of rolled homogeneous armor (RHA) steel, 2.54-cm thick, were fixed at an inclination of 60° and struck by a cylindrical rod of S7 tool steel. The rod had a mass of 65 g and a length-to-diameter (L/D) ratio of 10. For an impact at the ballistic-limit velocity (upper left) of 1282 m/s, the rod is totally consumed. In the process, it produces considerable damage to the target plate at the impact face and suffers a dramatic change in its flight path, being turned almost 60° within the target

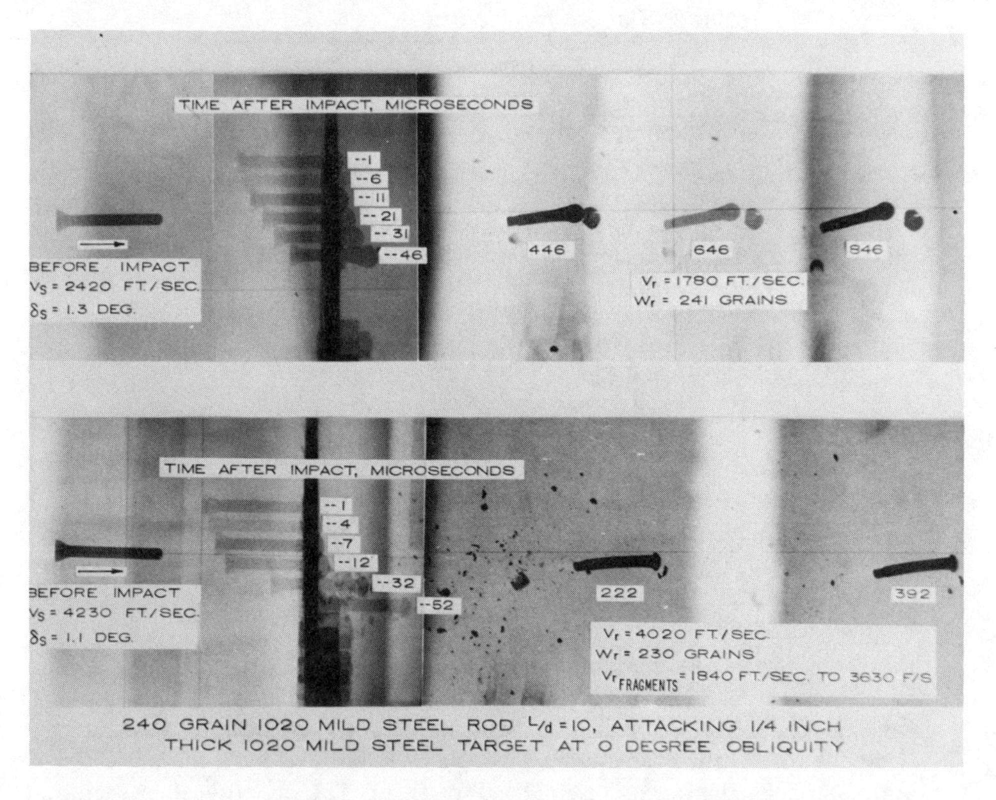

Figure 3 Plastic deformation in rod during perforation [Courtesy of C. Grabarek and A. Ricchiazzi].

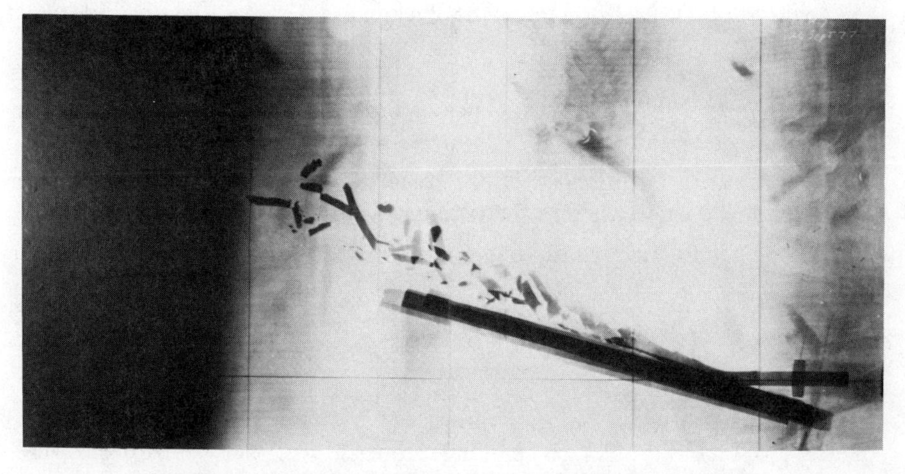

Figure 4 Rod deformation and shatter at high obliquity impact.

Table 3 Target Characteristics

Geometry	
Thickness	Shape
Thin	*Flat*
Single Plates	*Curved*
Composites	*Serrated*
Spaced Plates	*Irregular Surface*
Intermediate	
Barriers	
Thick	
Ground	
Deep Water	
Heavy Barriers	
Typical Materials	
Hard and Soft Metals	*Sand*
Concrete	*Snow*
Ceramics	*Plastics*
Rock	*Wood*

toward the normal to the rear face. As the impact velocity is increased, a greater portion of the striker survives while target damage progressively increases. The deviation from the flight path begins to decrease until, at a velocity about 1.5 times greater than the ballistic-limit velocity, the projectile path is undeviated from its original line of flight. The quantity of projectile surviving the impact (or residual mass) increases for velocities above the ballistic limit and reaches a limiting value for impact velocity about 20% greater than the ballistic limit. As the striking velocity increases beyond this point, projectile residual mass again begins to decrease.

Targets are best classified following Backman & Goldsmith (1978). A target is said to be:

1. *Semiinfinite* if there is no influence of the distal boundary on the penetration process.
2. *Thick* if there is influence of the distal boundary only after substantial travel of the projectile into the targets.
3. *Intermediate* if the rear surface exerts considerable influence on the deformation process during nearly all of the penetrator motion.
4. *Thin* if stress and deformation gradients throughout its thickness do not exist.

Impacted materials may fail in a variety of ways. The actual mechanisms will depend on such variables as material properties, impact velocity, projectile shape, method of target support, and relative dimensions of projectile and

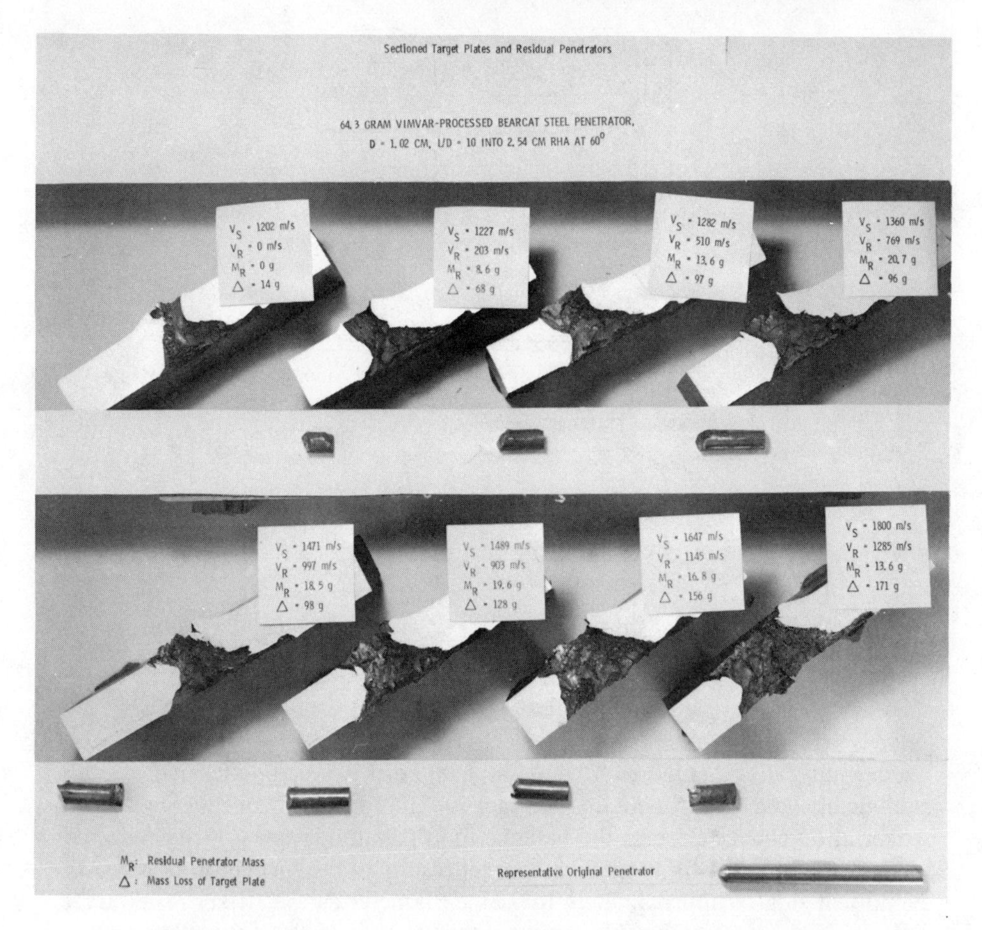

Figure 5 Sectioned target plates and residual penetrators [Lambert (1978a)].

target. Figure 6, adapted from Backman & Goldsmith (1978) and Backman (1976), shows some of the dominant modes for thin and intermediate thickness targets. Although one of these may dominate the failure process, they will frequently be accompanied by several other modes.

Spalling, tensile failure as a result of the reflection of the initial compressive wave from the rear surface of a finite-thickness plate, is a commonplace occurrence under explosive and intense impact loads, especially for materials stronger in compression than in tension. Scabbing is similar in appearance, but here fracture is produced by large deformations and its surface is determined by local inhomogeneities and anisotropies. Fracture, as a result of an initial stress wave exceeding a material's ultimate strength, can occur in weak, low-density targets while radial cracking is common in materials such as ceramics where the tensile strength is considerably lower than the compressive strength.

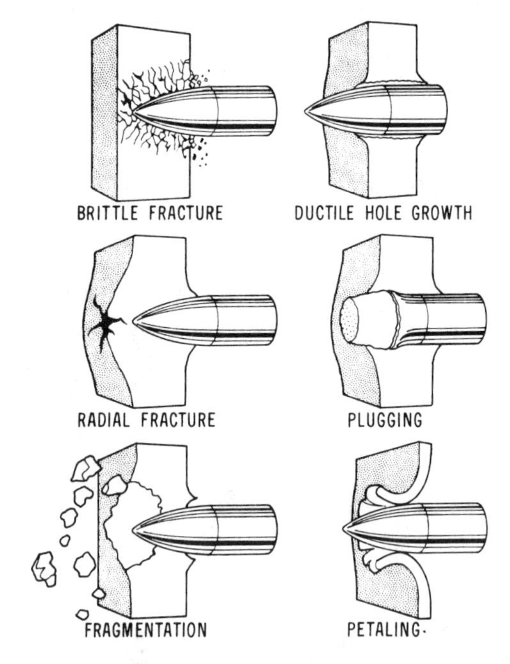

Figure 6 Failure modes in impacted plates [Backman (1976)].

Plugging failure (Figures 2 and 3) has been studied extensively, both analytically and experimentally. Impact by a blunt or hemispherical-nosed striker at a velocity close to the ballistic limit (the minimum velocity required for perforation) of a finite-thickness target results in the formation of a nearly cylindrical slug of approximately the same diameter as the striker. The target material is constrained by the geometry and motion of the projectile to move forward in the direction of penetration. Separation of the plug from the target may occur by a conventional fracture mode, that is, void formation and growth in shear, or by a mechanism known as *adiabatic shearing* which is characterized by the formation of narrow bands of intense shear. It is generally accepted that the adiabatic-shear instability develops at a site of stress concentrations in an otherwise uniformly straining solid. The work of plastic deformation is converted almost entirely into heat which, because of the localized high deformation rates, is unable to propagate a significant distance away from the plastic zone [Moss (1980) for example cites shear strain rates within adiabatic-shear bands of up to 10^7 s^{-1} and temperature within the band up to 10^5°C]. As a result, the temperature in the zone rises, encouraging additional local plastic flow and concentrating the local plastic strain still further. The process continues and results in the propagation of a narrow band of intense plastic strain through the material along planes of maximum shear stress or minimum strength until unloading occurs or the material fractures. For striking velocities exceeding the minimum perforation velocity by more than 5–10%, multiple fragments rather than an intact plug will result. Plugging failure is quite sensitive to the impact angle and nose shape of the projectile.

The significance of adiabatic-shear failure is more clearly seen for penetration by projectiles with sharp nose configurations. Woodward (1978a, b) has demonstrated that for these conditions, material in the target is generally displaced radially and no plug is formed. But if the material is susceptible to adiabatic shearing, a change in the mode of failure occurs and a plug is pushed out along intense shear bands irrespective of the projectile geometry.

Petaling is produced by high radial and circumferential tensile stresses after passage of the initial stress wave. The intense-stress fields occur near the tip of the projectile. Bending moments created by the forward motion of the plate material pushed by the striker cause the characteristic deformation pattern. It is most frequently observed in thin plates struck by ogival or conical bullets at relatively low-impact velocities or by blunt projectiles near the ballistic limit. Petaling is accompanied by large plastic flows and permanent flexure. Eventually, the tensile strength of the plate material is reached and a star-shaped crack develops around the tip of the projectile. The sectors so formed are then pushed back by the continuing motion of the projectile forming petals.

For thick targets impacted by malleable materials at velocities exceeding 1 km/s, there is a hydrodynamic erosion and inversion of the penetrator material against the receding face of the penetration channel in the target. The target material is forced aside much as though a punch were being pushed into it, although the channel is much bigger than the penetrator diameter (Figure 7).

A combination of ductile failure and spalling seems to be characteristic for the failure of thick plates of low or medium hardness.

5.1.2 Experimental Methods

In view of the complexity of penetration processes, it is not surprising that the bulk of the work in this area is experimental in nature. High-velocity impact test techniques, aside from routine proof tests, vary mainly in the degree of instrumentation provided and hence the amount of data retrieved. The most common types of testing have as their objective the determination of:

1. The velocity and trajectory of the projectile prior to impact.
2. Changes in configuration of projectile and target as a result of impact.
3. Masses, velocities, and trajectories of fragments generated by the impact process.
4. The ballistic limit.

High Velocity Impact Testing. High-velocity impact is a very energetic process. Some of the incident energy is converted into light that obscures the impact event (Figure 8). Further problems are caused by debris ejected at the

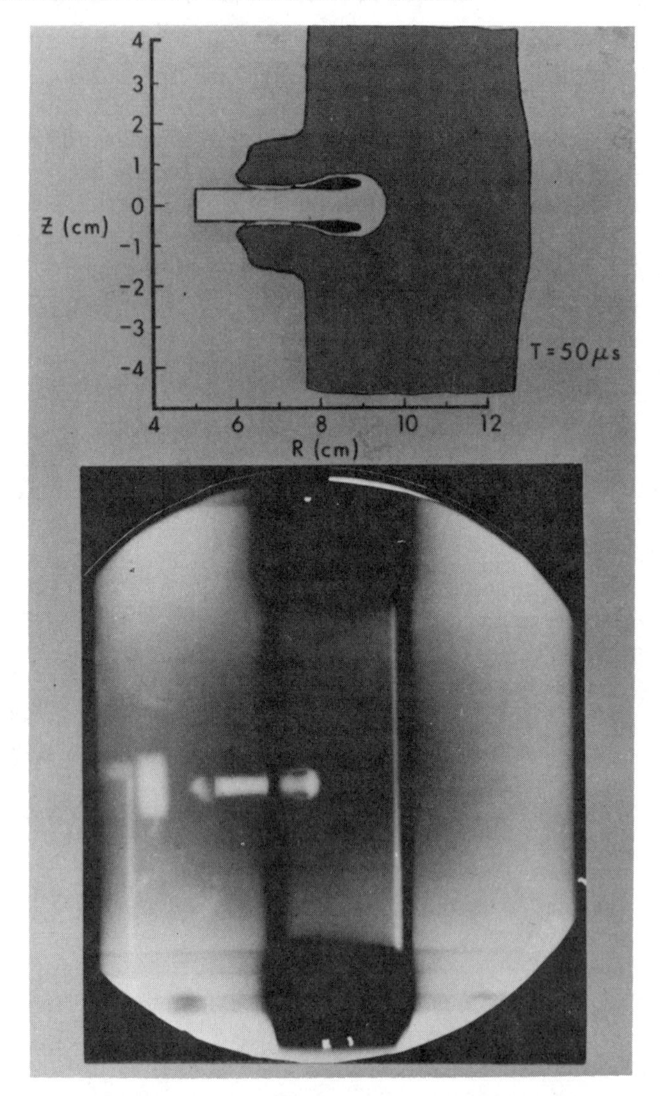

Figure 7 Rod erosion during penetration.

impact face and at the rear of the target once the projectile has broken through. Conventional optical techniques, such as high speed cameras, are therefore quite limited for such applications. This is evident in Figure 9, taken from Awerbuch and Bodner (1973), which shows a series of high-speed photographs of a lead rifle bullet impacting a 19 mm plate. Both debris and an energetic light flash obscure projectile and target deformation at the impact interface although with a wider field of view it would be possible to record projectile rear surface motion (therefore deceleration) for a limited time. The

Figure 8 Energy release at high-velocity impact [Courtesy of A. Ricchiazzi].

rear surface bulge is clearly observed with the high speed camera but the projectile is totally obscured on breakout. To overcome these problems, most experimental facilities rely on x-ray illumination of energetic interaction events. Frequently, both x-ray and optical methods are used to record impact phenomena.

Projectile trajectories may be determined in a number of ways; high-speed photography, orthogonal-flash radiography, or yaw-card measurements. Yaw cards are thin paper or plastic sheets located along the anticipated trajectory. Interaction of the projectile with the yaw cards does not generally affect its motion. The position of the perforations on the yaw cards determines projectile location in the plane perpendicular to the trajectory. The shape of the perforation allows flight orientation to be determined.

The striking velocity is determined from a measurement of transit times over fixed distances. The time of arrival at predetermined locations is established by the closing or opening of electrical circuits, interruption of light beams, synchronized photography, or flash radiography of the projectile.

An example of a test setup for retrieval of penetration data for solid (kinetic energy) projectiles is shown in Figure 10. It is a typical arrangement used for small scale (65 g or less) penetrators at the U.S. Army Ballistic Research Laboratory [Grabarek & Herr (1966)] and sufficiently flexible so that changes (such as the addition of x-ray tubes or high-speed framing cameras) can be

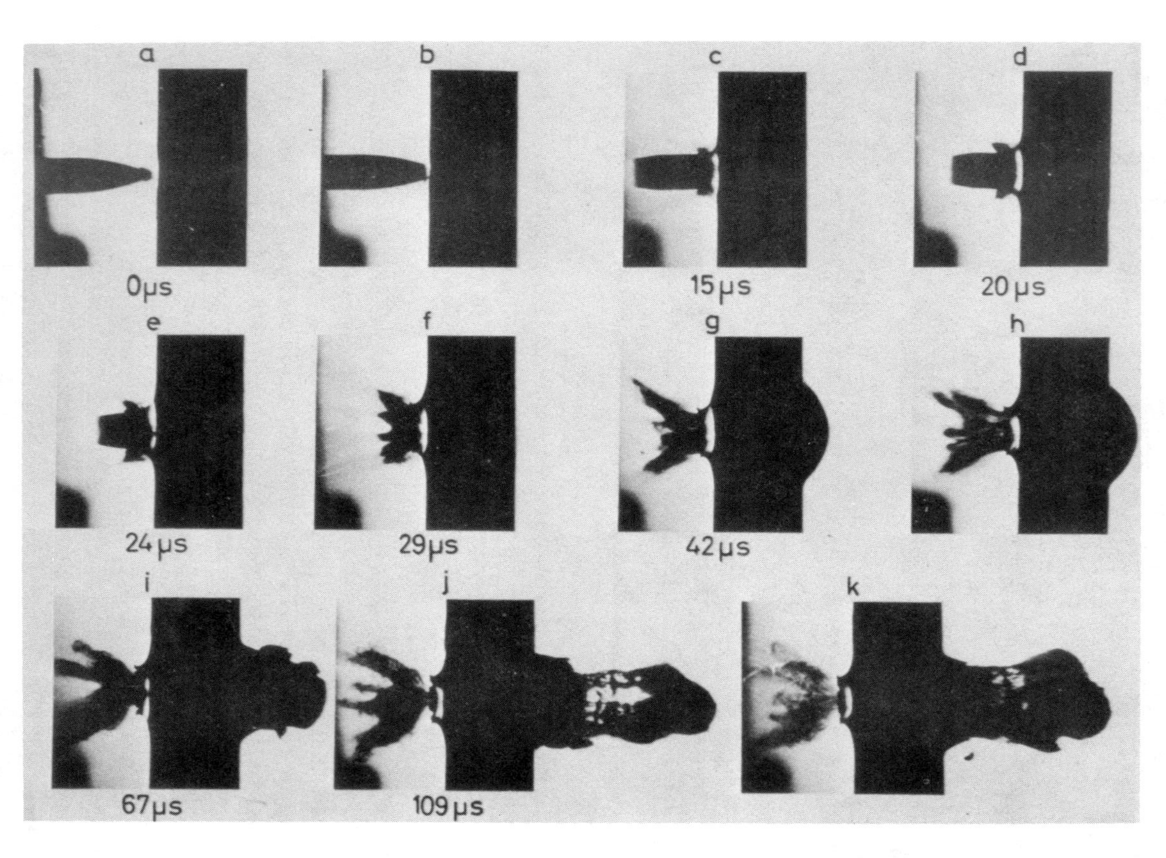

Figure 9 Photographic sequence for lead bullet penetrating aluminum plate [Awerbuch and Bodner (1973)].

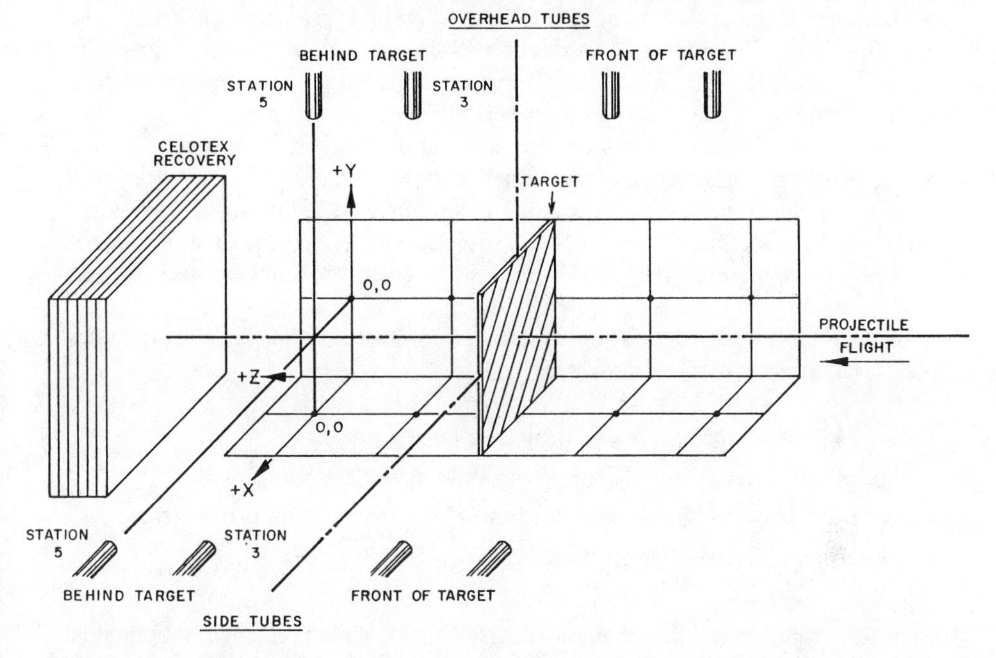

Figure 10 Experimental setup of x-ray instrumentation for projectile-penetration data acquisition [Grabarek and Herr (1966)].

made to accommodate a variety of test requirements and projectile types. The x-ray system consists of orthogonal pairs of x-ray tubes (105 or 150 kv) arranged as shown. A time delay generator in the system pulses each tube or set of tubes at preset intervals.

Four tubes, arranged in pairs, provide simultaneous orthogonal radiographs of the projectile in free flight before target impact. The projectile-striking velocity and orientation are measured from these radiographs.

A similar tube arrangement behind the target provides orthogonal radiographs to supply the data needed to determine the residual parameters. Additional tubes are usually added to view the projectile impacting and penetrating the target. Systematic procedures for acquiring and storing data from such experimental facilities are described by Lambert and Ringers (1978) while Herr and Grabarek (1978) provide procedures for evaluation of materials for high-velocity impact applications in such experiments.

In situations where fragment spray must be characterized for use in vulnerability analyses, recovery of fragments by procedures that inflict minimal damage to them is necessary. The most common recovery media are stacks of plywood or cane fiberboard. Wenzel and Dean (1975), among others, have considered alternatives. If there are many fragments, the task of measuring individual fragment masses, speeds, and directions of travel from multiply exposed x-ray films, and their correlation with recovered fragments is formidable because of the problems of identifying individual fragments on the orthogonal film pairs. Procedures for reduction of fragment data have been given by Arbuckle et al. (1973). Examples of typical results with small-scale (65 g) rods are to be found in the report by Lambert (1978a).

Post-mortem measurements on projectile and target include determination of the principal dimensions of the target crater such as depth, diameter, and crater volume (or entrance and exit diameters for a perforation) as well as the final length, diameter, and mass of the projectile and other massive fragments. Procedures for making the measurements are given by Lambert and Ringers (1978).

In summary, then, the data extracted from conventional high-velocity impact tests consists of the following:

1 Speed and orientation of the projectile prior to impact.
2 Speed and orientation of major projectile pieces after perforation.
3 Speed, mass, and spatial distributions of fragments behind the target.
4 Hole size and mass loss in the target.

Graphical representations of high-velocity impact data concern relationships among such variables as velocity, target thickness, angle of obliquity, total projectile yaw, impact-kinetic energy, impulse, force, and time. For such plots, physical and geometric characteristics of projectile and target (excepting target thickness) are held constant. Often plots are normalized using projectile diameter or length as a characteristic length, the limit velocity, material sound speed or similar as a characteristic velocity, limit energy as a characteristic energy and so on. Since plots are two-dimensional, the curves represent relationships between one dependent variable and one independent variable. Several curves may be plotted to show how these relationships vary with a third variable, but on any given plot, other variables must be held constant. Figure 11 from Backman & Goldsmith (1978) is an example of such a phase diagram portraying the behavior of a 6.35 mm ogival-nosed projectile striking a 6.35 mm aluminum alloy target. The various modes of projectile and target behavior are shown as functions of impact velocity and obliquity. For a given projectile and target situation, much more elaborate phase diagrams can be built up to serve as aids to designers.

The Ballistic Limit. One of the problems encountered in the study of impact phenomena is the determination of a velocity below which an object will fail to perforate a barrier or some type of protective device. This determination is of prime importance in the design of protective structures, in evaluation of the effectiveness of military vehicle armor, and in any problem area where an impact can cause damage. This velocity is commonly referred to as a critical-impact velocity or ballistic limit.

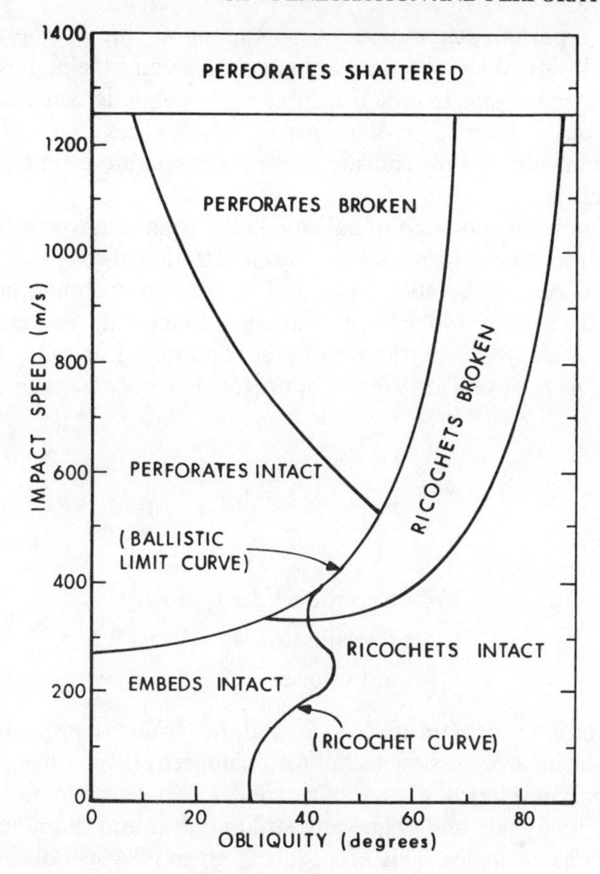

Figure 11 Phase diagram for projectile impact [Backman and Goldsmith (1978)].

The techniques available to determine this velocity can be classed as either deterministic or probabilistic. In the former category, a limit velocity is determined from physical principles (the conservation laws and material constitutive relations) but because of the complexity of the governing partial differential equations, simplifications are introduced that generally require empirical determination of one of two constants. In the probabilistic approach, models are built relying on a substantial base of data consisting of the object's striking velocity and either its residual velocity or a statement of either defeat or nondefeat of the barrier. The resulting critical velocity is most commonly expressed as a V_{50}, that is, a striking velocity for which there exists a 50% probability of perforation of the barrier. In V_{50} determinations, a statistical approach is employed where the response is quantal and a sensitivity test can be applied to the data. In the simplest approach, a V_{50} is determined by averaging six projectile-striking velocities that include the three lowest velocities that resulted in complete penetration and the three highest velocities that resulted in a partial penetration. A spread of 46 m/s or less is required between the lowest velocity for partial penetration and the highest velocity for a complete penetration. In practice, time and economic constraints limit the quantity of data obtained so that there is always less than that desired by either deterministic or probabilistic models for reliable estimation of critical impact velocities.

The deterministic approach to ballistic limits is an outgrowth of attempts to determine projectile performance during penetration through development of models based on conservation laws and assumptions about the mechanical behavior of the system. Much unpleasantness in modeling has been historically avoided by taking the penetrator to be nondeforming or rigid. Lambert and Jonas (1976) have observed that this approach generally leads to a form of the type

$$V_r = \begin{cases} 0, 0 \leq V_s \leq V_l \\ \alpha\left(V_s^p - V_l^p\right)^{1/p}, V_s > V_l \end{cases} \tag{1}$$

where

V_r = projectile residual velocity
V_s = projectile striking velocity
V_l = limit velocity

For nondeforming penetrators, $p=2$ and α, V_l are empirical parameters determined from a regression technique. Lambert (1978b) has extended this approach and developed a predictive model for determination of residual velocities of long rods and fragments striking steel and aluminum targets at arbitrary incidence angles. This procedure is given in Appendix A.

Limit-velocity curves (Figure 12a) define boundaries with a sharpness that is unreal. Actually, near these boundaries, terminal behavior tends to be probabilistic in nature, rather than deterministic. Figure 12b is a typical plot of the probability of complete penetration as a function of velocity for a certain missile and target combination. This type of plot is found by firing a great number of rounds so as to obtain a statistical sample. Any velocity in the range of mixed results can be used as a limit velocity. For example, the V_{10} and V_{90} limit velocities are those velocities at which there is a 10 and 90% probability of

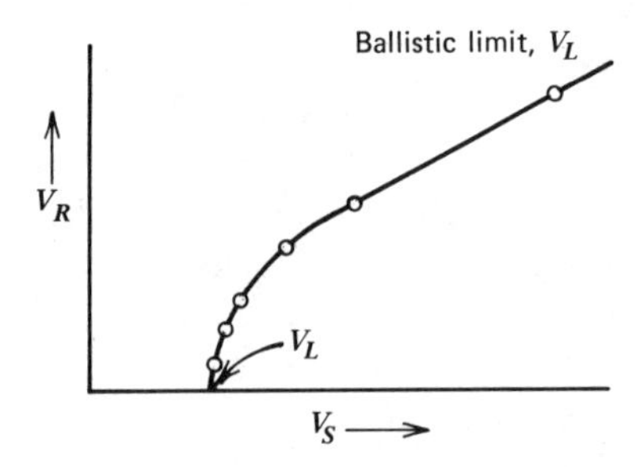

Figure 12a Limit-velocity curve.

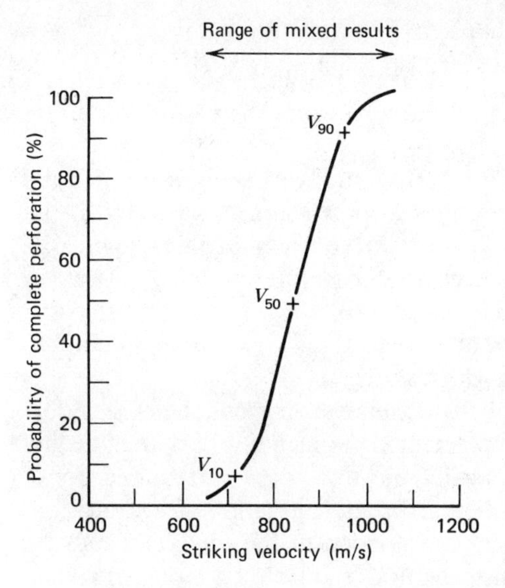

Figure 12b Penetration-probability curve.

a complete penetration, respectively. Usually the V_{50} limit velocity is used to represent limit velocity. Near the V_{50}, the slope of the probability curve is greatest and, thus, the V_{50} can be located with the greatest precision.

Before a limit velocity can be measured, two arbitrary definitions must be made. First, it is necessary to define a complete penetration (CP) (see Figure 13). Three such definitions are presented below as examples.

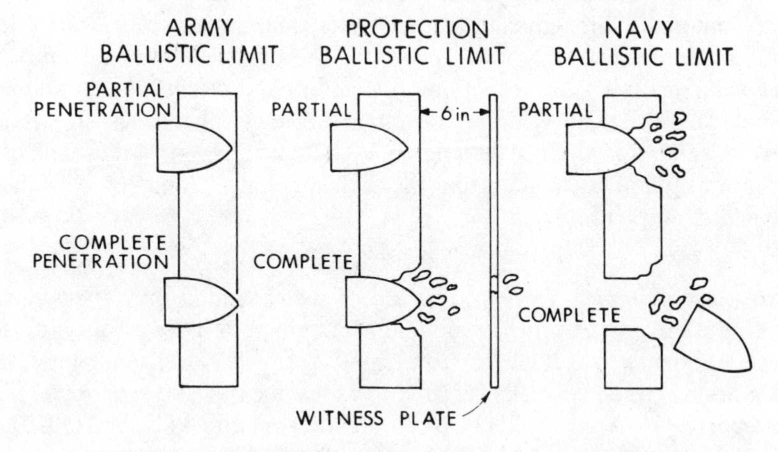

Figure 13 Various definitions for complete and partial penetration.

1 The target must be breached so that light will pass through it.
2 At least one-half of the missile must pass through the target.
3 A thin aluminum sheet placed behind the target must be perforated.

For a given encounter, the second limit velocity usually will be greater than the first. The third will be similar to the second except when material scabs off the back of the target. Note that only the second criterion pertains to missile behavior. Having defined a CP, it is necessary to define the limit velocity in terms of the probability of success. Usually, the V_{50} is utilized for this definition. A shot fired at a velocity that exceeds the V_{50} may not cause a CP, but the probability of a CP increases with velocity.

The (probabilistic) V_{50} measure of ballistic limit is very popular and widely used. It is fairly straightforward to implement experimentally and analytically. Its prime deficiencies are that a considerable data base is required for a statistically meaningful V_{50} and that there is little exploitation of information.

There are some obvious advantages associated with the (deterministic) V_l, principally a greater exploitation of information. Rather than binary information (yes or no as regards perforation), weighted values (velocities) are utilized in the case of perforations, which enhances precision and economy. Normally, a realistic V_s, V_r curve is generated along with a ballistic limit (V_l). This can be of value not only in penetrator design and correlation of experimental data with analytical or numerical results but also in consideration of multiple plate situations.

The only obvious, and sometimes significant, disadvantage inherent in V_l determination is the necessity of complex and costly instrumentation (currently x-ray radiography) for measuring residual velocity. Although the cost per round of an instrumented V_s, V_r test is greater than for a V_{50} test, the cost per determination of a ballistic limit (V_l) may well be less than that for a V_{50} since overall fewer shots will need to be fired. Furthermore, the determination of a V_l is better suited for additional research in advancing ballistic technology as regards material failure (spallation, penetrator fracture) and characterization of behind-armor debris.

The ballistic limit V_l is a function of geometry and orientation as well as the material properties of both striker and target. Grabarek of the Ballistic Research Laboratory has systematically investigated experimentally the effects that various parameters have on the ballistic limit. Some of this data is presented in Appendix B.

Instrumented Impact Testing. The use of instrumented impact tests is not new. Bluhm (1956) performed experiments in which thin plates were impacted against stationary projectiles instrumented with wire resistance-type strain gauges. Similar tests, normally referred to as reverse ballistic experiments, have been reported by Arajs (1971), Lasher, Henderson, and Maynard (1975), and others. Instrumented targets have also been used to investigate penetration by kinetic energy projectiles. Weirauch (1971) reported results of an extensive investigation employing radiographic and optical techniques to examine penetration into copper and Duralumin plates. Netherwood (1979) has recently used small cylindrical probes to measure rates of penetration. Currently instrumented impact techniques are being developed and extensively tested by George Hauver and his coworkers at the Army Ballistic Research Laboratory.

Hauver has been systematically examining the response of long rods during penetration by taking strain measurements with foil resistance gauges. Early experiments employed the reverse ballistic approach in which stationary rods were impacted by steel targets launched from a light-gas gun. Although satisfactory results were obtained for both normal and oblique impacts, the reverse ballistic approach strictly limited the target diameter and the target mass limited the impact velocity to about 700 m/s. To overcome these limitations, Hauver (1978) developed a technique for launching instrumented rods. Strain gauges, located at 20, 40, and 60 mm from the tip of the rod, are

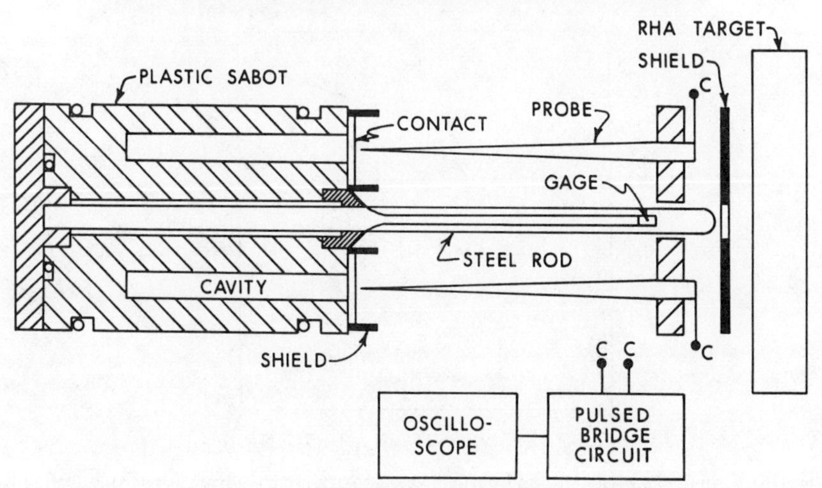

Figure 14 Projectile package for instrumented impact testing [Hauver (1978)].

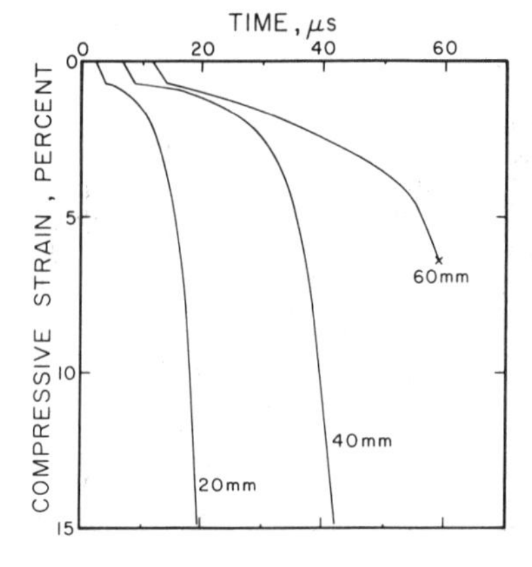

Figure 15 Experimental strain-time records from forward ballistic technique [Hauver (1978)].

connected to thin metal contacts located on the front surface of a plastic sabot (Figure 14).

Prior to impact, the contacts are pierced by stationary probes and a low-resistance electrical path is provided during approximately 100 mm of travel by the sabot. A series of impact tests was performed at normal incidence with steel rods impacting steel targets at a velocity of 1 km/s. Strain-time signals from these tests (Figure 15) were amenable to a simple wave analysis using the rate-independent theory discussed in Chapter 4 which provided relationships between stress and strain, and particle velocity and strain, to strain of 15% (Figure 16). Numerical simulations for this situation were also undertaken by Gupta and Misey (1980) and Misey et al. (1980) with the computer programs EPIC-2, HELP, and REPSIL.

Good agreement with experimental strain-time records at all gauge locations was obtained when dynamic materials data was employed in the calculations.

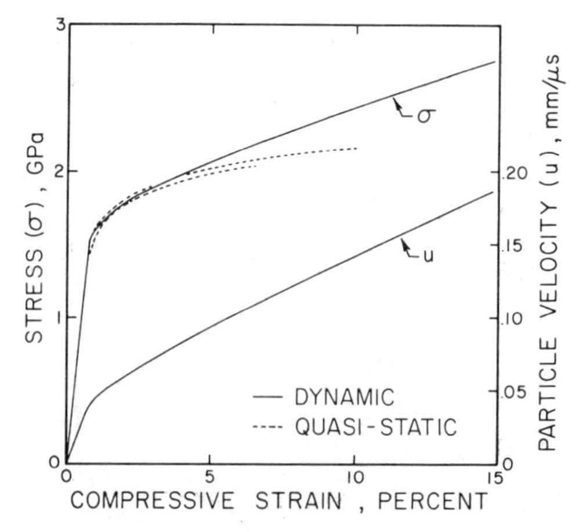

Figure 16 Dynamic stress-strain and particle velocity-strain curves from forward ballistic instrumented experiment [Hauver (1978)].

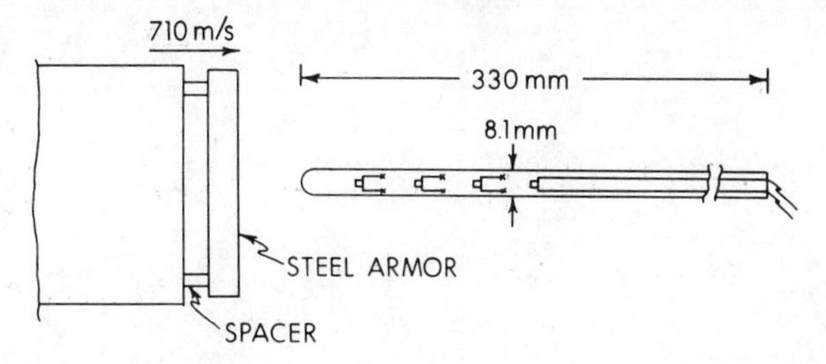

Figure 17 Experimental arrangement for reverse ballistic testing with instrumented rods [Hauver (1980)].

Most recently, Hauver (1980) has reverted to the reverse ballistic technique (Figure 17) since he finds that keeping the instrumented portion of the experiment stationary yields consistently superior data. In this arrangement, the stationary rod is impacted by a steel target launched from a light-gas gun. Rods are commonly instrumented with gauges which are diametrically opposed and are located 20, 40, 60, and 80 mm from the tip. The targets are made of rolled homogeneous armor. The rods were made of low-inclusion tool steel, Saratoga 7, prepared by vacuum-induction melting with vacuum-arc remelting.

As with the forward-ballistic technique, strain-time histories are obtained at the gauge locations. All gauges provide data to fairly high strain levels at low-impact velocities (<700 m/s) until they are destroyed by front-face debris (or spall) as they approach the target surface. At higher velocities (1 km/s) a complete strain-time record is obtained only at the 20 mm position. The other gauges are destroyed by front-face debris before large strain levels can be

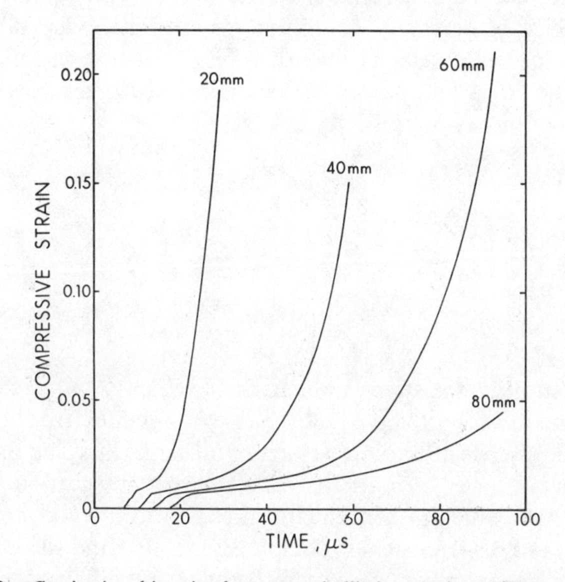

Figure 18a Strain-time histories for reverse ballistic experiment [Hauver (1980)].

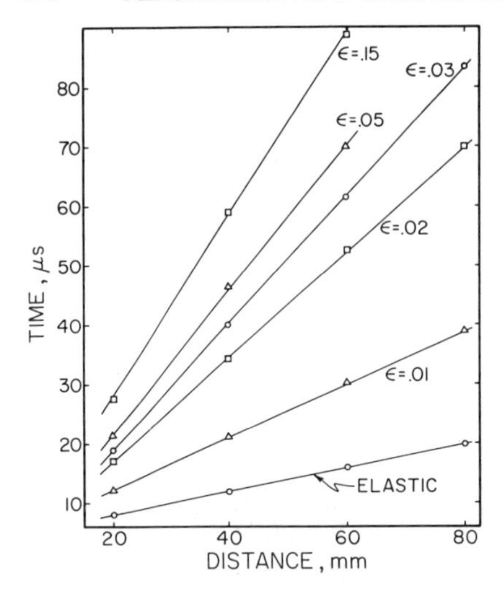

Figure 18b Lagrangian distance-time diagram [Hauver (1980)].

achieved. The data for these gauges can be extended to large values of strain using the data from the 20 mm position. This requires prior knowledge of the relationship between wave velocity and strain. This relationship is obtained with the rate-independent theory of dynamic plasticity from analysis of data for corresponding tests with smaller targets (target diameter $\approx$ 3 rod diameters). For these situations, debris trajectory diverges sufficiently to miss the incoming gauges.

The data of Figure 18 may also be represented as a Lagrangian distance-time plot that shows the arrival of different strain levels at the gauge locations. The linearity of the plot suggests that different strain levels propagate at a characteristic velocity, a necessary condition for the rate-independent theory of plasticity to hold. Wave velocity may then be obtained as a function of strain (Figure 19). The stress and particle velocity are then obtained from the strain rate-independent theory of plasticity in which

$$\sigma = \int_0^{\varepsilon} \rho c_p \, d\varepsilon \tag{2}$$

$$v = \int_0^{\varepsilon} c_p \, d\varepsilon \tag{3}$$

$$c_p = \left(\frac{1}{\rho} \frac{d\sigma}{d\varepsilon} \right)^{1/2} \tag{4}$$

where ε is the strain, σ the stress, v the particle velocity, and c_p the plastic wave velocity. Figure 20 shows the σ-ε and v-ε curves obtained from this procedure. The relationship between particle velocity and strain may be used to convert the strain-time histories to particle velocity-time. Integration then provides displacement time which is conveniently used to locate rod points in spatial coordinates. If rod displacement is completely defined and the trajectory of the rod-target interface is known, data on rod erosion are obtained.

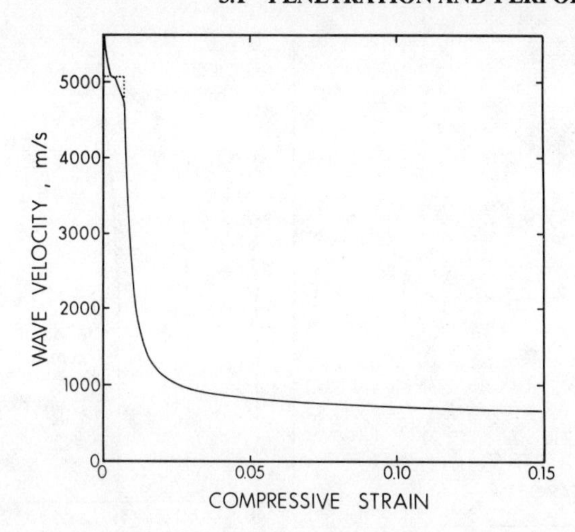

Figure 19 Wave velocity as a function of compressive strain [Hauver (1980)].

Displacement-time history also offers a means for examining the accuracy of strain measurements and the validity of the analysis. The experiment shown schematically in Figure 21 was performed to measure rod displacements after impact, providing data that were compared with displacements obtained from the analysis of strain data. The stationary, long steel rod in Figure 21 was chemically blackened except for highly reflective lines located 40, 60, and 80 mm from the tip. These lines were illuminated by an intense xenon light source and photographed by a rotating mirror-streak camera during impact by a plate of rolled homogeneous armor steel launched at 710 m/s. Results at the 40 mm location are shown in Figure 22. Circular data points from the optical

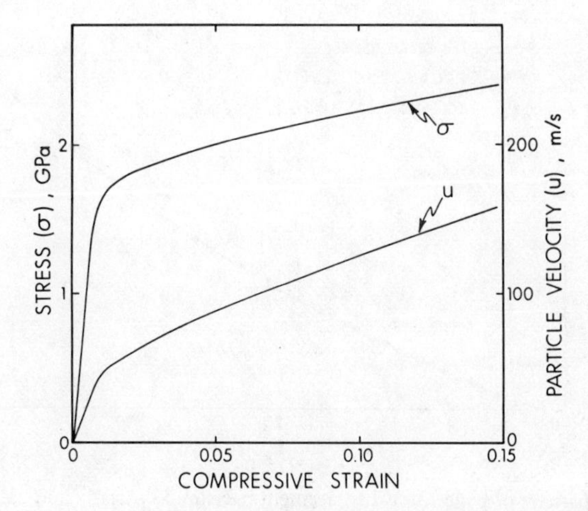

Figure 20 Stress and particle velocity as functions of compressive strain [Hauver (1980)].

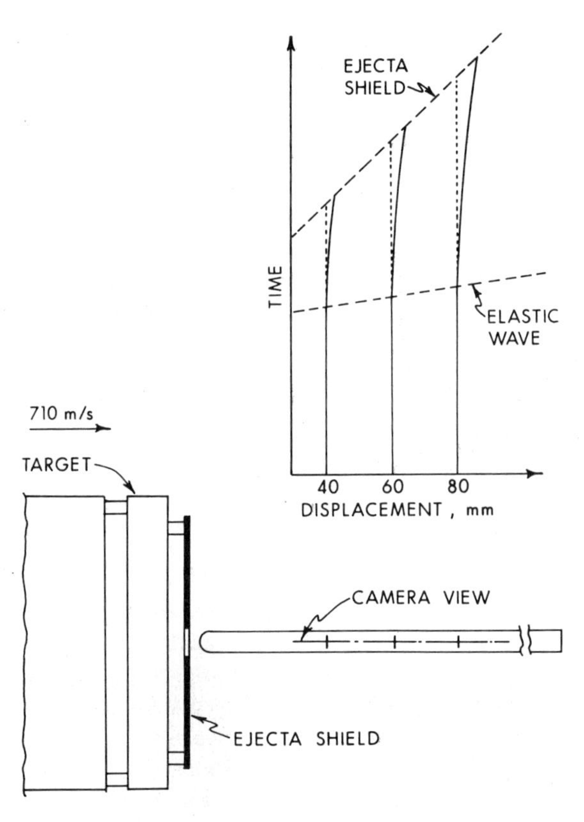

Figure 21 Experimental arrangement for the optical measurement of penetrator displacement, schematically showing a photographic record [Hauver (1980)].

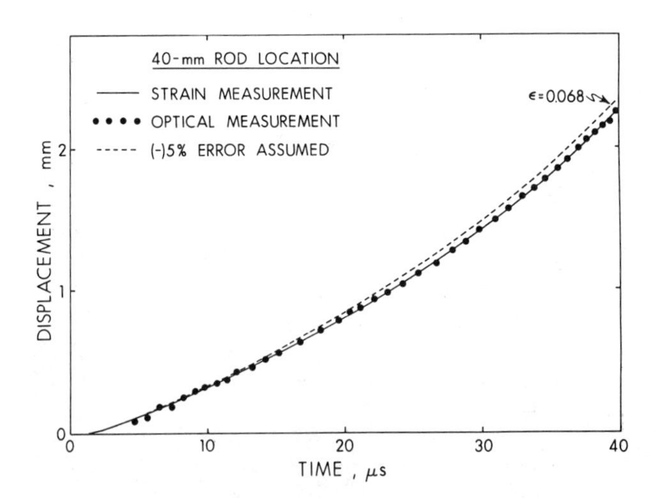

Figure 22 Comparison of penetrator displacement measured optically and calculated from strain data [Hauver (1980)].

measurement are in agreement with the curve obtained from the analysis of strain data. As an indication of sensitivity, measured strains were assumed to be 5% low. However, introducing a 5% correction shifted the calculated curve far out of agreement with data from the optical experiment. The agreement in this comparison tends to confirm both the accuracy of the strain measurements and the reliability of the simple-wave analysis, but only to a strain of approximately 7%. Strains measured during penetration attain much higher values, so that further confirmation experiments were deemed necessary. These confirmation experiments needed to be performed at lower impact velocities where the slowly moving high-strain levels can travel away from the target and into view for photographic measurements.

Hauver found it necessary to combine rod results with target data to construct the penetration-time path in the target. Several uncertainties are introduced in this approach; the principal ones being the validity of the target penetration-time measurements and the transformation of target data from material coordinates to spatial coordinates. A series of experiments was undertaken in which optical observations were made on target-front and rear-face behavior. Hole and bulge volumes were measured, with the Tate theory [Tate (1967, 1969)] being used to locate the depth of penetration when the back surface of the target begins to bulge. From these experiments, the target data of Netherwood (1979) and probe measurements, a distance-time diagram in spatial coordinates was constructed (Figure 23). Although several assumptions are implicit in this approach, the resulting penetration-time path appears to be both realistic and reliable. Additional experiments are being performed to assess the reliability of the methods involved.

Netherwood (1979) has investigated several methods for obtaining penetration-time data within steel targets for normal impacts with steel rods at velocities of 1 km/s. Steel on steel impacts present several difficulties for erosion and penetration rate measurements. In like material impacts, it is difficult to differentiate rod material from target material. High-amplitude

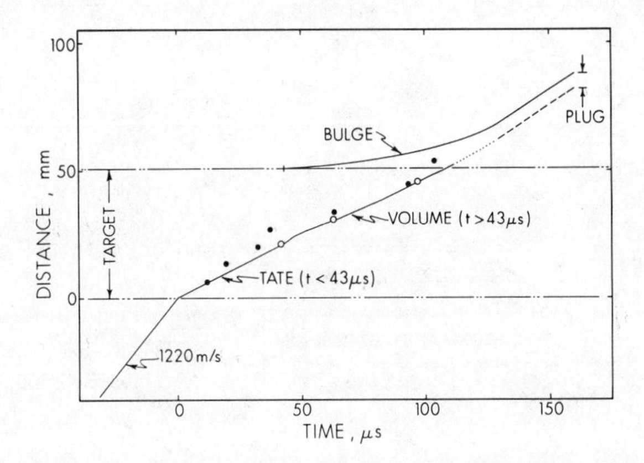

Figure 23 Distance-time diagram in spatial coordinates [Hauver (1980)].

stress waves and a plastic front will precede the projectile and may affect the target material and gauges. In addition, any alteration to a monolithic target may greatly alter the target response to impact. A laminated target is weaker than a solid one of the same thickness and the penetration mechanism for a thin target is different from that for a thick target. Nevertheless, the target must be altered in some fashion if measurements are to be made.

The most reliable results obtained thus far are with the experimental configuration in Figure 24. Insulated wires are placed in holes drilled in 2.54 cm rolled-homogeneous armor targets. As the projectile penetrates the target, wires at different depths are shorted out providing a distance-time record. Sapphire and Teflon were used to insulate the wires. Both materials are good insulators to high pressures but have very different mechanical behavior, sapphire being very hard and brittle while Teflon is very ductile but tough. Since no quantifiable difference in performance was measured between the two insulators in a number of tests, sapphire was used predominantly since it gave cleaner signals.

Results of several impact experiments are shown in Figure 25. As with previous investigations with thick targets at high-impact velocities [Baker (1969), Weirauch (1971)] the penetration response consists of an entrance phase where the rate of penetration is high, a steady-state phase where the penetration rate approaches a limiting value, and an exit phase. Netherwood's technique appears to be a viable one for the entrance and steady-state phases judging by the reproducibility of results. However, the technique fails when the rear surface of the target splits or plugs. The technique is being further

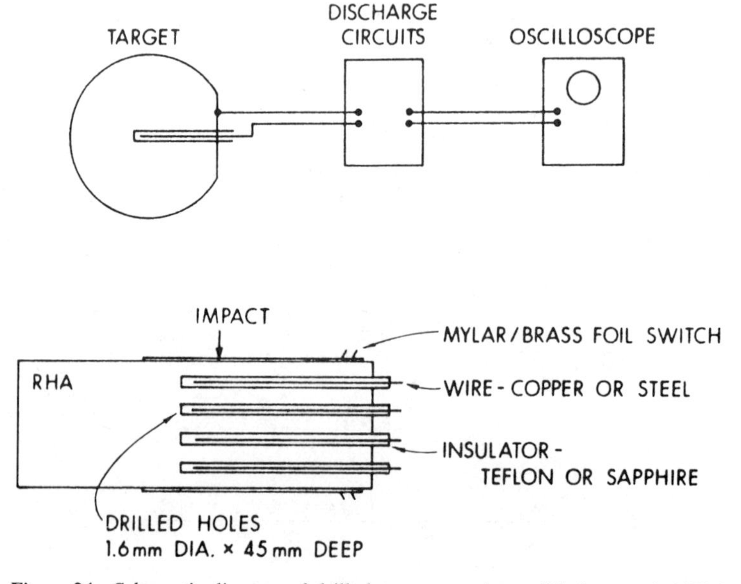

Figure 24 Schematic diagram of drilled target experiment [Netherwood (1979)].

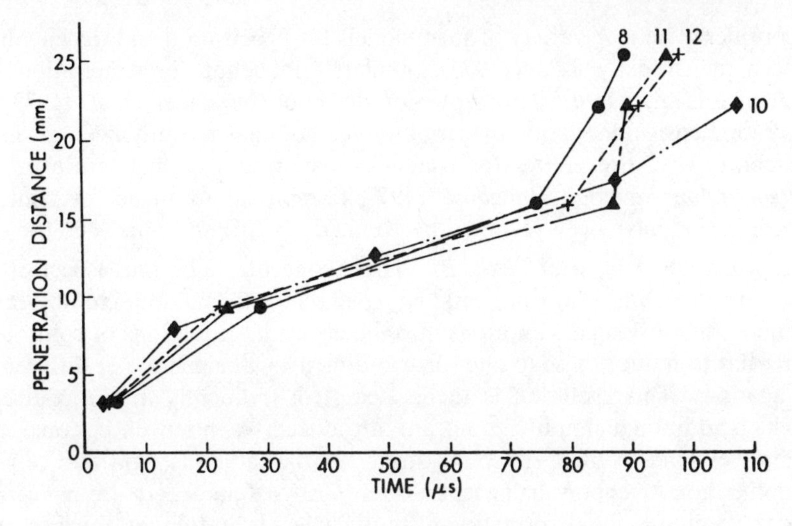

Figure 25 Penetration curves for long rod impact [Netherwood (1979)].

developed to include simultaneous measurement penetration rate and target motion. A similar technique is being studied by Pritchard (1980) to measure pressure amplitudes in impacted targets.

The PHERMEX facility [Venable and Boyd (1965)] at Los Alamos Scientific Laboratory provides another invaluable method for the study of penetration phenomena. Essentially a 6 MEV x-ray source, it is capable of generating extremely short duration pulses that can literally penetrate, or "see into" some 20 cm of steel, thus providing radiographs of projectile behavior within the target. Such information is invaluable for a better understanding of the complexities of material behavior under high-rate loading conditions as well as for providing a further check on numerical simulations. Jonas and Zukas (1978) provide comparisons of two- and three-dimensional computational results with radiographs obtained at the PHERMEX facility.

5.2 ANALYTICAL MODELS

Analytical approaches have tended to fall into three categories:

1 *Empirical or Quasianalytical.* Algebraic equations are formulated based on correlation with a large number of experimental data points and these are used to make predictions to guide further experiments. Such efforts are usually closely related to tests performed to discriminate between the performance characteristics of various materials or structures for a particular design objective. In general, these efforts do not significantly advance our understanding of material behavior and processes and will not be

considered here. A variety of such models for penetration and ricochet have been reviewed by Recht (1973). Similarity modeling for penetration mechanics is discussed in a chapter of the book by Baker et al. (1973). A comprehensive discussion of various force laws has recently been given by Dehn (1979). Procedures for fragment penetration of metals are in *The Penetration Equation Handbook* (1977). Empirical formulas for concrete penetration have been reviewed by Kennedy (1976) and Sliter (1980).

2 *Approximate Analytical Methods.* These concentrate on one aspect of the problem (such as plugging, petaling, spall, crater formation, etc.) by introducing simplifying assumptions into the governing equations of continuum physics to reduce these to one- or two-dimensional algebraic or differential equations. Their solution is then attempted, frequently in the course of which additional simplifications are introduced. With few exceptions, such analyses tend to treat either the striker or the target as rigid and rely on momentum or energy balance, or both. Only a few papers are concerned with predicting the deformation of both projectile and target. Furthermore, almost all such analyses either require some empirical input or rely on material parameters not readily available or measurable.

3 *Numerical Methods.* For a complete solution of impact problems, one must rely on a numerical solution of the full equations of continuum physics. Finite-difference and finite-element methods are capable of attacking the entire set of field equations, have greater flexibility than various algebraic equations, and can accurately model transient phenomena. They are still approximate in nature (one solves a set of discretized equations rather than the corresponding differential equations) but at present, errors associated with material properties are usually far greater than errors inherent in the numerical method. Numerical methods are treated in Chapter 10. The remainder of this chapter is concerned with a discussion of model solutions.

Most existing models consider either a single damage mechanism (plugging, hole enlargement) or conservation law (energy, momentum). A few allow multiple mechanisms, that is, a combination of such factors as compression, plugging, tearing, target inertia, dynamic pressure, friction, and drag. The models could also be subdivided as descriptive versus predictive depending on the degree of empiricism involved, but this point will not be pursued further.

Generally, analytical models tend to reduce the complexity of the problem by treating the projectile as rigid and nondeforming. This removes quite a bit of mathematical unpleasantness as well as reality from the analytical process, except for impacts involving quite thin targets. Several additional hypotheses are frequently invoked to model the target:

1 *Localized Influence.* It is assumed that only a small region with dimensions comparable to the projectile diameter is affected by the impact, the remainder of the target not influencing events.

2 *Rigid Body Motions Are Negligible.*

3 *Thermal Phenomena May Be Neglected* This includes neglect of frictional effects, shock heating, and changes in material behavior.

4 *The Target Is Initially Stress-Free.*

As noted by Backman and Goldsmith (1978) these assumptions generally prove useful except for certain situations near the ballistic limit. If there is an overall tendency toward analytical schemes, it is simplicity. The simplest approaches and failure criteria are frequently employed, primarily because it is not possible to make exact predictions of the complex behavior of materials in impact situations. There exists an amazing and sensible adherence to the Matuska Dictum that, loosely paraphrased, states: *if you are going to be wrong, be wrong as simply and cheaply as possible!*

Several excellent surveys of ballistic-impact modeling exist. Nicholas (1967) reviewed some 245 articles dealing with ballistic impact from a mechanics viewpoint while Goldsmith (1973) surveyed thin-target perforation analysis for normal impact situations. Material covered in these surveys will be mentioned here only briefly when required for the sake of completeness.

5.2.1 Penetration Into Semiinfinite Targets

Although penetration into semiinfinite targets at hypervelocity has received considerable attention, the literature for the same situation at ordnance velocities is quite limited. Brooks (1973) considers penetration and cratering processes for nondeforming projectiles impacting ductile targets. His model is based on an aerodynamic analogy, assuming a rigid-plastic "atmosphere." The model allows for an entry phase where the projectile tip is immersed in the target, a variation of flow stress with increasing penetration depth, an elastic limit below which permanent hole enlargement does not occur, and also allows penetrators with doubly tapered conical-nose tips. Predictions are made for the shape of the crater and the depth of penetration and comparisons made with experimental results for several projectile-target material combinations. The model agreed with crater profiles for steel and tungsten-carbide penetrators striking soft aluminum, and tungsten-carbide penetrators striking mild steel and 4340 steel targets. Among the principal findings of the study are the existence of a critical value of a nondimensional ballistic number (the ratio of the instantaneous dynamic inertial pressure to local flow stress) below which the crater diameter is equal to that of the projectile and the observation that for a semiapex angle greater than 55° there is virtually no difference in the hole profiles produced at a given velocity. The depth of penetration was found to vary strongly with semiapex angle, being some 50% greater for an angle of 10° than for 55°.

Brooks (1974) also advances a hypothesis to explain the behavior of ductile projectiles striking thick targets. Analytical techniques exist to treat impact of rigid projectiles and soft ones that flow hydrodynamically. However, for deformable projectiles where material strength is a significant factor, the deformation behavior is strongly dependent on the impact velocity, the dynamic properties of the projectile and target materials, and on the instantaneous shape of the projectile. At a particular impact velocity, the projectile deformation process will exhibit a dynamic instability that will change its behavior from an essentially elastic character to one that is essentially hydrodynamic. The velocity at which this occurs is termed the hydrodynamic transition velocity. On the basis of an extensive experimental program, Brooks proposes the following deformation sequence:

1 For all penetration velocities, target material is accelerated radially way from the axis of penetration by the passage of the projectile.
2 At low velocities, the kinetic energy imparted to this material as lateral motion is recovered as elastic strain energy such that the target material always remains in contact with the projectiles.
3 As the impact velocity increases, the lateral acceleration of the target material increases; the magnitude of the target-kinetic energy approaches the elastic-strain energy and hence the degree of hydrostatic support afforded to the projectile by the target decreases.
4 At the tip of the projectile, the slope of the surface relative to the trajectory is greatest, hence the radial acceleration of the target material near the tip of the projectile will be greater than at any other point of the ogive. This effect will be even more pronounced if the target material is of a type that hardens significantly when subjected to plastic deformation.
5 At the transition velocity, the kinetic energy imparted to the target material near the nose reaches a level at which it exceeds the elastic-strain energy and the target can no longer provide hydrostatic constraint to prevent the projectile from deforming laterally.
6 Once deformation of the point of the projectile is initiated, the process becomes unstable. As the point becomes spherically blunted, the radial acceleration of the target material increases and exceeds that which would allow adequate hydrostatic support for the next element of projectile behind the deformed point. The rate of deformation increases progressively and finally results in a total destruction of the ogive and the formation of a new, dynamically stable profile.

Ballistic tests with ogival projectiles showed that the transition velocity varied inversely with tip radius—a result supporting the hypothesis. It was also found experimentally that the transition velocity for a given shape can be forced to a higher level if tip deformation is inhibited by an appropriate selection of tip material. Many of the salient points in Brooks' hypothesis are also supported by Zukas and Jonas (1975) who numerically studied the effectiveness of ballistic caps of various materials for long rod penetrators.

Tate (1977) suggests that the hydrodynamic transition velocity depends on the relative rates of rod erosion and plastic wave propagation. He states that when the rod-erosion rate exceeds the rate of propagation of gross-plastic deformation, then all the gross deformation is constrained to occur very near the tip of the rod in a region of increased entropy and temperature resulting in a jet mode of penetration. Expressions are given for the speed of propagation of gross-plastic distortion and for the hydrodynamic-transition velocity for right-circular cylindrical rods striking thick targets at normal incidence in terms of the empirical dynamic strength of the rod and target and the dynamic work-hardening rate or dynamic large-strain tangent modulus of the rod material. A comparison with data for copper rods striking copper targets indicates qualitative agreement.

Tate (1967a, b) proposes a modification to Bernoulli's equation by including two strength parameters (stresses for rod and target above which each behaves as a fluid) to predict deceleration of a long rod after striking a thick target. The strength parameters are empirically determined quantities. Comparison with experimental data indicates fair to poor agreement and a high degree of sensitivity of predicted results to the assumed values of the two strength parameters. In additional developments along the above lines, Tate (1969) provides models to account for deformation of a soft rod striking a rigid target and the penetration of a rigid projectile into a soft target. He shows that it is theoretically possible to have a decrease in depth of penetration with increasing impact velocity, but predictions are not substantiated by the experimental results cited.

Most recently, Tate (1980) has developed a model for the ricochet of long rods. The problem of ricochet is a long-standing one. As yet, no definitive analytical treatment has emerged. Much of the early work involving ricochet of military projectiles from metal plates, soil, concrete, and water has been documented by Ipson et al. (1973) and Recht and Ipson (1973). Ricochet of nondeforming spheres from metallic surfaces has been studied by Backman and Finnegan (1976). Other studies of nondeforming projectiles striking various media at high obliquity are those of Johnson and Reid (1975), Hutchings (1976), Soliman et al. (1976), and Daneshi and Johnson (1977a, b, 1978).

Recht identifies three major factors that produce direction and velocity changes during high obliquity impacts:

1 During impact, the projectile and target are compressed. The elastic portion of the total strain energy absorbed is recoverable. Its release produces change in projectile motion.

2 The determined direction and magnitude of the resisting-force resultant acting on the projectile will be affected by the characteristics of target surfaces and material interfaces (since these influence geometric changes and stress-wave reflections) as well as the deformations produced by high-velocity impact.

3 Shear and frictional forces resist projectile motion and serve to reduce velocities. Furthermore, unless the resisting-force resultant always acts on a line that intersects the projectile's center of gravity, the projectile will experience rotating moments.

Except for extreme obliquities, it is difficult to get ricochet from thin plates. Hence, most models and empirical relationships take no account of plate thickness. This is one of the assumptions of Tate's model. He further neglects deceleration of the rod (which is assumed to have a square cross section for computational simplicity), allows for erosion of the rod surface, and assumes his modified hydrodynamic theory [Tate (1967)] to hold. These assumptions lead to an expression for a critical ricochet angle in the form

$$\tan^3\beta > \frac{2}{3}\frac{\rho_P V^2}{Y_p}\left(\frac{L^2+D^2}{LD}\right)\left[1+\left(\frac{\rho_P}{\rho_t}\right)^{1/2}\right] \tag{5}$$

In (5), the subscripts "p" and "t" refer to penetrator and target respectively. the angle β is the angle of obliquity (the angle between the projectile trajectory and the normal to the target plate), V represents striking velocity, L and D are the length and diameter of the projectile, ρ represents density, and Y is a characteristic projectile strength which is closely related to the Hugoniot elastic limit of the material. The critical angle above which ricochet will occur depends primarily on density, length-to-diameter ratio, and projectile strength. Because the tangent of the ricochet angle is a function of the cube root of these variables, the critical angle is relatively insensitive to variations of these parameters. This model provides good qualitative prediction of rod behavior at high obliquity encounters. If details such as target thickness and nose shape critically affect rod performance, then the analysis would most profitably be done through computer calculations.

Byrnside, Torvik, and Swift (1972) have had considerable success in studying penetration processes and their work merits further consideration for generalization and extension to the oblique impact case. Their approach is a modification of the rigid penetrator and deep penetration theories of Goodier (1965) to account for projectile strength in crater formation. It is assumed that a spherical projectile is not fragmented while being completely consumed. Experiments were performed with 7075-T6 aluminum projectiles striking 6061-H aluminum targets. Measured values of crater diameter and crater depth agreed quite well (0–8% deviation for mean crater diameter) with predicted values for velocities under 2 km^{-1}. For increasing striking velocity and projectile strength, discrepancies between predictions and data increased (13% for hypervelocity data).

Persson (1974) has developed a simple model for response of a relatively thick target normally impacted by a rigid sphere. Projectile motion is retarded by elastic-plastic friction and inertia forces with provision for reduction of retarding forces because of edge- and rear-surface effects. The model comes with no fewer than eight adjustable parameters which must be determined by separate experiments. In this regard, it is worth quoting Glenn & Janach (1977): "...the proliferation of parameters, with the aim of providing plausible physical explanation of discrepancies between theory and experiment, has its limitations. In this connection, we were recently reminded of a comment attributed to the French mathematician, Cauchy, to the effect that, given four parameters, he could draw a credible version of an elephant and given five, he could make its tail wiggle."

5.2.2 Penetration of Finite-Thickness Plates

Perforation of finite-thickness plates has achieved greater attention. Fugelso et al. (1961, 1962) and Fugelso (1964) survey at length the theoretical aspects of penetration and perforation and justify the use of linear-elastic solutions to perforation problems for very thin plates at striking velocities under 1.2 kms^{-1} and impact durations of less than 50 μs. Fracture is based on a critical octahedral shear stress. No comparison with experiments is made. Florence and Ahrens (1967) and Florence (1969) offer a linear-elastic analysis of stresses in metal projectiles and ceramic targets. The experimental data presented is impressive.

Hole growth in ductile targets has been considered by Bethe (1941), G. I. Taylor (1947), Freiberger (1952), and Kumari (1975). W. T. Thompson (1955) and Brown (1964) use a quasidynamic energy approach to study petaling of thin plates. Kucher (1967a, b) optimizes penetrator shape using Thompson's theory. These approaches are conceptually interesting and permit considerable mathematical manipulations but are not very useful since few armor designs use infinitely thin sheets. Goldsmith (1973) comments on analytical treatment of hole growth at some length.

In the energy balance analyses cited above, wave propagation effects, crack formation, friction, adiabatic heating, and strain rate are not considered. Zaid and Paul (1957) assumed that penetration effects propagate at a finite rate and proposed a "zone of action" within which the significant effects are confined. Through use of momentum conservation and an "effective mass" of the target plate, they determined the resisting force, deceleration and penetration of a nondeforming conical projectile striking a thin plate at normal incidence. The analysis was later extended to cover truncated conical and ogival projectiles [Paul and Zaid (1958)] and truncated cones at oblique impact angles [Zaid and Paul (1959)]. A pitfall of their otherwise elegant approach lies in the requirement that the deformation pattern for the target be assumed a priori, requiring a good deal of insight on the part of the user of their models. Computed velocity-distance histories compare fairly well with experimental data quoted (10–25% discrepancy).

Pytel and Davids (1963) consider deformation of a viscous plate by a rigid projectile. The plate is assumed to be acted on by an initial velocity over a circular area with a radially symmetric shearing stress, uniform across the plate thickness, being the only nonnegligible stress component. The theory requires a viscosity coefficient and strain rate for prediction of displacement fields. Minnich and Davids (1964) modified the theory by including an empirically determined "impact yield constant" that has units of stress. Below this value the target is assumed to act as quasirigid whereas above it viscous flow occurs. This modification to the model leads to finite-final deformed shapes but as the impact yield constant was found to vary with plug displacement the model is of no practical utility.

Awerbuch (1970) and Awerbuch and Bodner (1974a, b) have analyzed the normal perforation of projectiles into metallic plates. The penetration process is assumed to occur in three interconnected stages with plug formation and ejection being the principal mechanism of plate perforation (Figure 26). In the first stage, shearing is assumed not to occur. This stage is considered to be a compressive stage in which the forces acting on the projectile are an inertial force and a compressive force. The inertial force is because of the acceleration of the mass of the target material in contact with the projectile in the direction of motion. The compressive force on the projectile is due to the compressive strength of the target material in contact with the projectile. Another basic

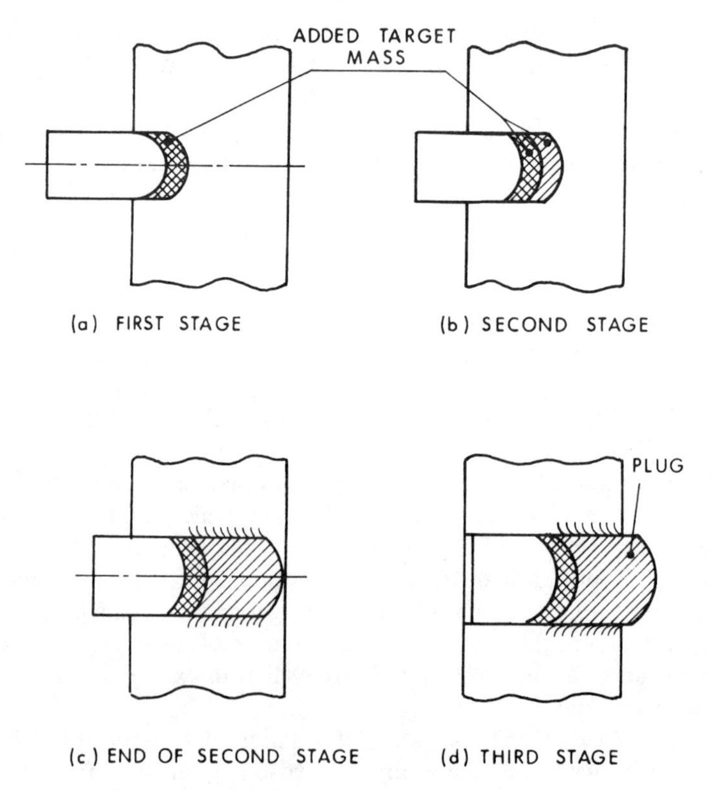

Figure 26 Three-stage perforation model [Awerbuch and Bodner (1974a)].

assumption for this stage is that mass from the target material is added to the projectile during the penetration process.

The second stage of penetration is the onset of shearing of a plug from the target plate. In this stage of insipient plugging, the projectile is acted on by the compressive and inertial forces of the first stage as well as a shearing force that results from the motion relative to the target plate of target material, which is accelerated by the projectile during this stage.

A third stage begins when plug and projectile move together as a rigid body with only a shearing force acting on the plug's circumference along its whole length. The theory allows computation of post-perforation velocity, force-time history, and contact time for perforation processes that include dishing, plug formation, and ductile cavity enlargement. However, several parameters in the analysis must be determined empirically, namely the target entrance and exit diameters, the plug length, the coefficient of target viscosity, and the width of the shear zone. For the latter, the authors cite the paper by Chou (1961) for an analytical expression. With the above quantities determined, experimental results for lead bullets striking steel and aluminum targets showed good agreement with predictions for post-perforation velocities and duration times.

Awerbuch and Bodner (1975) modified their normal perforation theory to include the effect of angle of impact for cases where perforation occurs without ricochet or projectile fracture. The primary modification consists of replacing plate thickness with an effective line-of-sight thickness for the target and adjusting force and momentum expressions in the analysis accordingly. Comparison of theory to experimental results for 0.22-caliber lead bullets striking aluminum targets ranging from 2–6 mm in thickness showed fair agreement.

Simpson (1972) proposes a model for penetration and perforation for a striking velocity range of 1.2–5 $\mathrm{km s^{-1}}$. The model envisions penetration/perforation of a finite-thickness plate to consist of an initial stage of penetrator hydrodynamic erosion. The second stage consists of continued penetrator erosion and onset of plate deformation assuming a plugging mechanism. In the remaining stage, the penetrator is assumed completely consumed and a plug ejected from the target plate. The target deformation stages closely follow the works of Minnich and Davids (1964) and Awerbuch (1970). The author was unable to find perforation data in the specified velocity regime and therefore presents comparisons only for thick-target penetration, comparing model predictions for penetration depth and hole diameter with experimental data for long rods and mass-focus slugs. Except for slug data, agreement is generally good. The report contains an extensive annotated bibliography and a computer program for the derived set of equations.

The deformation and perforation of thin plates resulting from the impact of spherical and conical projectiles has been examined by Goldsmith et al. (1965). Further work was reported by Calder and Goldsmith (1971) and Goldsmith and Finnegan (1971). The latter is of interest since an assessment is made of the relative magnitudes of dishing (plate bending) and plugging based on strain-gauge data acquired on each side of the target plate. It was found that at higher velocities the perforation mechanism changes from bending to punching.

Lethaby and Skidmore (1974) consider plugging near the ballistic limit. When the striking velocity is much greater than the limit velocity the target absorbs just enough energy to cause a punching-type failure and the rigid penetrator theories tend to give better results than in cases where the impact occurs at the ballistic limit. At such velocities, the target absorbs more energy which results in greater target deflection (bending). Their model, using the assumptions of membrane theory, also includes a plugging criterion based on a critical angle of deflection in the target. The model is limited to very thin plates and low-impact velocities. Predicted critical-projectile velocities agree to within 10% of those determined from experiments with mild steel cylindrical rods striking mild-steel plates at velocities of 38–170 ms^{-1}.

Woodward and deMorton (1976) compute critical velocities for plugging and a residual-plug thickness and velocity. The model is based on energy balance and assumes that shear and frictional forces resist penetration. Considerations of plug acceleration are also included. Comparison with experimental data showed good agreement for critical and plug-residual velocities and poor agreement for plug thickness. Since the projectile is assumed to be rigid, the model is better suited for computations with soft targets.

Woodward (1978) has further examined solutions to the failure of targets by hole enlargement and dishing when impacted by pointed projectiles. For the case of nondeforming penetrators, good estimates can be made of critical velocities to penetrate metal targets based on their mechanical properties. This treatment allows ranking of target material effectiveness by weight and cost.

Woodward (1977, 1978, 1979) has also examined various target failure modes when impacted by sharp-nosed projectiles. This has led to a novel treatment of adiabatic-shear plugging in terms of a natural geometric instability. A similar treatment can be used to explain the mechanics of scabbing failure.

Kowalski et al. (1976) employ one-dimensional wave theory and assume shear failure to compute graphically the minimum striking velocity necessary to eject a plug. Although a numerical example is given, no comparison with experiments is made.

Many researchers concern themselves with the minimum velocity necessary to perforate a plate and the residual velocity of the projectile after perforation. The simpler models rely on an assumed failure mode and energy or momentum balance (or both) as well as a few well-placed empirical constants. Nishiwaki (1951) proposed a residual-velocity model assuming that the total resistance to the motion of a rigid conical penetrator is a function of dynamic and static pressures. The assumption that displaced target material remains in contact with the projectile nose yields an expression for dynamic pressure. The static pressure is assumed to be a material constant. Gabbert (1970) modified the Nishiwaki theory by assuming that particles of target material are displaced in a direction along the projectile trajectory with a velocity equal to that of the projectile, rather than allowing target particles to displace normal to the projectile surface with a velocity equal to the component of projectile velocity in that direction. Both the Nishiwaki theory and Gabbert's modification were compared with a large body of experimental data. Both were found wanting.

Recht and Ipson (1963) develop a model for the residual velocity of a rigid penetrator using energy and momentum conservation and assuming plug formation to be the failure mechanism. The model requires an a priori knowledge of the minimum velocity required to perforate. They also suggest an expression for this velocity in terms of projectile diameter, length, density, and sonic velocity, and target shear strength, density, thickness, and sonic velocity. Giere (1964) offers a residual-velocity expression that differs little from that of Recht and Ipson. Ipson and Recht (1975) propose a means of determining minimum perforation velocities with a ballistic pendulum technique. Woodall et al. (1970) offer considerable data on plugging and perforation at velocities where plug shattering can be expected. There is limited confirmation of the model suggested by one of the authors (Heyda). Weidman (1968, 1969) presents several approximate methods for calculating ballistic limits. His analysis is restricted to short cylindrical projectiles perforating thin plates or sheets. He assumes perforation to take place when the magnitude of the strain rate is less than some critical value and the magnitude of the strain is greater than another critical value. Shearing is assumed to be the dominant failure mechanism and the critical value for strain and strain rate are determined graphically in terms of a mass-ratio factor. For thin sheets, perforation is assumed as a result of plugging, transverse shear stress is assumed constant through the plate thickness, and target material is treated as a viscoplastic Bingham solid. Projectile mass is assumed small in comparison to the mass of the resulting plug and failure occurs when the radius of the hole is equal to the radius of the projectile. Using a series expansion for strain and strain rate given by Chou (1961) the above failure criterion is incorporated and truncated at the second term to get an explicit formula for ballistic limit. The two term series shows excellent agreement with a full series solution (less than 5% discrepancy), especially at low velocities. No comparison with experimental data is made.

Heyda (1970) proposes a model for limit and residual velocities assuming plugging and resistance to projectile motion to be governed by two components of pressure, namely a high-intensity component computed from hydrodynamic theory since it is assumed that this pressure causes a thin fluid zone to be formed at the plug-projectile interface and a second component of pressure-resisting plug motion by shear. The shear stress is assumed constant through the plate thickness. It is further assumed that shear stress is the only component giving rise to mechanical work.

Leone (1975) develops a plugging model based on energy balance using an empirical relationship for limit velocity developed by Burch and Avery (1970).

He compares residual-velocity predictions of his own and nine other models ranging from the totally empirical to analytical. Two empirical models, his own and that of Recht and Ipson (1963) are found to give realistic results.

Lambert and Jonas (1976) have reviewed the penetration theories dealing with nondeforming projectiles. They find that the variety of models proposed, such as the resisting force approach of Poncelet-Morin, the energy-momentum analysis of Recht and Ipson, and other approaches attributed to Nishiwaki, W. T. Thompson, Zaid and Paul, almost invariable adhere to one basic form, namely

$$V_s = \begin{cases} 0, 0 \leqslant V_s \leqslant V_l \\ \alpha\left(V_s^2 - V_l^2\right)^{1/2}, V_s > V_l \end{cases} \tag{6}$$

Here V_s is defined as the striking velocity, V_r the projectile residual velocity, and V_l the ballistic limit,

$$V_l = \max\{V_s : V_r = 0\} = \inf\{V_s : V_r > 0\} \tag{7}$$

The various models reviewed ultimately differ insofar as do the formulations for α and V_l. Their examination of available data tended also to confirm that experimental results can often be well represented within the framework of the above form for residual velocity, particularly in situations where there is not excessive projectile deformation. Their report discusses in depth the characteristics of the form and generalizes it to

$$V_r = \begin{cases} 0, 0 \leqslant V_s \leqslant V_l \\ \alpha\left(V_s^p - V_l^p\right)^{1/p}, V_s > V_l \end{cases} \tag{8}$$

where α, p, and V_l are viewed as parameters to be subjected to optimal adjustment in a given situation. Lambert (1978b) has since provided equations for predicting the parameters α, p and V_l for perforation of steel and aluminum targets by long rods. (See Appendix A.)

Dunn and Huang (1972) have discussed impact on spaced plates, essentially summarizing existing equations for residual velocities based on plugging failure and estimate velocities and material parameters leading to projectile shatter based on simplistic shock-wave considerations. Zaid and Travis (1974) offer an intelligent discussion of the subject together with a review of the pertinent literature and data for impact of hardened steel cylindrical projectiles striking single and multiplate mild-steel targets at velocities up to 500 ms^{-1}.

Wilms and Brooks (1966) approximate the projectile as both a rotary and shear-beam column to obtain transient bending and shear stresses in the projectile on oblique impact. Closed form solutions are obtained via Laplace transforms and complex boundary conditions obviated by assuming that the nose is embedded in the target and loaded hydrodynamically—not very severe approximations if the target thickness to projectile diameter ratio is greater than 1 (i.e., thick targets). Variation of shear stress with obliquity is studied.

The authors find that the octahedral shear stress at 60° will be some 3.3 times greater than that at normal incidence.

Attempts to determine frictional or heating effects have confounded many researchers. Both analytical and experimental results are mixed. Wingrove (1972) experimentally obtained force-time curves for blunt, hemispherical, and ogival penetrators striking aluminum targets at velocities up to 240 ms^{-1}. He concludes that the maximum force is independent of geometry although the deformation pattern and nature of failure differ markedly for each of the penetrator shapes. He also concludes that frictional effects are probably negligible except in the vicinity of the ballistic limit and cites the results of Krafft (1955) who measured frictional adhesion between projectile and target during ballistic penetration with a torsion-type Hopkinson bar. Krafft concluded that sliding friction amounted to at most 3% of the projectile striking energy. This is in marked contrast with the findings of Mach et al. (1973), Mach (1975), and Weirauch and Lehr (1977) who attribute temperatures in excess of 1200°C to surface-friction effects for penetrators passing through thick aluminum targets. A previous determination of forces resisting penetration was made by Masket (1949).

Gordon (1973) attempted to use the Heyda (1970) model coupled with heat transfer to determine the temperature distribution at any point on an impacted plate. He assumes that heat is transferred only in the radial direction and that the velocity of the plug and projectile are the same. Comparison with experimental data indicated that the model is not a valid one.

Backman and Finnegan (1976) provide an analytical procedure for representing the behavior of a sphere-plate system solely in terms of the motion of the sphere, whose trajectory is modeled as a series of segments of constant speed and curvature. The resisting force is taken to be of the Poncelet form (quadratic in velocity) and failure of the target plate is modeled by removal of the force over predetermined areas once critical penetration depths are exceeded. Comparison with experimental data tends generally to support the model.

Qualitative discussions of adiabatic shear are given by, among others, Recht (1964), Stock and Thompson (1970), Wingrove (1973), Backman and Finnegan (1973), Wingrove and Wulff (1973), and Yellup and Woodward (1980). This subject is presently an item of intense research with analytical models anticipated shortly. The mechanics of the instability in penetration situations has been examined by Woodward (1978) and several reviews and discussions of the thermodynamic conditions responsible for the instability have appeared by Rogers (1974, 1979) and Bedford et al. (1974). Current work on containment of projectiles is that of Zaid, El-Kalay, and Travis (1973). The whole field of containment of ballistic fragments has been reviewed by Recht (1970).

Essentially all thin-plate perforation models deal with rigid, nondeforming penetrators. Two exceptions are the models proposed by Haskell (1972) and Recht (1976). Both assume plugging failure of the target. Haskell characterizes rod behavior in terms of one-dimensional plastic-wave propagation, a simple, maximum strain-failure criterion and a conservation of energy treatment of the perforation process. Recht follows concepts advanced by G. I. Taylor and treats mass loss as a multistage process characterized by differences between the relative velocity of deformed and undeformed portions of the projectile and the materials' plastic-wave velocity. The Recht model predicts mass loss caused by shock front, the lengths of deformed and undeformed portions of the rod, the shape of the deformed portion, and the extruded mass left behind during perforation. Both models predict residual mass and length reasonably well in comparison against finite-data sets.

Approximate models of the complex processes that occur during impact have their place. They serve to emphasize the primary mechanisms that govern the physical behavior of colliding solids. No less important is their function in providing a means by which experimental data may be ordered and partially analyzed. It must be realized however that they are quite limited in scope and frequently have embedded within them empirical constants masquerading as "material properties." Use of such models outside the approximations inherent in their formulation or the data bases for their empirical constituents frequently leads to conclusions that are both quantitatively and qualitatively incorrect.

Solutions of a more general nature must be obtained through numerical solution of the governing equations of continuum physics. Despite uncertainties associated with failure criteria, impressive results have been obtained with existing computer codes. The fundamentals of numerical modeling of impact phenomena and the characteristics of current three-dimensional computer codes for wave-propagation and impact studies are discussed in Chapters 10 and 11.

APPENDIX A: A RESIDUAL-VELOCITY PREDICTIVE MODEL FOR LONG ROD PENETRATORS

Lambert (1978b) has described a mathematical structure which, on the basis of experimental evidence and available theoretical analyses, seems appropriate for representing residual velocity as a function of striking velocity (and of projectile-target characteristics) for long rods impacting single-plate targets of rolled-homogeneous armor (RHA). Equations are provided for predicting the parameters a, p, and V_l in the context of the "standard" general form used to encompass V_s, V_r relationships:

$$V_r = \begin{cases} 0, 0 \leqslant V_s \leqslant V_l \\ a(V_s^p - V_l^p)^{1/p}, V_s > V_l \end{cases} \tag{1}$$

This form, its general characteristics, and adaptation to existing V_s, V_r data are discussed elsewhere [Lambert and Jonas (1976)]. Our interest here is confined to the predictive aspect.

For a given projectile-target situation, the following notation is used:

M = penetrator mass, grams
L = penetrator length, centimeters
D = penetrator diameter, centimeters
T = target thickness, centimeters
ρ = target density, grams per cubic centimeter
θ = incidence angle
M_r = penetrator residual mass, grams
V_l = limit velocity (or limit velocity estimate), meters per second
V_r = penetrator residual velocity, meters per second
V_s = penetrator striking velocity, meters per second

For purposes of this discussion, a "long rod" is essentially a cylinder having length-to-diameter ratio of least 4.

1.0 BACKGROUND AND DEVELOPMENT

Limit-velocity prediction is especially important and has been the focus of considerable effort for a long time. Because of the great complexity and consequent limited understanding of penetration processes, work in this area has been chiefly empirical: a frequent approach in formulating limit-velocity recipes has been to isolate and deal with the term MV_l^2/D^3, the "specific limit energy" (the numerator is twice the "limit energy," the penetrator kinetic energy expended when the velocity of impact is V_l).

A dimensionally awkward formula proposed by deMarre in 1886, for the case of normal incidence was:

$$\frac{MV^2}{D^3} = \alpha\frac{T^{1.4}}{D^{1.5}}, \qquad \alpha \text{ constant} \tag{2}$$

(See Curtis (1951) for an interesting and extensive account of early approaches to limit-velocity prediction.)

Formulae in which (for normal incidence) specific limit energy is proportional to a power of T/D have been used extensively and are generally referred to as modified deMarre formulae:

$$\frac{MV_l^2}{D^3} = \alpha(T/D)^{\beta}, \qquad \alpha \text{ and } \beta \text{ empirical constants} \tag{3}$$

In the case of nonnormal impact, T is usually replaced by $T \cdot g(\theta)$ where g is some function of the obliquity (e.g., the secant). The parameters α and β are typically set on the basis of past experience and fixed values for them tend in practice to be usefully applicable to only a small class of projectile-target situations; the "constant" is, it turns out, particularly vulnerable to necessary adjustment in going from situation to situation. A full range of values of α between 1 and 2, variously proposed to cover different situations, appear in the literature. Typically the parameter values are derived via a regression on available limit-velocity data sets.

Though empirical as generally postulated, (3) has also evolved from theoretical analyses of penetration mechanics. Bethe (1941), in a classical paper, used elasticity theory to analyze the action of armor plate in stopping penetrators. Considering the progressive expansion of a hole in the plate by lateral displacement of material, he determined limit energy to be proportional to TD^2, and thus devised formula (3) with $\beta=1$. A constant hydrostatic pressure α, resisting penetration, was taken to be a certain multiple of target yield stress (G. I. Taylor later modified Bethe's analysis and obtained a different value for α). Other early theories [Zener and Holloman (1942)] postulating different modes of target failure (such as petaling and plugging), inferred limit energy to be proportional to T^2D, thereby implying a version of (3) having $\beta=2$.

A series of early papers by H. P. Robertson provides a valuable descriptive analysis of the penetration cycle and an account of the pioneer work of Poncelet. The Poncelet assumption (1840) that the resistance encountered by a penetrator in passing through a plate is a linear function of the square of the velocity of the penetrator, has since been a recurrent feature of penetration analyses. A perceptive commentary by A. H. Taub and C. W. Curtis in an addendum to one of Robertson's papers [Robertson (1943)] discusses limit-velocity formulations inspired by the Poncelet and Bethe theories. In an analysis that supposes the Bethe theory to be valid while the penetrator is in the main body of the plate, but that the mechanism of failure changes to a petaling type near the backface of the plate, they derive the formula:

$$\frac{MV_l^2}{D^3}=\alpha\left(\frac{T}{D}+\beta\right), \qquad \alpha, \beta \text{ constant}, \qquad \beta<0 \tag{4}$$

Their development supposes the ratio of backface thickness (where petaling prevails) to penetrator diameter to be constant and β then turns out to be a quadratic function of that constant value. The basic reasoning behind (4) is appealing—it seems realistic to account for a change in the mechanism of plate reaction—but the supposed (nonzero) constancy of β cannot be strictly correct: in the limit, as $T\to 0$, for fixed M and D, the left-hand side of (4) should tend toward zero (clearly limit energy, and hence, specific limit energy, should approach zero) but the right-hand side tends to $\alpha\beta\neq 0$. Though not entirely satisfactory, as just observed, (4) does appear to be qualitatively in general agreement with experimental results (in better agreement over a broader spectrum of situations than other models) and so is used as a starting point.

First amend and modify equation (4) in three respects:

1 Replace the constant β by $e^{-T/D}-1$, a slowly varying, monotonically decreasing bounded function of T/D that tends toward zero as T does. This overcomes the discrepancy mentioned above.

2 Replace D^3 in the left-hand side of (4) by $D^{3-c}L^c$, c constant, the feeling being that this term should be partly responsive to penetrator volume (which is approximately proportional to D^2L). Equivalently, leave the left-hand side of (4) intact and multiply the right-hand side by $(L/D)^c$.

3 To accommodate arbitrary obliquity, θ, replace T by $T\sec^k\theta$, k constant. This is a customary way of dealing with obliquity and target thickness to produce an "effective thickness"; e.g., if $k=1$ then one has the length through the target of the projectile line of fire.

Let $z=z(k)=T\sec^k\theta/D$ and let $f(z)=z+e^{-z}-1$. Then, altering (4) in accordance with the above stipulations yields:

$$\frac{MV_l^2}{D^3}=\alpha\left[\frac{L}{D}\right]^c f(z) \tag{5}$$

or

$$V_l=\sqrt{\alpha\left[\frac{L}{D}\right]^c f(z)\frac{D^3}{M}} \tag{6}$$

The parameters α, c, and k require empirical determination. Working from a diverse collection of available experimental limit-velocity data for long rods into RHA,* taking velocity to be in units of m/s, using (6), and optimizing with respect to these three parameters in a least squares sense yields:

$$\alpha=(4000)^2, \qquad c=0.3, \qquad k=0.75.$$

The working formula then, for estimating limit velocity in situations involving long-rod penetrators and single-plate RHA targets is:

$$V_l=4000\left[\frac{L}{D}\right]^{.15}\sqrt{f(z)\frac{D^3}{M}}\ \text{(m/s)} \tag{7}$$

where

$$z=\frac{T(\sec\theta)^{.75}}{D}$$

$$f(z)=z+e^{-z}-1=\sum_{j=2}^{\infty}\frac{(-z)^j}{j}$$

Some notes on equation (7) and its usage:

1 The data collection that serves as the source of past experience involves straightforward penetrator, geometrical configurations—penetrators that are essentially right-circular cylinders with conical, flat, or hemispherical noses, and that have obvious values for D, L/D, and M. In practice, geometries will not always be so clear-cut and it may be necessary to estimate "effective" values for these input parameters. Additionally, the

*Lambert's limit-velocity data base contained limit velocities for 200 situations involving a wide range of masses ($\frac{1}{2}$ to 3630 g), diameter ($\frac{1}{5}$ to 5 cm), L/D's (4 to 30), target plate thickness ($\frac{3}{5}$ to 15 cm), obliquities (0 to 60°) and rod materials (densities vary from 7.8 to 19 g/cc).

data collection reflects relatively thick-target situations ($T/D>1.5$); it is suggested that usage of the formula be restricted to situations where target thickness exceeds 1.5 calibers.

2 The equation does not directly involve known material properties such as target strength and hardness; nor is account taken of rod-nose geometry. Generally the influence of such parameters is masked by overall experimental variability and there is little to be gained at this time by trying to isolate their effect on limit velocity. No doubt the degree of influence of nose shape is inversely related to target thickness and would need to be taken seriously for thin-target situations (i.e., T/D values less than, say, 2); it may also be that certain dynamic material properties, as they become available, can be effectively exploited.

The second phase in devising a predictive residual-velocity model [in the context of form (1)] is to account for estimation of parameters a and p. Here there is little in the way of traditional wisdom to provide guidance.

An essential premise of much V_s, V_r modeling is that the penetrator remains rigid during penetration and accordingly, that penetrator mass be conserved; this, together with further assumptions about energies dissipated in plugging target shearing (and perhaps otherwise distributed), facilitates statements of mass or momentum balance—that is—if the mass, M, is conserved then post-perforation penetrator momentum and energy are available as MV_r and $MV_r^2/2$ respectively, and, preimpact terms being apparent, one easily concocts the framework for a "theoretical" penetration model. The assumptions typically made, penetrator rigidity and otherwise, are simply not realistic in general and are almost never tenable in situations that tend to involve relatively thick target plates. For example, in a set of 38 firings involving 65-g rods fired into 1-in. and $1\frac{1}{2}$-in. plates at various obliquities and speeds [Lambert (1978a)] the residual mass of recovered rods was less than half of the original mass. It is probably more realistic to regard residual mass as an increasing function of striking velocity, rather than to avow its constant equality to striking mass.

Taking an interpretation that considers the totality of momenta of behind-target ejecta to approach the preimpact penetrator momentum as striking velocity gets large (i.e., supposing target absorption of momentum to become negligible), it is possible to write

$$M_rV_r + \sum_{i=1}^{n} M_iV_i \to MV_s \qquad \text{as } V_s \to \infty \tag{8}$$

where $\{M_i\}_{i=1}^n$ is the collection of debris masses (target and penetrator fragments) having respective speeds $\{V_i\}_{i=1}^n$. Generally, we have no real handle on most of the terms in (8). Let $M' = \rho\pi D^3 z/4$. Two basic assumptions will be made concerning limiting behavior as $V_s \to \infty$.

1 $\sum_{i=1}^{n} M_iV_i = hM'V_r$, h constant. M' is approximately (exactly in the case of normal impact) the mass of target material projected in front of the

penetrator (along its line of flight) prior to impact. $\sum_{i=1}^{n} M_i V_i$ is set proportional to the momentum of a hypothetical "plug" of mass M' having speed V_r. In traditional "plugging models" a term such as $M'V_r$ is frequently used to account for the nonpenetrator portion of residual momentum.

2 $M_r/M \to 1$. This is a considerably weaker assumption than the frequently asserted but faulty statement that $M_r = M$ for all striking velocities.

Incorporating assumptions 1 and 2 in (8) yields:

$$MV_s - hM'V_r - MV_r \to 0 \qquad \text{as } V_s \to \infty$$

or

$$\frac{V_r}{V_s} \to \frac{M}{hM' + M} \qquad \text{as } V_s \to \infty \tag{9}$$

But the right hand of (9) is then the slope of the asymptote to the V_s, V_r curve, the parameter a of (1). Empirical evidence suggests a value of $\frac{1}{3}$ for h. Thus

$$a = \frac{M}{M + M'/3} \tag{10}$$

Finally, there remains the parameter p. Traditional rigid-penetrator theory almost invariably leads to a model corresponding to form (1) with p fixed at 2, and such models have been highly successful in instances where the penetrator suffers little or no deformation. Lambert therefore requires p to be sensitive to (and an increasing function of) effective target thickness in calibers (i.e., to $z = T/D(\sec\theta)^{.75}$ and additionally to be near 2 for "thin" plates, having a limiting value of 2 as $T \to 0$. A simple appropriately behaved function that was found to be effective is:

$$p = 2 + \frac{z}{3}. \tag{11}$$

2.0 THE MODEL

In summary, collect ingredients and restate the predictive model. Let

$$z = \frac{T}{D}\sec^{.75}\theta$$

$$f(z) = z + e^{-z} - 1 = \sum_{j=2}^{\infty} \frac{(-z)^j}{j!},$$

$$M' = \rho\pi D^3 z/4 \ (\rho = 7.8 \text{ for RHA targets}),$$

$$a = \frac{M}{M + M'/3}$$

$$p = 2 + z/3$$

and $$V_l = u\left[\frac{L}{D}\right]^{.15}\sqrt{f(z)\frac{D^3}{M}}$$

where $u = 4000$ for RHA targets. Then

$$V_r\begin{cases} 0, 0 \leqslant V_s \leqslant V_l \\ a(V_s^p - V^p)^{1/p}, V_s > V_l \end{cases}$$

Remarks

1 For the case of penetration into aluminum target plates Lambert tentatively suggests using these equations with $u = 1750$ and $\rho = 2.74$. It is expected that the equations presented here will be useful for characterizing penetration into other target materials—other (empirical) values for u will need to be determined from experimental evidence and the appropriate density, ρ, inserted.

2 With little assurance (because of little experience to date) Lambert also proposes that these formulae might be used in representing the penetration of fragments. Supposing the material density of the fragment to be ρ(g/cc) and its presented area (the known or expected area projected by the fragment along its line of flight onto the target surface at impact) to be A (cm^2), let $D = 2\sqrt{A/\pi}$ and $L = \frac{M}{A\rho}$, and predict according to the preceding equations.

APPENDIX B: PARAMETERS AFFECTING THE BALLISTIC LIMIT

The ballistic limit for projectiles that penetrate solely on the basis of their kinetic energy is affected by many parameters. Among these are target and projectile hardness, density, and yield strength; projectile geometry (length, diameter, nose shape), target thickness, projectile yaw, and target obliquity. One of the most comprehensive investigations of these effects has been conducted by Grabarek (1973). His principal findings are summarized in this appendix.

1.0 EFFECTS OF MATERIAL HARDNESS ON BALLISTIC LIMIT

A series of tests were conducted against finite-thickness and effectively semiinfinite target plates to assess hardness effects. Table B1 lists the geometric and material parameters for the finite-thickness plate tests.

Figure B1 is a flash radiograph showing the rod before impact, during penetration, and after perforation of the target. The soft rod (BHN* 200)

*BHN—Brinnel Hardness Number.

Table B1 Test Conditions

Rod Characteristics
Material—AISI 1090 *Steel*, BHN200, 285, 600
Weight—7.78 g
Geometry—cylindrical, L/D=20
Nose Shape—50° *Conical Nose*

Target Characteristics
Material—Rolled-Homogeneous Armor (RHA), BHN 280
Thickness—6.35 mm

Initial Conditions
Striking Velocity—1052 ms
Impact Obliquity—0° (*Normal Incidence*)

remains intact after target perforation but suffers considerable plastic deformation at the impact end while losing approximately 7% of its initial mass during penetration. The intermediate hardness rod (BHN 285) suffers less plastic deformation but is more susceptible to fracture and loses 28% of its mass. The hard BHN 600 rod suffers even less deformation than the others but breaks up during penetration. The two largest pieces constitute approximately 66% of total initial rod weight. It can be seen that for approximately equal rod striking velocities, a 29% gain is obtained in the residual rod velocity by increasing hardness from 200 to 600 BHN, but this is achieved at the expense of rod integrity.

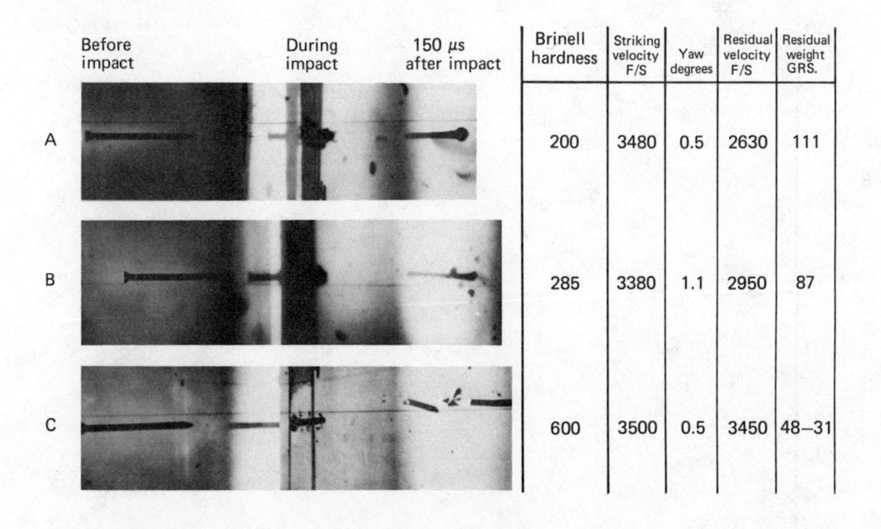

	Brinell hardness	Striking velocity F/S	Yaw degrees	Residual velocity F/S	Residual weight GRS.
A	200	3480	0.5	2630	111
B	285	3380	1.1	2950	87
C	600	3500	0.5	3450	48–31

120–Grain 1090 steel rod (L/D = 20) of different hardness atacking $\frac{1}{4}$ in. RH steel armor (BHN 280 ⊥ ± 20) at 0 degree obliquity

Figure B1 Radiograph of projectile penetration.

Similar tests at fixed striking velocity were performed with thick targets (AISI 6150 steel alloy, BHN 190, 76mm thick). Rods were of the same material and dimensions as in Table B1, but now hardened to BHN 220, 375, and 600. The nominal striking velocity was 625 m/s. The BHN 220 and 375 rods deformed significantly on impact with only slight penetration of the target (1.52 and 3.3 mm respectively). However, the BHN 600 rod did not deform

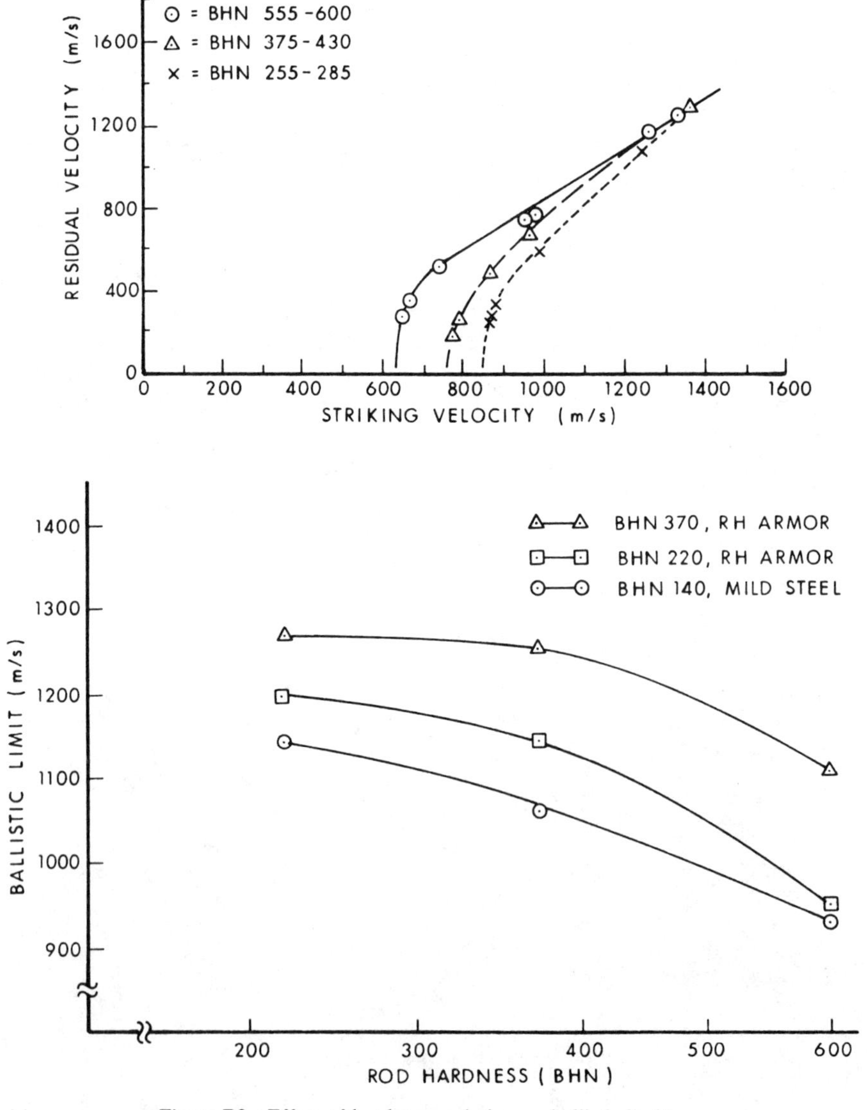

Figure B2 Effect of hardness variation on ballistic limit.

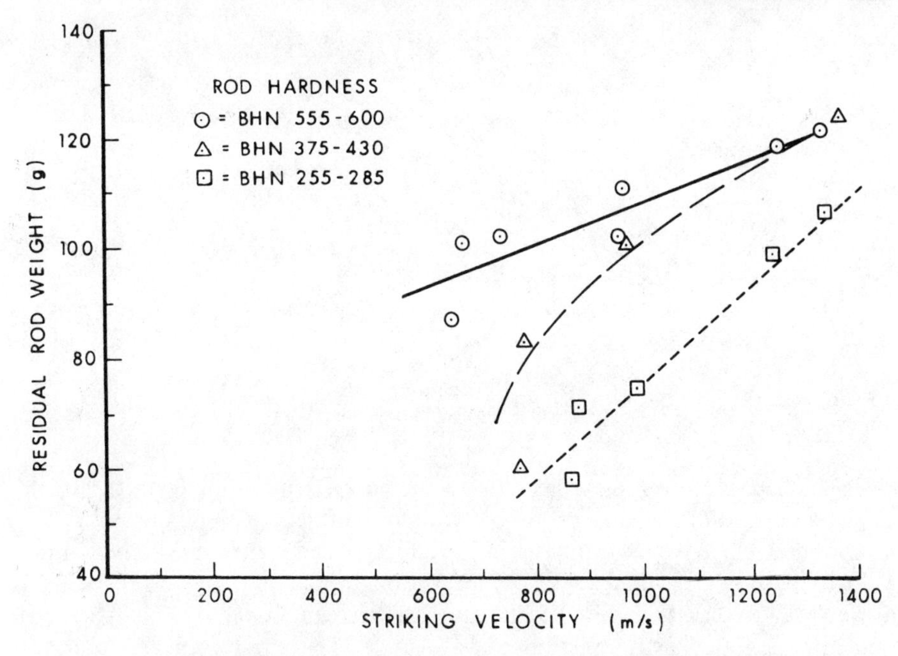

Figure B3 Effect of hardness on residual weight of projectile.

during penetration and was stopped at a depth of 19.81 mm. This penetration depth is 13 times greater than that achieved by the soft BHN 220 rod.

The effects of hardness variation on ballistic limit can be seen in Figure B2. Rods of AISI 1066 steel with hardness of BHN 255–285, BHN 375–430, and BHN 555–600 were fixed at 6.35 mm of rolled-homogeneous armor, BHN 380, at normal incidence. The rods had a length-to-diameter ratio of 10, a diameter of 5.6 mm and 50° conical nose. The tests served to measure target resistance to penetration as a function of the rod hardness over a range of striking velocities from 635 to 1375 m/s.

Figure B2 shows that a 25% reduction in ballistic limit can be achieved by increasing the rod hardness from BHN 255–285 to BHN 555–600. However, for striking velocities above 1200 m/s, rod hardness has little effect on the residual velocity. None of the rods shattered in these firings but their residual weight after plate perforation (W_R) showed a significant dependence on rod hardness. Figure B3 indicates an increase in residual rod weight as impact speed and hardness increase.

2.0 EFFECTS OF STRIKING YAW

Most projectiles arrive at the target with some perturbation of the flight path (Figure B4). Small angles of projectile yaw have little effect on the projectile's ability to penetrate a target, especially in situations where the target is grossly

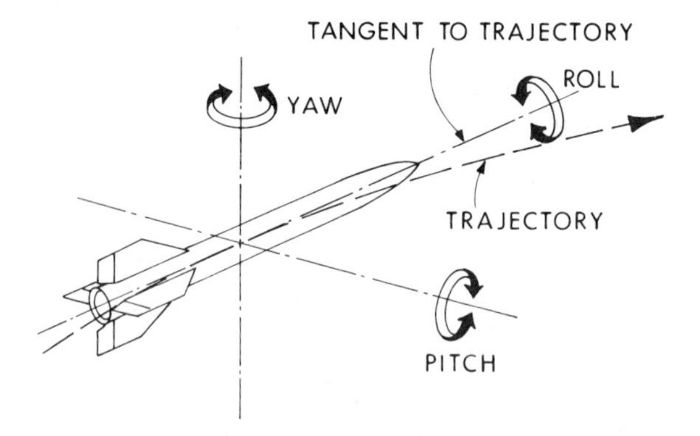

Figure B4 Projectile trajectory.

overmatched. However, as projectile yaw increases, penetration of thick targets is degraded and deformation and breakup of the projectile is enhanced.

Grabarek (1973) made estimates of the effects of rod yaw on the ballistic-limit velocity based on residual velocity data for long rods impacting armor plate at various yaw angles. In Figure B5, the estimated percent increase in ballistic-limit velocity is plotted as a function of yaw angle. The curve indicates that for normal incidence encounters with yaw angles up to about 3° there is less than a 1% increase in ballistic-limit velocity. For greater yaw angles however, the increase in the ballistic limit becomes significant. For high obliquity encounters, critical yaw angles may be less than 1°.

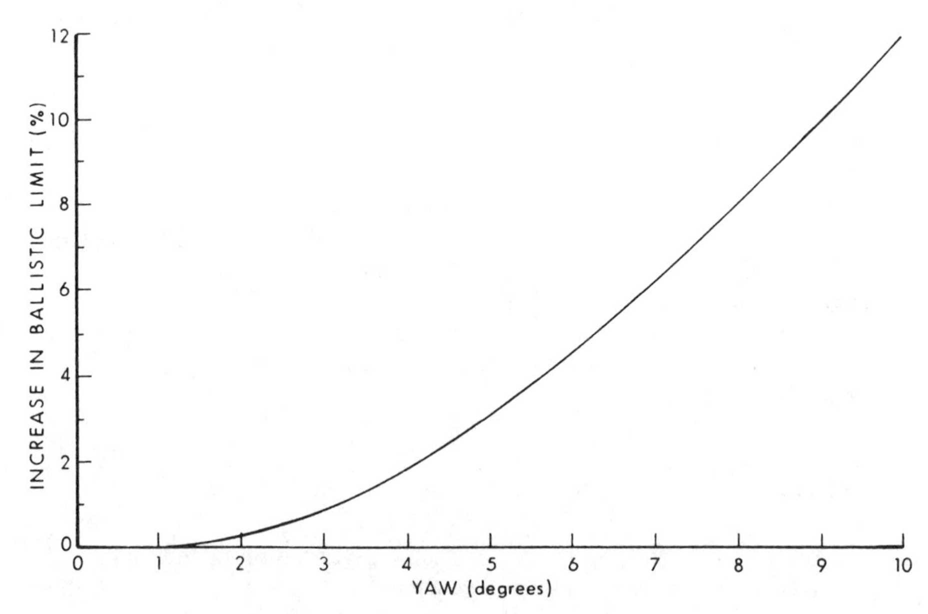

Figure B5 Estimated effect of striking yaw on ballistic limit.

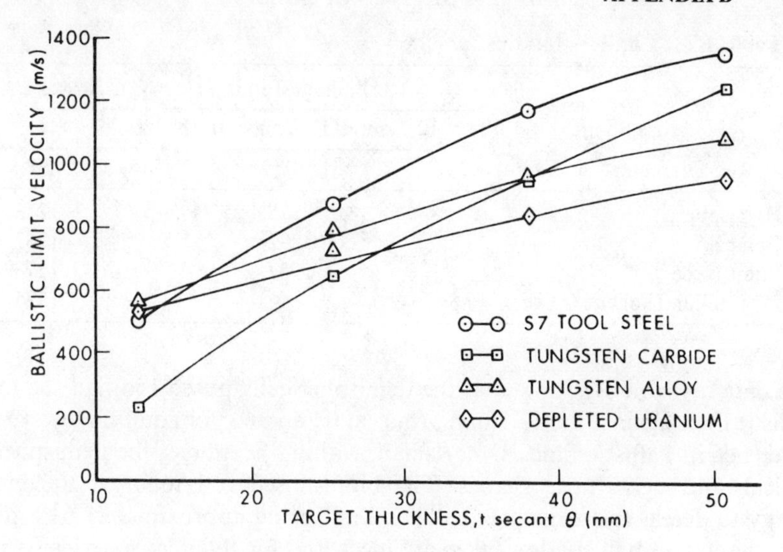

Figure B6 Effect of penetrator material on ballistic limit.

3.0 DENSITY EFFECTS

Material density has a significant effect on ballistic limit. Figure B6 shows results of a study using 65-g rods, L/D of 10, with hemispherical nose shape made of steel, tungsten carbide (WC), tungsten alloy ($W2$), and depleted uranium (DU) striking steel plates. The ballistic limit velocity is plotted as a function of line-of-sight target thickness. It is seen that for the higher striking velocities against thick target plates the higher density projectiles are more effective performers. However, at the lower striking velocities against thinner plates, the tungsten carbide and steel rods perform better. Rod hardness and compressive strength are the dominant properties here that lead to improved penetration capability with minimal rod deformation and breakup.

4.0 EFFECTS OF NOSE SHAPE

The nose shape of a projectile plays an important part in penetration of targets at velocities below the dynamic yield strength of the rod material; the more blunt the nose shape, the higher the ballistic-limit velocity. At target impact velocities that exceeds the dynamic yield strength of the rod material, the nose shape has a negligible effect on the ballistic limit velocity.

Fixed weight, 14.6 g, L/D of 10, steel rods of BHN 555, having nose shapes of a hemisphere, conical (40° angle), blunt, and 2.1-caliber tangent ogive were tested against 12.7 mm RH armor 380 at 0° and 60° obliquity. The results are given in Table B2.

Table B2 Limit Velocities

Nose Shape (Rod Diameter=6.35 mm)	Ballistic-Limit Velocity, m/s — 12.7 mm RH Armor, BHN 380	
	0°	60°
Hemisphere	875	1213
Conical	892	1262
Blunt Nose	942	1273
2.1-Caliber Tangent Ogive	—	1225

The data in Table B2 show that the hemispherically nosed rod had the lowest ballistic limit against the 12.7 mm armor at 0° and 60° obliquity. However, the difference in ballistic limit is very small, within 4%, among the hemispherical, conical, and ogive nose shapes. The blunt-nosed rod required the highest energy to defeat the target; its ballistic limit being approximately $6\frac{1}{2}\%$ higher than the lowest ballistic-limit velocity measured for the hemispherically nosed rod.

5.0 EFFECTS OF LENGTH-TO-DIAMETER RATIO ON BALLISTIC LIMIT

Grabarek (1973) conducted tests with steel and tungsten rods with two L/D ratios striking RHA. The rods had a fixed weight of 65 g, L/D ratios of 5 and

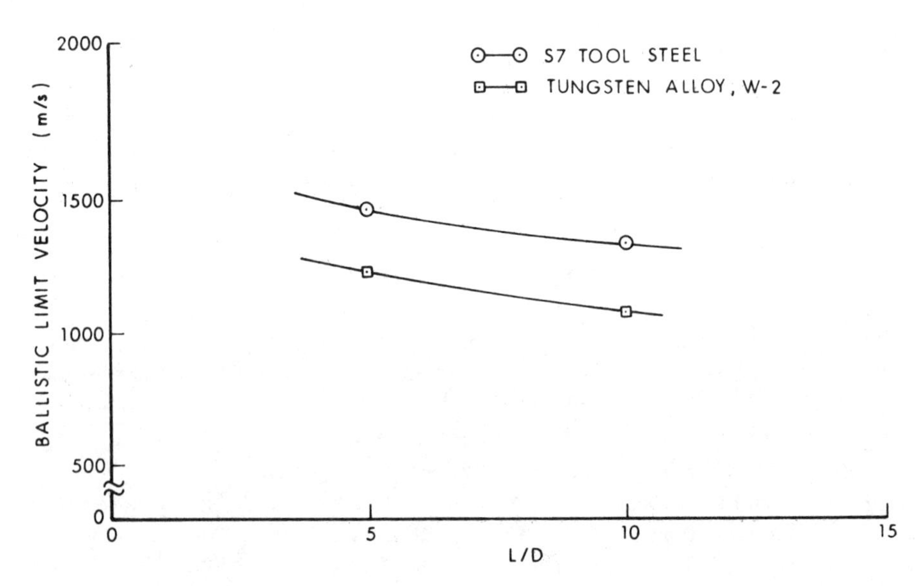

Figure B7 Effect of length to diameter and density on ballistic limit.

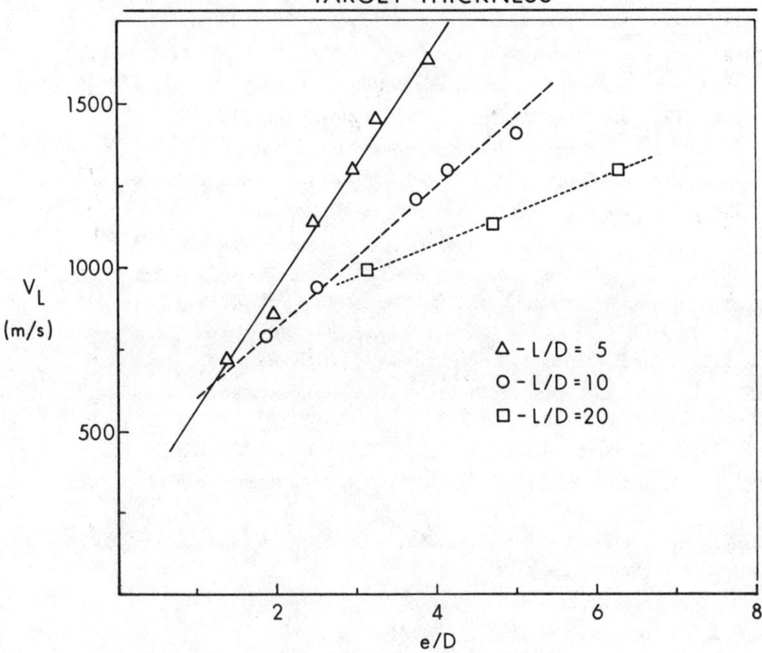

Figure B8 Effect of target thickness on ballistic limit.

10, and a hemispherical nose shape. The materials were S7 tool steel, BHN 380 and tungsten alloy ($W2$), BHN 290. The target plate was 2.54=cm thick, BHN 380 and was set at 60° obliquity.

Figure B7 shows the effects of both density and L/D on ballistic limit. The high-density (18.6 gm/cc) tungsten rods, when compared to the same weight and L/D of steel (7.8 gm/cc) resulted in an 18% lower ballistic-limit velocity. The L/D effect on ballistic limit is quite evident for both materials tested. For the tungsten rod there is a 13% reduction in V_l in changing the L/D from 5 to 10. For the steel rod, a reduction of 8% in V_l is obtained. Similar results from a much larger data base, involving steel, hemispherically nosed rods striking RHA plates of various thicknesses and obliquities (0°, 45°, and 60°) have been obtained by Lambert et al. (1976) and Lambert (1978a). Figure B8 shows the variation of ballistic limit velocity with normalized target thicknesses. The parameter e/D accounts for thickness (T) and obliquity (θ) and is given by

$$\frac{e}{D}=\frac{T}{D}\left(\frac{1+2\sec\theta}{3}\right)$$

REFERENCES

_____ (1955), *Proc. 1st Hypervelocity and Impact Effects Symp.*, Santa Monica, CA.

_____ (1957), *Proc. 2nd Hypervelocity and Impact Effects Symp.*, Washington, D.C.

_____ (1959), *Proc. 3rd Symp. on Hypervelocity Impact*, Chicago, IL.

_____ (1960), *Proc. 4th Symp. on Hypervelocity Impact*, Eglin AFB, FL.

_____ (1961), *Proc. 5th Symp. on Hypervelocity Impact*, Denver, CO.

_____ (1963), *Proc. 6th Symp. on Hypervelocity Impact*, Cleveland, OH.

_____ (1965), *Proc. 7th Hypervelocity Impact Symp.*, Tampa, FL.

_____ (1969), *Proc. AIAA Hypervelocity Impact Conference*, Cincinnati, OH.

_____ (1977), *Penetration Equations Handbook for Kinetic Energy Penetrators*, Joint Technical Coordinating Group for Munitions Effectiveness, 61 JTCG/ME-77-16.

Arajs, V. (1971), *An Investigation of Forces on a Projectile During Perforation of Thin Alumin Plates*, M.S. Thesis, Air Force Institute of Technology, Wright-Patterson AFB, Ohio.

Arbuckle, A. L., Herr, E. L. and Ricchiazzi, A. J., Ballistic Research Laboratory, BRL-MR-2264 (AD908362L).

Awerbuch, J. (1970), Technion-Israel Institute of Technology, MED Report No. 28.

Awerbuch, J., and Bodner, R. S. (1973), Technion-Israel Institute of Technology, MED Report No. 41.

Awerbuch, J., and Bodner, S. R. (1974a), *Int. J. Solids Struct.* **10**, 671; also (1973) Technion-Israel Institute of Technology, MED Report No. 40.

Awerbuch, J., and Bodner, S. R. (1974b), *Int. J. Solids Struct.*, **10**, 685.

Awerbuch, J., and Bodner, S. R. (1977), *Exp. Mech.* **17**, 147.

Backman, M. E. (1976), Naval Weapons Center, NWC TP 578.

Backman, M. E., and Finnegan, S. A. (1973), in R. W. Rhode et al. (Eds.), *Metallurgical Effects at High Strain Rates*, Plenum Press, New York.

Backman, M. E., and Finnegan, S. A. (1976), Naval Weapons Center, NWC TP 5844 (AD A030268).

Backman, M. E. and Goldsmith, W. (1978), *Int. J. Eng. Sci.*, **16**, 1.

Baker, J. R. (1969), Naval Research Laboratory, NRL 6920.

Baker, W. E., Westine, P. S., and Dodge, F. T. (1973), *Similarity Methods in Engineering Dynamics*, Hayden Book Co., Rochelle Park, N.J.

Baker, W. E. (1976), *Shock and Vibration Digest*, **7**, No. 7, 107.

Bedford, A. J., Wingrove, A. L., and Thompson, K. R. L. (1974), *J. Aust. Inst. Metals*, **19**, 61.

Bethe, H. A. (1941), Ordnance Lab., Frankford Arsenal, unnumbered report.

Bluhm, J. I. (1956), *Proc. Soc. Exp. Stress Anal.*, **13**, 167.

Brooks, P. N. (1973), Defense Research Establishment Valcartier, DREV R-686/73.

Brooks, P. N. (1974), Defense Research Establishment Valcartier, DREV 4001/74.

Brown, A. (1964), *Int. J. Mech. Sci.*, **6**, 257.

Brunton, J. H. and Rochester, M. C. (1979), in C. M. Preece (Ed.), *Treatise on Materials Science and Technology Vol. 16: Erosion*, Academic Press, New York.

Burch, G. T., Jr., and Avery, J. G. (1970), Air Force Flight Dynamics Laboratory, AFFDL-TR-70-115.

Byrnside, N. C., Torvik, P. J., and Swift, H. F. (1972), *J. Basic Eng., Trans. ASME*, **94**, 397.

Calder, C. A., and Goldsmith, W. (1971), *Int. J. Solids Struct.* **7**, 863.

Chou, P. C. (1961), in J. E. Lay, and L. E. Malvern, (Eds.), *Developments in Mechanics*, North-Holland, Amsterdam.

Daneshi, G. H., and Johnson, W. (1977a), *Int. J. Mech. Sci.*, **19**, 555.

Daneshi, G. H., and Johnson, W. (1977b), *Int. J. Mech. Sci.*, **19**, 661.

Daneshi, G. H., and Johnson, W. (1978), *Int. J. Mech. Sci.*, **20**, 255.

Davison, L., and Graham, R. A. (1979), *Phys. Re.*, **55**, 257.

Dehn, J. (1979), Ballistic Research Laboratory, ARBRL-TR-02188.

Dunn, W. P., and Huang, Y. K. (1972), Watervliet Arsenal, WVT 7267 (AD 756403).

Florence, A. L., and Ahrens, T. J. (1967), Army Materials Research Agency, AMRA-CR-67-05 (F).

Florence, A. L. (1969), Army Materials and Mechanics Research Center, AMMRC-CR-69-15.

Freiberger, W. (1952), *Proc. Comb. Phil. Soc.*, **48**, 135.

Fugelso, L. E., Arentz, A. A., Jr., and Poczatek, J. J. (1961), American Machine & Foundry Co., Mechanics Research Div., MRD 1127 (AD 272947), Vol. 1.

Fugelso, L. E., and Arentz, A. A. Jr., (1962), General American Transportation Corp., MRD 1127 (AD 421590), Vol. 2.

Fugelso, L. E. (1964), General American Transportation Corp., unnumbered report.

Gabbert, R. D. (1970), M.S. Thesis, Air Force Institute of Technology, Wright-Patterson AFB, Ohio.

Giere, A. C. (1964), *AIAA J.*, **2**, 1471.

Glenn, L. A., and Janach, W. (1977), *Int. J. Fract.*, **13**, 301.

Goldsmith, W. (1960), *Impact*, Edward Arnold, London.

Goldsmith, W. (1963), *Appl. Mech. Revs.*, **16**, 855.

Goldsmith, W., Liu, T. W., and Chulai, S. (1965), *Exp. Mech.* **5**, 385.

Goldsmith, W. (1967), in K. Vollrath and G. Thomer (Eds.), *Kurtzzeitphysik*, Springer-Verlag, New York.

Goldsmith, W., and Finnegan, S. A. (1971), *Int. J. Mech. Sci.*, **13**, 843.

Goldsmith, W. (1973), in *Joint Technical Coordinating Group for Munitions Effectiveness, JTCG/ME Working Party for KE Penetrators, Information Exchange Meeting*, Aberdeen Proving Ground, MD.

Goodier, J. (1975), in *Proc. 7th Hypervelocity Impact Symp.* Tampa, Florida.

Gordon, P. F. (1973), Frankford Arsenal, M73-6-1.

Grabarek, C., and Herr, L. (1966), Ballistic Research Laboratory, BRL-TN-1634 (AD 807619).

Grabarek, C. (1973) in *Joint Technical Coordinating Group for Munitions Effectiveness, JTCG/ME Working Party for KE Penetrators, Information Exchange Meeting*, 13–14 February 1973 at U.S.A. Ballistic Research Laboratories, Aberdeen Proving Ground, MD.

Gupta, A. D. and Misey, J. J. (1980), AIAA Paper 80-0403, presented at the AIAA 18th Aerospace Sciences Meeting, 14–16 January 1980, Pasadena, CA.

Haskell, D. F. (1972), Ballistic Research Laboratory, BRL-MR-2248.

Hauver, G. E. (1978), *Int. J. Eng. Sci.*, **16**, 871.

Hauver, G. E. (1980), in *Proc. 5th Int. Symp. on Ballistics*, Toulouse, France.

Herr, L, and Grabarek, C. (1978), Ballistic Research Laboratory, ARBRL-MR-02860 (ADA062101).

Herrmann, W., and Jones, A. H. (1961), Massachusetts Institute of Technology, Aeroelastic and Structures Research Laboratory, ASRL 99-1.

Heyda, J. F. (1970), General Electric Co., Tech. Memo. Rept. TM 70-002.

Hopkins, H. G., and Kolsky, H. (1960), in *Proc. 4th Symp. on Hypervelocity Impact*, Eglin AFB, FL.

Huszarik, F. A., Reichman, J. M., and Cheung, J. B. (1979), in W. F. Adler, (Ed.), *Erosion: Prevention and Useful Applications*, American Society for Testing Materials, ASTM STP 664.

Hutchings, I. M. (1976), *Int. J. Mech. Sci.*, **18**, 243.

Ipson, T. W., Recht, R. F., and Schmeling, W. A. (1973), Naval Weapons Center, NWC TP5607.

Ipson, T. W., and Recht, R. F. (1975), *Exp. Mech.*, **15**, 249.

Johnson, W. (1972), *Impact Strength of Materials*, Crane, Russak, New York.

Johnson, W., and Reid, S. R. (1975), *J. Mech. Eng. Sci.*, **17**, 71.

Johnson, W. (1979), in *Mechanical Properties at High Rates of Strain*, Conference Series No. 47, Institute of Physics, London.

Jonas, G. H., and Zukas, J. A. (1978), *Int. J. Eng. Sci.*, **16**, 879.

Jones, N. (1975), *Shock and Vibration Digest*, **7**, No. 8, 89.

Jones, N. (1978a), *Shock and Vibration Digest*, **10**, No. 9, 21.

Jones, N. (1978b), *Shock and Vibration Digest*, **10**, No. 10, 13.

Jones, N. (1979), in *Mechanical Properties at High Rates of Strain*, Conference Series No. 47, Institute of Physics, London.

Kennedy, R. P. (1976), *Nuc. Eng. and Des.*, **37**, 183.

Kinslow, R., Ed. (1970), *High Velocity Impact Phenomena*, Academic Press, New York.

Kornhauser, M. (1964), *Structural Effects of Impact*, Spartan Books, Baltimore.

Kowalski, S. J., Kolodziej, J. A., and Raniecki, B. (1976), *Nuc. Eng. and Des.*, **37**, 225.

Krafft, J. M. (1955), *J. Appl. Phys.*, **26**, 1248.

Kucher, V. (1967a), Ballistic Research Laboratory, BRL-R-1379 (AD 823537).

Kucher, V. (1967b), Ballistic Research Laboratory, BRL-R-1384 (AD 664138).

Kumari, S. (1975), *Int. J. Mech. Sci.*, **17**, 23.

Lambert, J. P., and Jonas, G. H. (1976), Ballistic Research Laboratory, BRL-R-1852 (ADA021389).

Lambert, J. P. (1978a), Ballistic Research Laboratory, ARBRL-TR-02072.

Lambert, J. P. (1978b), Ballistic Research Laboratory, ARBRL-MR-02828 (AD B 027660L).

Lambert, J. P., and Ringers, B. E. (1978), Ballistic Research Laboratory, ARBRL-TR-02066.

Lascher, F. R., Henderson, D. and Maynard D. (1975), Avco Systems Division, Wilmington, MA, AVSD-0306-75-RR.

Leone, S. G. (1975), Air Force Flight Dynamics Laboratory, AFFDL-TR-75-18.

Lethaby, J. N., and Skidmore, I. C. (1974), in *Mechanical Properties at High Rates of Strain*, Conference Series No. 21, The Institute of Physics, London.

Letho, D. L. (1972) Naval Ordnance Laboratory, NOLTR-72-274 (AD 755429).

Levy, S., and Wilkinson, J. P. D. (1976), *The Component Element Method in Dynamics*, McGraw-Hill, New York.

Mach, H., Masur, H., Muller, H., and Werner, U. (1973), Deutsch-Franzosischer Forschungsinstitut Saint-Louis, Report 32/73.

Mach, H. (1975), Bundesministerium der Verteidigung, BMVg-FBWT-75-7.

Masket, A. V. (1949), *J. Appl. Phys.*, **20**, 132.

Minnich, H. R., and Davids, N. (1964), Penn. State University Dept. of Eng. Mech Interim Tech. Rep. No. 3.

Misey, J. J., Gupta, A. D., and Wortman, J. D. (1980), in J. R. Vinson (Ed.), *Emerging Technologies in Aerospace Structures, Design, Structural Dynamics and Materials*, ASME, New York.

Moss, G. L. (1980), Ballistic Research Laboratory, Technical Report ARBRL-TR-02242.

Netherwood, P. H. (1979), Ballistic Research Laboratory, ARBRL-MR-02978.

Nicholas, T. (1967), Air Force Materials Laboratory, AFML-TR-67-208 (AD 820356).

Nishiwaki, J. (1951), *J. Phys. Soc. (Japan)*, **6**, 374.

Paul, B., and Zaid, M. (1958), *J. Franklin Inst.*, **265**, 317.

Persson, A. (1974), in *Proc. 1st Int. Symp. on Ballistics*, Orlando, FL.

Pitek, M., and Hammitt, G. F. (1966), University of Michigan, Nuclear Eng. Dept., Technical Report No. 1 (AD 803278).

Pritchard, D. (1980), Ballistic Research Laboratory, private communication.

Pytel, A., and Davids, N. (1963), *J. Franklin Inst.*, **276**, 394.

Rawlings, B. (1974), in *Mechanical Properties at High Rates of Strain*, Conference Series No. 21, Institute of Physics, London.

Recht, R. F., and Ipson, T. W. (1963), *J. Appl. Mech., Trans. ASME*, **30**, 384.

Recht, R. F. (1964), *J. Appl. Mech., Trans. ASME*, **31**, 189.

Recht, R. F. (1970), in *I. Mech. E., 3rd Int. Conf. on High Pressure*, University of Denver, Denver, CO.

Recht, R. F. (1973), in *Joint Technical Coordinating Group for Munitions Effectiveness, JTCG/ME Working Party for KE Penetrators, Information Exchange Meeting*, 13–14 February 1973 at U.S.A. Ballistic Research Laboratories, Aberdeen Proving Ground, MD.

Recht, R. F., and Ipson, T. W. (1973), Denver Research Institute, DRI 2025 (AD 274128).

Recht, R. F. (1976), in *Workshop on Mechanics of Impact and Penetration*, Ballistic Research Laboratory.

Recht, R. F. (1978), *Int. J. Eng. Sci.*, **16**, 809.

Robertson, H. P. (1943), National Defense Research Council, Armor and Ordnance Report A-227, OSRD No. 2043.

Rogers, H. C. (1974), Drexel University, unnumbered report for U.S. Army Research Office.

Rogers, H. C. (1979), *Ann. Rev. Mater. Sci.*, **9**, 283.

Ross, C. A., Strickland, W. S., and Sierakowski, R. L. (1977), *Shock and Vibration Digest*, **9**, No. 12, 15.

Simpson, R. A. (1972), Honeywell Govt. & Aeronautical Products Div., unnumbered report.

Sliter, G. E. (1980), *J. Struct. Div., ASCE*, **106**, 1023.

Soliman, A. S., Reid, S. R., and Johnson, W. (1976), *Int. J. Mech. Sci.*, **18**, 279.

Stock, T. A. C., and Thompson, K. R. L. (1970), *Metal. Trans.*, **1**, 219.

Summers, D. A. (1979), in C. M. Preece (Ed.), *Treatise on Materials, Science and Technology, Vol. 16: Erosion*, Academic Press, New York.

Swift, H. (1972), in P. C. Chou, and A. K. Hopkins, (Eds.), *Dynamic Response of Materials to Intense Impulsive Loading*, U.S. Govt. Printing Office, Washington, D.C.

Tate, A. (1967a), Royal Armament Research & Development Establishment, RARDE Memorandum 46/67 (AD 824293).

Tate, A. (1967b), *J. Mech. Phys. Solids*, **15**, 387.

Tate, A. (1969), *J. Mech. Phys. Solids*, **17**, 141.

Tate, A. (1977), *Int. J. Mech. Sci.*, **19**, 121.

Tate, A. (1980), *J. Phys. D: Appl. Phys.*, **12**, 1825.

Taylor, G. I. (1947), *Quart. J. Mech. and Appl. Math.*, **1**, 103.

Thompson, W. T. (1955), *J. Appl. Phys.*, **26**, 80.

Venable, D., and Boyd, T. J., Jr., (1965), in *Proc. 4th Int. Symp. on Detonation*, 12–15 Oct. 1965, Naval Ordnance Laboratory, White Oak, MD., ACR-126, U.S. Govt. Printing Office.

Vinson, J. R. (1968), University of Delaware, Dept. of Mechanical & Aerospace Eng. Tech. Rep. No. 89.

Weidman, D. J. (1968), *AIAA J.*, **6**, 1622.

Weidman, D. J. (1969), National Aeronautics & Space Administration, NASA-TN-D-5556.

Weirauch, G. (1971), Franco-German Research Institute, Saint-Louis, France, ISL Report 7/71.

Weirauch, G., and Lehr, H. F. (1977), in *3rd. Int. Symp. on Ballistics*, Karlsruhe, Germany.

Wenzel, A. B., and Dean, J. K. (1975), Ballistic Research Laboratory, BRL-CR-262.

Wilms, E. V., and Brooks, P. N. (1966), Canadian Armament Research & Development Establishment, CARDE-TR-551/66.

Wingrove, A. L. (1972), *J. Phys. D: Appl. Phys.* **5**, 1294.

Wingrove, A. L., and Wulff, G. L. (1973), *J. Aust. Inst. Metals*, **18**, 167.

Wingrove, A. L. (1973), *Metal. Trans.*, **4**, 1829.

Woodall, S. R., Heyda, J. F., Galbraith, H. J., and Wilson, L. L. (1970), Air Force Armament Laboratory, AFATL-TR-70-112.

Woodward, R. L., and de Morton, M. E. (1976), *Int. J. Mech. Sci.*, **18**, 119.

Woodward, R. L. (1977), *J. Aust. Inst. of Metals*, **22**, 167.

Woodward, R. L. (1978a), *Int. J. Mech. Sci.*, **20**, 349.

Woodward, R. L. (1978b), *Int. J. Mech. Sci.*, **20**, 599.

Woodward, R. L. (1979), *Metals Technol.*, **13**, 106.

Wright, J. P., and Baron, M. L. (1979), in K. Kawata, and J. Shioiri, (Eds.), *High Velocity Deformation of Solids*, Springer-Verlag, Berlin.

Yellup, G. M., and Woodward, R. L. (1980), *Res Mech.* **1**, 41.

Zaid, A. I. O., El-Kalai, A., and Travis, F. W. (1973), *Int. J. Mech. Sci.*, **15**, 129.

Zaid, A. I. O., and Travis, F. W. (1974), in *Mechanical Properties at High Rates of Strain*, Conference Series No. 21, The Institute of Physics, London.

Zaid, M., and Paul, B. (1957), *J. Franklin Inst.* **264**, 117.

Zaid, M., and Paul, B. (1959), *J. Franklin Inst.*, **266**, 24.

Zener, C., and Holloman, J. (1942), Watertown Arsenal, Report 710/454.

Zukas, J. A., and Jonas, G. H. (1975), AIAA Paper 75-749, presented at the AIAA/ASME/SAE 16th Structures, Structural Dynamics & Materials Conference, Denver, Colorado, 27–29 May 1975.

Zukas, J. A. (1980), in J. R. Vinson, (Ed.), *Emerging Technologies in Aerospace Structures, Design, Structural Dynamics and Materials*, ASME, New York.

6

HYPERVELOCITY IMPACT MECHANICS

Hallock F. Swift

The study of hypervelocity impacts has been conducted during the past three decades with varying degrees of urgency and excitement. The field has made major contributions in the past in antiballistic missile technology, protection of space vehicles from meteoroid impacts, and the study of materials response to ultra-high pressures. More recently, the field is making contributions in evaluating the performance of high-velocity ordnance weapons, and serious speculation is developing about using hypervelocity impacts to produce thermonuclear fusion within reactors.

An interesting point of ambiguity concerning the field is its definition. Originally, absolute velocities were proposed as minimums for the hypervelocity regime. More recently, favor has spread to considering onset of specific impact phenomena for determining the low-velocity limit of the hypervelocity regime. The most common of these is the complete pulverization of the projectile and target material in the immediate region of the original contact point. This definition has gained favor because it allows a significant simplification of the analysis of such impacts. The projectile and local target material may be considered fluids when the stresses induced by the impacts are many times the materials strengths. Thus, the principles of fluid mechanics may be used to at least start hypervelocity impact analyses. This simplification makes hypervelocity impacts rank only behind elastic impacts in their ease of mathematical analysis. As such, study of hypervelocity impacts is an extremely attractive starting point for developing quantitative understandings of more generalized impact processes.

The velocities at which the approximation of negligible projectile and target strength become valid vary widely with projectile and target material combinations. Velocities below 1 km/s will produce such phenomena in wax targets impacted by wax projectiles (and, of course, in fluid target-projectile combinations). Velocities of 1.5 to 2.5 km/s can produce hypervelocity impacts when soft dense metals such as lead, tin, gold, and indium are considered. Velocities of 5 to 6 km/s must be achieved before hypervelocity phenomena can be triggered in typical structural and stony materials such as aluminum, steel, quartz, and so on. Finally, velocities of 8 to 10 km/s and above are required to produce these phenomena in very strong, low-density materials such as berylium and boron metals, and hard ceramics like allumina, boron carbide, and finally diamond.

Fascinating impact situations arise when either the projectile or target material is shocked to pressures where it behaves like a fluid, but the other materials response is still controlled by strength-related phenomena. Here a broad variety of impact phenomenology can be produced where projectiles remain intact and tunnel deeply into soft low-strength targets, or projectiles splash on the surface of targets that either remain undeformed or are subjected to strength-related failures such as cracking and scabbing.

6.1 HYPERVELOCITY PENETRATION MECHANICS

The phenomenology governing hypervelocity penetration mechanics can be understood most effectively by considering:

1. Impacts into very thick targets where the side and rear walls play no significant part in the cratering process.
2. Penetration of intermediate thickness targets where shock reflections from the rear wall affect cratering.
3. Perforation of thin targets where the projectile bursts through the target plate and causes energetic debris to be launched rearward.

6.1.1 Penetration of Thick Targets

A hypervelocity projectile striking a target produces shock waves in the projectile and target materials whose peak stresses are many times any of the materials strengths. The materials in the projectile and target deform according to the laws of fluid mechanics which treat material compressibility and inertia. A universal characteristic of craters produced by "chunky" projectiles (all dimensions are comparable) traveling at hypervelocities is that the penetration is approximately one-half the crater diameter at the original target surface. Obviously the ultimate "chunky" fragment is a sphere, but cubes, cylinders 1-caliber long, blunt cones, and so forth, all meet this criterion and produce craters of similar shape when impacts are in the hypervelocity regime. Another interesting feature of these craters is that the crater volume, V_c, per-unit kinetic energy of the projectile, E_p, is nearly constant for each combination of projectile material and target material.

$$V_c = KE_p \tag{1}$$

Values for the proportionality constant, K, are dependent on both the shock and strength parameters of the projectile and target materials. The value for the constant is usually in the range of 0.5 to 2.0×10^{10} erg/cm^3. Equation (1) can be expanded by evaluating the crater volume under the assumption that the crater has a hemispherical shape and its volume is proportional to projectile kinetic energy to produce the important relationship:

$$\frac{P_c}{D_p} = \left(\frac{\rho_p K}{4}\right)^{1/3} U_p^{2/3} \tag{2}$$

Equation (2) produces the famous result that penetration depth, P_c, for hypervelocity-impact craters varies with the two-thirds power of projectile impact velocity. The absence of any term involving linear dimensions from (2) infers that the sizes of hypervelocity craters scale linearly with projectile dimensions. This scaling dependence has been investigated carefully and found to be nearly accurate, but crater size relative to projectile size increase with projectile diameter to the 0.06 power.

$$\frac{P_c}{D_p} = \left(\frac{\rho_p K}{4}\right)^{1/3} D_p^{.06} U_p^{2/3}; \qquad V_c = KE_p D_p^{.12} \tag{3}$$

This very small deviation from linear-size scaling dependence is unimportant when scaling over factors of 1:5 or 10, but can become important when scaling between micro and macro impacts.

It is instructive to consider the characteristics governing the onset of the hypervelocity regime as impact velocities are increased. Figure 1 shows a series of plots of the ratio between crater depth and diameter (crater shape factor) versus impact velocity for a number of projectile-target materials combinations. A crater shape factor of 0.5 represents classic hypervelocity crater shape.

When a projectile material is much more rigid and dense than target material (as in the case of tungsten-carbide projectiles striking soft aluminum targets), the craters rapidly become deep tunnel-like structures as impact velocities are increased from very low values. This behavior continues as long as the projectile remains a rigid penetrator but it terminates rapidly when impact-induced stresses cause breakup and finally pulverization of the projectile. Once projectile pulverization occurs, the shape of the crater changes rapidly as velocities are increased further toward the approximately hemispherical shape of a hypervelocity crater. The opposite effect occurs when projectile materials are softer, weaker, and less dense than the target material as is the case when plastics impact steel. This situation produces shallow broad craters at low velocities whose depth-to-diameter ratios increase toward 0.5 as the hypervelocity regime is approached. Projectile-target materials combina-

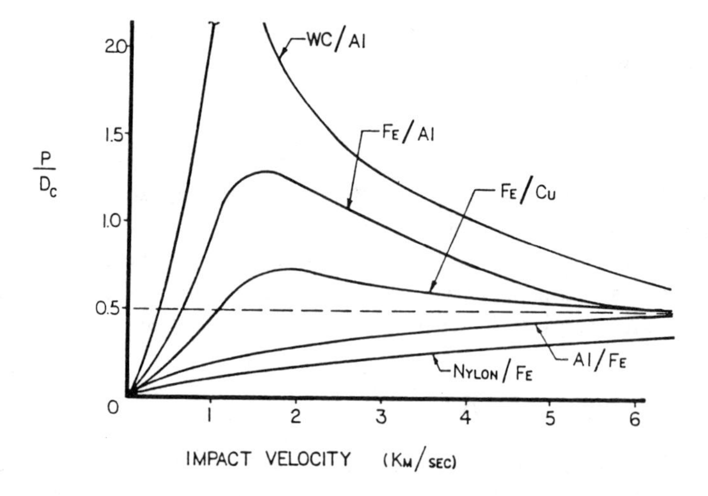

Figure 1 Crater shape factor versus impact velocity for several projectile-target materials combinations.

tions with intermediate strength ratios exhibit intermediate behavior between these two extremes.

The behavior characteristics described above lead to interesting target penetration results as are shown in Figure 2 where penetration normalized to projectile diameter is plotted against impact velocity. For cases where strong dense projectiles strike relatively weak targets of low density, penetration increases rapidly with velocity in the low-velocity regime where the penetrator remains rigid. Projectile breakup and pulverization is marked by a maximum

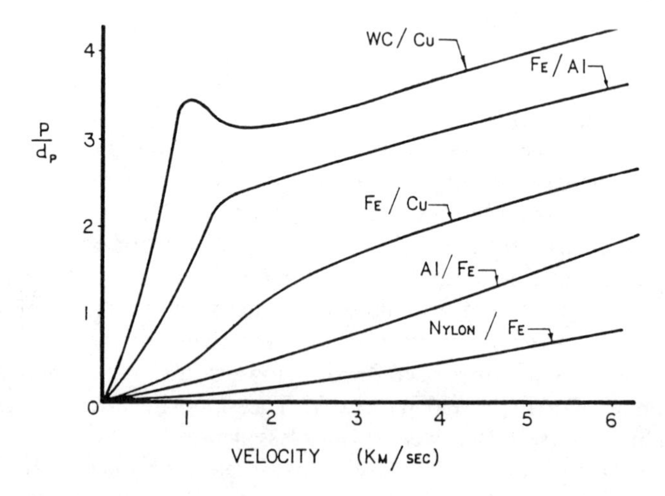

Figure 2 Impact penetration normalized to projectile diameter versus impact velocity for several projectile-target materials combinations.

in the penetration curve followed by a reduction in penetration effectiveness as impact velocity continues to increase into the projectile-pulverization regime. The curve again turns upward as the hypervelocity regime is approached, and increases with a two-thirds power of projectile velocity thereafter. The situation where low density, weak projectiles are launched against relatively strong and dense targets produced craters whose depths increase slowly with velocity at low velocities, and then accelerate to a two-thirds power-velocity dependence as the hypervelocity regime is approached. Target-projectile combinations with materials properties intermediate between these two extremes display intermediate behavior, although they all share the common characteristics that penetration depth increases with the two-thirds power of projectile velocity once the hypervelocity regime has been reached.

Long Rod Penetration. We must consider the case of long rod penetrators impacting thick targets at hypervelocities before we leave thick-plate hypervelocity impact mechanics. These impact situations are extremely important from an engineering viewpoint because long rods are the most efficient penetrators yet invented for attacking thick armor.

Basically the penetration of long rods into targets at hypervelocities occurs in three distinct steps. First, a cavity is produced at the impact site according to mechanics analogous to those that produce craters in targets impacted by "chunky" projectiles. The final crater diameter at the original surface approximates closely to what would have been produced by impact of a "chunky" projectile of the same diameter, material, and velocity as those of the rod. A small amount of material at the front end of the rod is consumed during this initial transitory phase. A steady-state situation then develops where rod material feeds onto the floor of the deepening crater and is disrupted and expelled up the crater walls. This steady-state situation is governed almost entirely by hydrodynamic considerations since the pressures produced are much higher than the materials strengths. The velocity of the incoming rod is also high enough to assure that the shock-wave controlling material deceleration adjacent to the crater floor remains fixed relative to the crater floor (a standing shock wave). Rod material behind the shock wave is not affected in any way by the impact process until it crosses the shock and is disrupted almost immediately at the interface with the target material at the crater floor. This factor is of extreme importance when considering the impacts of long, very slender rods because the lack of loading of the rod material behind the shock wave precludes the rod from bending or buckling. Extremely deep tunnel-like structures can thus be developed when very long rods impact very thick targets. The depth of the crater produced by the steady-state rod consumption is governed by the Eichelberger equation:

$$\frac{P_c}{D_p} = \left(\frac{\rho_p}{\rho_t} \right)^{1/2} \tag{4}$$

Perhaps the most important feature of (4) is that crater depth seems to depend only on rod length and the ratio of projectile-target materials densities. Greater depth is specifically independent of impact velocity and all other materials properties.

The steady-state regime is, of course, completed when the rear end of the rod passes through the standing shock wave. A crater termination phase then begins that leads to the formation of the final crater. Strictly speaking (4) only considers the crater depth generated during the steady-state penetration process and a small amount of penetration must be added to account for the initiating and terminating phases of the cratering process.

Figure 3 presents a plot of crater depth normalized to rod diameter as a function of impact velocity for high-strength steel rods impacting thick, steel armor. In this case, penetration rises rapidly with velocity in low-velocity regimes and then penetration depth asymptotes to a value somewhat greater than the value predicted by (4) as velocities are increased. This curve was originally generated to identify the large advantage that could be expected from launching kinetic energy penetrators at velocities up to double current ordnance capabilities. The curve also shows that no further penetration advantage can be anticipated if velocities are increased beyond this value. One approach for developing tank armor to withstand attack from kinetic energy

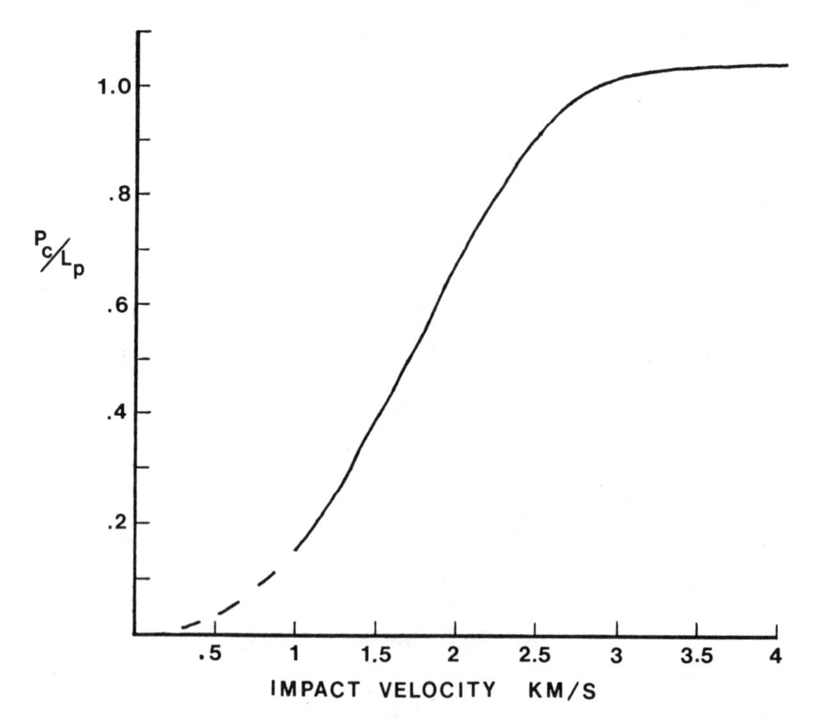

Figure 3 Normalized rod-penetration depth versus impact velocity for long-steel rods impacting thick-steel targets.

projectiles involves taking steps to increase the absolute velocity where the hypervelocity approximations become valid (or in other words, where the curve in Figure 3 approaches its final asymptotic value). Conversely, kinetic-energy round development emphasized technology for lowering this onset velocity.

6.1.2 Penetration of Intermediate-Thickness Targets

The cratering process described in the last section is initiated in an identical manner when the target is of intermediate thickness. Differences develop when the primary shock wave in the target material detaches from the expanding crater surface and reaches the free surface at the rear of the target as is sketched in Figure 4. This shock wave reflects as a tensile wave that propagates back into the target. The tensile wave is generated to satisfy the condition that the instantaneous normal stress at any free surface remains zero at all times. The backward-running tensile wave travels through material compressed by the later stages of the oncoming compression wave, and therefore travels at a velocity somewhat higher than the compressive wave velocity. The instantaneous pressure at any point within the target material being subjected to these waves, is simply the algebraic sum of the stresses produced by the forward-running compression wave and rearward-running tensile wave.

The process is shown schematically in Figure 4. The sketch of the propagating shock-rarefaction wave shows that this algebraic sum can reach negative (or tensile) amplitudes. At any point in the target material where the ultimate-dynamic tensile-yield strength is exceeded, a fracture will be formed (spall plane). These fractures typically occur along sectors of spheres whose center is

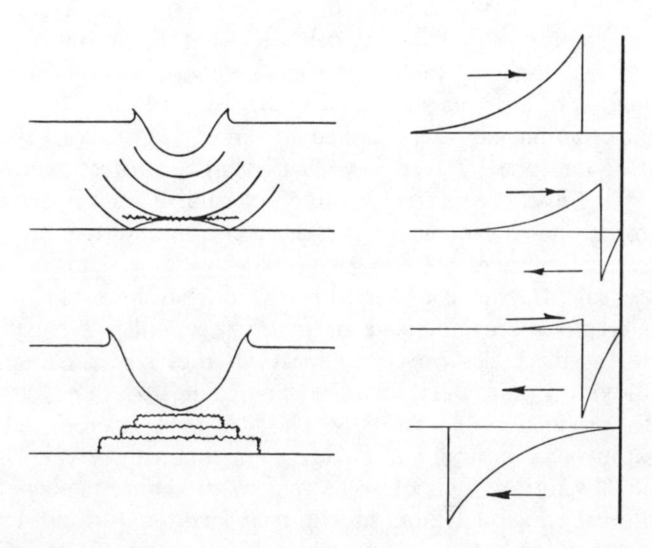

Figure 4 Conceptual sketch of a "chunky" projectile impacting a target of intermediate thickness and a demonstration of tensile wave production on reflection at a free surface.

located along the symmetry axis of the impact at a point one target thickness above the original impact point. Part of the momentum from the shock-rarefaction wave system is trapped in the material beyond the fracture. This momentum produces a rearward velocity in the material beyond the failure that tends to tear the material away and project it rearward. Velocities up to several hundred meters/second of the failed material can be achieved in this manner. Failure of the material forms one or more spall fragments and exposes a new rear surface of the target on which the entire process can be replayed producing a second spallation plane. This process can continue as long as the appropriate stress waves are available and, in general, will produce a pattern of spallation whose total depth approaches one-third of the target thickness as the ballistic limit conditions are approached.

Meanwhile, the crater growth continues normally until the rarefaction waves reflected from the rear surface of the target reach the moving crater floor. Here, they can act to extend total penetration slightly. When the crater reaches approximately two-thirds of the target thickness, its floor meets the uppermost spall plane resulting in target perforation. This situation is widely referred to as the "ballistic limit condition."

The precise target-impact conditions required to just perforate the target have a small but definite statistical component. For this reason, no specific set of target-impact conditions can be specified as the limit for target perforation. This factor has led to the definition of a term that specifies an idealized impact velocity where the probability of perforation is 50%, V_{50}. Target perforation is usually expressed in terms of impact velocity for a given penetrator-target combination but individual studies have used the same type of concept to specify target perforation limits in terms of target thickness, projectile size, and so forth.

Before leaving this subject, let us consider what circumstances constitute perforation of an impacted target. The most rigorous definition of ballistic limit is formation of a permanent bulge at the rear surface of the target. This ballistic-limit criterion has been applied to piping in a space vehicle where constriction of the tube by even a small amount can affect vehicle mission reliability. A slightly more relaxed definition of ballistic limit is detachment of a spall fragment. Such detachment represents a clear hazard to a fluid-handling system where solid fragments can cause blockages, or damage or destroy pumps. These spall fragments projected rearward from the armor of a military vehicle can create internal damage or injure occupants. One ballistic-limit criterion used by the U.S. Army is the perforation of a hard aluminum plate (2024-T3 alloy) 0.5 mm thick. Studies have indicated that this situation approximates production of a casualty. The next most rigorous definition of ballistic limit involves the ability of a target to maintain a gas seal. This limit is used traditionally for the hulls of space vehicles or vehicle tankage that must maintain internal pressure without indefinite expenditure of fluid. Finally, the most relaxed common definition of ballistic limit is that the target will just transmit light. The impacted target is simply held up to a light source and a gleam of light at the base of the crater constitutes perforation.

Long Rod Penetration. The impact mechanics of long rods perforating targets of intermediate thickness differs significantly from that of "chunky" projectiles striking similar targets. Careful reflection about the physics controlling spall formation shows that both the intensity of the primary compression shock wave and rate of its decay control spall formation. Decay rate is quite rapid when the primary wave is produced by the impact of a "chunky" fragment because shock wave termination is produced mainly by rarefaction waves emanating from the rear surface of the incoming projectile. Removal of impact pressure during a rod impact is produced by a completely different and much slower process. Pressure in the target adjacent to the crater floor is nearly constant during the steady-state penetration phase of the cratering process and removal of this stress at the termination of this process takes considerable time. Accordingly, the pressure in the primary shock wave drops slowly, thus inhibiting spall formation. Final crater perforation occurs when either the crater floor intersects the rear surface of the target or when the material between the crater floor and the target rear surface fails in shear—producing a plug.

A case of special interest occurs when the projectile is considerably longer than is required to perforate the target. A fixed amount of material is removed from the nose of the projectile by the initial transient process and material is consumed at an approximately fixed rate as the projectile negotiates the steady-state regime. This regime terminates abruptly when the target is perforated and a variety of material, including the remainder of the projectile, is projected behind the target plate.

The residual length of the projectile can be estimated with some accuracy by estimating the amount removed in the transient process, the amount removed during the steady-state process, and any residual material that is removed as stress within the projectile falls back to low values supportable by the material strength. The velocity of the residual projectile fragment is nearly that of the initial projectile because only a small amount of momentum contained in the elastic portion of the shock wave in the projectile material at the time of perforation is transmitted to the projectile segment after perforation has been achieved. This momentum is always much smaller than the directed momentum of the projectile and, therefore, does not change rod segment velocity significantly.

Much effort has been expended in the recent past defining the debris emanating from the rear surface of armor plate impacted by long rod penetrators because this information is directly usable for determining lethality of kinetic energy penetrators to armored structures such as tanks and bunkers.

6.1.3 Thin-Plate Perforation

Logical extension of our discussion of hypervelocity penetration mechanics requires consideration of the interaction of hypervelocity projectiles with very thin targets which require a small fraction of the projectile kinetic energy to achieve perforation. This area of the penetration mechanics field is extremely important for two reasons. First, impacts of this sort control the operation of dual-sheet hypervelocity armor used widely to protect spacecraft from meteoroid impact damage or destruction. Second, impacts with thin plates provide the only situation known for producing the fluid mechanical processes associated with early stages of all hypervelocity impacts without also producing phenomena associated with viscoplastic and elastic behavior of the materials that tend to mask results from the fluid mechanical phase.

Let us start this discussion by considering the basic physics of chunky projectiles striking thin plates. Stresses in the projectile and target that are computable using Hugoniot relationships are generated during initial contact. Shock waves carrying these stresses propagate rearward into the projectile and forward into the plate until they reach free surfaces where they reflect as tensile disturbances with equal and opposite amplitudes to the original compressive ones. Generally, the first free surface reached is the rear surface of the target. The dual effect of the oncoming compression wave and the rarefaction wave reflected from the rear surface is to project the target material under the projectile rearward at a velocity near that of the original impact. Meanwhile, the primary shock wave propagating rearward into the projectile intercepts the side and rear walls of the projectile. The net effect of this wave motion is to produce a velocity field in the projectile material that causes the great majority of it to follow the target material through the hole formed in the target into an expanding bubble of debris. The remainder of the material is projected rearward into a fine veil that expands conically away from the impact site. Meanwhile, the side walls of the hole in the target expand outward at a rapidly decreasing rate until the final hole diameter of a few times the projectile diameter is achieved.

This impact situation is extremely interesting from a fundamental physics viewpoint because virtually all of the projectile and target material exposed to the primary shock wave is launched into the debris plume in front of and behind the impacted target during periods when the stress levels are much greater than materials strengths. For this reason, we may safely assume that materials behavior is purely fluid-dynamic and that the results of this materials behavior are "stored" as the characteristics of the debris plumes. These characteristics may be measured over relatively long time periods after the impact by a variety of techniques. Thus, hypervelocity impact of thin plates becomes the only known method for studying fluid mechanical behavior of solid materials in the absence of complicating factors.

Meteoroid Shield. Thin plates have found extensive use as sacrificial armor for space vehicles subjected to meteoroid impact hazards. The basic concept involves placing a thin sacrificial plate some distance outboard from the surface to be protected. Incoming meteoroids strike the plate and are disrupted massively as they perforate it. The plume of debris that expands both transversely and longitudinally behind the plate intercepts the protected surface over a relatively broad area. Thus, the impulse intensity delivered to the second plate is reduced sharply below that characteristic of a primary impact. Weight-saving factors of 10 have been achieved experimentally between single plate and dual plate armor providing equivalent protection. Theoretical investigations indicate that weight savings of twice this value are feasible when considering impacts at typical meteoroid velocities in near-earth space.

The state of material in the debris cloud affects the penetrating ability of this cloud against the protected surface. Vaporous clouds provide simple blast loading for the impacted surface and the intensity of this loading falls rapidly as plate spacing is increased. Clouds made up of liquid droplets provide loading of the underlying structure somewhat similar to that provided by vaporous clouds. The liquid droplets subdivide continuously as they expand rearward until they reach very small sizes because they are held together by surface tension forces only. Thus, craters from individual droplet impacts are of negligible practical importance although they are several times the size of the impacting droplets. Experimental data shows that plumes of liquified debris expanding behind impacted plates are thin (i.e., all debris launched from the impact site in a particular direction is launched at the same time and travels at a single velocity so that it flys together). The resulting short local impact durations satisfy the conditions for producing spallation of the underlying plate. Indeed, experimental impacts of liquified debris clouds do produce severe spall failures of underlying plates! Solid debris particles within impact generated debris clouds produce individual impacts on the protected surface and no feasible increase in plate spacing reduces the intensity of this threat.

The large differences between the damages produced by vaporous, liquid, and solid clouds led investigators to consider in detail the physics of debris cloud formation. The mechanism identified as being responsible for establishing the state of debris cloud material involves the entropy injected into the projectile and target material by the primary shock wave from the impact. This entropy increase is trapped in the material as it relaxes back to low pressure because relaxation processes are virtually isentropic. The trapped entropy appears as internal energy in the material that elevates its temperature. If this internal energy exceeds the fusion energy of the debris material, it returns from the shocked condition as liquid. If this energy exceeds the materials sublimation energy, the material returns as a vapor. This model for describing material behavior is not completely accurate because it fails to account for a number of known processes. The model has been shown repeatedly to predict material state in the debris clouds accurately, however, which indicates that the unaccounted processes provide approximately balancing errors.

Table 1 contains information on the shock pressures required to melt and vaporize a variety of materials by entropy trapping from shock waves.

Let us now consider application of thin-target impact physics to the design of a two-plate meteoroid shield for a spacecraft.

The following is an analysis of such a system which demonstrates an elegantly simple approach to evaluating the characteristics of debris clouds

Table 1 Shock Conditions Required to Melt and Vaporize Materials by Entropy Trapping

Material	Melting				Vaporization			
	Incipient		Complete		Incipient		Complete	
	Pressure Mb	Al Impact Velocity km/s	Pressure Mb	Al Impact Velocity km/s	Pressure Mb	Al Impact Velocity km/s	Pressure Mb	Al Impact Velocity km/s
Magnesium	0.48	5.40						
Aluminum	0.70	5.60	1.00	7.0				
	0.67	5.50	0.88	6.6	1.67	10.2	4.70	
	0.61	5.10	0.85	6.5				
Titanium	1.30	7.60						
Iron (steel)	1.80	7.90	2.10	8.80				
Cadmium	0.33	2.50	0.46	3.20				
	0.40	3.0	0.59	3.9	0.88	5.2	1.80	8.1
	0.33	2.5	0.43	3.15	0.70	4.4	5.30	
Copper	1.40	6.60	1.84	8.00				
	1.40	6.60	1.84	8.00	3.40	12.6	34.00	
Nickel	2.3	9.00						
Lead	0.25	2.00	0.35	2.60				
	0.27	2.1	0.34	2.5	0.84	4.8	2.30	9.1

launched behind thin plates, and the effects these clouds may have on underlying structures. Variables for the analysis are defined in Figure 5.

The debris clouds behind an impacted plate has been modeled as a symmetrically expanding sphere whose center moves rearward behind the impacted plate. All of the materials within the sphere is assumed to be concentrated at its surface which corresponds accurately to a wide range of experimental observations.

Let us consider first the motion of the center of the expanding sphere. This motion can be estimated accurately by evaluating momentum conservation between the incoming projectile and the debris cloud material that consists of the incoming projectile and material from the impacted plate that is carried with it. This step requires the further assumption that negligible momentum is transferred to the remainder of impacted plate or to the debris plume projected from the front surface of the front plate. The only way momentum will reach the remainder of the plate is through mechanical shear at the edge of the impact crater. The shear strength of realistic target materials when considered with the times during which the shear forces can operate (near the time required for the projectile to pass the plate) leads to the conclusion that the negligibility approximation is valid. Studies using ballistic pendula have been carried out at several laboratories to evaluate the impulse contained within debris clouds launched from the front surfaces of thin plates subjected to impacts. These studies all concluded that less than 1% of the incoming impulse is distributed in this manner.

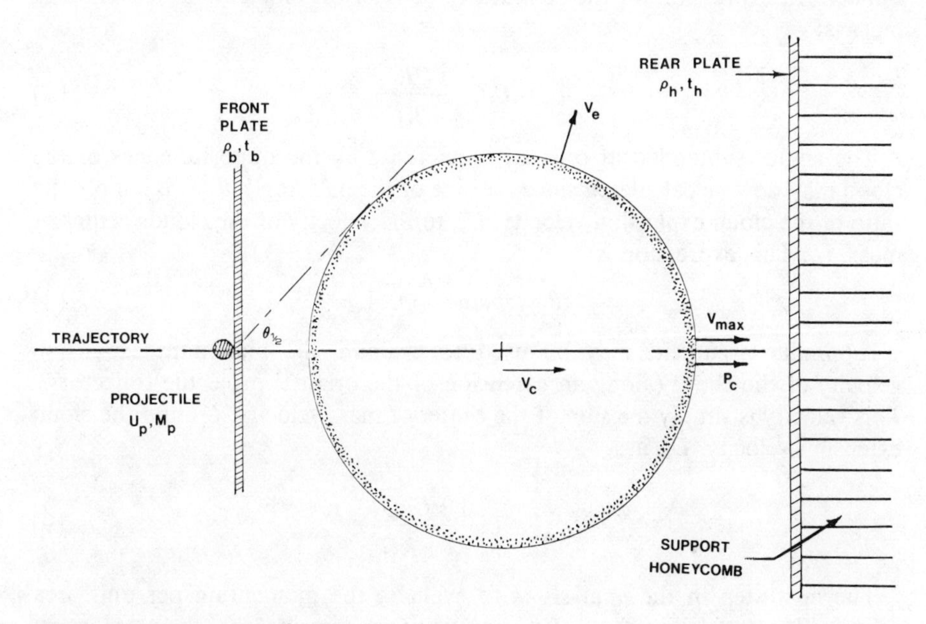

Figure 5 Schematic diagram of a model used to evaluate operation of dual-plate hypervelocity armor.

The velocity of the center of mass relative to that of the projectile was calculated from momentum conservation considerations is presented as:

$$U_c = \frac{U_p}{1+KG^2} \tag{5}$$

The parameter, K, is the ratio of the masses per unit area of the projectile and target plate; and G is the ratio of projectile diameter to the target diameter that contributes debris to the spherically expanding cloud.

Let us now consider the source of energy for causing the sphere of debris material to expand. The only source identified to date is the excess energy made available by the conservation of momentum described in (5). This energy is:

$$E_e = E_p \frac{KG^2}{1+KG^2} \tag{6}$$

This evaluation of available kinetic energy may now be used to calculate the expansion velocity of the sphere under the assumption that a fixed fraction, Q, of the available energy appears as directed kinetic energy. This fixed fraction assumption is probably not strictly valid because the mechanisms for inserting heat energy into the debris material are functions of peak shock pressures to which this material was subjected. Further development of this analysis can probably be accomplished by considering models for relating this energy partition term, Q, to peak shock pressures produced during the original impact. An expression for the outward velocity of the cloud about its center of mass is:

$$U_e = U_p G \frac{\sqrt{QK}}{1+KG^2} \tag{7}$$

The angles subtended at original impact site by the opposite edges of the cloud may now be calculated since the sine of the half-angle, $\theta_{1/2}$, is simply the ratio of the cloud expansion velocity, U_e, to the velocity of the clouds center of mass, U_c. This expression is:

$$\theta_{1/2} = \sin^{-1}\sqrt{QK} \tag{8}$$

A similar argument may be used to evaluate the maximum expansion velocity of the cloud (along an extension of the original projectile trajectory). This velocity is simply the sum of the center of mass velocity, U_c, and the cloud extension velocity, U_e, as:

$$U_{\max} = U_p \frac{1+G\sqrt{QK}}{1+KG^2} \tag{9}$$

The next step in the analysis is to evaluate the momentum per unit area contained in the debris cloud. For simplicity we describe only the momentum contained in the section of the debris cloud near its axis of symmetry (near the projection of the original velocity vector). The momentum per unit area, P_m, is:

$$P_m = \frac{M_p U_p \left(1 + G\sqrt{QK}\right)^3}{4\pi x^2 K G^2} \tag{10}$$

This very basic analysis or straightforward extensions of it define the important debris-cloud parameters for analyzing the operation of dual-plate meteoroid shields. The remainder of the analysis presented here is concerned with the response of underlying structures to impingement of the advancing debris cloud. We have chosen to consider a rather specialized structure consisting of a thin homogeneous plate supported by expanded honeycomb material for this analysis. The reason for choosing this particular structure is that it is typical of many structures used in current or planned spacecraft and it has the unique feature that honeycomb material crushes at constant pressure independent of displacement when it is loaded quasistatically or dynamically at velocities up to a few hundred meters per second.

The momentum per unit area evaluated in (10) may be used to evaluate the displacement produced in the front face of the honeycomb-supported panel by: first establishing the instantaneous rearward velocity imparted to the panel when it absorbs momentum from the debris cloud; and then calculating the kinetic energy per unit area developed in the plate. This kinetic energy is absorbed in the honeycomb as the plate moves rearward under constant resisting pressure. The distance required to absorb the kinetic energy in the plate then becomes the final deflection we are seeking, δX.

The instantaneous velocity reached by the rear plate on axis, U_h, is evaluated in (11); the kinetic energy per unit area associated with this movement, E_h, is presented in (12); and the displacement required to arrest this motion by honeycomb crushing, δX, is evaluated in (13).

$$U_h = \frac{W M_p U_p}{4\pi X^2 \rho_h t_h} \left(1 + G\sqrt{QK}\right)^3 \tag{11}$$

The parameter, W, defines the momentum multiplication at the cloud-rear plate interface.

$$E_h = \frac{W^2 M_p^2 U_p^2 \left(1 + G\sqrt{QK}\right)^6}{32\pi^2 X^4 \rho_h t_h} \tag{12}$$

$$\delta X = \frac{W^2 M_p^2 U_p^2 \left(1 + G\sqrt{QK}\right)^6}{32\pi^2 X^4 P_h \rho_h t_h} \tag{13}$$

Equation (13) may be used as the principal design tool for developing dual-plate meteoroid shields to protect against a specific threat particle since it contains all of the parameters that can be adjusted. These parameters include the parameter, K, which relates the masses per unit area of the threat projectile and the front plate; ρ_h and t_h determine the mass per unit area of the second plate; P_h is the crushing strength chosen for the honeycomb material; and X is the spacing between the front and rear plates. The remaining parameters describe aspects of the physics of the problem and cannot be controlled. The dimensionless parameters G, Q, and W are: the ratio of projectile diameter to front-plate hole diameter providing material to the debris field; the ratio of kinetic energy to heat energy in material of the debris cloud; and the momentum multiplication occurring when the debris cloud impinges on the rear plate, respectively.

Perhaps the most striking individual observation about (11) is the enormous effect separation between the two plates, X, has on the effectiveness of dual-plate hypervelocity armor.

6.2 HYPERVELOCITY LAUNCHERS

The equipment used for launching macroscopic projectiles to hypervelocities may be subdivided conveniently into staged light-gas guns, explosive devices, and electrically powered devices. Each of these areas has produced subcategories with individual launch characteristics as discussed in the following paragraphs.

6.2.1 Staged Light-Gas Guns

Staged light-gas guns are analogous to conventional powder guns in that a reservoir of energetic gas is used to produce pressure at the rear of a launch package contained within a tube. As the package is accelerated, the velocity achieved, U_p, may be expressed as:

$$U_p = \left(\frac{2A}{m} \int_0^{X_0} P\,dx \right)^{1/2} \tag{14}$$

The pressure specified in (14) is that which appears at the base of the package. This pressure is, in general, less than the gas reservoir pressure because the gas must expend energy accelerating its own mass to the instantaneous projectile velocity. Energy expended in accelerating itself is not available for projectile acceleration. This energy loss is reflected as a pressure gradient along launch barrel from its base at the reservoir to the rear of the moving projectile. The extent of the pressure drop between the reservoir and the projectile is dependent on projectile velocity and the characteristics of the gas, particularly its sonic velocity as is presented in (15).

$$P = P_r \left(\frac{U_p}{a_0} - 1 \right)^{1/2}$$

$$a_0 = \left(\frac{\gamma R T}{M} \right)^{1/2} \tag{15}$$

The two-stage light-gas gun was invented to take advantage of the low molecular weights of helium or hydrogen to provide the ultrahigh-velocity gas flows needed to achieve hypervelocities with useful sabot-projectile packages. The operation of a typical light-gas gun is shown in Figure 6. A conventional powder gun with a smooth bore is installed with its muzzle encased in a massive steel block that contains an extension of the powder gun tube, a transition section to a smaller diameter tube, a cavity for containing a pressure-actuated valve mechanism, and a receptacle for a relatively small diameter tube that serves as a launch barrel for the second stage gun. The projectile-sabot package is mounted at the base of the launch barrel and the now-closed barrel of the relatively large powder gun is evacuated and back-filled with either hydrogen or helium gas. In operation, the powder charge is ignited driving the first projectile (now known as a pump piston) into the charge of low molecular weight gas. The velocity of this piston remains well below the sound speed in the light gas charge so the compression remains nearly adiabatic, although pressure disturbances are generated that propagate up and down the shortening gas column. The pressure increases rapidly at the base of

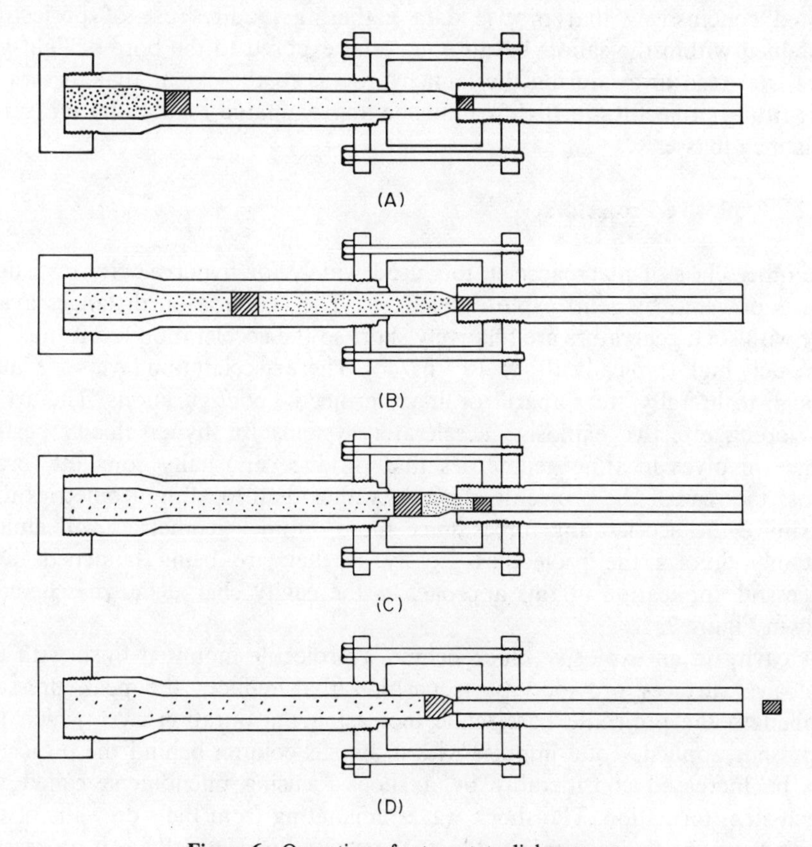

Figure 6 Operation of a two-state, light-gas gun.

the launch barrel until the piston enters the central section and the gas pressure reaches the critical value needed to actuate the valve mechanism. A complex gas flow situation then ensues where the gas volume is decreased by motion of the compression piston while it simultaneously is increased by motion of the projectile down the launch barrel. Pressures in the gas reservoir rise from a few hundred megapascals when the value mechanism opens, to above one gigapascal and then fall rapidly as the projectile nears the muzzle of the launch barrel and the piston expends the last of its kinetic energy deforming itself into the transition section.

Light-gas guns are used routinely to produce velocities of projectile-sabot packages up to 7.5 km/s. In isolated cases, sabot packages have been launched to velocities above 8.5 km/s. Homogeneous slugs of low-density plastic material have been launched to velocities just over 12 km/s. Light-gas guns are the only means currently available for launching projectiles of predetermined shape to hypervelocity regimes of projectile-target materials combinations of engineering interest. As such, they have found extremely wide use for hypervelocity impact investigations.

Studies conducted in the early days of hypervelocity investigations demonstrated conclusively that precise data gathering requires use of projectiles contained within the sabots because material exposed to the bore of light-gas guns is eroded away during the launch process to the extent that masses of bore-fitting projectiles in flight cannot be approximated reliably by measured prelaunch masses.

6.2.2 Explosive Projectors

The other class of macroaccelerators used widely for hypervelocity investigation is powered by solid explosive charges. The acceleration distances available with such generators are relatively short so the acceleration levels must be extremely high (typically 10^9 to 10^{11} m/s^2). These acceleration levels are large enough to literally "tear apart" ordinary projectile configurations. The art of developing effective explosive accelerator systems for hypervelocity performance involves locating geometries that produce unusually constant forces across the rear of the projectile elements rather than locating geometries that maximize the accelerating forces since these optimal geometries will almost certainly deform the projectiles grossly as they are being launched. One successful application of this approach is the cavity charge shown schematically in Figure 7.

A cavity in an explosive block behind a projectile mounted flush with the explosive surfaces provides a gas cushion that reduces the peak pressure applied to the projectile base while increasing the duration over which the impulse is applied. Total impulse within the gas column behind the projectile can be increased considerably by a shock-focusing phenomena called the Mach-stem formation. The shock waves emanating from the side walls of the cavity compress the gas projected from the cavity floor laterally as it progresses

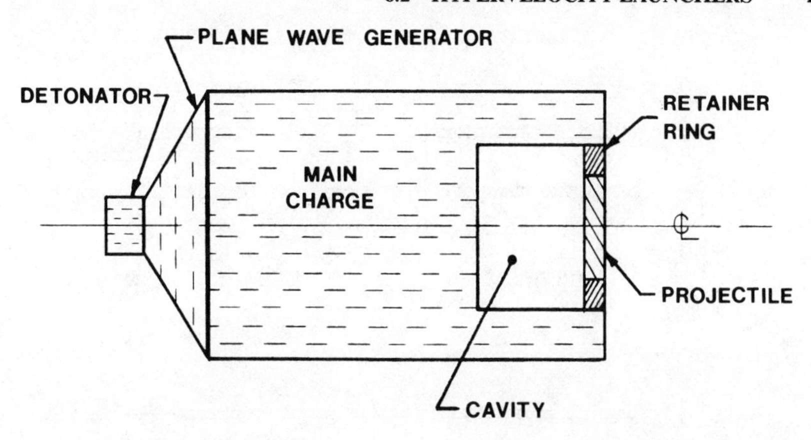

Figure 7 Cavity explosive charge used for launching metallic discs.

up the cavity. The diameter of the projectile to be launched is set approximately equal to the theoretical diameter of the Mach-stem. The projectile is encased in a retainer ring of identical material whose outer diameter equals that of the cavity. This rim is subjected to a steep impulse gradient with the inner diameter receiving more impulse per unit area than the outer diameter. Radial-traveling compressive waves produced in the projectile by its violent longitudinal acceleration propagate across the projectile-rim interface and reflect from the outer edge of the rim as tensile disturbances which propagate inward radially as they focus. The rim-projectile interface cannot sustain a tensile disturbance so the reflected tensile wave rereflects at this interface and is trapped in the rim. The stress gradient and the radial running waves combined to tear the rim apart and project the fragments radially away from the projectile which, it is hoped, proceeds forward intact.

Cavity-charge explosive devices can launch discs with aspect ratios as high as, perhaps, 0.3 to velocities near 6 km/s. Each configuration must be designed carefully and then optimized experimentally to produce the desired effects. These devices have proven very useful for extensive impact studies where many similar impacts are required, but they have not found wide use where impact conditions must be varied systematically.

Another very interesting class of explosive projectors employs the very high pressures available from detonating explosives to forge projectile material from one geometric shape to another in the process of launching it to hypervelocities. The most interesting and useful of these devices involve explosively collapsing a metal linear material surrounding a cavity onto the cavity axis to produce a material jet that travels at extremely high velocity. Figure 8 demonstrates this process applied to a cylindrical liner. Upon detonation from one end, the explosive accelerates the linear material toward the center of the cavity at high velocity. After a short period, a collapse point is defined that represents cavity closure. This collapse point propagates in the direction of

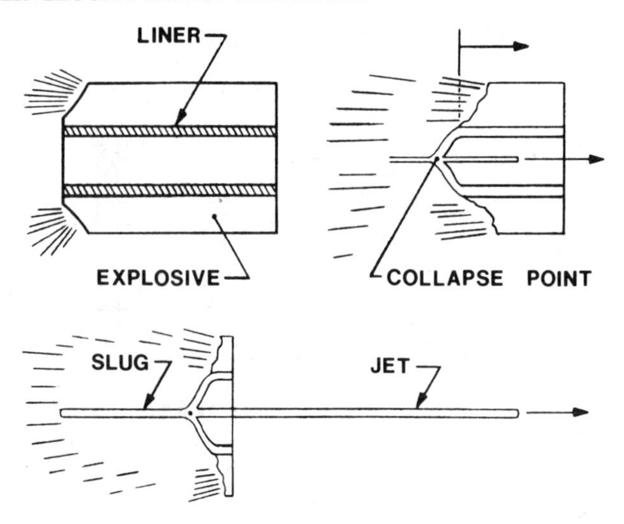

Figure 8 Operation of a cylindrical liner-shaped charge.

explosive detonation at the detonation velocity. The linear material at the collapse point divides between material that propagates forward into a jet and material that remains nearly stationary which forms the "slug." The jet moves away from the collapse point at a velocity approximately twice the closure velocity (detonation velocity) subject to the condition that the closure velocity is lower than the shock wave speed in the shocked linear material.

These devices produce truly amazing results! Intense explosives with detonation velocities above 7 km/s used in conjunction with liners made from beryllium whose shock speed can exceed 7 km/s, have produced practical jets with lengths approaching one to one and one-half times the liner length that travel at velocities in excess of 12 km/s. Copper jets traveling at speeds approaching 10 km/s have also been demonstrated.

Clearly, none of these devices produce projectiles that can be examined in the laboratory before launching so instrumentation must be provided for estimating the mass and material condition of the projectiles in flight if precise ballistic studies are to be conducted.

A famous variant of the cylindrical cavity charge is the conical cavity charge where the forged element is a hollow cone. Explosive collapse of such a cone produces an axial collapse point that moves forward at a maximum velocity as soon as it is formed and then decelerates steadily. The resulting material jet contains a velocity graident along its length that causes the jet to extend to many times the cone length during transit over a considerable distance. Eventually, these graidents cause the jet to fracture into a sequence of individual projectiles traveling along a single trajectory with progressively decreasing velocities. When ductile materials such as copper are used for the liner, charges only 10 cm long can produce continuous jets with lengths up to

40 cm. We pointed out in Section 6.1.1 that such jets can perforate steel armor nearly as thick as the jets are long (i.e., a charge weighing only a few pounds can perforate 40 cm of steel armor plate)!

6.2.3 Electrical Accelerators

We conclude this discussion of hypervelocity launchers with a consideration of a new accelerator technology with some demonstrated capability for achieving hypervelocities and much potential for further development. The technology area is "dc electromagnetic rail guns" whose basic operation is presented in Figure 9. An electrical current propagates down one parallel rail, crosses the armature, and returns up the opposite rail. This current path produces a magnetic field at the armature that interacts with current passing through the armature to produce a force directed along the rails as:

$$F_p = L'I^2 \tag{16}$$

Clearly, the critical parameters are the inductance per unit length, L' and the current, I. Earlier activities in these areas produced poor results which have recently been traced to difficulties with the current source. As with other gun-type projectors, the effectiveness of the device is closely related to the ratio between average and peak accelerations experienced by the projectile during launch. Very high peak accelerations disrupt the projectile-sabot packages when total impulses required to achieve hypervelocities are applied. It is clear from (16) that optimum performance for such a device occurs when the current through the device is constant. This characteristic has been hard to achieve because the impedance of the devices, Z, is proportional to armature velocity as:

$$Z = Z_0 + L'U_p \tag{17}$$

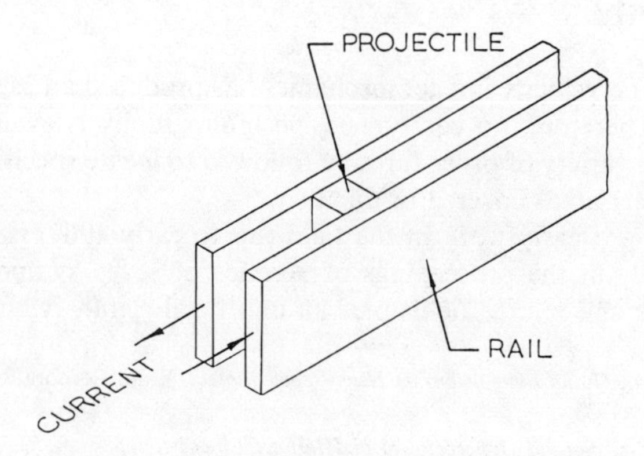

Figure 9 Basic operation of a DC electric-rail gun.

This equation shows that the voltage applied across the gun terminals must increase orders of magnitude during the acceleration cycle if currents are to remain nearly constant.

No high current source was available that met these criteria until a specialized power supply system was invented at the Australian National University (ANU). Basically, a homopolar generator, which is a low-voltage high-current power source, was used to charge an inductor to currents of several hundred thousand amperes. Once the critical current was reached, this inductor was switched across the terminals of the dc electrical rail gun where it delivered the same current with which it was charged. An inductor in this configuration is essentially a constant current source that automatically adjusts the voltage across its terminals to maintain the current almost regardless of the impedance of the load circuit. In the case of the ANU experiment, voltages at the breech of the electric gun rose from approximately 100 V to over 15,000 V as the projectile completed its acceleration. Peak velocities of 6.5 km/s have been achieved with payloads weighing up to 3 g. Projectile acceleration profiles, electrical characteristics of the firing cycle, and final projectile velocities were maintained within a few percent of predicted values during these experiments.

Analyses indicate that velocities of 15 km/s and above can be achieved in a more or less straightforward manner using this basic technology. Three application areas for this technology have attracted interest. The first is development of an ordnance weapon that can launch useful payloads to velocities approaching twice current capability. The second is the use of high-performance electric guns (velocities over 10 km/s) as reaction propulsion systems for vehicles operating in space. Recoil from sequences of pellets fired rearward provide average propulsion forces many times greater than those achievable using currently available ion engines. The third area is a meteoroid simulator that can achieve velocities high enough (>12 km/s) to assure impact vaporization of projectile and target materials that are appropriate simulants of actual encounters in space.

BIBLIOGRAPHY

The field of hypervelocity impact mechanics has produced an estimated 10,000 publications. Therefore, no comprehensive bibliography is available. We recommend that a variety of procedures be followed to locate specific information regarding subject areas covered in this text.

First, important early work in the field (up to early 1969) is covered quite comprehensively in the proceedings of a series of seven symposia on hypervelocity impact and related fields plus an unofficial eighth symposium:

1 *Proceedings of the Rand Symposium on High-Speed Impact*, Rand Corporation, Santa Monica, California, May 1955.

2 *Proceedings of the Second Hypervelocity and Impact Effects Symposium*, U.S. Naval Research Laboratory, Washington, D.C., May 1957.

3 *Proceedings of the Third Hypervelocity Impact Symposium*, Armor Research Foundation, Chicago, Illinois, October 1958.

4 *Proceedings of the Fourth Hypervelocity Impact Symposium*, Eglin Air Force Base, Florida, April 1960.

5 *Proceedings of the Fifth Symposium on Hypervelocity Impact*, Colorado School of Mines, Denver, Colorado, April 1962.

6 *Proceedings of the Sixth Symposium on Hypervelocity Impact*, Firestone Tire and Rubber Co., Cleveland, Ohio, April/May, 1963.

7 *Proceedings of the Seventh Hypervelocity Impact Symposium*, Martin Company, Tampa, Florida, November, 1964.

8 *AIAA Hypervelocity Impact Conference, A Volume of Technical Papers*, Cincinnati, Ohio, April/May, 1969.

Two other technical meetings have (or will) produce proceedings of specific importance. The Office of Research and Advanced Technology (ORAT) of NASA held a conference in late 1968 that produced a classic proceedings covering information specifically relevant to space vehicle protection from meteoroid impact threats. The European Space Agency held a specific workshop during the spring of 1979 to consider protection of a probe to fly by Comet Halley during its next appearance (in the spring of 1986) that attracted recent contributions on the same subject. These two volumes should form a compendium of the most up-to-date information available.

1 *Compilation of Papers Presented at the ORAT Meteoroid Impact and Penetration Workshop*, Johnson Manned Spacecraft Center, Clearwater, Texas, October 1968.

2 *Proceedings of the Comet Halley Micrometeoroid Hazard Workshop*, European Space Research and Technology Center, Noordwijk, Netherlands, April 1979.

More specific information can be located at a special library of Impact Mechanics compiled by the University of Dayton Research Institute. This library contains approximately 5000 volumes that cover the field from the mid 1950s to the present. The appropriate contact is Dynamics Mechanics Laboratory, University of Dayton Research Institute, Dayton, Ohio 45469.

The field of hypervelocity impact mechanics is addressed in two fairly recent books which present broad ranges of information in a more orderly and comprehensive manner than is available from conference proceedings.

1 *High-Velocity Impact Phenomena*, Edited by Ray Kinslow, Tennessee Technological University, Academic Press, New York City, 1970.

2 *Dynamic Response of Materials to Impulsive Loads*, Edited by Pei Chi Chou and Alan K. Hopkins, Air Force Materials Laboratory, Wright-Patterson Air Force Base, Ohio, August 1972.

These sources also serve well for locating information on hypervelocity launchers. Two other references are of special interest. Dr. A. E. Siegel produced a comprehensive discussion of gas gun operation many years ago which remains a classic. More recently, Dr. J. P. Barber described the operation of modern electric guns in his Ph.D. thesis and continuations of his work are presented in a recent journal article.

1 Siegel, A. E., "The Theory of High-Speed Guns," *AGARD-O-graph*, **91**, May 1955.

2 Barber, J. P., "The Acceleration of Macroparticles in a Hypervelocity Electromagnetic Accelerator," Ph.D. Thesis, The Australian National University, Canberra, Australia, 1972.

3 Rashleigh, S. G., and Marshall, R. A., "Electromagnetic Acceleration of Macroparticles to Hypervelocities," *J. of Appl. Phys.*, **49**, 1978, pps. 2540–2542.

SYMBOLS

A	Cross section area of projectile-sabot package
D_p	Equivalent projectile diameter
E_e	Kinetic energy of explosion
E_p	Projectile kinetic energy
G	Ratio of projectile diameter to hole diameter in thin plate that contributes material to the debris cloud
I	Electrical current
K	Ratio of a mass-unit area of meteoroid and bumper shield
L'	Rail inductance per unit length
M	Molecular weight
M_m	Mass of meteoroid
P	Gas pressure behind sabot
P_c	Crater penetration depth
P_m	Peak momentum intensity produced by cloud
P_r	Pressure in gas reservoir
P_s	Shield material density
Q	Fraction of energy available to debris-cloud formation expended or directed kinetic energy
R	Universal gas constant
T	Absolute temperature
U_c	Velocity of debris cloud, c.g.
U_e	Outward velocity of debris bubble
U_h	Instantaneous rear-plate velocity
U_{max}	Maximum rearward-cloud velocity
U_p	Meteoroid-projectile velocity
V_c	Crater volume
X	Position along projectile-launch tube, separation between bumper and rear plate
X_0	Launch-tube length

Z	Rail-gun impedance
Z_0	Residual impedance of a rail gun with a stationary projectile in firing position
a_0	Sound speed in perfect gas
d_m	Equivalent meteoroid diameter
f_p	Force accelerating projectile package
k	Ratio of projectile kinetic energy to crater volume
m	Mass of projectile-sabot package
t_h	Rear-plate thickness
t_s	Bumper thickness
x	Distance behind meteor-bumper plate
δ_x	Deformation distance of rear plate
γ	Ratio of gas-specific heats
ρ_b	Density of the front-plate material
ρ_h	Rear-plate density
ρ_m	Meteoroid-material density
ρ_p	Projectile-material density
ρ_r	Target-material density
$\theta_{1/2}$	Half-angle of debris bubble from original impact point
L_p	Projectile length along direction of travel

7

IMAGE FORMING INSTRUMENTS

Hallock F. Swift

Cameras and related image forming devices are the most widely used instruments for studying dynamic mechanical events such as impacts and explosions. The reasons for this wide usage are: investigator familiarity with related equipment (conventional cameras), availability of usable equipment with applicable capabilities, and the enormous information storage rates achievable with cameras.

Cameras are, basically, "parallel information recorders" (i.e., each point on the image format is recorded at the same time as every other point). For instance, a conventional 35 mm SLR camera used with fine-grain color film can gather data at maximum rates exceeding 10^{13} bit/s—a data recording rate exceeding the fastest electronic technology currently available. While it is true that no data recovery systems have yet been developed that make full use of camera's data-storing capabilities, the recording rates are formidable and are highly useful.

Data recording rates of conventional cameras have been considerably exceeded by those of high-speed cameras but the principal advantages offered by high-speed imaging technology are in the area of extending our time-resolving capabilities. Cameras with demonstrated resolution times approaching 10^{-13} s (100 fs) have been reported and cameras are commercially available with resolution times near 10^{-12} s (1 ps). These performances are the shortest time-resolution capabilities of any form of instrumentation currently available, and they also represent the greatest extension of any of our human senses yet achieved. (Human beings can resolve 5×10^{-2} s visually without aid, which is 500 billion times slower than the time-resolution limit of current photoinstrumentation.)

Most high-speed photographic systems are analogous to conventional single-frame (snapshot), and ciné (movie) cameras. They are capable of visualizing objects moving at high velocities and of observing characteristics of their motions just as conventional cameras do for objects moving at ordinary velocities. A third type of high-speed camera is available that has seen little application at conventional speeds—the smear-streak camera. These units produce image information that represents the observed object only indirectly but visualizes object motion in the form of graphical plots revealing position-versus-time information. These cameras have the advantage of observing objects continuously rather than intermittently and they are capable of the best time resolutions currently available.

This chapter develops the optical characteristics of high-performance cameras needed to gather specific information from dynamic mechanical events and describes the technology currently available for taking single-frame images, ciné sequences, and smear-streak records of them.

7.1 IMAGING REQUIREMENTS FOR OBSERVING DYNAMIC EVENTS

The principle difficulty in observing dynamic events with imaging equipment is the motion blurring caused by object motion during the exposure period. This blurring destroys spacial resolution in the direction of object motion and must be countered, if usable images are to be recorded, by making the exposure times short enough to reduce motion blurring to manageable proportions. The very short exposure times, in turn, reduce the total radiation energy available to form images.

7.1.1 Exposure-Time Requirements

The critical factors determining exposure-time requirements for producing usable images is the time for the observed object to cross the view field. The ratio between this time and the exposure time determines the resolution limit of the resultant image imposed by motion blurring as

$$R_o = \frac{U_o X_i}{\tau_i} \tag{1}$$

where R_o = the total resolution of the observed object (in line pairs); U_o = object velocity; X_i = the width of the observed area; and τ_i = the exposure time. Should the term, R_o, exceed 1200, high-quality images (such as those produced by a 35 mm SLR camera) can be produced without any degradation as a result of motion blurring. Reduction of this term to 500 allows photographs to be taken with image qualities equaling those of the finest black-and-white television pictures (using the American television system). Images recorded with R_o as low as 3 contain just enough information to determine basic geometric shapes of the images such as circles, triangles, squares, and so forth. Further reduction of R_o to unity produces images with only enough quality to establish the presence of an object but contain essentially no information about its shape.

Figure 1 is presented to provide an idea of motion-blurring limitations occurring during the imaging of some specific dynamic events. We have assumed here that the object being photographed effectively fills the field of view of the imaging system. Note that, the critical parameter for determining image quality, under these conditions, becomes the time required for the object to move its own length along its direction of travel, L_o.

$$R_o = \frac{U_o L_o}{\tau_i} \tag{2}$$

The graph in Figure 1 plots this time-versus-image exposure time, and demonstrates the important fact that the critical timing parameter is proportional to the ratio between object size and object velocity. Several candidate objects are presented along the ordinate. Note that the fastest of these objects (a planet

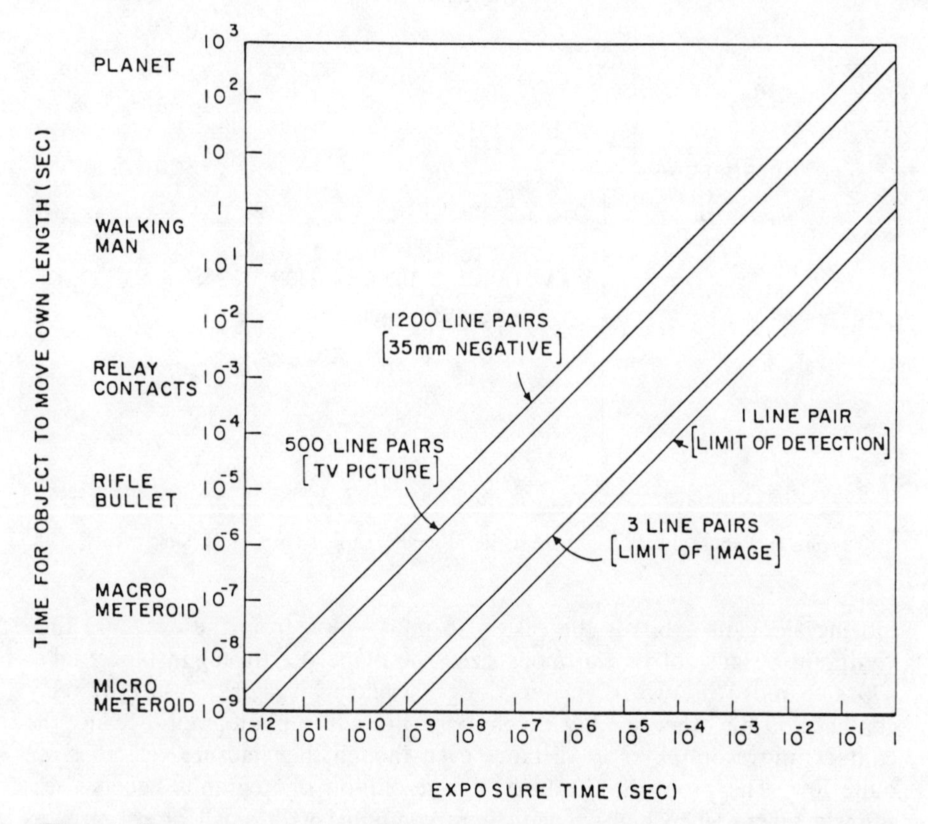

Figure 1 Exposure-time requirements for producing various quality images of moving objects.

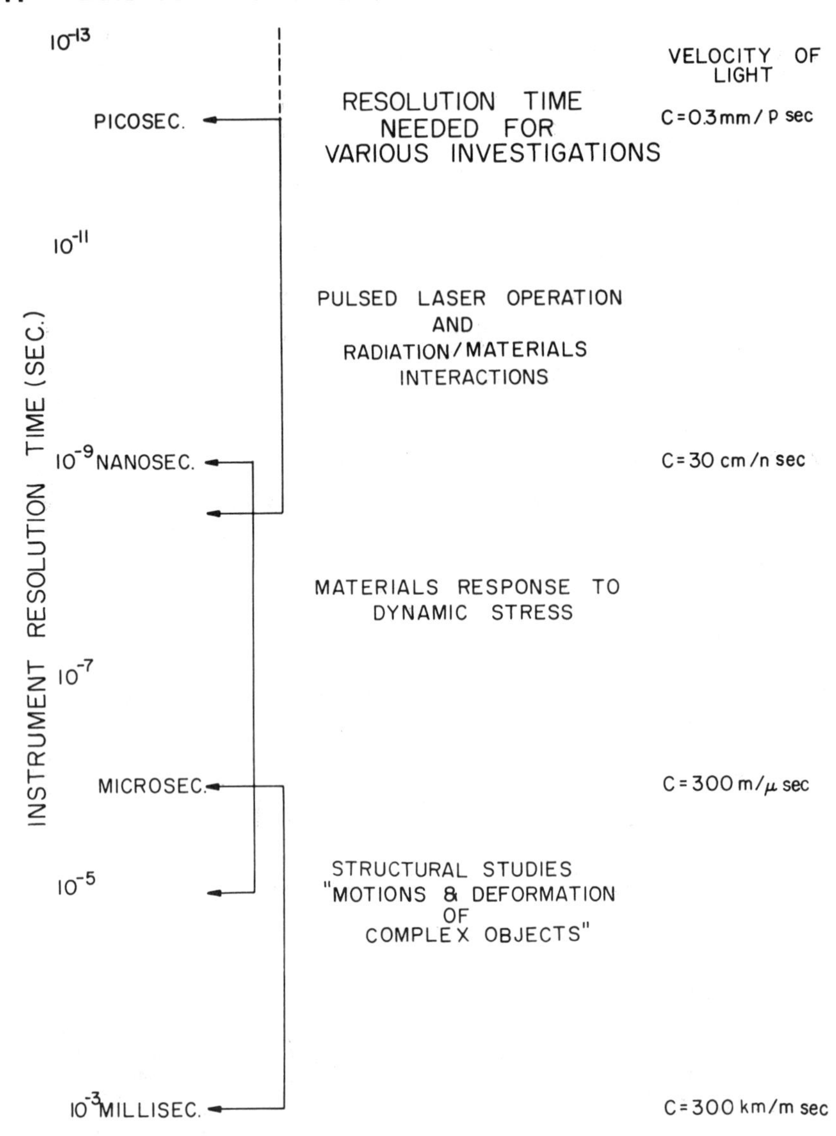

Figure 2 Exposure time requirements for observing various types of dynamic events.

moving along its orbit) is the easiest to photograph from the exposure time viewpoint because of its enormous size. The planet Earth, for instance, takes 470 s to move its own diameter along its velocity vector. Closure of relay contacts, on the other hand, is much more difficult to photograph because the contacts move only a short distance even though their closure velocities are quite low. The problem of taking high-resolution photographs becomes extremely severe when high velocities are combined with small object sizes as occurs when small projectiles launched to ultra-high velocities are observed.

The diagonal plots in Figure 1 represent the various picture qualities described earlier. The abscissa defining the exposure-time requirements provide strong evidence that exposure-time capabilities down to 10^{-12} s are of direct use for imaging small and energetic mechanical events.

The exposure times required for effectively observing various classes of dynamic mechanical events are presented in Figure 2. Investigations of structures subjected to dynamic loads can generally be conducted with imaging equipment capable of producing exposure times as short as 10^{-6} s (1 μs). Studies of materials response to shock loading requires more advanced equipment with exposure time capabilities down to 10^{-9} s (1 ns) and studies of radiation interaction with materials such as is required for research into laser-induced thermonuclear fusion require exposure times down to 10^{-12} s (1 ps) and possibly below. The scale at the extreme right of Figure 2 shows the distance that light travels (and other forms of electromagnetic radiation) during the exposure times considered.

7.2 SINGLE-FRAME CAMERAS

The simplest and oldest method for observing dynamic events is to take a "snapshot" with an exposure time short enough to effectively "freeze" the object motion. Two approaches to achieve this end have been developed successfully. The photosensitive material (film) can be exposed to the event continuously and the light source can be flashed to provide the exposure; or the camera can be shuttered at the required speed to observe an event that is more or less illuminated continuously. Figure 3 is a synopsis of the important imaging capabilities of technologies to current single-frame photographic technology. Framing time in seconds is presented versus spacial resolution of the resultant images in total line pairs.

7.2.1 Unfocused Shadowgraph

The oldest single-frame technique is the unfocused shadowgraph shown schematically in Figure 4. An open sheet of film is placed close to the object being observed (such as a rifle bullet in flight) and a small, diameter-pulsed light source such as a fast electrical spark gap is mounted some distance on the far side of the object. Triggering the spark source at the appropriate time exposes a silhouette image of the event on the film that can be subsequently developed and observed at leisure. Commercially available spark generators can produce effective exposure times as short as 5×10^{-9} s and very simple spark circuits can readily produce exposure times of 10^{-6} s duration.

Recently, pulse lasers with expanded beams have been used to take simple shadowgraphs with durations as short as 10^{-11} s.

Another important application of unfocused shadow imagery occurs when the pulsed light source is replaced by a pulsed x-ray source. Dynamic x-radiographs with exposure times down to approximately 5×10^{-8} s can be

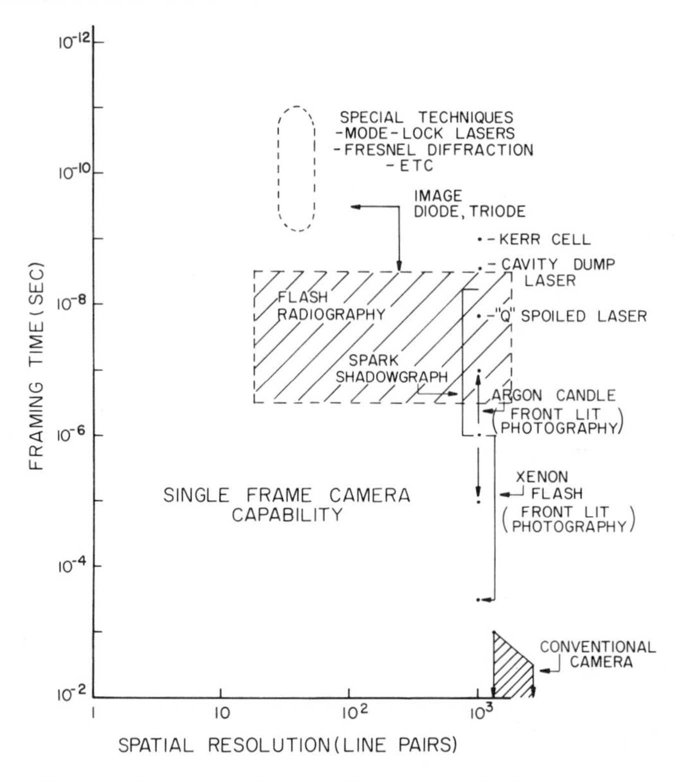

Figure 3 Current capabilities of high-speed, single-frame cameras.

produced to observe a wide variety of energetic mechanical events using commercially available equipment. This capability allows the effects of smoke and dust that prohibit photography of many interesting events such as explosions and impacts to be circumvented. The internal operations of opaque devices can also be visualized. Finally, flash radiography can be used to determine instantaneous shifts in local material densities within objects that are produced by propagating shock waves. Section 7.2.6 describes flash x-ray equipment in more detail.

7.2.2 Focused Shadowgraph

The simple shadowgraph equipment is severely limited by at least two factors. First, experiments must be carried out in nearly total darkness which is, at best, an inconvenience. This limitation can easily preclude using this technique for data-taking if the object under observation emits light. Light from the object simply fogs the film and produces no image information. Finally, placing the film very close to the object can be a problem if the event under study is violent enough to cause film damage.

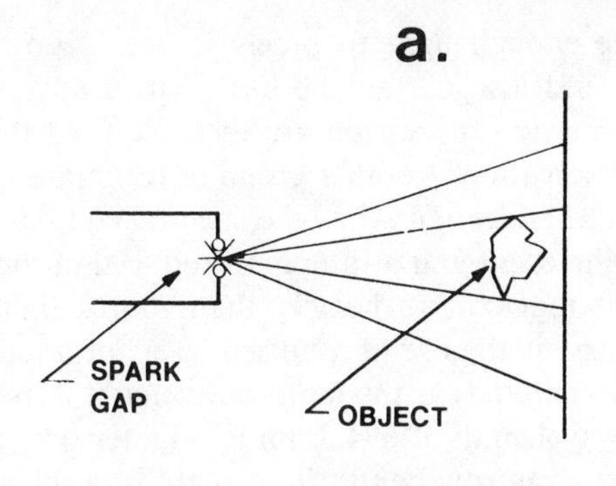

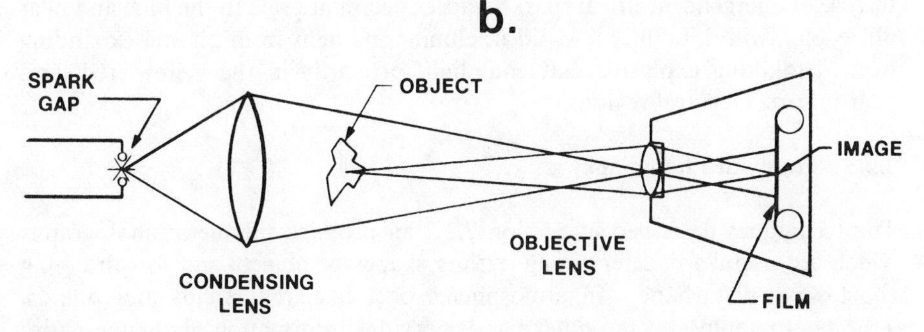

Figure 4 Shadowgraph setups for taking silhouette images of moving objects: A, unfocused; B, focused.

All of these difficulties may be circumvented by using a focused shadowgraph system also shown schematically in Figure 4. The film is contained in an ordinary camera which is focused on the event. A condensing lens mounted beyond the event images a small light source on the iris of the camera lens after the light has passed through the object area. Both the camera and the condensing lens may be placed some distance from the event so that they may be protected if necessary.

Note that the effective diameter of the camera objective lens is the diameter of the light-source image focused by the condensing lens rather than the iris diameter. Since this image is quite small, the effective F number of the camera lens is high, which produces a substantial depth of field for the overall system. The efficiency by which light is gathered from the source is determined by the F number of the condensing lens and the magnification of the light-source image and is not affected by the aperture setting of the camera objective lens. Closing down the aperature of the camera objective lens, on the other hand, sharply reduces the camera's efficiency in collecting light from the area of the event under observation so a capability is developed for selectively suppressing light from the event area with respect to light from the source.

Some events produce enough light to preclude use of unfocused shadowgraph systems entirely and also curtail the use of focused systems. Stronger means than simple geometric suppression are then required to eliminate light from the event under observation. Another group of techniques for suppressing event light involves spectral selectivity (i.e., the spectral content of the source is separated from that of the event and a filter is introduced in the optical system to attenuate event light much more heavily than source light). An extreme example of this technology is the use of a pulsed laser for a light source and a very narrow band filter centered on the laser wavelength. A pulsed ruby laser that emits energy at a wavelength of 694.3 nm is brighter over a spectral band of 0.5 nm defined by a narrow-band filter than is a black body whose temperature is 5×10^6 °K! Less extreme applications of this technology involve the use of energetic electrical sparks whose spectra are rich in the blue and near ultraviolet, with blue filters to aid in eliminating light from plasma expanding from detonating explosive that emit light primarily in the yellow, red, and near-infrared spectral regions.

7.2.3 Front-Lit Photography

The techniques described in Section 7.2.2 all produce silhouette photographs which are useful for determining outline shapes of objects and for observing shock-wave disturbances in atmospheres or transparent solids and liquids. True photography, on the other hand, provides information about thé entire front surface of the event under observation. Taking high-speed photographs is more difficult than taking silhoutte photographs because the event under view must be bathed in light and the camera system must gather light that has been scattered and/or reflected to form an image—a rather inefficeint process. Potential exists for efficiently illuminating small view fields with pulsed light sources through use of condensing lenses and mirrors. Little potential exists, however, for using geometric constraints for allowing the camera system to absorb light incident upon the event that originates from the source more effectively than light emitted near the event. Thus, an experimenter can easily get into a conflicting set of requirements where the aperture of the camera's objective lens must be opened to gather sufficient light for photography, but must also be closed down to eliminate fogging from light produced by the event or to produce enough depth of field to meet an experimental requirement.

Spectral selectivity can, of course, be used to aid in illuminating difficult events, or the event may be photographed in its own emitted light if the duration of this light output is short enough or if adequate shuttering of the camera optics is available.

Considerable interest has recently occurred in taking high-speed color snapshots by using sensitive color film with nearly white-light illuminators such as Xenon flash lamps. Results are often spectacular artistically and the color may provide additional information concerning the event such as the original identity of launched particles, the instantaneous condition of surfaces, or local material densities.

Light sources designed for taking single front-lit frames at high speeds are available commercially. The larger light output required for front-lit photography generally precludes use of fast sparks, but very intense sparks of relatively long duration may be used for lighting an event when shuttering is available to produce short exposure times. Xenon flash lamps have proven especially useful for front-lit photography when operated at high speed. The relatively high efficiency with which Xenon flash lamps convert electrical energy to optical energy enables such systems to operate effectively with a equivalent photographic exposure times down to several tenths of a microsecond while illuminating a large event area.

Pulsed lasers such as *Q*-spoiled ruy units and frequency-doubled Neodynimum YAG or glass devices can produce light pulses with less than 10^{-7} s duration containing energies up to several joules. This light may be expanded to cover several dozen square meters of event area to produce high quality front-lit photographs. The spatial coherence of light from lasers causes light scattered from adjacent points in the field-of-view event to interfere constructively and destructively at the image plane. This intereference pattern causes local variations in illumination intensity (image speckle) which is, at best, objectionable and, at worst, may destroy significant amounts of the information content of the resultant images. Designers of laser-illumination photographic systems go to considerable lengths to destroy spacial coherence of the laser beam to eliminate speckle from the resultant photographs. Cells filled with carbon disulfide liquid through which the unexpanded beam travels for several tenths of a meter have proven useful for eliminating beam coherence as has bouncing the beam off several diffusely reflective surfaces before allowing it to reach the event.

7.2.4 High-Speed Shutters

The equipment described thus far determines the effective exposure time for producing images by pulsing the light source. This technology cannot be employed when one wishes to photograph self-luminous events using their own light or when exposure times shorter than those available with pulsed-light sources are required. Mechanical shutters have been developed to operate at exposure times down to 10^{-4} s. Exposure times shorter than this limit require use of electronic shutters.

Perhaps the oldest high-performance electronic shutter is the Kerr cell, which relies on certain organic liquids becoming birefringent when subjected to intense electric fields. An optically birefringent material splits linearly polarized light incident on it into two parts polarized perpendicular to one another. This feature allows shuttering action by containing the fluid in a cell mounted between crossed-optical sheet polarizers. Light incident on the first polarizer sheet is divided between a component whose polarization is parallel to the sheet that passes the polarizer and the other component whose polarization is perpendicular, which is reflected and absorbed. The polarized light then passes through the unactivated cell without change in polarization state and impinges on the second polarized sheet whose polarizing direction is perpendicular to that of the first. Since the light incident on the second sheet contains no polarization component parallel to its polarized direction, the light is reflected or absorbed and the shutter is closed.

Electroides mounted along two opposite lateral walls of the cell produce a strong electrostatic field across the organic fluid when they receive a high-voltage pulse, which causes the incoming polarized light to split into two polarized components one of which is parallel to the second polarizer direction and passes through it causing the shutter to open. A schematic diagram of a Kerr cell is presented in Figure 5.

Nitrobenzene is the organic fluid used in almost all Kerr cells because of its high Kerr constant. A cubic cell 5 cm in a side can be built into a shuttter that is fully opened with 40 kV applied across its electrodes (1600 V/cm). The birefringence produced by the Kerr effect is a prompt electrooptical phenomenon that is produced within 10^{-12} s after local application of the electrostatic field and which decays in a similar time after field removal. Kerr-cell cameras have been built commercially with exposure times as low as 10^{-9} s. One famous experiment was carried out using the Kerr phenomenon to photograph a laser light pulse traveling through dilute milk with an equivalent exposure time of less than 10^{-11} s.

Kerr cells used as optical shutters have the enormous advantage of producing virtually no image degradation, and cameras using them have developed a justified reputation for fine photographic quality. Kerr-cell optical shutters possess one notable weakness—poor optical efficiency. The optical polarization process on which Kerr-cell operation depends expends at least 50% of the light incident on the front surface of the first polarizer sheet even when the polarizers operate with perfect efficiency. Practical polarizers of large size that can produce useful on/off ratios are seldom more than 70% efficient, thus introducing another factor-of-two reduction and optical efficiency. Finally, nitrobenzene is opaque to blue and ultraviolet light even in its most pure form.

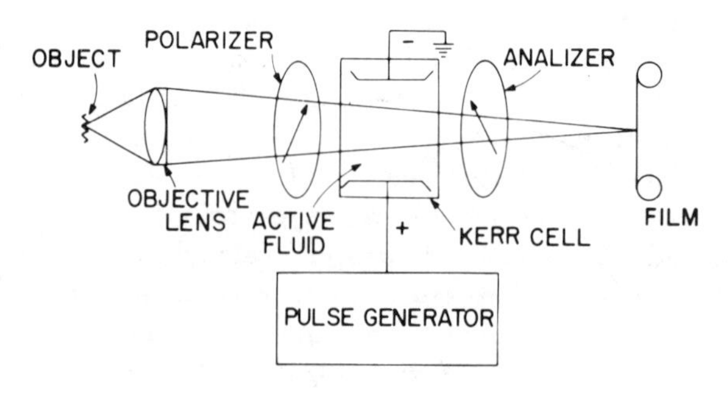

Figure 5 Schematic diagram of a Kerr-cell optical shutter.

Thus, Kerr cells cut off optically in the spectral range that is otherwise the most useful for high-speed photography because very bright thermal light sources are all rich in the blue and ultraviolet. In general, a realistic Kerr cell introduces an attenuation of at least 10 X into an optical system when the cell is open and attenuations of, perhaps, 10^7 when the cell is closed.

Other electrooptical phenomena have been used to produce electrooptical shuttering. A phenomenon closely associated with the Kerr effect is produced in some transparent crystals when they are subjected to axial magnetic fields. The theoretical efficiency of such cells are somewhat better than Kerr cell because the basic polarization process has an ideal efficiency of 70% rather than 50%. In addition, the crystals employed are transparent throughout the visual spectral range. Overall photographic efficiencies four or five times those achievable with Kerr cells have been demonstrated with commercially available equipment. The principal problem with magnetic electrooptical shutters is that the required magnetic field is much more energetic than the electrostatic field used with Kerr cells and hence requires longer times to deploy and remove. Typically magnetic shutters produce exposure times of 10^{-6} s and longer.

7.2.5 Electron-Image Tubes

An entirely different approach to producing electronic shutters is the use of electron-image tubes. These vacuum tubes consist of a photocathode where electronic analogues of optical images focused on it are produced continuously and a phosphor screen against which the electrons are accelerated by high electrostatic potential. The screen produces a visual image that is a more or less faithful rendition of the image originally focused on the photocathode. When used in high-speed single-frame cameras, image tubes are shuttered by applying voltages to intervening electrodes that inhibit electron flow from the photocathode to the phosphor screen except during a brief instant when flow is permitted. The phosphor screen is excited during this instant but may continue to emit light for durations extending from tens of nanoseconds to seconds. This light is usually recorded on a photographic film but may also be recorded on the photocathode of a television camera tube for electronic storage and/or electronic image processing.

Electron image tubes can be classified by the method used for focusing electrons generated at the photocathode onto the output phosphor screen. The simplest approach diagrammed in Figure 6 is to mount the phosphor screen a few millimeters behind the photocathode and switch the tube by applying a high-positive potential (5 to 20 KV) between the cathode and screen when tube operation is desired (proximity-focused image tube). The electrostatic field in the gap is almost precisely normal to both surfaces and the low-energy electrons emitted from the photocathode travel along the field lines to the screen to produce an image that is virtually devoid of conventional distortion. Small lateral velocities imparted to electrons as they exit the photocathode lead to slight blurring of the image on the phosphor screen since these electrons do

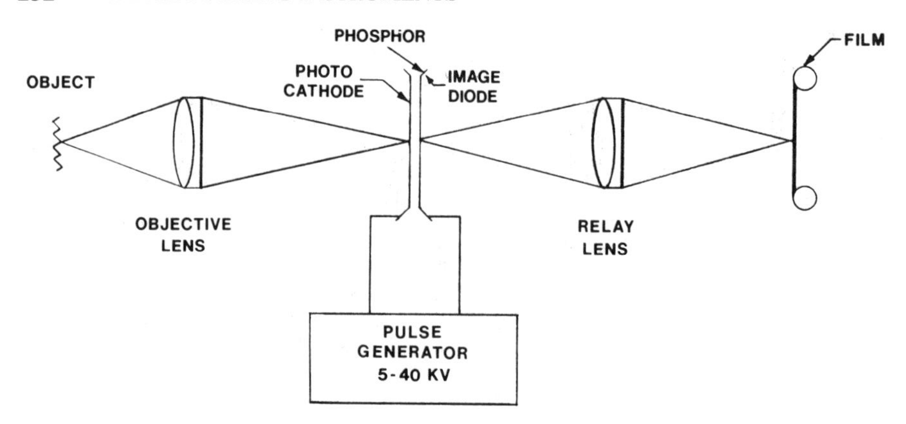

Figure 6 Single-frame electronic camera using a proximity-focused image diode.

not follow the field lines precisely whereas other electrons emitted from the same point with lesser lateral velocities do. This blurring increases with separation between the photocathode and screen, but decreases with applied voltage. The net effect of these two opposing tendencies is that the blurring increases proportional to the square root of the applied voltage when tubes with different cathode-screen spacings are operated at similar voltage stresses (i.e., applied voltage divided by plate separation). When a tube whose phosphor screen is mounted 3 mm behind its photocathode is operated at 20 KV, the images produced have spacial resolutions limited to slightly over 20 line pairs per mm (lp/mm) by this phenomenon.

The second type of electron-image tube separates the photocathode from the phosphor screen by some distance and employs electrostatic lenses to provide required electron focusing (electrostatically focused image tubes). Electrostatic lenses can provide fine spacial resolution to the output images (better than 50 lp/mm) but they tend to produce limited amounts of barrel or pincushion distortion. Image resolution tends to vary over the output format with the finest resolution produced at the center and the poorest at the edges. These tubes can be shuttered either by switching the full accelerating voltage or by installing electrodes near the photocathode that can accomplish switching at voltages as low as 10% of the applied accelerating voltage, thus greatly simplifying the gating requirements for recording high-speed images. Image resolution developed by electrostatically focused tubes is essentially independent of applied voltage.

Perhaps the finest quality images from electron image tubes are produced by magnetically focused units. The phosphor screen is placed some distance behind the photocathode and the electrons are magnetically focused on it by a constant-axial magnetic field either produced by permanent magnets or a solenoid coil (magnetically focused image tubes). Electrons exiting the photocathode with slight lateral velocities spiral around the magnetic lines of force with a rotation rate proportional to the field intensity as

$$W_0 = \frac{Be}{m} \tag{3}$$

where W_0 is the rotation frequency of electrons in a magnetic field (Larmour frequency), B is the magnetic field intensity, and e and m are the charge and mass of an electron respectively. The electron's flight time down the tube may be controlled by adjusting the accelerating voltage between the photocathode and the phosphor screen. When this flight time is a precise multiple of the electrons' rotational period, precise focusing occurs (i.e., all electrons leaving a point on the photocathode arrive at a single point on the phosphor screen regardless of the direction or magnitude of their lateral velocities). Typically, magnetically focused tubes are operated at conditions that produce one to four electron rotations during their travel.

These magnetic-focused tubes have the advantage of forming nearly perfect images when they are inserted into a nearly ideal magnetic fields. Generally, the best magnetic fields are produced by solenoidal coils but these coils must dissipate considerable electrical power to produce strong enough fields. This problem is exacerbated for most camera systems because fields high enough to effectively eliminate distortions caused by the Earth's magnetic field must be achieved.

Development of electronic image-tube technology is especially exciting to the field of high-speed photographic instrumentation because several important advantages are offered, the most important of which is potential for light amplification. The image produced at the tube output screen may be brighter than the image delivered to the tube. The potential for light amplification is enormous because it offers a major step toward the solution of the fundamental problem of providing adequate lighting for recording images at very high speeds.

Electron-image tubes also have some basic disadvantages with regard to alternative shuttering devices. The internal operations of the tubes can limit spacial resolution of the transmitted images and can produce image distortion. They are also not yet capable of transmitting color information so that the advantages described earlier from using color photography for aesthetic reasons or to promote image analysis are denied to cameras using electron image tubes.

Let us now consider potential for light amplification afforded by the use of electron-image tubes. Typical photocathodes produce one emitted electron per five incident photons from the input image (an average quantum efficiency of 20%). An efficient phosphor on the output screen will produce approximately 200 photons for each incident electrons on it if the electrons are accelerated to 20 kV. The number of photons emitted is approximately proportional to the accelerating voltage in the range of 5 kV to, perhaps, 40 kV as

$$N_p \cong 0.01\ V_a (5\ \text{kV} \leqslant V_a \leqslant 40\ \text{kV}) \tag{4}$$

where N_p is the number of photons emitted from a phosphor screen per-incident electron and V_a is the accelerating voltage across the tube. Thus, the tube has a potential light amplification capability of $0.2 \times 200 = 40$ X when it is operated at 20 kV.

Early cameras built with proximity-imaging tubes failed to achieve this gain because they used a relay lens to image the output images onto photographic film. The efficiencies of these relay lens systems where typically 2.5% so that the overall camera system had an optical gain of approximately one.

The first step in improving this situation was to replace the relay lens with a coherent fiber-optic end plate at the rear of the photo tube. Light from the phosphor screen passes inot the ends of individual fibers that conduct it to the film clamped against the rear tube surfaces. Net efficiencies of this image transport mechanism are 60 to 70%. This factor, when considered with a basic optical gain of the tube (40 X) lead to total optical gains for the camera system of 20 to 30 X. This advance significantly aids the problem of illuminating high-speed events for detailed observation.

A much more powerful technological advance occurred with the development of microchannel plates. These plates are arrays of electron multipliers that are capable of increasing the number of electrons in clouds emitted by the photocathode by factors in excess of 10,000 X. Subsequent acceleration of these multiplied electron clouds against phosphor screens provide enough amplification to produce overall optical gains for the tubes between 100,000 and 500,000 X! Successful exploitation of microchannel plate technology has a potential of producing high-speed cameras that can operate with exposure times down to 10^{-9} s using even illuminations typical of ordinary photography.

The microchannel plate, which is the heart of the new optical amplification technology, consists of a glass plate approximately 1 mm thick through which are pierced a vast number of small diameter holes (see Figure 7). Typically, 10 μm diameter holes are spaced at the corners of 20 μm squares. The surfaces of the plate are electroplated to form conducting paths and the walls of the holes are treated to produce very high electrical resistance (typically 1000 V across the faces of the microchannel plate produces a current of 10 $\mu a/cm^2$). When an electron cloud is accelerated to several hundred volts before impinging on the front surface of the plate, electrons enter the holes in the plate and strike the internal walls. There, each electron impact liberates two to four secondary electrons that are accelerated down the hole by the electrostatic field and strike the hole wall. Each of these impacts produces two to four additional electrons. This process is repeated 20 or 30 times as the electron cloud proceeds toward the rear surface of the plate. On average, each electron entering the front of the microchannel plate produces tens of thousands of electrons emitted from the rear surface. The output electrons may then be focused on the output phosphor using proximity, electrostatic, or magnetic focusing.

Clearly, the image information of the incident electron beam is retained at the output of the microchannel plate but the spacial resolution of the resultant image is limited by the separation between adjacent holes.

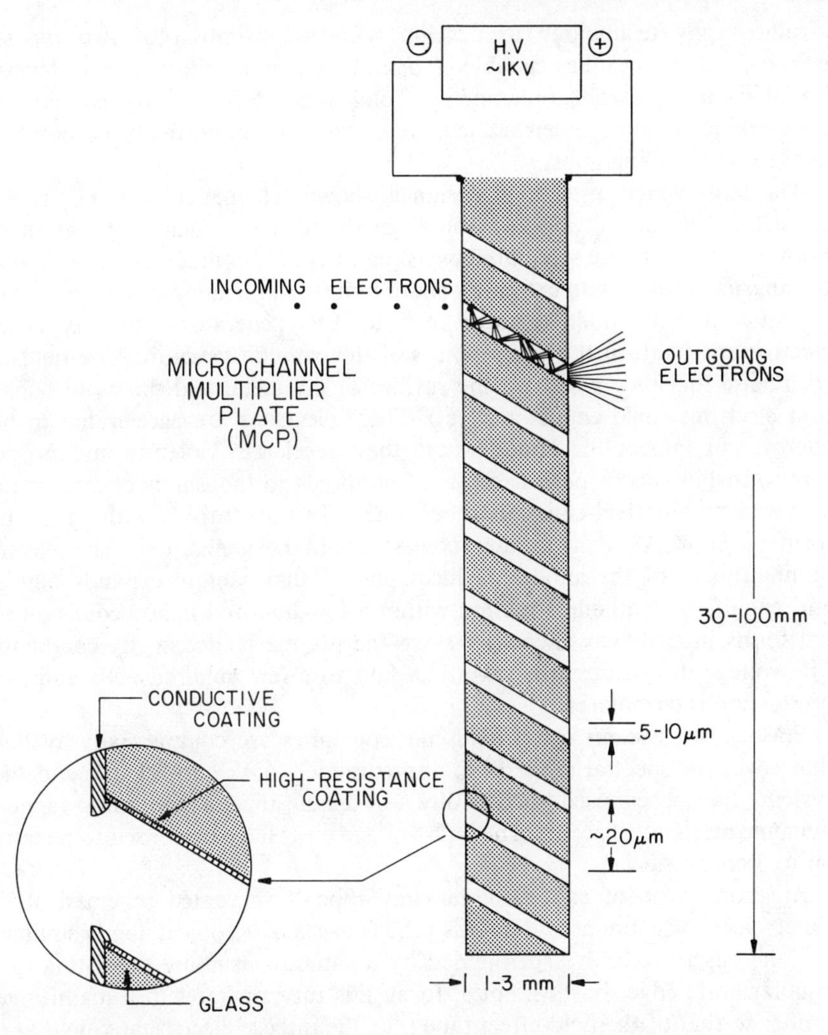

Figure 7 Operation of a microchannel plate used for amplifying electron-image clouds.

7.2.6 Flash Radiography

Flash radiography was described originally as a specific form of unfocused shadowgraph imaging. Its obvious power is that it can effectively "see through" dust and smoke associated with energetic mechanical events and can also visualize internal operations of solid devices made from opaque materials. Finally, flash radiograph is the only technique currently available for determining the dynamic local-density profiles of materials subjected to impulsive loading.

Typically, x-radiographs are exposed for periods ranging between large fractions of the second to tens of hours because conventional x-ray sources have intrinscially low intensities. The breakthrough that allowed high-speed x-radiography technology to develop was the invention of two types of cold-cathode x-ray tubes capable of operating at momentary current levels of hundreds to thousands of amperes. Total x-ray doses from each pulse are comparable to, but somewhat less than, total doses normally used to take conventional radiographs.

The most widely used tube design is shown schematically in Figure 8. It consists of a conical anode made from dense metal such as tungsten surrounded by a cathode structure consisting of combs containing large numbers of tungsten needles with extremely sharp points. Application of a high, positive potential to the anode (greater than 65 kV) generates extremely intense electrostatic fields at the sharp points of the cathode structure. The fields are so intense that they penetrate the surface of the metal and draw out conduction electrons (field-emission effect). These electrons are accelerated to high energy and impact the anode where they decelerate violently and produce x-rays. Instantaneous peak currents of hundreds to thousands of amperes are achieved within field-emission tubes when they are driven with fast pulse-forming networks that usually consist of Marx generators. The electron bombardment of the anode produces plasma that rapidly expands outward and engulfs the cathode structure within a few hundred nanoseconds and an arc forms immediately thereafter since the plasma is electrically conducting. The voltage drop across the tube then falls to a few hundred volts and x-ray production is terminated.

Flash x-ray systems using field-emission tubes are commerciálly available that cover the spectral range from approximately 65 kV to 2.1 mV and these systems have been used successfully for diagnosing an enormous range of dynamic mechanical events. The high voltage units have been used to penetrate up to 3 cm of steel.

A second type of cold-cathode x-ray tube is presented schematically in Figure 9, the vacuum-arc tube. This tube consists of a conical anode similar to the one described above surrounded by a cathode structure consisting of an annular knife edge. In its simplest form, this tube operates in a manner very similar to that of the field-effect tube (i.e., the intense electrostatic field at the

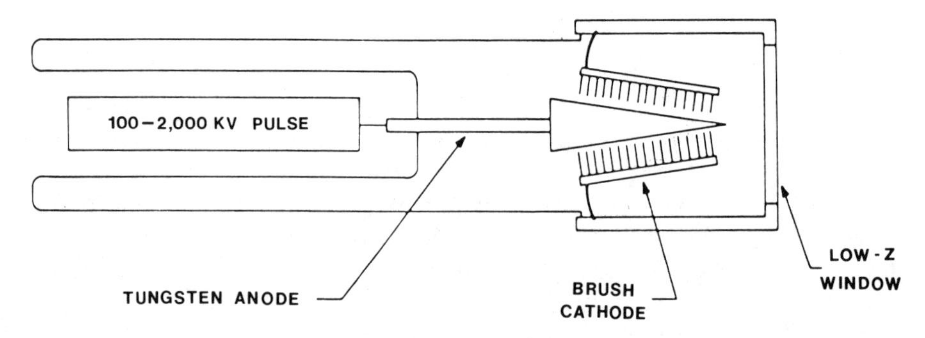

Figure 8 Schematic drawing of a field-emission flash, x-ray tube.

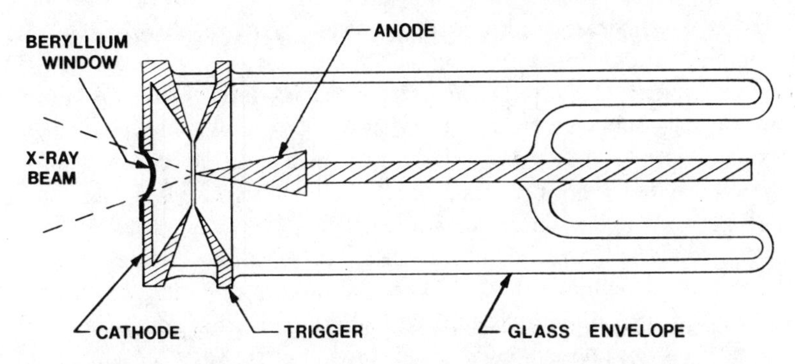

Figure 9 Vacuum-arc, flash x-ray tube with triggering capability.

knife edge produces an electron flow from the cathode that produces x-rays at the anode surface). Currents achievable with these tubes are similar to those available from field-effect tubes. The tubes are much simpler to design and construct but are not quite as stable in operation as tubes employing multiple sharp points.

A major advantage with the vacuum-arc tube is that it can be operated at voltages well below the minimum need to stimulate a field effect discharge by introducing a third electrode immediately adjacent to the cathode structure. Operation in this mode is achieved by connecting the tube across a pulse-forming-network that produces a steady potential between the cathode and the anode that is below the 80 to 90 kV that causes spontaneous tube breakdown. The trigger electrode is then excited with a 15 to 20 kV pulse between it and the cathode. The very short gap between these two electrodes (typically 100 μm) produces a strong enough electrostatic field to trigger electron emission that develops almost immediately into a vacuum arc. Electrons from the arc plasma are accelerated across the gap to the anode and produce the x-ray pulse. Plasma from both the trigger arc and the anode meet in the anode-cathode gap within a few hundred nanoseconds after tube operation commences to form a vacuum arc that terminates x-ray production. These triggered tubes have been used to produce x-ray pulses with voltages as low as 10 kV.

Thus, the combined use of the two x-ray tube technologies can provide pulsed x-ray sources with durations less than 2×10^{-7} with characteristic spectra ranging from below 10 kV to above 2 mV. Doses are sufficient to allow practical x-radiographs to be taken of many dynamic mechanical events.

These radiographs may be taken directly with photographic film when the x-ray spectra are maintained below 50 kV but require the use of intensification screens when higher voltage spectra are used because the film emulsion cannot intercept enough energy from the more penetrating radiation to achieve effective exposures. The highest spacial resolutions are achieved with screens made of thin lead foils clamped in intimate contact on both sides of the film.

Incident x-rays liberate electrons in the lead that are absorbed in the film emulsion and expose individual grains. The lead screens are rather "slow" (i.e., provide inefficient exposure or film). Less spacial detail is available but much greater speed is achieved from intensifier screens made with rare-earth phosphors mixed with high-density salts. These screens produce copious amounts of light when excited by x-rays and the light is transferred to film emulsions clamped between them. There is a trade-off to some extent between intensifier screen efficiency and spacial resolution of the resulting radiographic images. The screens must be made thicker to absorb more x-ray energy but the thicker screens tend to spread the light emitted from the adsorbed radiation. Typically, high quality radiographs taken with an intensifier screen may contain up to 5 lp/mm spatial resolution and radiographs taken where imaging sensitivity was maximized are limited to 2 to 4 lp/mm spacial resolution.

The most important parameter for controlling design of an experiment using x-ray imaging for the observation medium is the wavelength of x-ray spectrum. In general, higher voltage spectra penetrate objects under investigation more effectively than lower voltages but produce less image contrast. The best images are usually produced by using the softest (lowest voltage) radiation to produce high contrast that is consistent with providing enough penetrating power to produce a usable image. An advantage of working with extremely soft x-rays (below approximately 50 kV) is that the film emulsion itself becomes an efficient x-ray absorption medium and may be used without any form of intensification. This ability remarkably improves spacial resolution, contrast, and dynamic range of the radiographic images. A concomitant disadvantage of working at low x-ray voltages is the tendency of the x-ray beam to scatter from material in the event area that produces background fogging of the image.

In general, very soft radiation is optimum for investigating objects made exclusively from materials with low atomic weights. Voltages below 50 kV have proven effective for observing fiberglass cords in automobile tire treads. Intermediate voltages appear to be best for general light-duty use and especially for penetrating dust and smoke associated with violent mechanic events. Higher voltages are used only when deep penetration of high-density material is required. Again, the lowest voltage sources that can produce adequate radiographic images are always preferred.

7.3 HIGH-SPEED CINÉ TECHNOLOGY

High-speed ciné cameras visualize the motion of energetic mechanical events just as ordinary ciné cameras visualize the motions of events occurring at ordinary velocities. The critical parameters for determining cinemagraphic capability needed to achieve desired results are: the framing rate, the exposure time for each frame, and the total spacial resolution of the images format. Exposure time and spacial resolution may be treated in a manner identical to that suggested for single-frame cameras. The framing rate establishes the

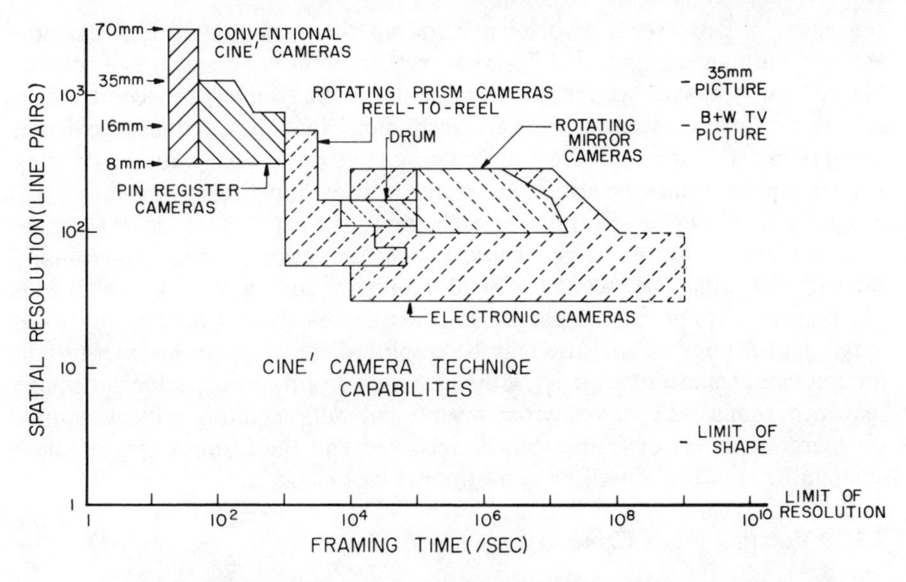

Figure 10 Capabilities of currently available high-speed ciné cameras.

amount objects under observation move in the view field between frames and hence determines the precision with which their motions can be determined. Nearly constant velocities can be evaluated to precisions of better than 1% if 10 position-time coordinates are available. Thus, interframe times of 10% of the period required for objects to cross the camera's field of view will produce excellent results for simple motion analyses. Many high-speed ciné cameras have fixed ratios between exposure time and interframe time. Thus, one must chose the higher framing rate consistent with proper image motion across the field of view and motion blurring of the individual images. Framing rate must also assure that the total time during which the event under investigation must be observed is available. Many of the faster ciné cameras produce only small numbers of images (typically 5 to 150) so that the framing rate must be set to allow the entire period of interest to be spanned.

The capabilities of available ciné camera concepts important for analyzing dynamic mechanical events are presented in Figure 10. Spacial resolution for each frame is plotted against time between frames for the technologies of interest.

7.3.1 Intermittent Motion Cameras (Pin-Registered Cameras)

Conventional ciné cameras operate with a mechanism that rapidly advances the film one frame at a time and stops it while a rotary shutter produces an exposure. The process is then repeated until the camera is turned off or the film supply is exhausted. The earliest efforts for obtaining high-speed ciné sequences of fast events involve speeding up the conventional ciné cameras action—with amazing results! Today several cameras are commercially available that can take ciné sequences at speeds up to 500 frames per second. It is a tribute to modern mechanical-design technology that the complex activity of accelerating a film from rest and bringing it to rest again while a rotating shutter exposes it may be accomplished repetitively in periods as short as 2 ms.

Typically, the exposure time for each frame is near 20% of the interframe time so that 400 μs duration exposures may be accomplished with framing rates of 500 frames per second. The film advance and shuttering mechanisms place little restraint on the use of objective lenses so that almost any focal length and *F* number aperture may be employed. The high-speed intermittent motion cameras are often used with color film to produce spectacular image sequences that can be viewed with conventional motion picture projectors since very large numbers of frames can be recorded and the formats can be made compatible with those used by conventional ciné cameras.

7.3.2 Rotating Prism Cameras

Although the performance of intermittent-motion ciné cameras is spectacular from a mechanical design viewpoint their performance is inadequate to observe the majority of high-speed mechanical events. The next stage in development of higher speed cameras was replacement of intermittent film motion with continuous film motion. Such replacement can be accomplished if the image being exposed on the film is made to move in synchronization with the film during the exposure time so that no sensible relative motion occurs. This motion compensation is usually accomplished by directing the camera light through a glass prism as shown schematically in Figure 11. Rotation of the prism in synchronization with film motion allows the light passing through the prism to be displaced laterally while maintaining parallelism of the individual rays. Thus, the entire image is moved downward at a velocity equal to that of the film motion that produces motion compensation.

Cameras using this technology have framing rates limited by the speed with which film can be moved through the camera and centrifugal burst strength of the rotating prism. Cameras are available commercially that take 16 mm

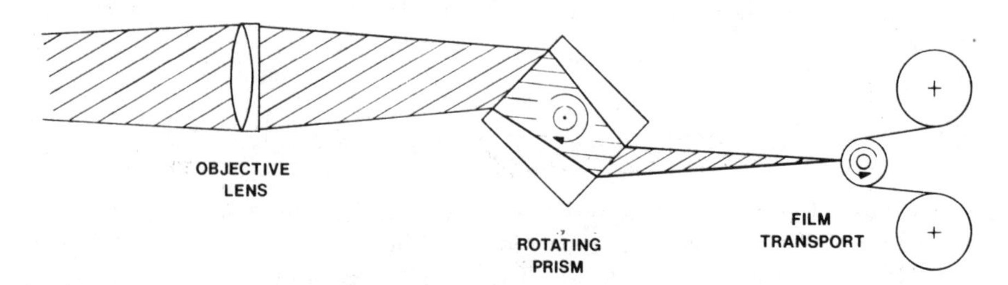

Figure 11 Rotating prism-camera technology.

frames at rates up to 11,000 frames per second and 35 mm frames at rates up to 5000 frames per second. Half-height and quarter-height 16 mm frames have been taken routinely at rates up to 22,000 and 44,000 frames per second respectively for special-purpose application. Shuttering in one form or another is used to maintain fixed ratios of exposure time to interframe time. Several modern designs of rotating prism cameras allow different shuttering discs to be installed to vary this ratio from 40% down to below 10% (for full-height frames). Thus, exposure times as low as 10 μs can be achieved with 16 mm ciné sequences.

The prism assembly places some restrictions on the minimum F numbers and back-focal lengths of objective lenses used with rotating prism cameras. Especially lenient designs can accommodate lenses with 50 mm back-focal lengths with F numbers as low as 2.0. Other designs require back-focal lengths up to 80 mm and restrict F numbers to values above 4.0.

Film strips from full-frame rotating prism cameras can be projected with conventional motion picture equipment since formats can be made compatible with standards for normal-speed equipment, and sufficient frames are available for extended presentations. Color film is often used with rotating prism cameras and aesthetically spectacular sequences are often achieved although rotating prism cameras lack the photographic crispness and brilliance of fine-quality, intermittent-motion cameras.

7.3.3 Rotating Mirror-Drum Cameras

The limitations on the speeds of the rotating prism cameras caused by peak film velocities and centrifugal burst strengths of the prisms can be overcome to some extent by mounting a relatively short strip of film on the inside surface of the high-speed rotating drum and replacing the prism mechanism for image motion compensation with a rotating mirror assembly. The drum containing the film is accelerated to maximum velocities several times higher than those achieved by reel-to-reel mechanisms during several tens of seconds of torque application. The small components needed for a rotating mirror-motion compensator are also capable of operating safely at velocities considerably higher than those achievable with relatively massive glass prisms used in rotating prism cameras. Sequences of 16 mm frames may be taken along individual strips at rates up to 25,000 frames per second and individual exposure time as short as 4 μs can be achieved. An elegant extension of this technology is available in a camera that uses a more complex image-motion compensator to take four separate framing sequences in quadrature with one another (i.e., the four-strip sequences operate independently from one another except that they are synchronized to produce frames at four times the rate achieved by any one strip). The four strips of 16 mm frames are exposed on a single strip of 70 mm film to produce several hundred images taken at rates up to 100,000 frames per second with individual exposure times near 4 μs.

The capabilities of rotating mirror drum cameras are clearly formidable and one example stands along as a practical photographic instrument capable of operating at taking rates near 100,000 frames per second. Rotating drum-framing cameras have several notable disadvantages that are inherent to their operation. First, the number of frames that can be taken during an individual operation is limited to at least some extent. Second, the format does not lend itself to motion picture projection without laborious reprinting of the individual frames onto a single strip of film. Third, the F number of the optics system and the back-focal distance of the lens are restricted even more severely than for rotating prism cameras. Finally, the use of multiple mirrors in the optical path reduces the optical sensitivity of the camera below that predicted by the F number and has a subtle but real degrading effect on image quality.

7.3.4 Rotating Mirror Cameras

Rotating mirror cameras were invented before World War II and provided an enormous boost to the field of high-speed photography by allowing framing rates approaching 10 million frames per second with exposure times down as low as 5×10^{-8} s to be achieved. These capabilities were exploited massively during World War II when rotating mirror cameras became the principal instruments supporting the development of explosive components for use in nuclear weapons.

The basic operation of a rotating mirror framing camera is presented schematically in Figure 12. The camera objective lens forms an image on the face of a mirror rotated to high speed by a gas turbine. From 5 to 144 individual cameras are mounted so that each of their objective lenses image the rotating mirror onto a single strip of photographic film. Each individual camera receives light only when the rotating mirror is positioned at an angle to reflect light from the main objective lens into it. Thus, the rotating mirror performs as a high-speed switch directing light from camera to camera to produce a ciné sequence.

Rotating mirror framing cameras have the obvious disadvantage that the total number of frames per taking sequence is severely limited. Special multiple reproduction techniques are needed to produce motion picture sequences suitable for production. Another severe limitation is that the internal complexity of the camera limits F numbers of the optical systems to at least 10 and more typically 15–25. The light gathering efficiencies of the camera are curtailed even more severely than the F number restriction by the multiple reflection required of the light beams. In general, rotating mirror framing cameras require especially intense illumination to produce adequate high-speed photographic operation.

Two types of rotating mirror cameras have been produced. The first or “synchronous” type contains secondary cameras arranged over only 60 to 90° of the circle surrounding the rotating mirror. The camera thus becomes sensitive as the mirror moves into position for illuminating the first frame and

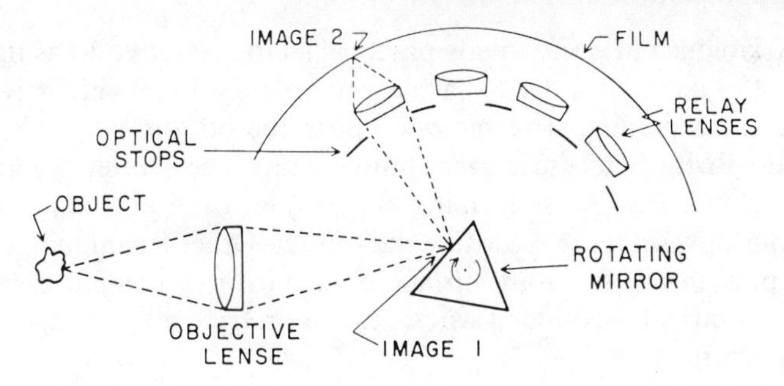

Figure 12 Operation of a rotating mirror ciné camera.

remains sensitive until the last frame is exposed. The camera then becomes insensitive for periods up to four times the duration of the sensitive period before the process is repeated. This camera design approach allows use of a flat-plate rotating mirror shape whose reflective face may be located very close to the axis of rotation. This feature greatly simplifies the design of the camera and enables this design to be perfected to the point where truly spectacular ciné sequences can be recorded in either black and white or full color. The literature abounds with brilliant visualizations of energetic events such as exploding wires, actuation of explosive devices, hypervelocity impacts, and so forth.

The primary disadvantage of the synchronous design approach is that means must be provided to trigger the event under observation as the camera becomes sensitive to assure that the camera can operate properly.

The second design or "continuous-access" rotating mirror camera employs secondary cameras over a wider angle surrounding the rotating mirror and a multisided mirror to assure that the first frame in the sequence is exposed immediately after the last frame of the preceding sequence has been exposed.

This camera design possesses the major advantage that it can be employed to observe events whose timing cannot be controlled accurately as long as event initiation can be detected early enough to turn on appropriate lighting.

Continuous access cameras have three major disadvantages. First, the optical quality of the images is degraded somewhat by the necessity of mounting the reflective surfaces of the rotating mirror some distance from the mirror axis. Second, the cameras are much more costly than similar synchronous cameras. Third, extreme measures must be taken to prevent rewriting a second sequence of images over previously exposed frames since no "dead time" is available for extinguishing the scene lighting.

A number of extreme mechanical devices have been employed as capping shutters to blind continuous access cameras quickly enough to prevent rewrite. One design that has been sold commercially employs four small blasting caps to shatter a thick glass block contained between two thin plastic sheets that prevent dispersal of the resulting debris. Crazing waves produced by the explosive charges in the blasting caps can "close" a 75 mm aperature within 50 μs. A more frivolous but occasionally practical shutter referred to as the "Betty Crocker" shutter contains a dead space approximately 1 cm wide between two glass plates. An exploding wire mounted near the bottom edge of the dead space is covered with flour during the shutter setup. The shutter is actuated by discharging an electrical pulse-forming network into the exploding wire that blows the flour upward thereby closing the shutter. Other techniques, including electrical explosion of the front surface of a mirror, the rapid dispersal of plasma from electrically exploded wires, and so on have all been used successfully as blast shutters.

7.3.5 Cranz-Schardin Spark Camera

The Cranz-Schardin spark camera like the rotating mirror cameras opened the possibility for taking ultra high-speed ciné sequences many years ago, and these units contributed significantly to the study of high-speed dynamic mechanical events. The Cranz-Schardin camera is essentially a ciné version of focused shadowgraph cameras described in Section 7.2.2. A number of spark gaps are mounted in a plane (see Figure 13). The condensing lens images each of these gaps at a separate point in the plane containing the camera objective lens. A special camera with a separate objective lens located at each spark-gap image point is used to record the camera output. Sequential firing of the spark gaps produces a ciné sequence of images of the event under observation. Cranz-Schardin cameras containing as many as 121 (11^2) shadowgraph channels have been constructed and used successfully.

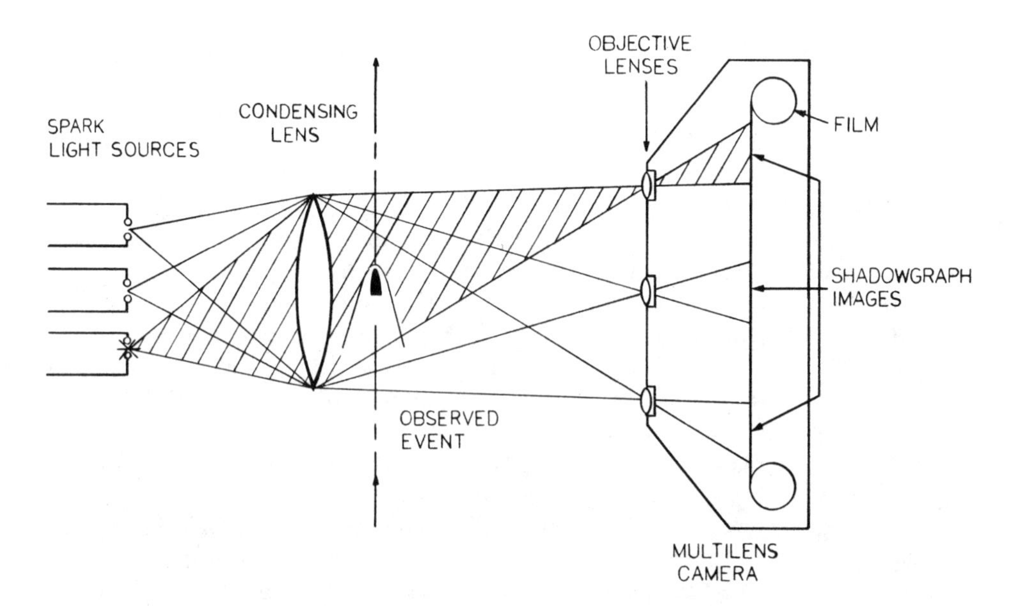

Figure 13 Cranz-Schardin multiframe spark-shadowgraph camera.

This type of camera possess the enormous advantage that the framing rate may be adjusted independently of the exposure time for each frame. In principle, the framing rate need not be confined to a single value but may be tailored specifically to the experimental requirements. For instance, a Cranz-Schardin camera might be set to observe a rod penetrating a target of medium thickness. The first few frames are taken with relative long durations between them to observe the velocity and orientation of the incoming projectile. A rapid burst of frames might then be recorded to observe the initial impact under high temporal resolution. The camera might then be shut down for the time required to perforate the target and turned on again at high speed to observe rear-surface breakout. The final frames may then be taken with relatively large separations between them to analyze the motions of debris particles projected behind the target.

Perhaps the single most important disadvantage of Cranz-Schardin camera technology is that it is strictly limited to taking shadowgraph images. Two other disadvantages of varying importance are: each frame has a field of view slightly different from the others, which complicates extraction of quantitative motion data from the records to some extent; and the framing format is not suitable for motion picture projection unless the relatively arduous task of reprinting the frames in a standard strip sequence is accomplished.

7.3.6 Electronic Ciné Cameras

The ciné cameras described thus far are mechanical or quasimechanical instruments. Their framing rates span most of the range of interest for taking ciné sequences of macroscopic mechanical events. Their principal limitations are in the areas of light-gathering efficiency and the fact that some mechanical events involving very small objects moving at very high velocities are beyond their recording capabilities.

Electronic ciné cameras have become important during the past two decades primarily because of their greater efficiency in use of light and their ultra-high framing rate capabilities.

High-speed electronic ciné cameras use an electron-image tube similar to the one sketched schematically in Figure 14. The tube operates as the one described in Section 7.2.5 except that two sets of opposing deflection plates are used to deflect the relatively small images of the photocathode onto different regions of a large phosphor screen. Movement of these images from point to point on the screen surface has been accomplished at rates required to produce image sequences at 3×10^9 frames per second. An elegant technique has been used to double this framing rate by splitting the camera viewing format into two parts and extending the pathlength of one of these parts 50 cm with respect to the other. The light that forms the image on the photocathode travels at 30 cm/ns, so delaying the beam 1.6 ns causes the half-frame to be recorded midway between the electronic interframe time, thus producing a doubling of the framing rate.

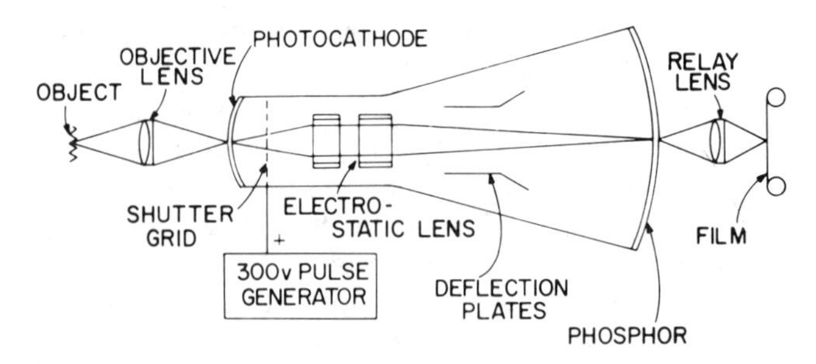

Figure 14 Electrostatically focused image tube with image-deflection capability.

Modern electrostatically focused image tubes employ coherent fiber-optic end plates that allow optical gains of up to 100 through the cameras. Essentially no limitation is placed on the back-focal length or F number of the objective lens. Thus, ciné camera configurations with formidable light-gathering capability can be produced.

Recently microchannel plate technology has been applied to enormously increase the light-gathering capabilities of electronic ciné cameras. At present, the microchannel plate is enclosed in a separate proximity-focused image tube with fiber-optic input and output windows. This second (amplifier) tube is "rung" onto the output of the ciné-imaging tube so that the two form a tandem pair. Greater efficiency will probably be achieved in the near future by adding microchannel plates in the cavity of the framing tube itself at a position immediately in front of the phosphor screen. Incoming electron-image analogues would then be multiplied in the channel plate before receiving final acceleration on to the phosphor screen.

Work is also underway on the development of a medium-speed electronic framing camera that uses magnetic focusing and deflection in a very large electron-image tube to provide frames of unusually high spacial resolution and low geometric distortion. This camera contains a microchannel plate that will allow it to operate at gains over 100,000 X. The camera is expected to take pictures at rates up to 250,000 frames per second with the spacial resolution of fine-quality television frames.

In general, the electronic ciné cameras' weakest points are that frame size and resolution limit information content of individual frames relatively severely. Such cameras are also basically black-and-white instruments with no color rendering capability.

7.3.7 X-Ray Cinematography

X-ray cinematography is a technology that extends the x-ray shadowgraph technology discussed in Section 7.2.6 to the production of ciné sequences that allow detailed motion analyses to be conducted through opaque view fields. Two basic approaches are available for pursuing this activity. Both approaches have been investigated experimentally, and both of them require the extreme-light amplification capabilities of modern-image converter cameras.

The first approach employs a quasisteady x-ray source to illuminate the photocathode surface of a large x-ray intensifier tube. Tubes are currently available commercially with photocathodes up to 30 cm diameter. Incoming x-rays are first converted to visible light in a scintillator screen mounted on the front surface of the tube and the visible light is then converted to an electron-image analogue at the photocathode surface. The electrons from the photocathode are focused electrostatically onto a small diameter output phosphor after being accelerated to 30 kV. The image on the output phosphor is approximately 10,000 times brighter than the image falling on the photocathode because of the geometric demagnification and the effects of energetic electrons striking the phosphor. The output phosphor from the x-ray intensifier tube is observed with a high-speed electronic ciné camera that provides enough light amplification to effectively eliminate the losses caused by the relay lens optics between the intensifier tube and the photocathode of the ciné camera.

This x-ray ciné system suffers from three principal limitations. First, the view field is limited by the diameters of available x-ray intensifier tubes (typically 30 cm). Second, the observation of energetic events necessitates a difficult trade-off between safety of the expensive x-ray intensifier tube that requires separating the photocathode from the event and introducing protective layers between them, and spacial resolution of the event that requires the photocathode to be mounted close to the event. Third, a limitation exists on maximum framing rates that can be achieved. These framing rates are usually limited by the light-output characteristics of the x-ray intensifier screen and the phosphor screen at the output of the image-intensifier tube. Clearly one image must disappear from both screens almost completely before the second image can be recorded if "ghosting" is to be avoided. To date this limitation has prevented framing rates from exceeding $5 \times 10^4 \ s^{-1}$. Finally, the use of a continuous or quasicontinuous x-ray source severely limits the dose rate that can be produced without overheating the x-ray tube.

The second approach involves using flash x-ray generators similar to those described in Section 7.2.6 to produce the x-ray illumination. A free-standing scintillation screen is used to convert the x-ray images to visible ones. An electronic framing camera with microchannel plate amplification is used to observe the screen in synchronization with actuating each of the flash x-ray generators.

This approach allows the relatively inexpensive scintillation screen to be positioned close to energetic experiments. The criteria for experimental success shifts from screen destruction to destruction of the screen before the last frame of information is recorded. The great intensity of x-ray beams from flash sources eliminates difficulties associated with source brightness. The principal disadvantage of this approach is that the sources must be separated by at least small distances, which insures that each frame has a viewing angle slightly different from the others. This disadvantage may be converted to an advantage by positioning each of the sources to provided its own best angle for minimizing parallax between the event under observation and the x-ray sensitive surface.

7.3.8 High-Speed Television

All of the equipment described thus far employs photographic film for data recording. Photographic film technology is mature and wide varieties of emulsion sensitivities, spectral responses, contrast ranges, and dynamic ranges are currently available in both black and white and full color. The difficulty with film is that it must be developed before it can be observed and it is relatively expensive if used in large quantities. Both of these problems can be solved for many situations by substituting television cameras for film cameras.

The problem with this approach until recently has been that the serial nature of television framing has severely limited the maximum-achievable recording rates. This difficulty is rapidly disappearing. Units with framing rates up to 400 frames per second have been developed into complete motion-analysis systems and a television camera-recording system capable of operating to rates up to 2000 frames per second has been announced. These systems greatly enhance their photographic capabilities by being employed with high-speed pulsed Xenon flash lamps that can be operated in synchronization with camera operation. The light flashes produce exposure times of a few microseconds to effectively eliminate motion blurring even when the cameras are operating at relatively modest framing rates. Systems operating at hundreds of frames per second can be synchronized with operation of repetitive equipment such as machinery used for manufacturing small components so that the flash lamps can serve as stroboscopes. Introduction of slip in the synchronization path allows complete high-speed repetitive operations to be examined in detail with equipment employing very modest framing rates.

7.4 SMEAR AND STREAK PHOTOGRAPHY

Smear and streak photography concepts have been applied almost exclusively in the fields of high-speed photography and hence are little known to the general technological public. They produce specialized images of the object under investigation that are essentially position-versus-time plots of motion along an accurately predetermined line across the object plane.

The simplest of the devices in this series is a smear camera which, in its basic form, consists of an objective lens that images the object of interest onto the surface of a continuously moving film transport (see Figure 15). If the object is some light-emitting device such as cathode ray tube (CRT) in an oscilloscope with the sweep circuitry disabled, the film motion can provide the sweep action so that voltage variations with time (which produce movement of the beam across the CRT face) can be resolved for extended time periods. One of the

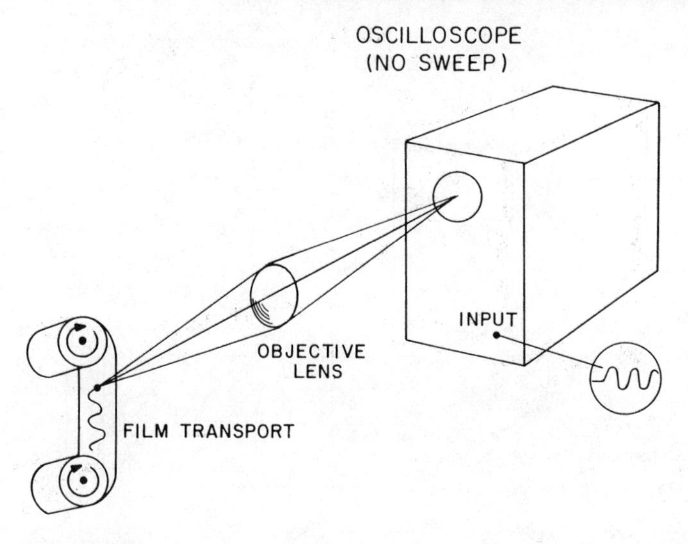

Figure 15 Smear-camera concept for recording extended sweep durations from an oscilloscope.

most important factors considered in the operation of such a camera is that it observes the object continuously rather than for a brief instant as is the case for a single-frame camera or as series of individual instances as in the case of a ciné camera.

The streak camera represents full development of the smear-streak concept. Here an objective lens images the event under study onto an opaque surface through which is cut a narrow slit. A relay lens images the slit across a film transport so that continuous illumination of the slit produces a stripe from one end of the film to the other as is shown in Figures 16 and 17. If the slit is positioned to intercept the image of a projectile traveling along its trajectory so that the images move along the slit, an unexposed strip will be generated across the exposed stripe as appears in Figure 16. The tangent of the angle the strip makes with a film velocity vector, θ_s, is a function of the velocity of the projectile, U_o, the velocity of the film, U_i, and the optical magnification of the system, M.

$$\tan\theta_s = \frac{MU_o}{U_i} \tag{5}$$

A completely different image-formation process occurs if the slit is positioned so that the image of the object under observation crosses it rather than travels along it as shown in Figure 17. Here a high resolution image of the object is wiped on to the film. These images have several interesting characteristics. First, the magnification in the spacial direction, M_s, and magnification in the temporal direction, M_v (the directions perpendicular and parallel to that

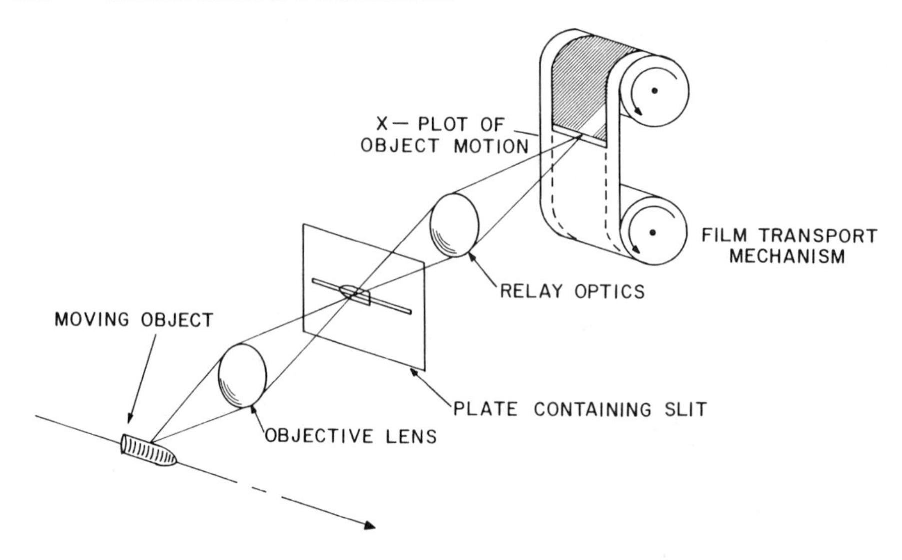

Figure 16 Streak-camera concept operating in "parallel slit" mode.

of the film motion respectively) are independent of one another as:

$$M_v = \frac{U_o}{U_i} \tag{6}$$

$$M_s = M \tag{7}$$

Magnification in the spacial direction is simply the systems optical magnification but the magnification in the temporal direction is the ratio of the velocity of the object to that of a slit image across the film as can be realized by noting

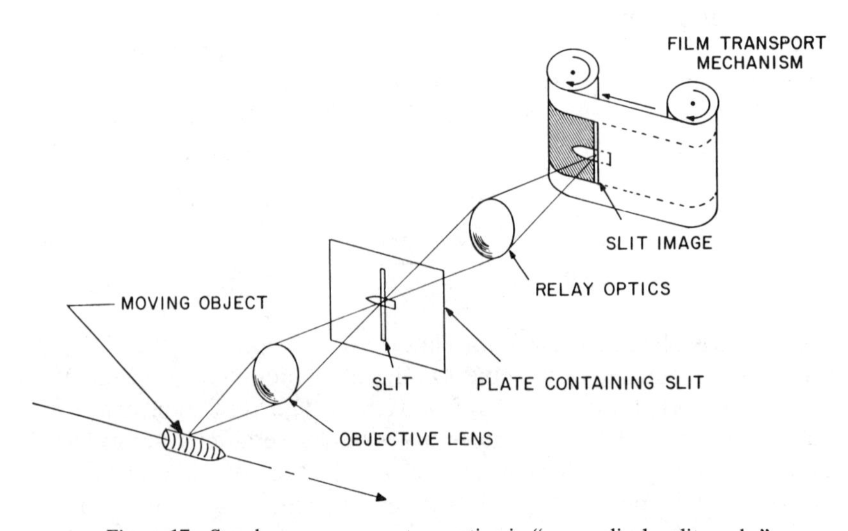

Figure 17 Streak-camera concept operating in "perpendicular slit mode."

that the time required to lay down the image is identical to the time required for the object to pass a line in space. The second and probably most important characteristic of a streak-produced image is that the camera remains sensitive continuously during its entire operating period (i.e., no blind periods are interposed between the sensitive periods as the camera observed its line in object space).

7.4.1 Reel-To-Reel Streak Cameras

The simplest smear and streak cameras are essentially motor-driven film transports with objective lens mounted in front of the film gate. No intervening optics are present between the film and the objective lens. An addition of a slit in front of the objective lens on which it is focused and the second objective lens, which images the event under investigation onto the slit surface, completes the design of a simple streak camera.

The resolution time of such a camera is simply the time required for the slit image to traverse the resolution limit of the relay lens operating in conjuncion with the film. A high performance reel-to-reel camera can reach ultimate resolution times as short as 2×10^{-7} s! Higher velocity and proportionally shorter resolution times are achieved by substituting a rotating drum-film transport system for the reel-to-reel one. Typical peak speeds near 300 m/s can be achieved with such devices that produce resolution times as short as 7×10^{-8} s.

7.4.2 Rotating Mirror-Streak Cameras

Rotating mirror-streak cameras can produce time resolutions down to a fundamental limit of approximately 10^{-9} s. The operation of a rotating mirror-streak camera is shown schematically in Figure 18. The relay lens images a slit onto a length of stationary film via the rotating mirror that is spun at high speed by a gas turbine. In principal, the speed at which the slit image can be wiped across the film can be made as large as desired by

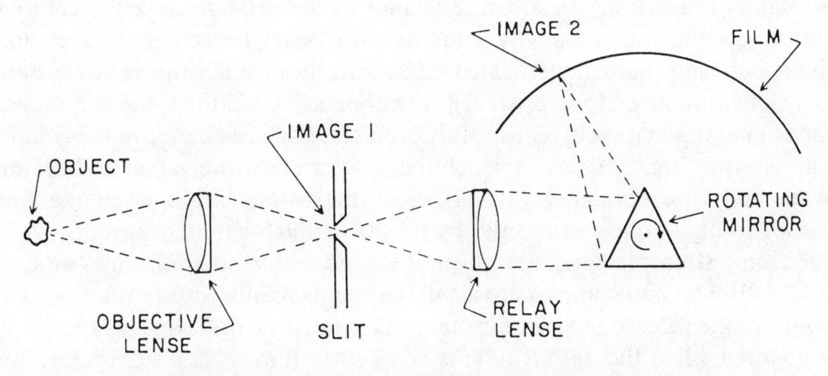

Figure 18 Operation of a rotating mirror-streak camera.

lengthening the distance between the rotating mirror and film plane indefinitely. Typically, top speeds for the slit image moving across the film plane are limited to 30 mm/μs (km/s) because system defocusing caused by unavoidable mirror distortion develops at a rate rapid enough to offset greater-potential time resolutions produced by increase slit image speed. The capability of resolving a single nanosecond with an all mechanical system still remains a formidable development, however.

Rotating mirror streak cameras are limited to using high F number optics ($\geqslant F/10$) because of the large separations required between the relay lens and the film plane. This limitation makes the cameras relatively insensitive optically so that intense lighting is required to produce adequate film records.

7.4.3 Electronic Streak Cameras

Electronic streak cameras are used extensively for two basic reasons. First, they are capable at operating at much lower light levels than mechanical cameras because they use fast optics and have optical amplification, and they can be made to resolve times far less than 1 ns.

Electronic streak cameras use electron-image tubes very similar to those used for electronic ciné cameras (see Figure 14). In fact, ciné-camera tubes are often used for streak-camera applications and several combinations of streak and ciné electronic cameras are available commercially. The electron cloud produced at the photocathode is accelerated toward the oversize phosphor screen and is focused on it by an electrostatic lens system. It then passes between one set of deflection plates on which a linearly increasing voltage ramp is applied. The deflection of the image across the phosphor screen is continuous and travels at nearly a constant velocity, which satisfies the conditions for a smear camera. Restricting the cameras view field to a slit with either external optics or baffles extends the camera system to the streak mode.

Specialized streak electron-image tubes have been developed whose internal geometries are optimized for high-speed operation. One of the more interesting specializations is replacement of the areal photocathode by a thin strip cathode, which is all that is required for streak operation. This adaptation significantly reduces the electrical resistance of the cathode since the cathodic material has medium resistivity and the area near the center of a circular cathode can only be communicated electronically through the resistive outer areas. Substitution of low-resistance interconnective coatings for the unused cathode material markedly reduces this problem. High resistance in the cathode circuit tends to limit the rate at which the tube can be switched on and off and also can produce voltage gradients across the cathode face when the tube responds to high-optical intensities by producing high-cathode currents.

Electronic streak cameras have been designed with resolution times as short as 10^{-13} (100 fs) and cameras are available commercially with resolution times down to single picoseconds. The cameras have intrinsic optical gains of up to a few hundred when the output tube is coupled to film or other photosensitive media with a coherent fiber-optic plate. Microchannel plates have been added to electronic streak camera systems both inside the tube (near the phosphor screen) and also in a second tube operated in tandem with the streak tube.

An interesting modern innovation of electronic streak cameras has been replacement of photographic film in small electronic streak cameras with a television camera tube that allows the resultant image to be read out of the instrument shortly after it is operated as an electronic signal. The result may be viewed immediately, stored, or processed electronically for enhancing image definition or computing engineering data.

The principal technical disadvantage of electronic streak cameras is that their recording times are relatively small multiples of their resolution times (typically 200 X to 600 X). Thus, synchronizing an electronic streak camera with an event under investigation may become an extremely complex task. High-performance electronic cameras generally do not have precisely constant deflection speeds so that relatively complex data analysis procedures must be followed to extract detailed quantitative data from the records.

A related problem to the short recording time is the limited amount of spatial information (information along the direction of the slit image) that can be recorded. Typically only 100 to 300 line pairs are available in the spatial direction of high-performance camera records.

Research is currently under way to alleviate this problem significantly for medium-performance electronic streak cameras by employing the potentially greater spatial resolution of magnetically focused and deflected image tubes. Streak records with over 1000 line pairs of spatial resolution and 600 to 1000 line pairs of temporal resolution are expected.

The difficulty with this approach is the higher energy densities of required magnetic fields beyond electrostatic fields with similar electron-deflecting performance reduces the peak deflection speeds available. Magnetic cameras have potential for resolving, perhaps, 10^{-10} s rather than fractions of 1 ps resolvable with electrostatic counterparts.

Generally, the technologies of high-speed photography and the general means of data recording are now adequate to meet the needs of researchers. This balance is very dynamic and has been lost several times during the past two decades. An extreme example is the recent intense activity to improve the performance of electronic streak cameras to provide time resolutions needed for observing radiation-material interactions driving laser-induced thermonuclear fusion experiments.

Current developments in the field are concentrating on exploiting technologies that have been generated already such as the use of a variety of flash x-ray and electronic camera developments to achieve a practical high-speed x-ray cinematography capability.

It is always difficult to predict future developments—especially in a field of technology that has produced so many breakthroughs and innovations as has high-speed photography. I do feel safe in predicting: the advent of practical x-ray cinematography systems, the appearance of single-frame ciné and streak cameras that fully exploit the optical amplification capabilities of microchannel plates, and the extension of high-speed television capability and applications—especially in industrial fields where it will be used to observe operations of reciprocating equipment.

BIBLIOGRAPHY

High-speed photographic technology and instrumentation techniques have produced a vast body of literature. Much of this literature is buried in texts of technical reports whose titles and principal subjects reflect end results of the investigations.

Notable exceptions are proceedings of two series of international congresses that treat instrumentation technology specifically. The first is the International Congress on High Speed Photography (and Photonics) and the International Congress on Instrumentation in Aerospace Simulation Facilities (ICIASF). References for recent proceedings for each of these series follow.

1 *Proceedings of the 1969 International Congress on Instrumentation in Aerospace Simulation Facilities*, published by IEEE Transactions on Aerospace and Electronic Systems, Farmingdale, New York (USA), May 1969.

2 *Proceedings of the 1971 ICIASF* published by IEEE/AES Rhodes Saint-Genese, Belgique, June 1971.

3 *Proceedings of the 1973 ICIASF* published by IEEE/AES Pasadena California (USA), September 1977.

4 *Proceedings of the 1975 ICIASF* published by IEEE/AES, Ottawa, Canada, September 1975.

5 *Proceedings of the 1977 ICIASF* published by IEEE/AES, Shrinveham, England (UK), September 1977.

6 *Proceedings of the 1979 ICIASF* published by IEEE/AES, Monterey, California (USA).

7 *Actes du 10'e Congres International Congress on High Rapide*, Nice, France, September 1972.

8 *Proceedings of the 11th International Congress on High Speed Photography*, London, England (UK), September 1974.

9 *Proceedings of the 12th International Congress on High Speed Photography*, Toronto, Canada, August 1976.

10 *Proceedings of the 13th International Congress on High Speed Photography and Photonics*, Tokyo, Japan, August 1978.

A third source of extremely valuable information are the proceedings of 29 meetings of the Aeroballistic Range Association (ARA). The ARA meetings are attended by technical staff members from most facilities conducting impact research who present papers on measurement technology, instrumentation, ballistic range operations, and so forth. All of the papers are technique-oriented—which is rare. Unfortunately, the ARA is a closed organization and their proceedings are unavailable to nonmembers. I recommend that you contact me or a organization member with specific requests for information. We can put you in touch with appropriate authors who may or may not make available copies of their ARA manuscript.

Finally, all general instrumentation areas are covered rigorously in texts available in any good technical library. Vendor organizations for high-speed photographic equipment are also excellent sources of general information as well as specific facts about their products.

SYMBOLS

B	Magnetic field intensity
L_o	Object length in direction of motion
M	Optical magnification
M_s	Image magnification in the spatial direction
M_v	Image magnification in the temporal direction
N_p	Number of photons emitted from a phosphor per-incident electron
R_o	Image resolution
U_i	Velocity of slit image across film
U_o	Object velocity
V_a	Accelerating voltage
W_o	Electron rotation frequency
X_i	Width of observed area
e	Electronic charge
m	Mass of electron
β	Magnetic-field intensity
τ_i	Exposure time
θ_s	Angle between streak record and the image-velocity vector

8

MATERIAL BEHAVIOR AT HIGH STRAIN RATES

Theodore Nicholas

The solution to problems in impact mechanics requires the application of the basic laws of mechanics and physics, as well as a description of the behavior of the material being considered. The mathematical description of the relationship between the stresses, strains, and their time derivatives is referred to as the constitutive equation of the material. Materials are often broken up into categories such as linear-elastic, nonlinear elastic, viscoelastic, viscoplastic, and so on. Each description refers to a mathematical representation for a class of idealized material behavior.

In describing the relationships between stress and strain and their time derivatives for a material, note that both stress and strain are point functions or tensors, thus a constitutive relation relates stress and strain at a point. A homogeneous material is defined as one having the same constitutive equation at all points. Here, constitutive relations that involve stress or strain gradients or spatial derivatives, will not be considered, that is, stress at a point will not be considered to be related to strain at other than the same point.

In dealing with stress and strain at a point we are dealing with tensors, each of which have six independent components. In the most general case, therefore, the relations between six stresses (three normal stresses and three shear stresses) and six strains (three normal strains and three shear strains), their time derivatives, as well as any other functions such as state variables, histories, or integral functions necessary to describe the material behavior, must be considered. In general, the constitutive relationship necessary to completely and accurately describe a material's behavior can be extremely complex and mathematically intractable. Thus, simplified models, restricted classes of material behavior, and mathematical idealizations are utilized in the solution of engineering problems.

For a linear-elastic material, the most general relation between the six stresses and six strains involves 36 constants; symmetry considerations reduce this number to 21 independent constants. For an isotropic material, one that has a constitutive equation independent of the orientation of the coordinate axes in the material, the number of independent constants reduces to two. The discussion in this chapter is confined to isotropic material behavior for simplicity.

In describing the behavior of materials, it is common to break up the total strain or strain rate into elastic and plastic components. In addition to describing the rate-dependent or flow properties of the material, we must also describe the yield behavior of the material when unloading or reversed loading occurs. Thus, we generally refer to the subject as dynamic plasticity. For simplicity, and in accordance with experimental observations, the stress and strain components are decomposed into their hydrostatic or spherical and deviatoric parts. The spherical parts, that is, average or mean stress (pressure) and average or mean strain (volume or density) are considered to be related elastically at lower pressures, and to be related through an equation of state at higher pressure levels where thermodynamic influences must be considered. The deviatoric components, or shear stresses and strains, include the description of the plastic behavior or yield function of the material, as well as any strain-rate effects. Shock wave or hydrodynamic theory primarily deals with the hydrostatic components because of their dominance, while yield surfaces and strain-rate effects are primarily associated with the shear components. The coupling or interactions between shear components and hydrostatic components are usually second-order effects and are generally ignored for mathematical or computational simplicity.

The problem that faces the materials scientist or experimentalist is how to determine the constitutive relationship that best describes a given material or class of materials. Because of the large number of components of stress, strain, and their time derivatives that may be involved in the formulation, experimental investigations tend to focus primarily on one-dimensional states of stress or strain. In dynamic testing, inertia and wave-propagation effects complicate the stress and strain states and often must be considered. This chapter concentrates on some of the problems and assumptions in determining dynamic material behavior at high rates of strain where wave-propagation effects may either be ignored or treated in a simple fashion. Chapter 4 discusses wave-propagation phenomena and their relation to dynamic material behavior. Dynamic properties of materials will be treated here from a purely phenomenological point of view. No attempt will be made to treat the fundamental mechanisms that control dynamic deformation. This subject, while vital to the understanding and developments in dynamic plasticity, is treated extensively in the literature and is referenced where appropriate.

The topic of the behavior of materials at high rates of strain has been one of considerable interest since World War II when dynamic plasticity and plastic-wave propagation first received attention. Since that time, a number of conferences and symposia have been held dealing with the subject and an extensive amount of literature has been generated on the subject area. No attempt is made here to reference all of the significant publications that have appeared. Rather, selected references are cited to illustrate specific points and, in some cases, to establish a historical perspective. A general discussion of impact phenomena can be found in Goldsmith (1960). The proceedings of the various conferences and symposia on behavior of materials at high rates of strain are to be found in Shewmon and Zackay (1961), Huffington (1965), Lindholm (1968a), Harding (1974), Kawata and Shioiri (1978), and Harding (1980). Metallurgical aspects of high strain-rate behavior are treated in Rohde et al. (1973), while a detailed review of testing techniques at high rates of strain can be found in Lindholm (1971).

8.1. CONSIDERATIONS IN DYNAMIC TESTING

8.1.1 Mathematical Description of Material Behavior

The most general form of a material-constitutive equation should cover the description of material behavior under the total range of strain rates that may be encountered. However, this can be extremely difficult, even for uniaxial stress, and therefore the majority of constitutive equations generally cover only a narrow range of strain rates. This is not inconsistent with the physics of the problem, since different mechanisms govern the deformation behavior of materials within different strain-rate regimes. Some of the considerations in dynamic testing have been summarized by Lindholm (1971) as shown in Figure 1. At strain rates of the order of 10^{-6} to 10^{-5} s^{-1} the creep behavior of a material is the primary consideration, usually at elevated temperatures for metals, and creep-type laws are used to describe the mechanical behavior. At higher rates, in the range 10^{-4} to 10^{-3} s^{-1}, the uniaxial tension, compression, or quasistatic stress-strain curve obtained from constant strain-rate tests is used to describe the material behavior. Although the quasistatic stress-strain curve is often treated as an inherent property of a material, it is a valid description of the material only at the strain rate at which the test was conducted. As higher strain rates are encountered, the stress-strain properties may change, and alternate testing techniques have to be employed. Constant strain-rate tests can be performed with specialized testing apparatus at strain rates up to approximately 10^4 s^{-1}. The range of strain rates from 10^{-1} to 10^2 s^{-1} is generally referred to as the intermediate or medium strain-rate regime. It is within this regime that strain-rate effects first become a consideration in most metals, although the magnitude of such effects may be quite small or even nonexistent in some cases. Strain rates of 10^3 s^{-1} or higher are generally treated as the range of high strain-rate response, although there are no precise definitions as to strain-rate regimes and care must be taken in evaluating data to note the actual strain rates rather than the terminology.

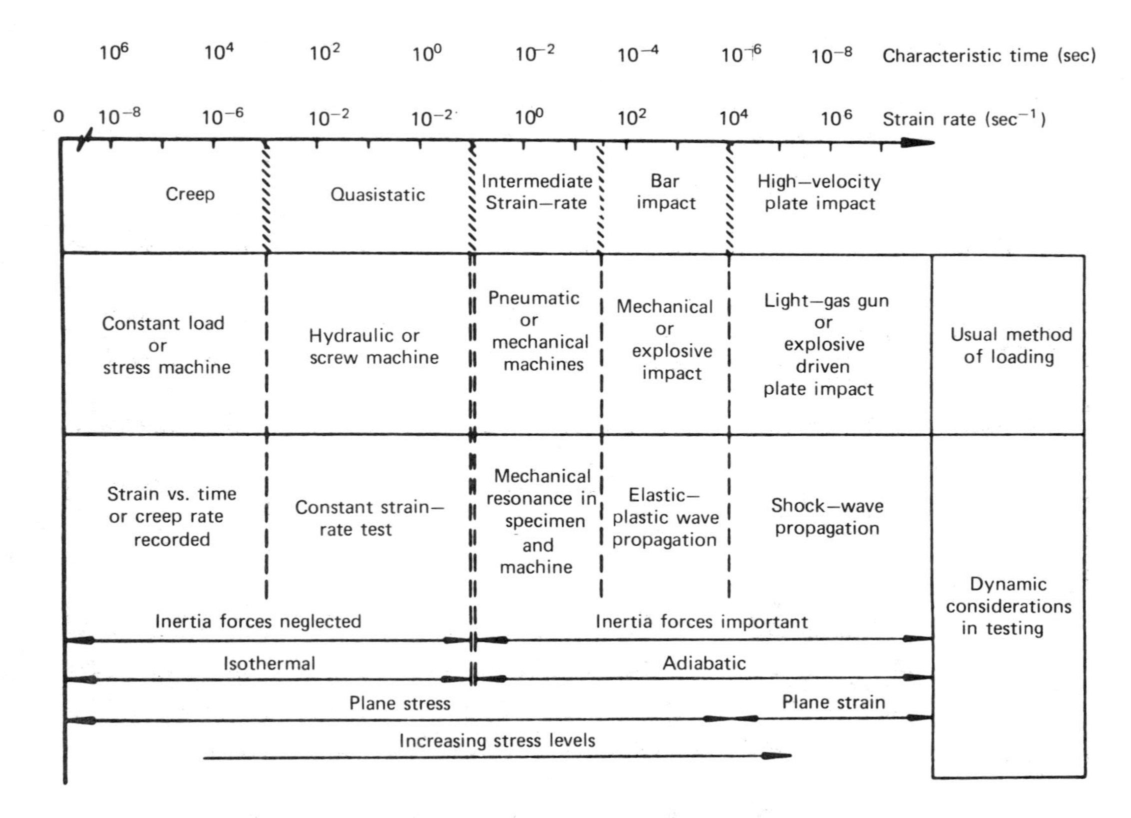

Figure 1 Dynamic aspects of mechanical testing [Lindholm (1971)].

It is within the high strain-rate regime that inertia and wave-propagation effects first become important in interpreting experimental data. At these high rates, care must be taken to distinguish between average values of stress and strain and local values that may be the result of one or more high-intensity stress wave propagating through a material. At strain rates of $10^5\ s^{-1}$ or higher, we are generally dealing with shock waves propagating through materials that are in a state of uniaxial strain. At these very high rates and the associated very short time scale involved, thermodynamic considerations become important. At these high rates we transition from nominally isothermal conditions to adiabatic conditions.

In performing wave-propagation experiments to study dynamic properties of materials, in addition to the inherent problem that the approach is an indirect one (see Chapter 4), the strain rates that occur during the test are primarily a function of the materials own constitutive behavior. Thus, for example, in uniaxial stress-wave propagation in long rods, the dispersive nature of plastic waves precludes achieving strain rates much above $10^2\ s^{-1}$ in most materials. If higher amplitude impacts are attempted, the stress state achieved will gradually change from uniaxial stress to uniaxial strain or complete lateral constraint as higher strain rates or shorter rise times are achieved. Higher impact velocities are generally used, therefore, in configurations such as the flat-plate impact (see Section 4.2). At very high strain rates, uniaxial strain states with corresponding high hydrostatic pressures occur, making it difficult to study the deviatoric stress behavior because of the dominance of the hydrostatic terms. These subjects are discussed in detail in Chapters 1 and 4. In this chapter, emphasis is placed on the delineation between uniform states of stress and strain and wave-propagation phenomena. It is important to distinguish between these two situations. In the former case, it is important to verify that average values are occurring, while in the latter case the analysis must be based on wave-propagation theory and analysis. This chapter primarily deals with situations where stress- and strain-averaging assumptions are a fundamental part of the analysis. Since the wave-propagation speed of a material cannot be controlled, configurations that involve the averaging assumptions generally cover the so-called intermediate and high strain-rate regimes of material behavior up to rates of approximately $10^4\ s^{-1}$.

8.1.2 Uniaxial Tests at High Strain Rates

The simplest method for determining information on the strain-rate sensitivity of a material is to increase the speed of a uniaxial tension or compression test. Hydraulic and pneumatic machines have been developed with ever-increasing controlled-speed capability. The tension or compression test appears to be ideal because the state of stress is purely uniaxial, or at least assumed so. The most important question that must be answered is how fast a uniaxial test can be performed before we are unable to obtain a valid stress-strain curve for a material. Consider a specimen of initial length l, fixed at $x=0$ and subjected to a uniform velocity v_o starting at time $t=0$ (see Figure 2). This represents a test in a constant crosshead velocity-testing machine or a drop-weight type of test where a large mass impacts one end of the specimen. Letting $u(x, t)$ denote the displacement of any point in the x-direction and assuming purely uniaxial motion, that is, neglecting radial inertia effects, the equation of motion is:

$$\frac{\partial^2 u}{\partial t^2}=c^2\frac{\partial^2 u}{\partial x^2} \tag{1}$$

where

$$c=\left(\frac{E}{\rho}\right)^{1/2} \tag{2}$$

is the longitudinal wave velocity in a bar or rod; E is Young's modulus, and ρ the mass density. Applying the initial conditions of zero displacement and velocity and the boundary conditions of the left end fixed and the right end moving with constant velocity v_o, the solution is found:

$$u(x,t)=\frac{v_o l}{2c}\left[f\left(\tau+\frac{x}{l}\right)-f\left(\tau-\frac{x}{l}\right)\right] \tag{3}$$

where $\tau=tc/l$ is a dimensionless time and $\tau=1$ represents the time it takes a wave to propagate the length of the specimen, l. The function $f(\tau)$ is shown in Figure 3. Strain can be obtained from $\varepsilon=\partial u/\partial x$ and stress from $\sigma=E\varepsilon$. Introducing the dimensionless variables:

$$\xi=\frac{x}{l} \tag{4a}$$

$$v^*=\frac{v_o}{c} \tag{4b}$$

plots of stress and strain can be constructed as a function of time. Figure 4 shows strain normalized with respect to v^* against dimensionless time τ at an arbitrary position ξ along the bar. The dashed line shows the average strain in the bar that we normally compute from total displacement divided by bar length. Figure 5 shows the normalized stresses at both ends of the bar. It can be seen from both figures that the stresses and strains are made up of numerous waves propagating back and forth in the bar. Note that the solution to the mathematical problem has assumed an instantaneous jump in velocity at $t=0$, whereas some finite rise time usually occurs because of imperfect impact or machine response. Nonetheless, if the number of wave propagations is large during the duration of a test, then the use of average stresses and strains seems

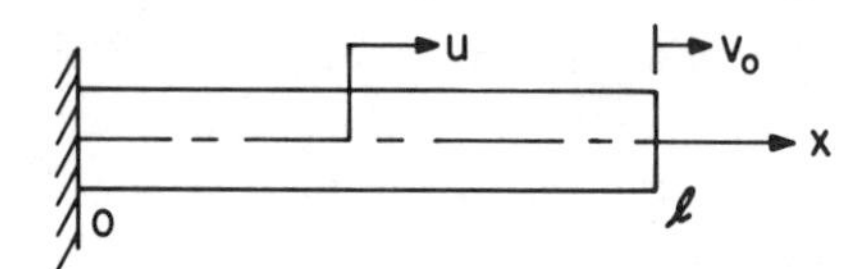

Figure 2 Schematic of uniaxial-tension test.

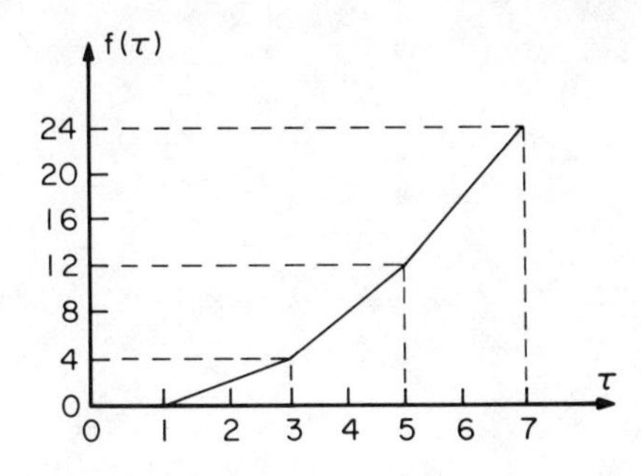

Figure 3 Graph of the function $f(\tau)$.

justified. However, if the velocity is large, then only a few wave reflections may occur. If this is the case, then individual wave propagations must be considered, average values alone cannot be treated. Inspection of the strain-time response of Figure 4 shows that the average strain rate is different than the actual strain rate if only a few wave reflections are considered during the entire test. For only a few reflections, strain rates can be as high as allowed by the rise time of the applied velocity. Quoted strain rates above the order of $10^5\ s^{-1}$ generally refer to the rise time of the propagating strain waves in a specimen unless very small specimens are being used, thus wave-propagation phenomena must be considered. Note that this analysis considers the case of a material that is linear-elastic and assumes a zero rise time in the applied velocity. Also, note that in determining the number of wave reflections or the transit time of the waves, the elastic-wave velocity has been used. In dealing with a material deformed into the plastic region, the plastic-wave velocity can generally be an order of magnitude smaller.

As a numerical example, consider a steel bar of 25 mm length. For steel, $c=5\times10^3$ m/s and $E=200$ GPa. The stress generated at the end of the bar on

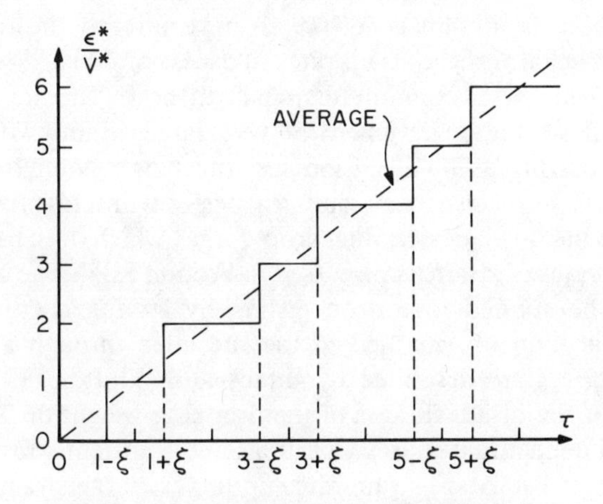

Figure 4 Nondimensional strain-time history.

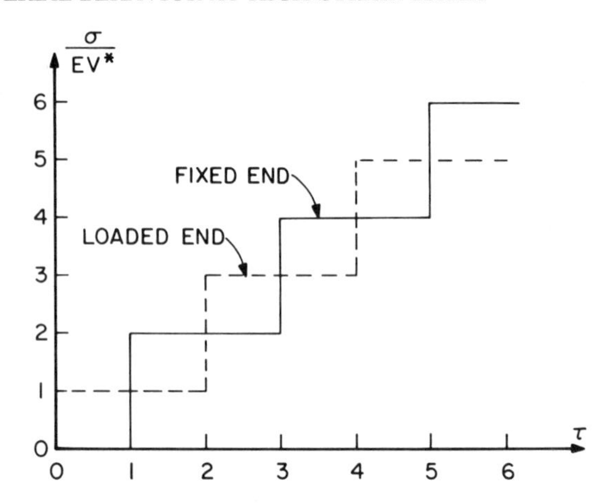

Figure 5 Nondimensional stress history at ends of bar.

first impact is $\sigma = \rho c v_o$, thus for an impact velocity of 2.5 m/s, the first stress pulse is 100 MPa and the average strain rate is 100 s^{-1}.

8.1.3 Intermediate Rate Tests

Various types of machines are available for performing tests at intermediate or medium strain rates using crosshead velocities up to the order of 1 m/s or greater. Strain rates up to approximately 10^2 s^{-1} can be achieved with these machines in tension or compression. Hydraulic or pneumatic devices are utilized to rapidly accelerate a driving ram to a constant velocity and then sustain that velocity for the duration of the test. In using hydraulic test machines in the medium strain-rate regime, the compliance of the machine must be considered when trying to achieve constant strain rates [Cooper and Campbell (1967)]. Lindholm et al. (1980) have utilized the torsion mode of deformation to achieve shear-strain rates in excess of 300 s^{-1} with a hydraulic machine designed for maximum torsional stiffness. The torsional mode of deformation allows the achievement of very large strains without geometric instabilities occurring as in a tension test (necking). Numerous drop-weight types of devices have also been built to perform intermediate-rate tests in compression. One such device, the drop forge, which has been used up to relatively high rates of strain, is discussed in Section 8.3.4. The dropping-weight principle can be applied to torsion testing by storing energy in a rotating flywheel that is suddenly coupled to the specimen through a clutch engagement. Such devices are described by Bitans and Whitton (1971) and Culver (1972). Some of the disadvantages of the dropping-weight or flywheel concept are the sudden impact that causes vibrations and transient stress waves and the inability to achieve constant strain rates up to very large strains.

A device that has been designed to achieve constant, true strain rates in compression is the cam plastometer, the basic principle of which was first proposed by Orowan (1950). It uses the energy stored in a rotating flywheel to drive a logarithmic shaped cam that applies the compression load to the test specimen. Some of the earliest experiments using a plastometer with a rotating cam are those of Loizou and Sims (1953) on lead, and Alder and Phillips (1954) on aluminum, copper, and steel over a range of temperatures. In the latter, reductions in height up to 50% were achieved at constant, true strain rates from 1–40 s^{-1}. A modified version of the rotating-flywheel cam plastometer was first used by Hockett (1959). With this device, constant, true strain rates in compression up to 240 s^{-1} are reported to be achieved regularly. Experiments at elevated temperatures up to 0.95 of melt, on several materials are described by Bailey and Singer (1963), who achieved strain rates over 300 s^{-1} at natural strains in excess of 2. In these latter experiments, flat-plate specimens were deformed in plane-strain compression as opposed to uniaxial compression. In either case, the use of appropriate lubricants is a critical part of the test procedure to eliminate end friction and to insure that the assumed stress state is achieved.

In addition to the wave-propagation considerations that can invalidate the assumptions of uniform stresses and strains across the specimen as very high strain rates are attempted, consideration must also be given to the measurement of stress. Stress is generally determined through the use of some type of load cell in series with the specimen and often some distance away from the end of the specimen. At high velocities of impact, one must consider the finite time it takes a wave to propagate from the end of the specimen to the load recording device, the wave propagation through the load cell and attachment fixtures, and the inertia loads generated by the load cell and the fixtures themselves.

Several methods are available for determining the average or local strains in the specimen. Average values can be determined from crosshead displacements, but one must be able to establish the effective gage length. Optical extensometers that record displacement between two points or targets on the specimen allow the determination of local strain values with high-frequency resolution. Electrical resistance-strain gages attached directly to the specimen allow resolution of local strains averaged over the length of the gage. Measurements of diameter changes in cylindrical specimens coupled with an assumption of incompressibility can also be used to determine dynamic strains. But the final determination of the relation between stress and strain requires one either to relate stress and strain at the same position and time or to use average values over a specimen, which again raises the necessity of considering wave-propagation effects when attempting ever-increasing rates of strain in uniaxial testing.

8.1.4 Wave-Propagation Experiments

An alternate approach for determining dynamic properties, and the only method available for extremely high rate testing, is to consider the detailed propagation of inelastic waves in the material. There is, however, no direct method for obtaining dynamic property information because one must use wave-propagation theory to study the phenomena which, in turn, requires a constitutive equation. But it is the constitutive equation that one seeks in the first place. Only an indirect or iterative procedure can be used, and the results cannot be uniquely established. The procedure one must follow, which is discussed in Chapter 4, is as follows:

1. Assume the form of the constitutive equation(s) and the material constants in the equation.
2. Perform a wave-propagation experiment and establish or record the input or boundary conditions as well as information on the wave-propagation phenomena such as strain-time profiles, free-surface velocity measurements, wave velocities, and so on.
3. Make analytical predictions of observed phenomena from the appropriate wave theory and assumed constitutive relation.
4. Compare predictions with experimental observations and revise constants if necessary. Repeat steps 1–4.
5. Try another experimental configuration or other experimental observations using the same equation and constants to verify the predictions.

As can be seen, this is a trial-and-error procedure which may not lead to a unique solution. An important aspect of this procedure is to start with a wave-propagation theory and form of constitutive equation with a sound physical basis.

Plastic waves of uniaxial stress in bars or rods in most engineering materials are dispersive in nature and attenuate as they propagate down the rod. Although short rise times and high strain rates can be achieved near the impact end, the waves attenuate rapidly. Thus, measurements of plastic-wave phenomena at various stations along the length of a bar away from the impact end generally involve strain rates not much greater than the order of 100 s^{-1}. Furthermore, there is often only a small amount of variability in the strain rates occurring during a test. Thus, any constitutive equation that predicts the proper flow stresses over the narrow range of strain rates encountered stands a good chance of predicting the wave-propagation phenomena observed. The plastic-wave experiment in rods or bars is thus, generally, not a very effective method for obtaining high-rate constitutive equations because of the lack of uniqueness and the inability to achieve very high strain rates. On the other hand, very high strain rates can be achieved with states of uniaxial strain, rather than uniaxial stress, but these stress states are primarily hydrostatic in nature, and preclude the determination of strain-rate effects since the contribution of the shear stresses is essentially obscured in that configuration. In the following sections, attention will be devoted to the study of stress-strain–strain-rate relations in materials utilizing experiments that involve nearly uniform states of stress and strain at high rates of strain and where wave-propagation analyses are not required to obtain data.

8.2 THE SPLIT HOPKINSON PRESSURE BAR

One of the most widely used experimental configurations for high strain-rate material measurements is the split Hopkinson pressure bar or Kolsky apparatus developed by Kolsky (1948). The concept of the Hopkinson bar involves the determination of dynamic stresses, strains, or displacements occurring at the end of a bar through observation of the effect some distance away. If a long bar remains elastic, then disturbances at the end are propagated undistorted (except for very high frequency components) down the bar at the elastic-wave velocity, $c=(E/\rho)^{1/2}$. A strain gage at the center of a bar thus provides a record of the force-time history at the end of the bar, but is delayed in time. The concept of utilizing two bars on either side of a specimen to achieve high loading rates was developed by Kolsky. The split Hopkinson bar or Kolsky apparatus for compression testing consists of a striker bar, incident bar, and transmitter bar and associated instrumentation for recording data as shown schematically in Figure 6. The incident and transmitter pressure bars are mounted in teflon or nylon bushings to assure accurate axial alignment while permitting stress waves to pass without dispersion, while the specimen is sandwiched between the two bars as shown. The apparatus, instrumentation, and analysis have been documented frequently in the literature [see, for example, Lindholm (1964)].

The striker bar is accelerated by using the energy stored in a spring or utilizing a small gas-gun apparatus. Several methods are available for measuring the velocity of the striker bar. One technique is to use grooves machined in the bar at a known spacing that pass through a magnetic pickup, generating two pulses that are, in turn, fed into a time-interval counter. Another method is to utilize two photocells and light sources at a known spacing. When the striker

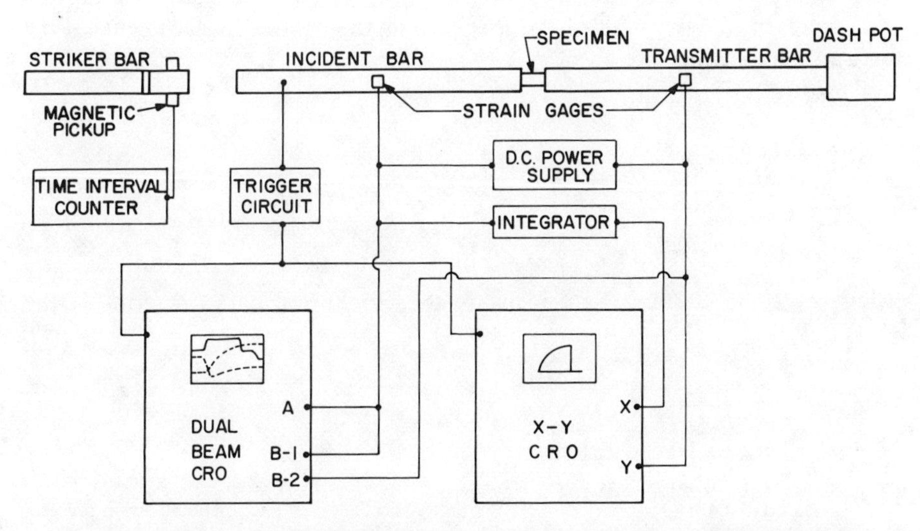

Figure 6 Schematic of apparatus and instrumention for compression split Hopkinson bar.

bar impacts the incident bar, constant amplitude compressive pulses are generated in both bars. The length or duration of the compressive pulse generated in the incident bar is twice the wave-transit time in the striker bar and the magnitude is directly proportional to the striker velocity. When the compressive pulse in the incident bar reaches the specimen, part of it is transmitted through the specimen and part is reflected because of mismatch in the cross-sectional area and acoustic impedance between the bar and the specimen. The area and mechanical behavior of the specimen determine the exact shape of the reflected and transmitted pulses. Knowledge of the strain-time history of the traveling pulses at strain-gage stations on the bars allows one to deduce the strain-time history at the ends of the bar. This basic principle of the Hopkinson bar depends only on the knowledge of the longitudinal wave speed to shift the pulse in time and the assumption of nondispersive elastic waves in the pressure bars.

8.2.1 Analysis

The equations for the analysis of the compression split Hopkinson bar assume that stresses and velocities at the end of the specimen are propagated down the bars in an undispersed manner. By assuming that the wave-transit time in the short specimen is small compared to the total time of the test, many wave reflections can take place back and forth in the specimen. The stress and strain are then assumed to be uniform along the specimen. By utilizing input and transmitter bars of the same material and cross-sectional area, a set of relatively simple relations are derived for the stress, strain, and strain rate in the specimen.

Figure 7 is a schematic of the specimen and Hopkinson bars and shows the incident, reflected, and transmitted pulses, ε_i, ε_r, and ε_t. Using subscripts 1 and 2 to represent the two ends of the specimen as shown, the displacements at the ends of the specimen are given by

$$u_1 = \int_o^t c_o \varepsilon_1 \, dt \tag{5a}$$

$$u_2 = \int_o^t c_o \varepsilon_2 \, dt \tag{5b}$$

where c_o is the elastic-wave velocity in the Hopkinson bars. In terms of the

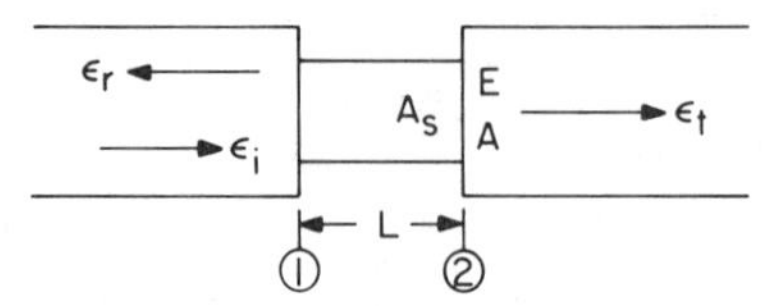

Figure 7 Schematic of specimen and strain pulses.

incident, reflected and transmitted pulses,

$$u_1 = c_o \int_o^t (\varepsilon_i - \varepsilon_r)\, dt \tag{6a}$$

$$u_2 = c_o \int_o^t \varepsilon_i\, dt \tag{6b}$$

where stresses and strains are assumed positive in compression. The average strain in the specimen is

$$\varepsilon_s = \frac{u_1 - u_2}{L} \tag{7}$$

or, in terms of the strain pulses

$$\varepsilon_s = \frac{c_o}{L} \int_o^t (\varepsilon_i - \varepsilon_r - \varepsilon_t)\, dt \tag{8}$$

where L is the length of the specimen. The forces at the ends of the specimen are obtained from

$$P_1 = EA(\varepsilon_i + \varepsilon_r)$$

$$P_2 = EA\varepsilon_t \tag{9}$$

where E and A are Young's modulus and the cross-sectional area of the Hopkinson bars. The average force is calculated from

$$P_{av} = \frac{EA}{2}(\varepsilon_i + \varepsilon_r + \varepsilon_t) \tag{10}$$

If it is assumed that $P_1 = P_2$, that is, that the forces are equal at both ends of the specimen, then from (9) and (8)

$$\varepsilon_i + \varepsilon_r = \varepsilon_t \tag{11}$$

$$\varepsilon_s = \frac{c_o}{L} \int_o^t (\varepsilon_t - \varepsilon_r - \varepsilon_r - \varepsilon_t)\, dt \tag{12}$$

For the specimen of cross-sectional area A_s, the strain, stress, and strain rate in the specimen become:

$$\varepsilon_s = \frac{-2c_o}{L} \int_o^t \varepsilon_r\, dt \tag{13}$$

$$\sigma_s = E\frac{A}{A_s}\varepsilon_t \tag{14}$$

$$\dot{\varepsilon}_s = \frac{-2c_o}{L}\varepsilon_r \tag{15}$$

It is important to note that the stress, strain, and strain rate are average values, and that they are calculated from a uniaxial stress-state assumption.

8.2.2 Instrumentation and Calibration

In performing the test, strain gages are placed on both the incident and transmitter bars equidistant from the specimen so that the reflected and transmitted waves arrive at the respective gages at the same time. The bars are long enough to permit observation of the entire loading event without interruptions caused by wave reflections from the free ends. The strain-gage bridges generally consist of two active gages to cancel bending. Data can be recorded on oscilloscopes or transient recorders.

For direct recording of stress-strain curves, the reflected pulse is passed through an electronic integrator to obtain a signal that is directly proportional to strain in the specimen. The signal is fed to the X axis of the recording instrumentation, while the output from the transmitter bar, proportional to stress, is fed to the Y axis.

It is recommended that the system be calibrated dynamically by passing a known stress wave past the gages on both the incident and transmitter bars that are butted together without a specimen. The magnitude of the strain pulse set up in the bars is $v_o/2c_o$ where v_o is the measured velocity of the striker bar and c_o the longitudinal wave speed.

The output from the strain gages on the incident bar can be electronically integrated and produces a voltage that is directly proportional to the area under the strain-time curve. The load calibration is determined by recording the amplitude of the strain on the transmitter bar.

An alternative method of calibration is sometimes used to check the dynamic calibration. This merely involves shunting a known calibration resistor across one arm of the strain-gage bridges. This produces a simulated strain of

$$\varepsilon_{\text{sim}} = \frac{1}{2G.F.} \frac{R_g}{(R_c + R_g)} \tag{16}$$

where $G.F.$ is the gage factor and R_g and R_c are the gage and calibration resistances. This method verifies the dynamic calibration to within several percent.

Since the strain rate is directly proportional to the reflected pulse, ε_r, the compression Hopkinson bar test is not a constant strain-rate test. Although the input pulse, ε_i, is of constant magnitude to the specimens, as the material strains the cross-sectional area of the specimen increases, making the specimen appear stiffer. The specimen material also work-hardens as deformation proceeds, making it stiffer as strain increases. Since the reflected strain, and hence strain rate, depends on the material, the strain rate generally decreases during the test. The strain-rate figures quoted from Hopkinson bar tests are, therefore, generally average values.

8.2.3 Tensile Tests

The split Hopkinson bar apparatus can also be adapted for use in tensile testing. Harding, et al. (1960) were one of the first to demonstrate a tension version of the Hopkinson bar technique that involved generating a compression pulse in a tube surrounding a solid inner rod. A similar setup has been described by Hauser (1966) and Christman et al. (1971). The setup is shown schematically in Figure 8, which compares the Hopkinson bar geometry in tension and compression as well as in shear. The tube and rod are connected by a mechanical joint. When the compression pulse in the outer tube reaches the joint, which is a free end, it reflects back through the solid inner rod as a tensile pulse. A threaded tensile specimen is attached to the inner rod to provide the mechanical connection necessary to transfer the tensile pulse through the specimen and into a second rod. A tensile version of the split Hopkinson bar is thus achieved. Strain rates of over 1000 s^{-1} were achieved in such an apparatus where loading eccentricity was minimized sufficiently to achieve stress accuracies of 5% [Harding et al. (1960)]. Although this is a relatively simple and direct method for generating high strain-rate data in tension, the technique suffers from the inability to generate tensile waves

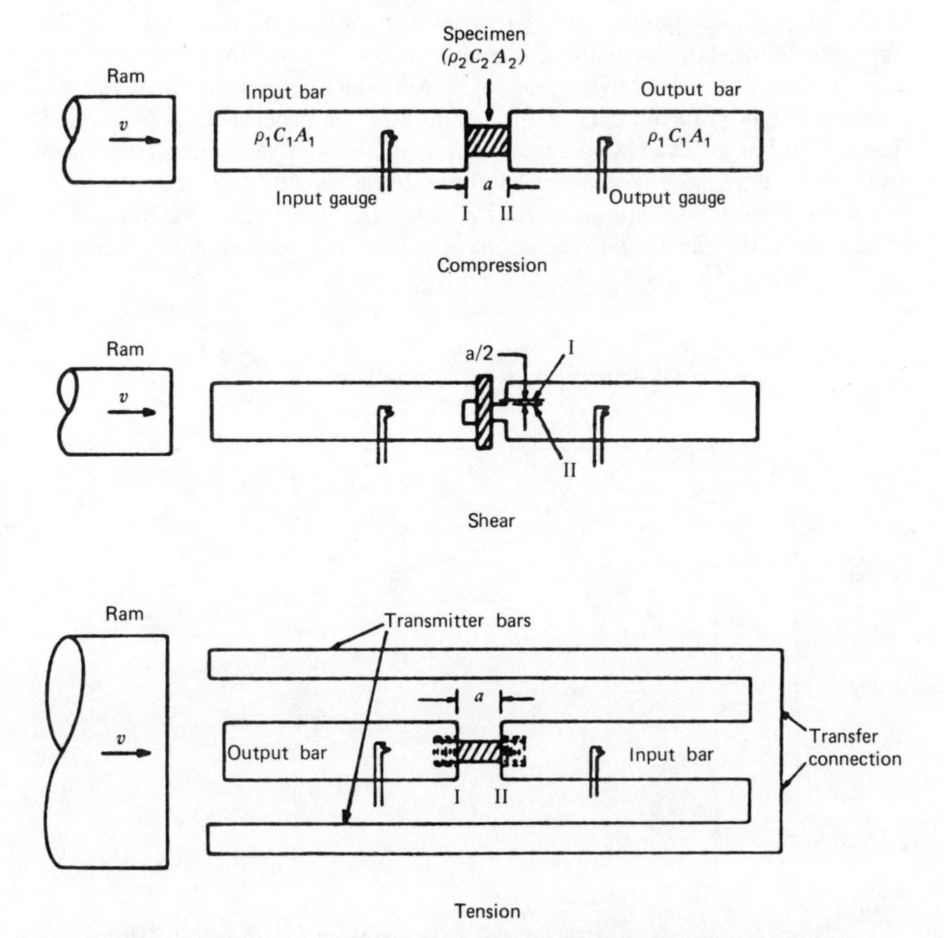

Figure 8 Examples of compression, shear, and tension-test arrangements [Hauser (1966)].

having very short rise times because of the wave dispersion at the mechanical joint.

Lindholm and Yeakley (1968) devised an alternate type of tension test based on the split Hopkinson bar technique. They utilized two Hopkinson bars as in the compression test, one being a solid rod and the other a hollow tube. Tension was achieved through a complex hat-type specimen design which is essentially a set of four very small tensile bars in parallel. Although the test is easy to perform, the specimen design requires considerable machining. More recently, Albertini and Montagnani (1974) utilized both an explosive loading device and rapid fracture of a clamp in a prestressed bar to generate the tensile pulses in a split Hopkinson bar setup. They reported rise times in the impact pulses of approximately 25 μs. A threaded tensile specimen is utilized as in most of the previous work. Only the difficulties of generating pulses with explosives prevents this from being a convenient and easily used laboratory apparatus for high-rate tensile testing.

Another version of a tensile Hopkinson bar utilizing a threaded specimen and a unique split-shoulder arrangement is described by Nicholas (1980a). The apparatus consists principally of a striker bar and two Hopkinson pressure bars. Figure 9 is a Lagrangian $x-t$ diagram which shows the details of the wave propagation in the bars and indicates how the experiment is performed. The striker bar is accelerated against bar 1, the impact generating a compression pulse whose amplitude depends on the striker velocity and whose length is twice the longitudinal elastic-wave transit time in the striker bar. The pulse travels down the bar until it reaches the specimen. A threaded tensile specimen

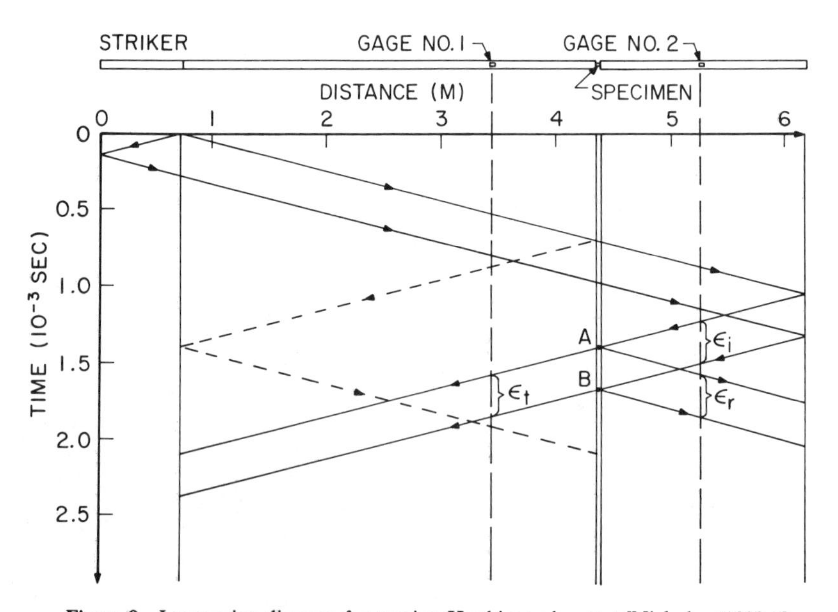

Figure 9 Lagrangian diagram for tension Hopkinson bar test [Nicholas (1980a)].

is attached to the two pressure bars and a split shoulder or collar is placed over the specimen. The compression pulse travels through the composite cross section of shoulder and specimen in an essentially undispersed manner, as if the specimen were not present. Referring again to Figure 9, the compression pulse continues to propagate until it reaches the free end of bar 2. There, it reflects and propagates back as a tensile pulse, shown as ε_i, as it passes gage 2. The tensile pulse, upon reaching the specimen at point A, is partially transmitted through the specimen (ε_t) and partially reflected back into bar 2 (ε_r). The shoulder, which carried the entire compressive pulse around the specimen, is unable to support any tensile loads because it is not fastened in any manner to the bars.

The experimental setup at the instant the tensile pulse is reflected from the free end of bar 2 and starts to propagate back down the bar is identical to the compression split Hopkinson bar apparatus except for the change in sign of the loading pulse and the use of a threaded joint to attach the specimen as opposed to the use of a cylindrical compression specimen. Data are reported for a number of different materials at strain rates up to $10^3\ s^{-1}$ by Nicholas (1980b). Figure 10 shows typical results for AISI 304 stainless steel. The data

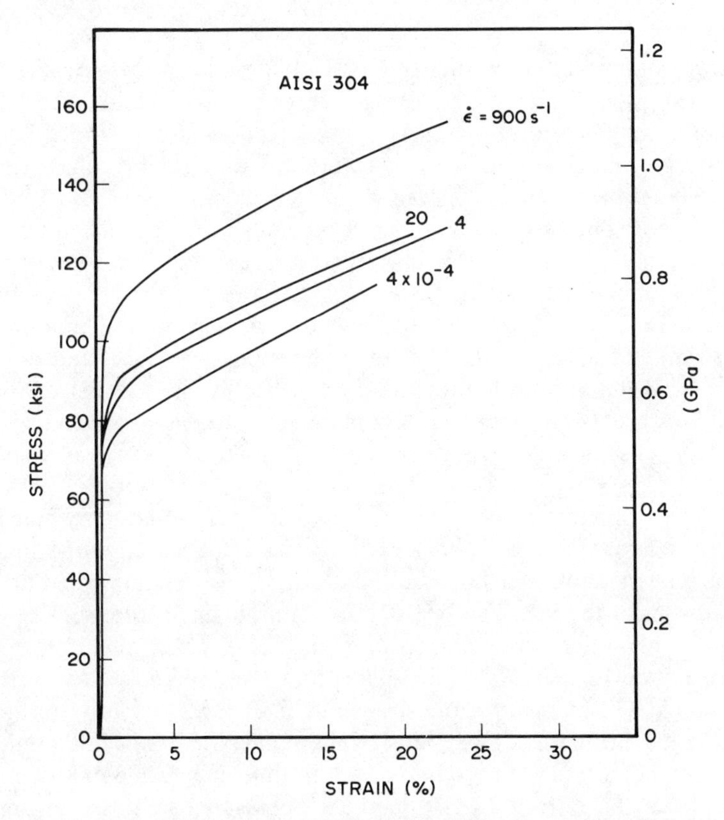

Figure 10 Tensile stress-strain data for AISI 304 stainless steel.

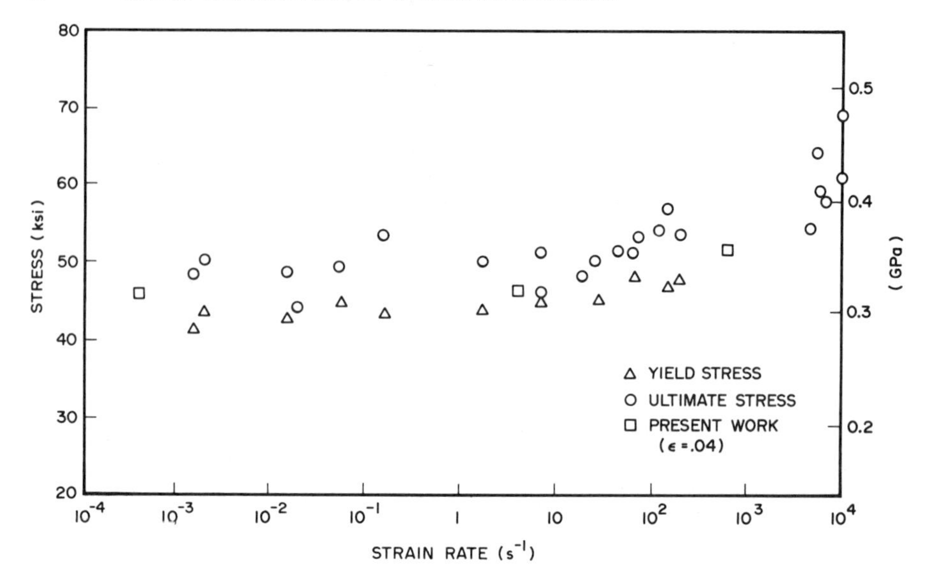

Figure 11 Tensile test data for 6061-T6 aluminum [after Jiang and Chen (1974)].

at the lower strain rates were obtained with a hydraulic servo-controlled testing machine using the identical specimen geometry as in the Hopkinson bar tests to facilitate comparison of data. This investigation also provided high-rate tensile stress-strain data on several structural aluminum alloys that are generally regarded as being relatively strain-rate independent. These data are plotted along with those obtained in a number of previous investigations in tension on 6061-T6 aluminum in Figure 11. It is interesting to note that surveys such as those by Lindholm and Bessey (1969) and Jiang and Chen (1974) show that although there is some degree of strain-rate dependence in pure aluminum, the higher strength aluminum alloys tend to be strain-rate independent over a range in strain rates from approximately 10^{-4} to 10^3 s^{-1}. Most of the data reported, however, were obtained in compression. The comparisons with data obtained in tension are discussed by Nicholas (1980a), as well as in Section 8.4.1.

The inherent limitations of the split Hopkinson bar method preclude its use for determining material behavior in the elastic region because of stress-wave reflections, stress nonuniformity, and large variations in strain rate during the initial portions of the test. The stress-strain data obtained from such a test are thus valid only after some degree of stress- and strain-rate uniformity is achieved. Figure 12 presents typical data from a tension split Hopkinson bar test on Ti-6Al-4V. In this test the strain rate is changing rapidly for approximately 25 μs. During this time, a total strain of over 1% has been accumulated. A plot of load or stress versus strain, based on time integration of strain rate as obtained from the reflected pulse in the Hopkinson bar, gives an apparent

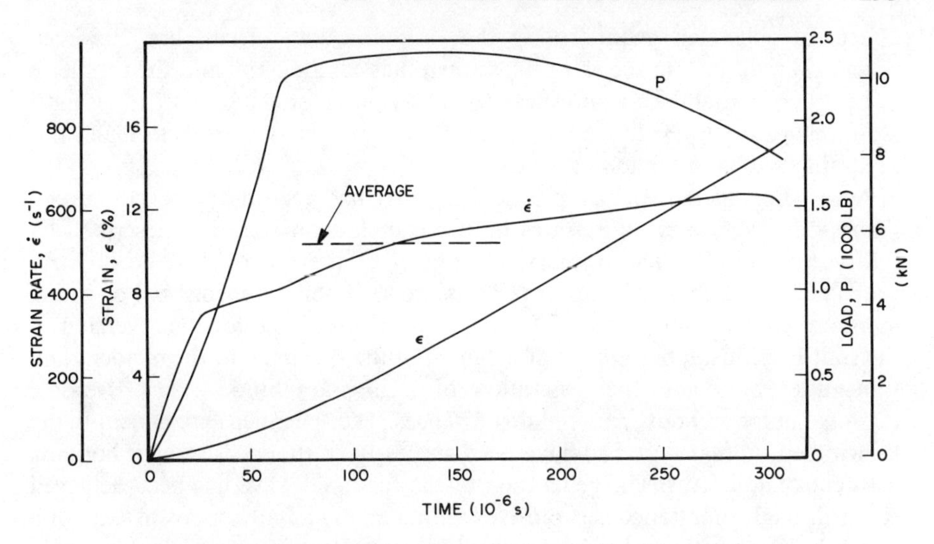

Figure 12 Typical tensile Hopkinson bar data for Ti 6Al-4V specimen.

elastic modulus much lower than the known values for titanium. This merely demonstrates that Hopkinson bar data can only be used to obtain flow-stress data outside of the elastic region or where the fundamental assumptions of the Hopkinson bar are satisfied.

Figure 12 also shows the typical variation of strain rate during the Hopkinson bar test. The average value of strain rate for this test is shown as a dashed line and is merely an estimate. These typical results show that strain rates are not constant in this type of test. They are controlled, primarily, by the dynamic response of the material. In compression tests, especially with high-strength materials, the combination of strain hardening and increase of cross-sectional area at large strains often leads to significant decreases in strain rate during the test. In some tests, the strain rate can decrease to zero if the amplitude of the incident pulse does not provide sufficient force to further deform the specimen.

8.2.4 Other Configurations

There has been increased emphasis, especially in recent years, to achieve ever-increasing strain rates using the Hopkinson bar apparatus. Within the constraints and limitations of the physical apparatus and the assumptions, one of the most direct methods for achieving higher strain rates is through the use of smaller test specimens. Edington (1969) was one of the first to achieve strain rates up to 10^4 s^{-1} by using 2 mm long by 12.5 mm diameter specimens that were fabricated using extremely accurate machining. Lindholm (1978) achieved rates up to 10^5 s^{-1} on specimens of 1100-0 aluminum and copper by reducing specimen lengths to 4 mm. He was careful, however, to preserve a constant ratio of length to diameter and thus avoided any additional questions of the effects of specimen geometry on the resultant data. There are, however, practical limitations to the size of specimen that can be used, and the problems of neglect of radial inertia and shear are always a factor. These difficulties have led investigators to explore other methods of subjecting materials to high rates of strain such as in torsion or in shear.

A variation of the split Hopkinson bar that has been developed at several laboratories to achieve high rates of strain in torsion was first described by Duffy et al. (1971). The primary advantages of the torsional mode of wave propagation are that it is nondispersive and that three-dimensional or radial inertia effects are not present. The major problems in the torsional version of the split Hopkinson bar are the attachment of the specimen to the incident and transmitter bars and the generation of a high-amplitude, short rise-time torsional pulse without any axial disturbances. The specimens are generally the thin-walled tubular type and have been successfully attached through bonding with epoxy adhesives. The generation of the torsional wave has been achieved either through simultaneous explosive loading on diagonally opposite sides of a bar or by pretorquing a bar and then suddenly releasing a clamp. Increasingly higher strain rates can be achieved by higher input torques or by using shorter gage-length specimens. Specimens of the order of 1 mm gage length have been employed to achieve strain rates in excess of $10^4\ s^{-1}$ by several investigators. Nicholas and Lawson (1972) demonstrated that data from extremely short specimens could be reliably reproduced with longer gage-length specimens. The torsional version of the Hopkinson bar was first used by Baker and Yew (1966). Alternate versions of the apparatus and its application are given by Campbell and Dowling (1970) and Lewis and Campbell (1972). The apparatus has found extensive application in incremental strain-rate tests described in Section 8.3.7. Further development in this area has considerable potential because of the inherent advantages of torsional testing, including the absence of hydrostatic components of stress.

The split Hopkinson pressure-bar apparatus has been modified by Bhushan and Jahsman (1978) to achieve nearly uniaxial strain conditions. A very tight fitting, rigid collar was utilized to restrain radial motion of the specimen in a compression Hopkinson bar setup using large diameter bars. However, since the collar is of an elastic material and a perfect fit between collar and specimen cannot always be assured, the achievement of purely uniaxial strain conditions (no lateral displacements or strains) is difficult. Furthermore, the yield strength of the Hopkinson bars becomes a limiting factor because the dynamic yield stress in uniaxial strain is higher than in uniaxial stress. Thus, the specimen in uniaxial strain may be stronger than the Hopkinson bars being used to test it that are in a nominally uniaxial stress state.

Other modifications of the split Hopkinson pressure bar include the superposition of a static radial confining pressure on dynamic compression of specimens. Chalupnik and Ripperger (1966), using a Hopkinson bar and "bomb" to get confining pressures up to 0.7 GPa, and Lindholm et al. (1974) performed such modifications to study the effect of hydrostatic pressure on the dynamic strength of materials. Ripperger (1965), in similar experiments, reported that the hydrostatic pressure component has an appreciable effect on the strain-rate sensitivity of high-purity aluminum, copper, and iron.

Sturges et al. (1980) report on a method of generating data at high strain rates and high hydrostatic pressure under torsional loading in a "torsion bomb." They used a heavy high-speed flywheel-clutch device to generate a mechanical impact. In this setup, no stress-wave effects were considered as in the Hopkinson bar. By using very short gage-length specimens of approximately 1.6 mm, they were able to achieve strain rates of $10^3\ s^{-1}$ with confining pressures up to 1.7 GPa on a range of materials from soft to very high strength steels. On ductile *fcc* metals such as aluminum and copper, the effect of pressure was to enhance the strain to failure. The application of very high strain rates was found not to reduce this effect.

The concept of the original Hopkinson bar, which involves measurements at one end or at some position along the bar to deduce what is occurring at the other end, has been applied in an attempt to achieve extremely high strain rates and very large strains in compression. Samanta (1971) utilized a direct impact version of a Hopkinson bar by firing a projectile directly against the specimen, thus eliminating the input bar as such. This procedure allowed him to obtain very large compressive strains in aluminum and copper. Load was determined in the conventional manner from strain gages on the output bar, while an optical method was employed to observe strain directly. Wulf and Richardson (1974) reported measurements of strain rates up to $10^5\ s^{-1}$ and true strains up to 2.0 using a hardened projectile impacting the specimen directly in place of the input bar. In their procedure, strain was measured directly using a coaxial capacitor. Their report describes this new circuit for measuring rapid changes of capacitance that is an improvement of a capacitance-type gage for measuring strains in a Hopkinson bar setup described earlier by Wingrove (1971). Gorham (1980) describes a modified Hopkinson bar system also using the direct impact of a striker bar against the test specimen. He uses a high-speed camera with a novel optical system to achieve very high radial-displacement resolution from which to calculate strain. In tests of 8 μs duration that subjected 1 mm diameter by 0.5 mm long tungsten-alloy specimens to strains of 30%, a strain rate of $4 \times 10^4\ s^{-1}$ was achieved. These high strain rates could generally not be achieved in a conventional split Hopkinson bar experiment because of limitations on the strength of the bars and because of the large forces necessary to deform very high strength materials.

The concept of the Hopkinson bar can also be extended to other test procedures. Instrumented Charpy impact tests or dynamic three-point bend tests can be conducted using this technique. Nicholas (1975) conducted three-point bend tests on beryllium using a system that consisted of a single bar instrumented with strain gages at the center and machined to a standard Charpy loading tup at one end as shown schematically in Figure 13. The support fixture for the specimen is rigidly attached to a base to simulate a fixed

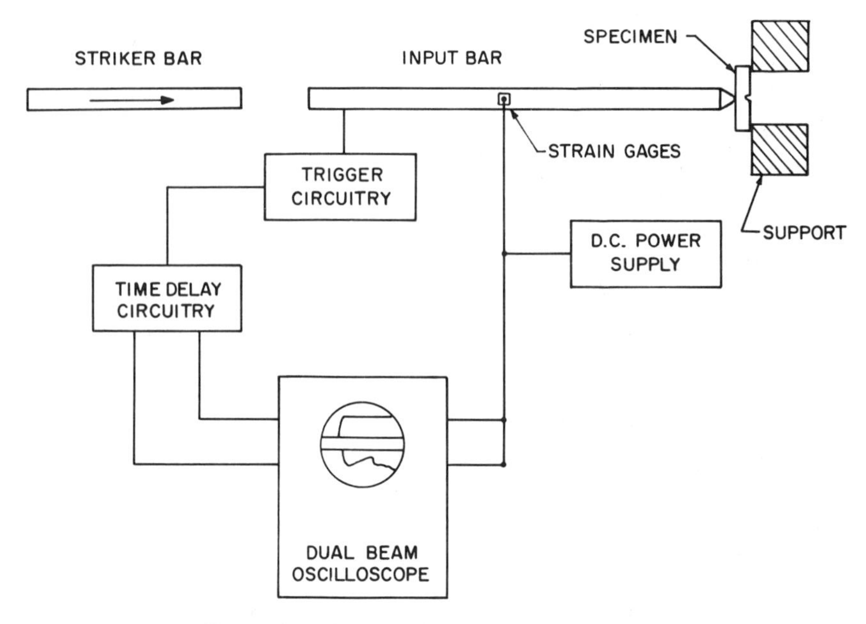

Figure 13 Schematic of Hopkinson bar-bend test.

boundary condition. Figure 14a depicts the transient signal at the strain gage on the input bar after it is impacted by the striker bar, while Figure 14b shows schematically the time-shifted signals as they would appear at the end of the input bar in contact with the specimen. Since the specimen supports are rigid, all beam displacements are those occurring at the loading tup on the end of the input bar. The equations for the data reduction for force, displacement, and velocity, are also shown in the figure. Since the entire history of force and displacement at the end of the bar can be reconstituted from the input and reflected signals measured at the strain gage, a method is available for performing a fully instrumented bend test. Figure 15 shows load-deflection histories for specimens of P-1 beryllium at three different velocities. It is to be noted that at the higher velocities, considerable "ringing" is apparent. This is not what is often referred to as load-cell ringing, since the data are obtained from longitudinal waves in a long bar that are assumed to propagate essentially undispersed. Rather, this effect is due to induced natural-bending vibration modes in the test specimen.

Tatro et al. (1979) have devised another modification of the split Hopkinson pressure-bar technique to permit performance of tension tests at strain rates exceeding $10^4\ s^{-1}$. The increase in strain rate is achieved by using a notched specimen. Direct strain measurements are achieved by photographing the specimen during deformation with a high-speed framing camera, while load measurements are made in a conventional manner with strain gages on the output bar. Synchronization of load and strain is somewhat difficult if the

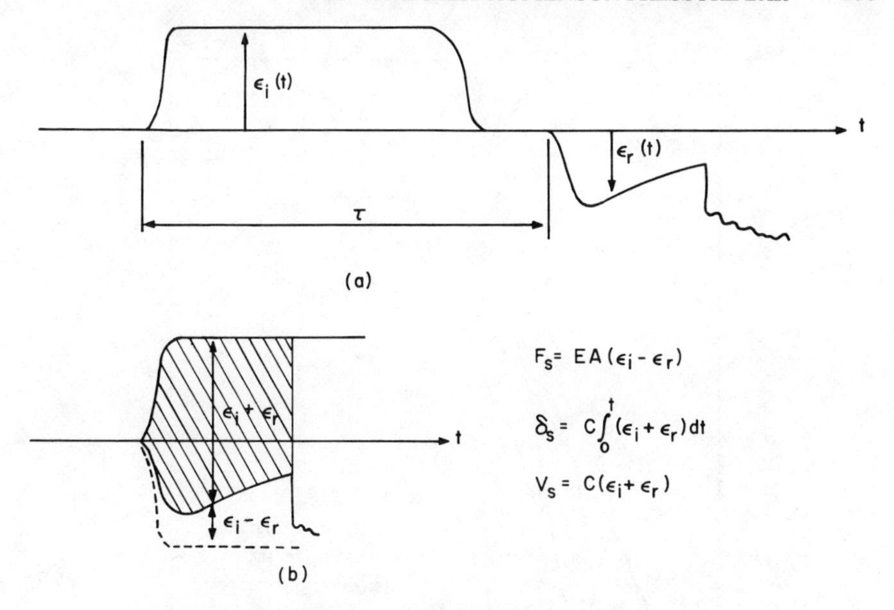

Figure 14 Data-reduction schematic for Hopkinson bar-bend test.

load-time and strain-time traces are not smooth and extrapolatable back to zero. The stress state at the notched cross section is not uniaxial, however, when sharp notches are introduced into the specimen. The stress distribution across the notch can be calculated from the notch geometry using the approximate analysis of Bridgman (1964). Assuming that the radial displacements at the minimum cross section are a linear function of the radius, r, and that the material obeys a Mises yield condition and Mises flow law, the axial and radial stresses at the notch cross section are

$$\sigma_z = F\left(1 + \log\frac{a^2 + 2aR - r^2}{2aR}\right) \tag{17}$$

$$\sigma_r = F\log\frac{a^2 + 2aR - r^2}{2aR} \tag{18}$$

where a is the minimum radius of the specimen at the notch and R is the radius of curvature of the notch. The flow stress σ_z at the surface $r=a$ is denoted by F. The average axial stress, $\bar{\sigma}_z$, is given by

$$\bar{\sigma}_z = F\left(1 + 2\frac{R}{a}\right)\log\left(1 + \frac{a}{R}\right) \tag{19}$$

The analysis of Bridgman has been criticized as being incorrect [Clausing (1969)] although a finite-element analysis of a simulated notched-tensile Hopkinson bar test has shown that the stress distribution across the notch postulated by Bridgman (17), is approximately correct [Zaslawsky et al.

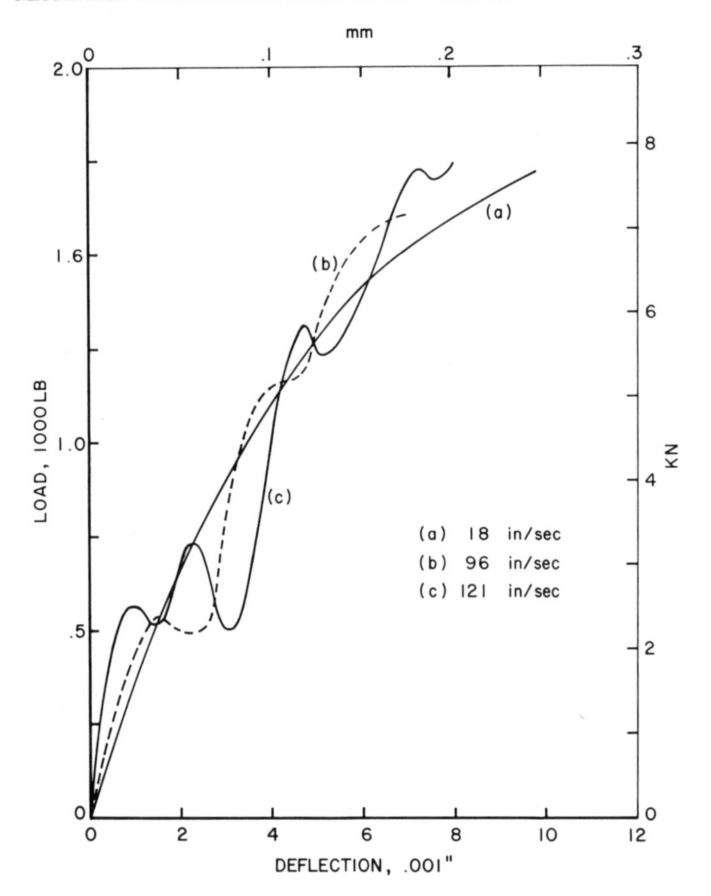

Figure 15 Typical load-deflection curves for dynamic bend tests on P-1 beryllium.

(1980)]. Additional analyses would appear to be warranted to firmly establish the stress state at the notch for the dynamic loading condition. The use of notched specimens thus allows the investigation of dynamic material behavior under triaxial stress states, assuming that an analysis of the notch configuration under transient loading is available. This technique also allows achievements in strain rate beyond those available with smooth bar specimens in tension.

The split Hopkinson bar has also been applied to measurements of fracture properties of materials through suitably designed specimen geometries. Klepaczko (1980) has used a wedge-loaded specimen in a conventional split Hopkinson bar arrangement to obtain dynamic load-displacement data from which dynamic fracture-initiation properties of a number of materials are determined. Another modification of the Hopkinson bar for dynamic fracture studies has been made by Costin et al. (1977) in which a section of a long bar is notched and prefatigued and subsequently loaded by a tensile pulse. The loading tensile pulse is generated by the detonation of an explosive charge against a specially shaped loading head, while conventional strain gages are used to monitor the incident and transmitted pulses. The crack opening displacement at the bar surface is measured against time with the aid of an optical interference fringe technique. Stress intensity rates in excess of 10^6 MPa-m$^{1/2}$ s^{-1} have been achieved with this setup that is an order of magnitude higher than can be achieved in a conventional Charpy impact test.

8.2.5 Analytical Studies

In addition to its fairly widespread use, the split Hopkinson bar has been the subject of extensive analytical studies to evaluate the assumptions and to establish the ranges of validity for its use. Perhaps no other dynamic test technique has been so thoroughly scrutinized as the split Hopkinson bar. The first critical analysis of this experiment was by Davies (1948) who roughly examined the effects of wave dispersion on the propagation of a pulse in a Hopkinson pressure bar. He concluded that a pressure bar is incapable of measuring pressures that are subject to changes in time of the order of 1 μs. Davies and Hunter (1963) derived an inertial correction factor from kinetic energy considerations, taking into account the nonuniform triaxial stress field induced by inertia. They concluded that optimum specimen geometries given by

$$\frac{a}{h} = \frac{2}{\sqrt{3}} \tag{20}$$

where a/h is the radius-to-thickness ratio providing minimal correction or error due to inertial effects as well as minimizing the problem due to interfacial friction. Conn (1965) was one of the first to emphasize the difference between mechanical effects, that is, wave-propagation phenomena, and intrinsic material properties. In a one-dimensional, rate-independent wave-propagation analysis of the split Hopkinson bar arrangement, assuming the dynamic stress-strain curve of aluminum, he showed that the rate-independent theory and a single nonlinear dynamic curve was able to explain data of Hauser et al. (1961) who had performed their own analysis using a bilinear approximation. Hauser et al. (1961) had concluded that the rate-independent elastic-plastic theory could not explain the results because the stress-strain curves at dynamic rates were higher than the corresponding static curves. Although the analysis was carried out for a very short specimen involving many wave reflections and interactions, the conclusions bear a tremendous similarity to those obtained by investigators using elastic-plastic waves in long rods as described in Chapter 4. Hauser (1966), in a review of experimental techniques for measuring stress-strain relations at high rates of strain, pointed out the difficulties and shortcomings of various techniques and presented a critical evaluation of the accuracy of the split Hopkinson bar. Further experimental studies by Bell (1966) tended to generate some doubt as to the validity of the Hopkinson bar technique,

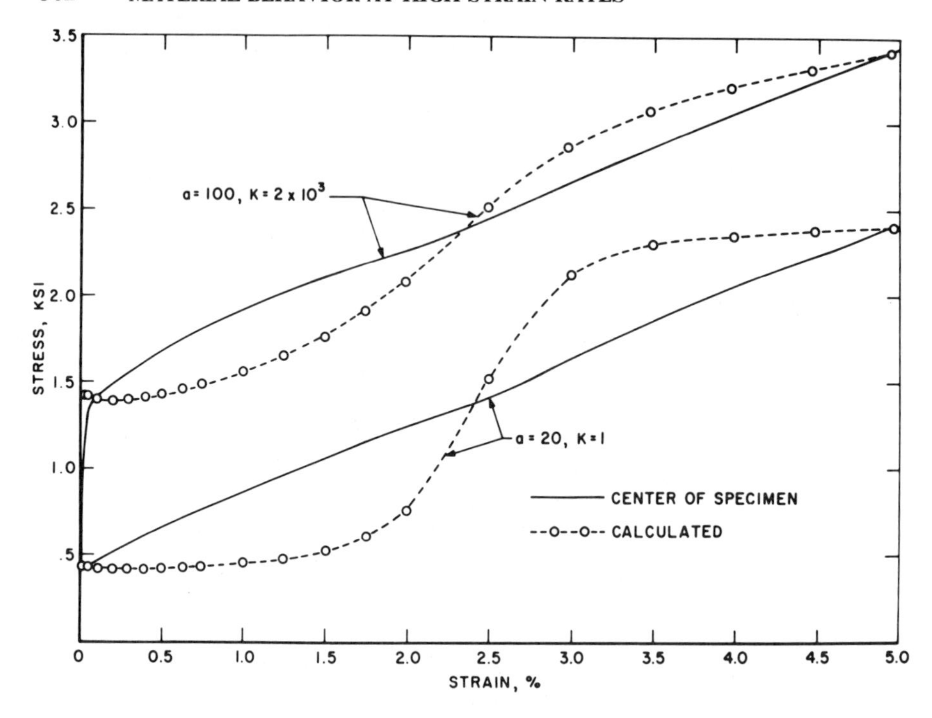

Figure 16 Calculated and actual stress-strain behavior in Hopkinson bar-jump test.

although the end conditions used by Bell involved friction that is usually not present in as large a magnitude in carefully performed Hopkinson bar tests.

Jahsman (1971) conducted a detailed one-dimensional wave-propagation analysis of the Hopkinson bar specimen using a rate-independent material model. He found slight discrepancies between the "reconstituted" curve and the original stress-strain curve, although the analysis was carried out only for short times and, hence, small plastic strains. Nicholas (1973) performed a similar type of analysis for a torsion Hopkinson bar specimen using a rate-dependent material model. It was concluded that for smooth stress-strain curves and moderate strain rates less than $10^4\ s^{-1}$ the technique appears to be reliable and accurate. For materials with a sharp yield point, it was demonstrated that slow plastic waves may tend to distort the apparent behavior of a material. However, for numerical examples that modeled a statically prestressed material subjected to dynamic stress in the Hopkinson bar technique, it was shown that significant errors could arise in the apparent incremental stress-strain curve. Figure 16, from that investigation, shows the actual and calculated stress-strain response for two different rate-dependent constitutive models and an assumed prestress into the plastic region. The calculated values are those that would be obtained theoretically if these materials were tested in a split Hopkinson bar test using the standard equations to reduce the experimental data. Note that these are increments of stress and strain superimposed on a quasistatic prestress. Numerical results of this type suggested that the assumption of equal stresses at both ends of the specimen appeared to be violated more in the incremental test than in other cases investigated.

One of the most thorough analyses of the Hopkinson bar technique was presented by Bertholf and Karnes (1975) using a two-dimensional, elastic-plastic finite-difference computer code. They investigated the effects of end friction as well as length-to-diameter ratio. Figure 17 shows comparisons of calculated stress-strain response with experimental data obtained using lubricated, dry (friction), and bonded end conditions. The calculations and experimental data show the importance of lubrication to eliminate friction between the specimen and the Hopkinson bars as the specimen tends to expand radially under axial compression. Although inertia, and especially friction, can produce an apparent strengthening effect in a material because of the additional constraint, Bertholf and Karnes concluded that the split Hopkinson bar experiment is very accurate and can be reliably used for the measurement of dynamic properties. They found that the inertia correction of Davies and Hunter was reasonable and that $a/h=1$ is a good criterion for specimen design. Their results also showed that friction can greatly increase the degree of

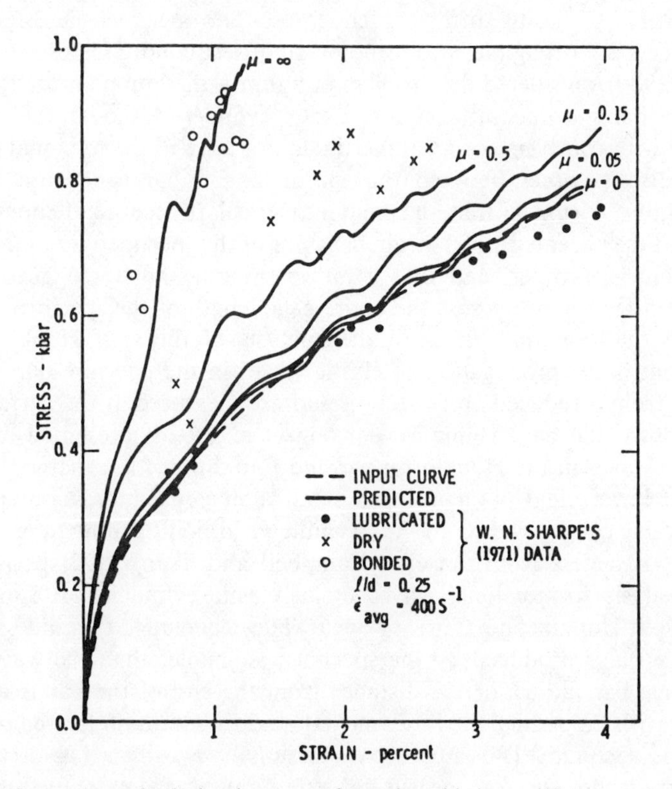

Figure 17 Comparison of calculated response for various friction coefficients and experimental data for different interface conditions [Bertholf and Karnes (1975)].

specimen stress and strain nonuniformity and concluded that Bell's (1966) conclusion that the uniform stress and strain assumption was invalid is correct only for the case of bonded specimens, but that it is false if reasonable care is taken to lubricate the ends of the specimen. Klepaczko and Malinowski (1978) presented the results of a simplified theoretical analysis to estimate the effects of both friction and inertia in the split Hopkinson bar using an energy approach. They derived the relationship

$$\sigma_0 = -P_{\text{av}}\left(1-\frac{2\mu a}{h}\right)-\rho\left(\frac{a^2}{8}+\frac{h^2}{12}\right)\ddot{\varepsilon}+\rho\left(\frac{a^2}{16}-\frac{h^2}{12}\right)\dot{\varepsilon}^2 \qquad (21)$$

where σ_0 is the true axial stress, P_{av} is the average of the forces on either side of the specimen, μ is the coefficient of friction between specimen and bar, and a and h are the specimen radius and length, respectively. For the case of no friction, this approximation reduces to that of Davies and Hunter (1963) when strain rates are not excessive. The approximation is in overall qualitative agreement with the two-dimensional, more exact numerical analysis of Bertholf and Karnes (1975). Their results, however, appear to be more generally applicable for estimations in actual test situations. Malatynski and Klepaczko (1980) conducted a series of experiments on lead in compression up to strain rates of $2\times10^3\ \text{s}^{-1}$ using different slenderness-ratio specimens and found that inertia effects are negligible within the range of test conditions used.

Leung (1980) considered the problem of a thin-walled tubular, torsional split Hopkinson bar specimen attached to a solid cylinder. Although the problems associated with transverse or axial inertia do not arise in the torsional case, the sharp reentrant corner between the specimen and bar can cause localized plastic strains. Using a finite-element numerical procedure, Leung demonstrated that the observed stress-strain behavior of the specimen is not influenced by the reentrant corner, and that a rather smooth and homogeneous shear deformation takes place across the entire gage length of the specimen.

Another question that arises in the analysis of the split Hopkinson bar arrangement is the propagation of elastic waves in the pressure bars that are generated from a reduced cross section and are measured at the surface some distance down the bar. Yeung Wye Kong et al. (1974) presented an elastic solution to waves in the Hopkinson bar and found that for a material with an upper and lower yield point, using a small diameter compression specimen, surface strains corresponding to the dynamic yield point can be in error as far as 16 bar diameters from the end. Campbell and Tsao (1972) performed a similar analysis for torsional waves in an elastic cylinder that simulates a torsional split Hopkinson bar arrangement. They concluded that unless the rise time of the pulse produced by the specimen is smaller than the wave-transit time for one bar radius, or the distance from the end of the bar is not large compared with the radius, the fundamental mode assumption for the propagating wave is adequate. Thus, the use of torsional waves in a bar or tube is a satisfactory technique for measuring stress pulses from a torsional split Hopkinson bar specimen, even when the rise times are very short.

There have also been experimental investigations to validate the use of the split Hopkinson bar. In these studies, attempts have been made to measure both stress and strain directly at one point or as close together as possible on a specimen and to compare the true stress-strain behavior with that deduced from the conventional method of obtaining split Hopkinson bar data. Karnes and Ripperger (1966) measured stress directly at one end of a Hopkinson bar specimen using an X-cut quartz crystal, while strain was measured using very short etched-foil strain gages mounted as close to the specimen end as possible. Chalupnik and Ripperger (1966) used a similar setup involving quartz disks at both ends of the specimen sandwiched between the specimen and the pressure bars and foil gages bonded directly to the specimen. In the latter investigation, high hydrostatic pressures were superimposed on the specimen in some tests. Both investigations tended to verify the basic assumptions of the split Hopkinson bar within the ranges of strain rates achieved. Watson and Ripperger (1969) extended the direct measurement technique to elevated temperatures through the use of a strain gage attached with flame-sprayed ceramic. Wasley et al. (1969) conducted experiments on brittle and low strength materials using a split Hopkinson bar instrumented with a set of piezoelectric quartz-crystal pressure transducers and concluded that this direct method of instrumentation provided valid and more accurate stress and modulus information at very low strain levels.

The split Hopkinson bar apparatus can be used for testing materials at elevated temperatures. However, the effect of the temperature gradient in the pressure bars between the specimen and the location of the strain gages must be considered to ascertain the effect of a modulus or density gradient on the elastic-wave propagation. The first work to account for such an effect was by Chiddister and Malvern (1963) who replaced the temperature gradient in the bar with a five-step discontinuous distribution assuming a uniform temperature between steps. From this discrete model, they worked out the equations for the finite wave reflections and interactions. Approximate temperature-correction factors relating the strain at the specimen to the strain at the gages were developed by Francis (1966) and Francis and Lindholm (1968) assuming an exponential form for the temperature profile.

The analysis of longitudinal elastic-plastic waves in a bar whose properties vary with distance along the bar is given by Perzyna (1959). Campbell and Ferguson (1970), and more recently, Malvern (1980) have treated the problem for an arbitrary temperature profile using the method of characteristics for both zero and finite rise times in the pulse. The propagation of a disturbance along a Hopkinson bar in which there is a temperature gradient is governed by

$$\frac{\partial \sigma}{\partial x} = \rho(x)\frac{\partial v}{\partial t} \tag{22}$$

$$\frac{\partial \varepsilon}{\partial t} = \frac{\partial v}{\partial x} \tag{23}$$

$$\sigma = E(x)\varepsilon \tag{24}$$

If it is assumed that ρ is a constant, that is, density does not vary significantly with temperature, the equations reduce to

$$\frac{\partial \sigma}{\partial x} = \rho \frac{\partial v}{\partial t} \tag{25}$$

$$\frac{1}{E(x)} \frac{\partial \sigma}{\partial t} = \frac{\partial v}{\partial x} \tag{26}$$

These equations can be solved using the method of characteristics. The relations in the characteristic plane are

$$d\sigma = \pm \rho c(x)\, dv \qquad \text{along} \quad dx = \pm c(x)\, dt \tag{27}$$

where

$$c(x) = \frac{E(x)}{\rho} \tag{28}$$

The following expression is derived assuming that the wave propagating down the bar has a zero rise time and using the differential relations in the characteristic plane [Malvern (1980)]:

$$\frac{\sigma}{\sigma_i} = \left(\frac{c}{c_i}\right)^{1/2} \tag{29}$$

where subscript i denotes the end of the Hopkinson bar, $x=0$, adjacent to the specimen and $c=c_0$, the elastic-wave velocity in a bar at ambient temperature. Denoting by subscript o the position of the strain gage at $x=x_0$, which is at ambient temperature, and thus where $c=c_0$, and using Hooke's law, it can be shown that

$$\frac{\varepsilon_i}{\varepsilon_0} = \left(\frac{E_i}{E_0}\right)^{-3/4} \tag{30}$$

$$\frac{\sigma_i}{\sigma_0} = \left(\frac{E_i}{E_0}\right)^{1/4} \tag{31}$$

If the modulus E is linearly dependent on temperature,

$$E_i = E_0(1 - c_\alpha) \tag{32}$$

where

$$c_\alpha = (T_i - T_0) \tag{33}$$

then the temperature correction factors are obtained:

$$\frac{\varepsilon_i}{\varepsilon_0} = (1 - c_\alpha)^{-3/4} \tag{34}$$

$$\frac{\sigma_i}{\sigma_0} = (1 - c_\alpha)^{1/4} \tag{35}$$

which relate the true stress or strain at the specimen to that measured at the gage. For the case of zero rise time, the correction is identical to that of Francis and Lindholm, independent of the shape of the temperature gradient.

An alternate approach to conducting elevated-temperature Hopkinson bar tests that avoids both spurious wave reflections and the necessity for correcting experimental data using the above formulas was developed by Eleiche and Duffy (1975). In torsional split Hopkinson bar tests, they employed bars that were tapered appropriately so that their mechanical impedance in torsion remained constant in spite of the temperature gradients. In torsion, the mechanical impedance is ρcJ, where ρ is the density, c the velocity of torsional waves, and J the polar moment of inertia. Assuming ρ is constant, the wave speed is given by

$$c(x)=\left[\frac{G(x)}{\rho}\right]^{1/2} \tag{36}$$

where G is the shear modulus and the variation with x is governed by the variation of G with temperature which, in turn, depends on the shape of the temperature gradient. The value of J depends only on the geometry (diameter) of the bar and can be designed appropriately to provide a constant mechanical impedance for any given temperature gradient. Although a different set of bars is required for any particular temperature of specimen, no corrections to the data are necessary. The setup was used successfully by Eleiche and Duffy up to specimen temperatures of 250°C and it was demonstrated that no spurious wave reflections occurred in a properly designed bar for temperatures as high as 500°C. This concept of a constant mechanical impedance could also be employed for tension or compression testing although no such application has been reported to date.

The effect of heating rate may be important in characterizing the elevated-temperature dynamic properties of materials. To study this phenomenon, Lipkin (1974) developed a method for dynamic testing of materials subjected to extremely high heating rates. Using electron-beam pulse heating, samples of 6061-T6 aluminum were heated in times of the order of 0.1 μs. A very short region of a long bar was subjected to this high heating rate, while the remainder of the bars remained at ambient temperature. A propagating stress wave generated within a few microseconds of the heating provides the high strain-rate loading. The central portion of the bar that is heated is the test specimen, while the unheated portions of the bar on either side act like the conventional split Hopkinson pressure bars. By performing the experiment rapidly, heat flow away from the central test area is minimized.

In all these analyses, density variations with temperature are neglected and only the variation of modulus with temperature is considered. The results indicate that the split Hopkinson bar experiment can be conducted at elevated temperatures using the appropriate corrections with a high degree of reliability.

8.3 OTHER EXPERIMENTAL TECHNIQUES

8.3.1 The Taylor Cylinder

A technique for determining the dynamic yield stress of a material was developed by Taylor (1948) and Wiffin (1948). The method involves impacting a right circular cylinder against a rigid target and making post-impact measurements of the deformed shape. Taylor used a very simple analysis that assumed rigid, perfectly plastic material behavior and simple one-dimensional wave-propagation concepts. Figure 18a shows the cylinder at some point during deformation. The deformed region is propagating away from the rigid wall at a velocity c_p, while the undeformed portion of the cylinder whose instantaneous length is h is traveling at a decreasing velocity v. It is assumed that the material behavior is rate-independent, $\sigma=\sigma(\varepsilon)$, and rigid-plastic, that is, elastic strains are negligible. Denoting the yield stress by σ_y and the original cross-sectional area by A_0, the areas and stresses on either side of the plastic wave front are as depicted in Figure 18a. The stress immediately ahead of the wave front is at incipient yield, while the material behind the wave front is at rest. Denoting engineering stress and strain by σ and ε, respectively, and assuming incompressible plastic flow,

$$c_p A=(v+c_p)A_0 \tag{37}$$

$$\frac{dh}{dt}=v+c_p \tag{38}$$

The strain directly behind the plastic wave front is calculated, using (37),

$$\varepsilon=\frac{A-A_0}{A}=\frac{v}{v+c_p} \tag{39}$$

Conservation of momentum leads to

$$\rho(v+c_p)v=\sigma-\sigma_y \tag{40}$$

because when the wave front passes through a distance $dx=-dt(v+c_p)$, the velocity of this element vanishes. The equation of motion of the nondeforming

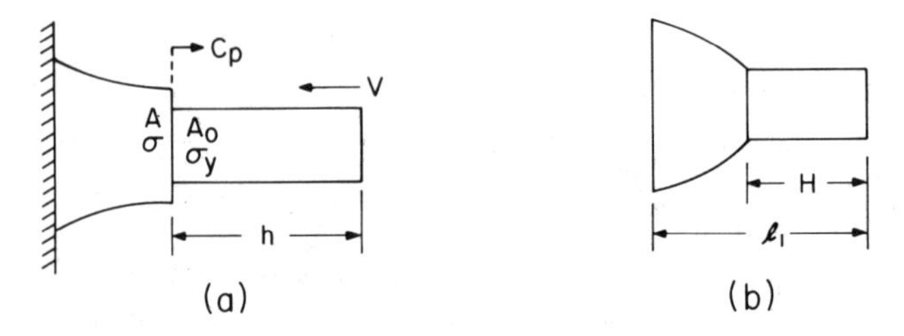

Figure 18 Schematic of Taylor cylinder; (a) during deformation, (b) after deformation.

part is

$$\rho h \frac{dv}{dt} = -\sigma_y \tag{41}$$

From (38) and (39) it can be shown

$$\tfrac{1}{2} d(\rho v^2) = \sigma_y \varepsilon d(\ln h) \tag{42}$$

and from (39) and (40) we obtain

$$\rho v^2 = \varepsilon(\sigma - \sigma_y) \tag{43}$$

Then, from (42) and (43),

$$d[\varepsilon(\sigma - \sigma_y)] = 2\sigma_y \varepsilon d(\ln h) \tag{44}$$

Letting $h = l_0$, $v = v_0$, and $\varepsilon = \varepsilon_0$ at time $t = 0$ and integrating,

$$\ln \frac{h}{l_0} = \tfrac{1}{2} \int_{\sigma_0}^{\sigma} \frac{d[\varepsilon(\sigma - \sigma_y)]}{\sigma_0 \varepsilon} \tag{45}$$

We can determine ε_0 from (43) and noting that $\sigma_0 = \sigma_0(\varepsilon_0)$:

$$\varepsilon_0 = \frac{\rho v_0^2}{\sigma_0 - \sigma_y} \tag{46}$$

The Taylor formula is derived by assuming $c_p = \text{constant}$, using (39) and (41)

$$\frac{dv}{dh} = \frac{\sigma_y}{\rho h(v + c_p)} \tag{47}$$

Rearranging and integrating,

$$\left(\frac{\sigma_y}{\rho}\right) \ln\left(\frac{h}{l_0}\right) = \tfrac{1}{2} v^2 - \tfrac{1}{2} v_0^2 + c_p v - c_p v_0 \tag{48}$$

We denote by H the final length of the undeformed portion as shown in Figure 18b that occurs after $v = 0$. Thus

$$\left(\frac{\sigma_y}{\rho}\right) \ln\left(\frac{H}{l_0}\right) = -\tfrac{1}{2} v_0^2 - c_p v_0 \tag{49}$$

If it is further assumed that the rear of the specimen decelerates at a uniform rate, then it can be seen that the plastic wave propagated a distance $(l_1 - H)$, where l_1 is the final total length of the cylinder, in a time

$$t = \frac{l_1 - H}{c_p} \tag{50}$$

But the time of deceleration is also approximately, assuming *uniform* deceleration,

$$t = \frac{2(l_0 - l)}{v_0} \tag{51}$$

Equating (50) and (51),

$$c_p = \frac{v_0}{2}\frac{(l_1 - H)}{(l_0 - l_1)} \tag{52}$$

and from (49)

$$\frac{\sigma_y}{\rho v_0^2} = \frac{(l_0 - H)}{2(l_0 - l_1)} \frac{1}{\ln\left(\dfrac{l_0}{H}\right)} \tag{53}$$

The yield stress or flow stress is easily obtained from (53) and involves only measurements of the undeformed portion of the cylinder and the impact velocity. It is to be noted that the Taylor experiment is limited to impact velocities below the plastic wave velocity in the cylinder, for otherwise shock waves would develop which are not considered in the analysis. It is also to be noted that strain rates are not constant in this experiment and, furthermore, that the values of the strain rates are not readily determined from the simplified analysis.

Taylor (1948) introduced a correction factor into the analysis since the deceleration of the cylinder is not necessarily uniform. Letting $\bar{\sigma}_y$ be the corrected value of σ_y as determined from (53), he derives the correction formula

$$-\frac{\bar{\sigma}_y}{\sigma_y} = \frac{l_0 - l_1}{l_0 - H} \frac{\ln\left(\dfrac{l_0}{H}\right)}{\left(K - \dfrac{c_p}{a}\right)^2} \tag{54}$$

where the quantities a and K are defined as

$$a^2 = \frac{2\sigma_y}{\rho} \tag{55}$$

$$K = \frac{v_0 + c_p}{a} \tag{56}$$

Taylor presents results of calculations in the form of plots of H/l_0 against l_1/l_0 for constant values of $\bar{\sigma}_y/\sigma_0$.

The Taylor cylinder experimental configuration has been analyzed in detail by Wilkins and Guinan (1973) using a two-dimensional finite-difference numerical scheme from which they found that an average dynamic flow stress can reasonably be obtained. Papirno et al. (1980), however, obtained computer code results that suggest that the Taylor formula for dynamic flow stress, when applied to high strength materials, yields results that are much higher than the actual flow stresses. Lee and Tupper (1954) presented modifications to the Taylor formulation by including elastic strains in the analysis. Raftopoulos and Davids (1967) used the Taylor experiment to determine the dynamic behavior of metals, particularly the dynamic yield stress. They extended the analysis to include elastic, work-hardening, and nonlinear material behavior and postulated various forms of constitutive equations in their analysis. They found that the assumption of one-dimensional motion was adequate and that up to impact velocities of approximately 300 m/s there was no deformation of the target. A rigid, work-hardening material model provided the best correlation with experimental data, and they concluded that work hardening was important to include in the material model. Recht (1978) extended the use of the Taylor model to determine the dynamic stress-strain behavior in situations where the impacted target is not rigid.

Other variations of the Taylor cylinder experiment include a reversed Taylor experiment that involves impacting a finite mass against a stationary cylinder [Gust and Young (1980)]. Another geometry involves impacting identical cylinders against each other to simulate the rigid boundary condition by symmetry considerations and, furthermore, to avoid questions about the friction between the radially deforming cylinder and the rigid wall. This latter geometry, however, presents experimental difficulties at elevated temperatures and loses some of its advantages in that situation. In any case, it should be recognized that the impacting cylinder experiment involves complex two-dimensional, axial-symmetric wave propagation and, although one-dimensional analysis has presented a simple method for obtaining material property data, the experiment lends itself to a more detailed analysis and more complete instrumentation to track the transient deformation profile. Such an approach, using the symmetric impact configuration, is being utilized by Curran et al. (1979) where the test is computationally simulated in iterative calculations with a two-dimensional computer code. The dynamic yield surface parameters are varied until the calculations and the measurements of the mushrooming using high-speed photography agree. This approach holds great promise for deducing material properties at very high strain rates under nonuniaxial stress conditions but suffers the disadvantages of the necessity of an iterative wave-propagation analysis.

8.3.2. The Expanding Ring

The expanding ring or cylinder provides another method for obtaining high strain-rate properties of materials. The ring that provides a state of uniaxial plane stress, and the cylinder that provides a state of plane strain, provide a geometry that can achieve a uniaxial stress or plane strain state at high rates through the application of a symmetrical radial pressure. Clark and Duwez (1950) appear to be the first to use the expanding ring or cylinder geometry for high-rate testing. They developed a method of expanding a tube using mercury as a driving fluid and achieved strain rates of approximately 200 s^{-1}. A piston driven at the desired velocity into the assumed incompressible mercury was used to generate a known internal fluid pressure in experiments on steel thin-walled hollow cylinders. Dynamic deformations of a ring have been achieved through the use of electromagnetic forces by Niordson (1965) or by detonation of an explosive charge by Hoggatt et al. (1967) and Hoggatt and Recht (1969). The expanding cylinder experiment has been described by Fyfe (1968) in which discharge of a capacitor into a wire is used to obtain an exploding wire. An optical displacement measuring system using a laser light source and photomultiplier detector, described by Swift and Fyfe (1970), is utilized to measure radial displacements. Walling and Forrestal (1973) used a magnetic pressure pulse to expand rings dynamically. They achieved a pulse of less than 10 μs duration that provided clean signals after 20 μs in their strain gages that were used for measuring circumferential strain in the ring. Wesenberg and Sagartz (1977) used the same technique of generating a magnetic pressure pulse from a capacitor-bank discharge to study the dynamic fracturing process in aluminum cylinders. Strain rates of up to $10^4\ s^{-1}$ were achieved. Fyfe and Rajendran (1980) used the thin cylinder configuration accelerated by an exploding wire system and laser-photomultiplier system for displacement measurements to study strain-rate and strain-rate history effects on the fracture of metals. Forrestal et al. (1980) have developed a new explosive loading technique for producing nearly uniform expansion of thick-walled cylinders. As in several of the other investigations cited, stress-strain data were not obtained, but strain-rate and strain-to-failure data were determined. In experiments on AISI 304 stainless steel cylinders, strain rates in excess of $4000\ s^{-1}$ were achieved.

If the ring or cylinder is impulsively loaded, that is, the time duration of the loading is very short, then the dynamic stress-strain relations of the material are obtained from consideration of the free flight deceleration of the ring or cylinder that achieved an initial radial velocity as a result of the impulse. For a thin ring, the engineering hoop stress in the absence of a driving pressure is easily shown to be

$$\sigma = -\rho R \frac{d^2 R}{dt^2} \tag{57}$$

where R is the radius, and ρ the mass density. If the forcing function, for example, an electromagnetic force or hydraulic or pneumatic pressure, is maintained during the experiment, it must be measured directly or calculated accurately and incorporated into the equation of motion. The strain in the specimen is obtained from measurement of the instantaneous diameter or radial deformation of the ring or cylinder. The stress, however, is obtained from (57) which shows that the calculation of stress depends on the knowledge of the second temporal derivative of radial displacement or the first derivative of radial velocity. In either case, accuracy is lost in differentiating experimentally measured displacement or velocity records, especially in double-differentiating displacement data. Daniel et al. (1980), in evaluating data from composite material rings loaded by an internal pressure pulse applied explosively through a liquid, obtained second derivatives using highly complex smoothing operations and approximations on strain-gage data in digital form. Warnes et al. (1980) utilized a laser velocitometer to obtain highly accurate velocity-time records that have to be differentiated only once to obtain radial acceleration. Limited success has been achieved to date, however, in the use of the expanding ring or cylinder for dynamic property measurements because of the lack of accuracy in calculating stress. On the other hand, very high strain rates of the order of $10^4\ s^{-1}$ or greater can be achieved with this technique. As with other dynamic test techniques, limitations in the maximum strain rate that can be achieved are reached when wave-propagation effects become important. In the case of the ring, this occurs when radial waves reflecting through the thickness of the ring become significant and the assumption of a purely circumferential stress state becomes invalid.

8.3.3 Dynamic Shear Tests

In addition to the torsional split Hopkinson bar, other techniques have been utilized to subject specimens to states of pure shear at high rates of strain. Dowling et al. (1970) developed a dynamic punch test that uses a Hopkinson bar system with a die tube and punch bar, a system previously suggested by Hauser (1966) and shown in Figure 8. Using flat test specimens, they were able to attain strain rates up to $10^4\ s^{-1}$ in shear. The apparatus is described in detail in Dowling and Harding (1967). In those tests, the effective gage length depended on the clearance between the punch and die and led to much variation in results in actual use. A more promising approach was a similar setup using a double notch type of shear specimen described by Campbell and Ferguson (1970). Using this specimen in a drop weight version of the Hopkinson bar, they were able to achieve strain rates up to $4 \times 10^4\ s^{-1}$ in shear in mild steel. Harding and Huddart (1980) have used the same apparatus and modified the double-notch shear-specimen geometry to obtain results that are in apparent better agreement with those from high-rate tensile tests on the same material. There are still questions as to the uniformity and magnitude of shear strain in these types of tests, but the use of pure shear coupled with the ability to achieve relatively high strain rates makes this technique one of the more promising for future studies in dynamic plasticity.

8.3.4 Drop Forge Tests

For testing in the medium strain-rate regime and higher, and especially for large plastic deformations, the drop forge test has been employed by a number of investigators. Although results obtained by this method are not always considered in reviews of the literature in dynamic plasticity, this is yet another method of achieving dynamic deformations under controlled conditions. The drop forge is a mechanical device that can subject a specimen to dynamic compression at strain rates of the order of $10^2\ s^{-1}$ and higher. In this test, a loading tup is raised by motor to a predetermined height and dropped. The tup is guided by carefully machined railings to insure axial alignment and loading. In general, the machine has an excess of tup energy to deform the specimen and can thus achieve somewhat constant velocity of deformation. For very large strains or insufficient tup energy, a constant velocity is difficult to achieve. For the case of nearly constant strain rate, the strain is easily calculated. Force measurements, however, require careful consideration of wave-propagation effects, especially as higher strain rates are attempted. Woodward and Brown (1976) present a method using short-ring piezoelectric load cells for force pulse measurements in experiments involving strain rates up to $10^4\ s^{-1}$. The effect of dynamic ringing, however, was evident at these high strain rates. Lengyel and Mohitpour (1972) used a high-energy forging press to subject specimens to small increments of deformation at high loading rates through a series of repetitive tests. In this manner, they tended to avoid any rapid temperature build up or the possibility of nonconstant strain rate at large strains and high strain rates. They achieved strain rates of up to $1000\ s^{-1}$ in these incremental tests. Using a single test, however, the stress-strain results fell well below those of the incremental test. They attributed this to the substantial temperature rise at large strains as well as the decrease in strain rate during the test. It is interesting to note, however, that strain-rate history effects were not considered as a contributing factor.

A sophisticated application of the drop forge test is described by Holzer (1978) in which short load cells are used for transient force measurements and a fiber-optics displacement transducer is used for transient strains. Force measurements are obtained by a computer-aided analysis of the force-time data that allows for correction for the dynamic characteristics of the load cell. The actual force of deformation is obtained by a fast Fourier transform of the signal from a piezoelectric load cell. The signal is then corrected in the frequency domain. This involves complex division of the frequency domain signal by the predetermined load-cell frequency-response function. The subsequent inverse transformation then yields the corrected true force signal. This procedure was employed by Holzer and Brown (1979) to obtain compression stress-strain data on two steels at strain rates up to $10^3\ s^{-1}$. The results show that the measured flow-stress value is influenced by specimen geometry, with stress increasing as the initial height over diameter ratio decreases. Although radial inertia effects might be suspected, these observations were made at both high and low strain rates. The effects of end friction, as discussed extensively in connection with the split Hopkinson bar analysis, would appear to be a contributing factor in the interpretation of these experimental results. It is important to note here the similarity between the drop-forge technique and the direct-impact Hopkinson bar configuration. In both cases, the assumptions and limitations of the split Hopkinson bar technique are equally important when trying to achieve very high rates of strain.

8.3.5 Dynamic Bend Tests

In discussing dynamic test techniques, a few remarks are in order regarding the Charpy impact or dynamic bend test. The Charpy impact test involves the determination of the total energy to failure of a notched beam in three-point bending under impact conditions. Its primary function is to screen materials, to compare similar types of materials, and to establish such quantities as the ductile to brittle transition temperature in steels through the measurement of total energy under impact. The use of a notched specimen and the recording of only the total energy makes the test of no value in the determination of constitutive relations or quantitative failure criteria for materials. The advantages of the test, however, are that it is extremely simple and quick to perform and allows one to subject a material to impact conditions rather easily, if that is the object of the material evaluation.

In recent years, an instrumented version of the Charpy test has been developed in which the actual load-displacement history of the specimen may be obtained [ASTM (1974)]. Load is determined through the use of strain gages placed directly on the loading head or tup of the impact pendulum, while displacement is obtained through a knowledge of the initial velocity and calculations of velocity decreases of the loading tup from energy considerations. The resultant load-displacement curve can be integrated to obtain an energy profile as a function of either time or displacement. The information from this type of test provides a better understanding of the dynamic notched-bend behavior of a material, and has been used to evaluate fracture toughness by numerous investigators but it still does not provide an easy method for determining material constitutive behavior.

To avoid the complexities of analyzing a notched beam in bending, several investigators have used unnotched or smooth bars in the instrumented Charpy test. In this instance, the use of beam theory allows one to make some approximate calculations regarding actual material response. For linear-elastic materials such as certain composites, or by restricting the analysis to the linear portion of the load-displacement curve, numerical results can be obtained for maximum stress to yield or failure under dynamic loading. When used in conjunction with quasistatic three-point bend tests on identical specimens, the instrumented (unnotched) Charpy test can provide useful information on the strain-rate sensitivity of materials in bending up to strain rates of the order of $100\ s^{-1}$.

As in the case of other test techniques, there are fundamental limitations to the strain rates that can be achieved because of wave-propagation or inertia effects. At impact velocities above approximately 1.5 m/s, inertia effects in Charpy tests become important, and bending-wave transit times across the specimen become large compared to the total duration of the test. The result is that as higher velocities are attempted, the recorded load-time history is increasingly distorted as a result of inertia and the vibrations of the specimen (see Figure 15). As in the case of the Hopkinson bar, the equivalence of load at the beam supports with that at the central loading point becomes a poorer assumption as increased loading velocities are attempted. Thus, there are severe limitations to performing Charpy tests at ever-increasing impact velocities.

One additional point must be made regarding the determination of failure criteria from the instrumented Charpy test. The standard Charpy geometry (unnotched) will lead to a tensile failure at the outermost fibers in three-point bending for most homogeneous, isotropic materials. For anisotropic materials, and especially laminated composites, this same geometry specimen will often fail first in interlaminar shear. Thus, in comparing loads to failure and corresponding energy absorption for different materials, it is extremely important to be cognizant of the first mode of failure that the material undergoes [Krinke et al. (1978)]. The use of relatively thin bend specimens has been recommended by several investigators to assure a tensile mode of failure under three-point bending in the Charpy test. If properly utilized, the instrumented Charpy test on unnotched specimens can be a valuable tool for additional evaluation of the strain-rate sensitivity of materials, especially for composite materials where specimen geometry and fabrication procedures make it difficult to employ other test techniques. Direct comparison of load-deflection traces from quasistatic and impact tests on identical specimens can provide an easy method for assessing the strain-rate sensitivity in bending of materials at strain rates up to the intermediate regime of approximately $10^2\ s^{-1}$.

8.4 EXPERIMENTAL RESULTS

8.4.1 General Remarks

The literature is filled with results of investigations on the dynamic properties of materials, which are too numerous to mention here. Several survey articles such as those by Lindholm and Bessey (1969) and Jiang and Chen (1974) have compiled data from a number of investigations and presented them in graphical or tabular form. Figure 19 shows a method of comparing data on a number of aluminums in terms of a rate-sensitivity parameter versus the static flow stress [Holt et al. (1967), Green et al. (1970), Lindholm (1974)]. The parameter is the increase in flow stress from a static test to a dynamic test at a given strain divided by the static flow stress and the log of the difference in strain rates. It represents the percent increase in stress per unit of log strain rate. The ratio obtained is useful for comparing materials only if the stress versus log strain-rate curve is essentially linear, otherwise the ratio can be influenced by the maximum strain rate obtained. An alternate method for presenting data in this format is to use two fixed strain rates, one for the dynamic test and one for the static test, for example 10^3 and $10^{-3}\ s^{-1}$, respectively. This, however, depends on the ability to obtain data at the given dynamic rate for all materials being compared.

As shown in Figure 19, for a number of aluminum alloys, the degree of rate sensitivity is seen to increase as material strength is decreased, or as purity increases. Duffy (1980) presents data from both constant strain rate and jump tests (see Section 8.4.3) in shear for a number of materials from various investigations reported in the literature. He uses a strain-rate sensitivity parameter that is calculated from the difference between the dynamic and

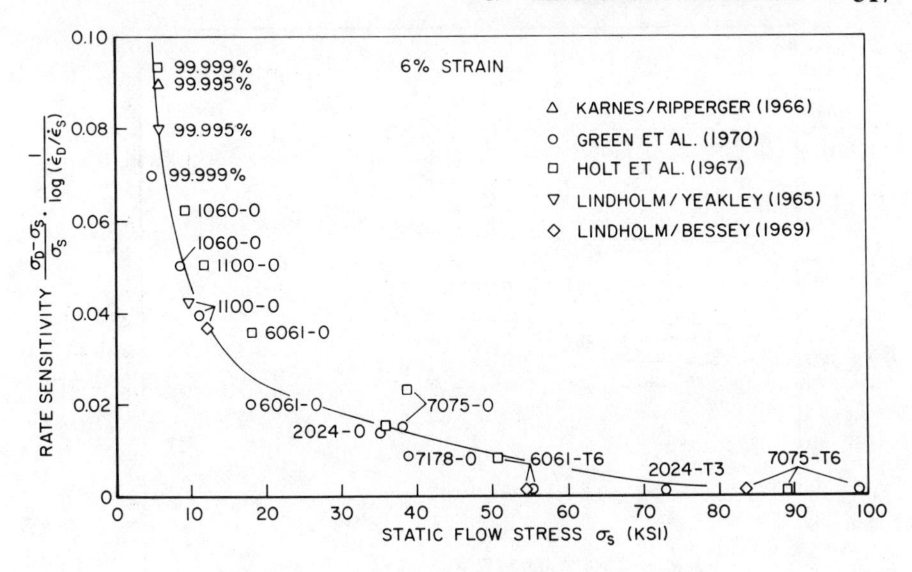

Figure 19 Strain-rate sensitivity of aluminum alloys.

static flow stresses in the constant rate tests, or the flow stress immediately following an increment in strain rate less the quasistatic value in the jump tests. An alternate method for presenting data is to plot flow stress against log strain rate for a given value of strain, the yield point, or the ultimate stress. A typical plot for a number of grades of structural steels is shown in Figure 20, which is based on data obtained in tension Hopkinson bar tests [Nicholas (1980a)]. Data from tensile tests on 6061-T6 aluminum is shown in a similar form in Figure 11.

The behavior of materials under dynamic loading has been reviewed extensively from various points of view and for various classes of materials. Metallurgical aspects of strain-rate sensitivity as well as examples illustrating the degree of rate sensitivity in various metals and alloys are presented in NMAB (1978). Holzer (1979) presents a tabular summary of the experimental work in dynamic plasticity of metals over the last three decades covering medium to high strain-rate tests performed using the cam plastometer, experimental drop forges, and the split Hopkinson pressure bar. Green et al. (1970) discuss the behavior of the *fcc* metals, aluminum and its alloys, copper, lead, and nickel and present an explanation of the observed behavior on the basis of dislocation theory. Malatynski and Klepaczko (1980) review the work on the plastic properties of lead as well as present data on strain-rate effects in technically pure lead in compression. Eleiche (1971) presents a survey of the available literature on the effects of strain rate at elevated temperatures including quantitative information on rate dependence in both tabular and graphical form. Nicholas (1980b) presents data from tension Hopkinson bar tests on a large number of materials along with quasistatic data for compari-

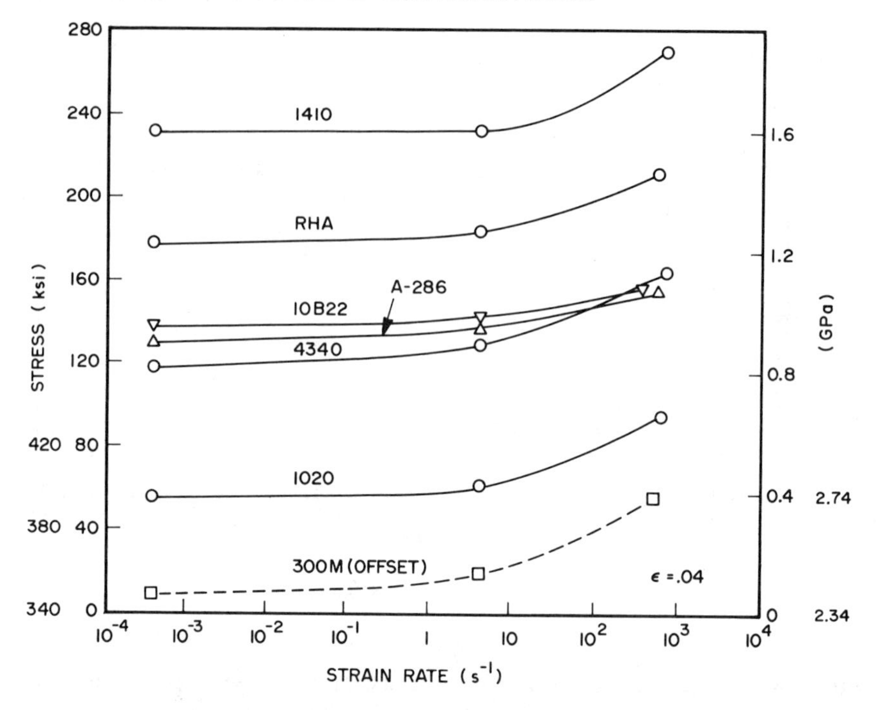

Figure 20 Effect of strain rate on the flow stress of several grades of steel.

son. The reader also is referred to the various conference and symposia proceedings previously referenced for specific information.

The results of various investigations using different, and in some cases the same experimental techniques are not always consistent with each other. There has been a continuing debate for years over the extent of the strain-rate sensitivity of metals, especially at higher rates of strain in the range above 10^3 to 10^4 s^{-1}. This is the range where the fundamental assumptions in the experiments can start breaking down, for example, the existence of uniaxial stress states, the neglect of radial-inertia effects, the uniformity of stress and strain in the specimen, and the lack of wave-propagation effects. As an example, examine data from experiments on strain-rate effects in copper, Figure 21, which show that at strain rates of approximately 10^3 s^{-1}, results of several different investigators show dramatic increases in flow stress over the quasistatic values. Compare these with the results of Lindholm (1978), Figure 22, on high-purity copper that show little or no increase in flow stress up to strain rates of nearly 10^5 s^{-1}. It should be noted that Lindholm took great care to preserve the length-to-diameter ratio in achieving the highest strain rates while some of the others did not.

As an example, pure aluminum, particularly 1100-0, has been a widely used material in dynamic plasticity investigations. It has generally shown a small amount of strain-rate sensitivity, with the rate sensitivity increasing with

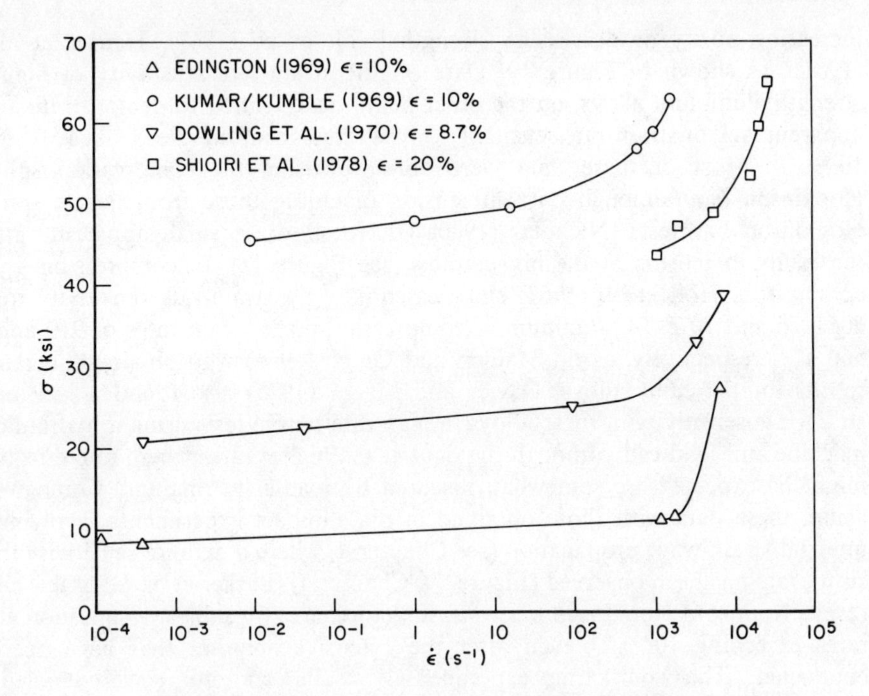

Figure 21 Effect of strain rate on the flow stress of copper.

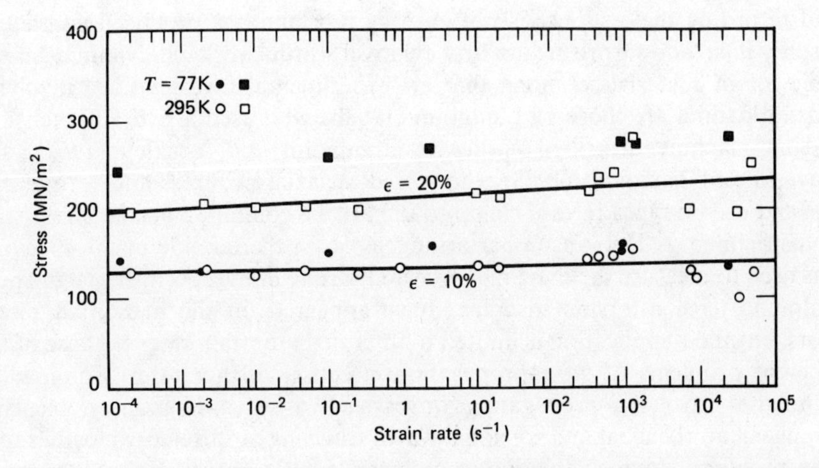

Figure 22 Effect of strain rate on the flow stress of 99.999% copper [Lindholm (1978)].

increasing purity or decreasing strength [Holt et al. (1967), Hauser et al. (1961)], as shown in Figure 19. Data on the strain-rate sensitivity of high-strength aluminum alloys, on the other hand, have tended to demonstrate an apparent lack of strain-rate sensitivity over a range in strain rates from 10^{-4} to 10^3 s^{-1}. Most of these data were obtained using the compression split Hopkinson bar, although data in tension including those from tensile split Hopkinson bar tests [Nicholas (1980a)] have shown a small apparent rate sensitivity in tension at the higher rates (see Figure 11). In compression, by comparison, Holt et al. (1967) show essentially no strain-rate sensitivity for 6061-T6 and 7075-T6 aluminum in compression up to strain rates of 910 and 560 s^{-1}, respectively, as do Maiden and Green (1966) who obtained similar results for the same alloys. Davies and Magee (1975) also found a lack of strain-rate sensitivity in these alloys in high-rate tensile tests using a hydraulic machine and load cell, although the data at the higher rates where rate effects might be expected are somewhat obscured by load-cell "ringing." Compare, again, these data with those obtained in plate-impact experiments involving uniaxial-strain wave propagation (see Chapter 4) where a definite sensitivity to strain rate has been observed (Figure 34, Chapter 4) [Barker et al. (1964)]. The results from split Hopkinson bar tests, therefore, are still subject to question at rates exceeding 10^4 s^{-1} even after the extensive analyses that have been conducted. The conflicting experimental results on rate sensitivity led Lindholm (1978) to conclude that the widely used Hopkinson bar methods must be carefully scrutinized when applied for testing at strain rates above 10^4 s^{-1} where the fundamental assumptions in the analysis become critical. He notes, however, that alternate techniques are not yet available for large dynamic plastic deformations.

8.4.2 Biaxial Testing

Since subjecting a specimen to high rates of strain in uniaxial stress conditions and recording the response is not an easy task and has produced conflicting results, it is not surprising to find relatively little work in dynamic biaxial behavior of materials. Among the few experimental investigations involving biaxial loading are those of Lindholm (1968b) who used a hydraulic tension-torsion machine described earlier by Lindholm and Yeakley (1967), and Hayashi and Tanimoto (1978) who stored an axial prestress and torque in a bar and used a rapid release clamp to impart the combined biaxial stress state to a specimen. A Hopkinson bar arrangement on the far side of the specimen was used to measure the dynamic force and torque in the specimen. High-speed hydraulic tension-torsion machines have appeared on the market in recent years, but their application is limited to intermediate strain rates because of the inherent problems of wave-propagation effects at higher rates. For loading techniques involving propagating waves such as in a Hopkinson bar-type arrangement, the axial and torsional waves traveling at different velocities may be a problem, although the two waves propagate in an uncoupled manner in the elastic bars.

The results reported by Lindholm (1968b) and Hayashi and Tanimoto (1978) show that the expansion of the yield surface with increasing strain rate is isotropic for annealed aluminum and that a Mises-type yield criterion is adequate to describe the observed behavior. Both investigations involved proportional loading paths. The corresponding analytical problem was addressed by Meguid et al. (1979) who considered the biaxial tension-torsion of a solid rod of a rate-dependent material. Their numerical results show that a very good approximation to the rate-dependent stress profiles can be obtained by simply increasing the yield or flow stress of the rate-independent material model from the quasistatic value to a value determined from a tensile test at the appropriate dynamic strain rate. Note the analogy to similar observations in uniaxial-stress wave propagation experiments in long rods, that is, the use of a single, dynamic stress-strain curve, described in Chapter 4, and to the results of a one-dimensional analysis of a split Hopkinson bar specimen by Conn (1965) who also demonstrated the use of a single dynamic curve. There has been no reported work using nonproportional loading paths, primarily because of the experimental difficulties in achieving such a state at very high strain rates.

The very few experimental data under dynamic biaxial loading and the general lack of demonstrated capability to model such behavior is primarily due to the lack of available experimental techniques. Recent investigations indicate that the use of a thin plate sandwiched between two hard, high-impedance elastic plates will provide a method for generating data at strain rates of the order of $10^5\ s^{-1}$ under states of combined pressure and shear [Clifton (1980)]. The configuration is based on the split Hopkinson bar concept and provides a biaxial stress state through the skew impact of a driver plate against the elastic plate that generates a wave of combined pressure and shear. Elastic-wave analysis coupled with free-surface velocity measurements provide sufficient data to obtain the high strain-rate response of metals in shear under a high confining pressure. This technique holds great promise for providing data in combined stress states at high strain rates which have heretofore been unobtainable.

8.4.3 Strain-Rate History Effects

In addition to the effects of strain rate and multiaxial stress states on the plastic flow of metals, the history of loading can affect the flow stresses at a given strain and strain rate. A number of investigators have examined loading history to determine its contribution to the mechanical behavior characteristics. A technique that has achieved popularity over the last decade is the jump test or more properly the incremental strain-rate test. Used mainly in torsion testing, a specimen is subjected to a slow rate of loading followed by a very high loading rate. Contrary to the incremental plastic wave experiments in which the wave speed and stress profile in the propagating wave are observed, this experiment provides the incremental stress-strain curve from the point where the dynamic stress was applied. Figure 23 shows typical results obtained

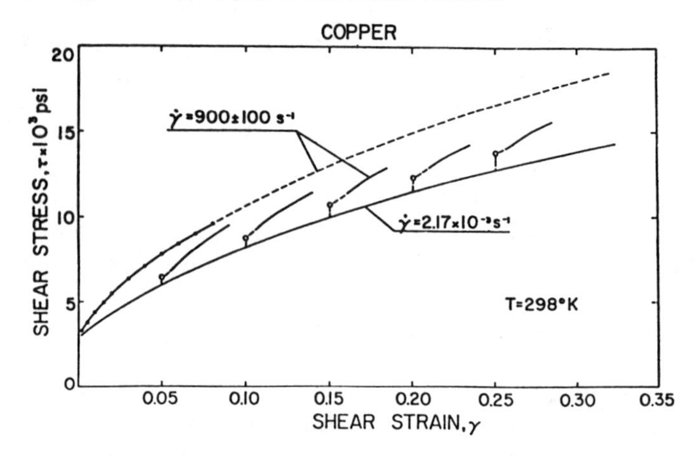

Figure 23 Results of incremental strain rate experiments with polycrystalline copper [Klepaczko et al. (1977)].

in this type of test. In addition to constant strain-rate stress-strain curves, incremental dynamic curves are obtained starting at various magnitudes of static prestrain. The experiment, which has been used extensively to study the effects of strain-rate history in materials, is described by Campbell and Dowling (1970) in which a rapid release clamp in a torsional split Hopkinson bar apparatus is used to provide torque and by Duffy et al. (1971) in which explosive loading is used for the dynamic torque. A comprehensive summary of work in this area is presented by Duffy (1980). Klepaczko (1968) and Klepaczko (1975) have attempted to relate the observed macroscopic behavior in jump tests to the dislocation behavior.

The earliest work on strain-rate history effects is that of Lindholm (1964) who demonstrated that the stress in dynamic loading in aluminum following static preloading was not the same as that obtained in dynamic loading starting from an unstressed condition. The difference was attributed to strain-rate history. Subsequent investigations by Klepaczko (1967), Klepaczko (1968), Campbell and Dowling (1970), Nicholas (1971), and Frantz and Duffy (1972) all tended to confirm the existence of strain-rate history effects in various materials by comparing the dynamic flow stress in an unstressed material to that obtained at the same dynamic strain rate in a material initially subjected to a much slower rate of loading. Figure 24 shows typical data for pure titanium. The solid line shows the path in the jump test where the material is subjected to a slow rate of straining to less than 4% shear strain and then suddenly subjected to a superimposed dynamic rate. The dashed lines show the reference constant strain-rate results at both the low and the high rate. In this particular material, the dynamic rate superimposed on a slow rate results in a flow stress that is below the constant dynamic-rate stresses for the same total strain. Among the most extensive series of strain-rate history or jump tests are those of Eleiche and Campbell (1976b) and Campbell, et al. (1977) who used a

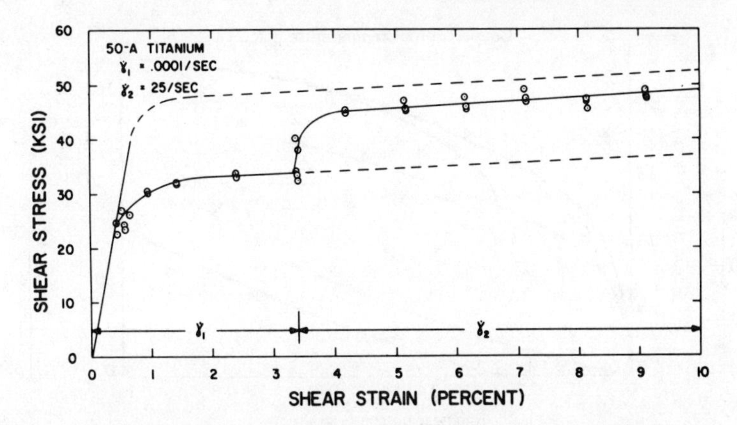

Figure 24 Results of variable strain-rate tests for 50-A titanium.

torsional split Hopkinson bar to achieve dynamic strain rates up to 3000 s^{-1} in shear. In experiments on copper, titanium, and mild steel they applied rapid changes in strain rate up to six orders of magnitude over a range of quasistatic shear strains up to 0.06. The results indicate that a history effect is present for all three materials, but that the dynamic stresses may be greater, as in mild steel (Figure 25), or less than, as observed previously, in titanium (Figure 24) and copper (Figures 23 and 26) than those obtained in constant rate tests. In their investigations, they examined the effects of adiabatic heating as a possible contribution to the observed behavior. The adiabatic temperature rise was calculated on the assumption that all the plastic work is converted into heat. Typical results shown in Figures 25 and 26 indicate that the temperature

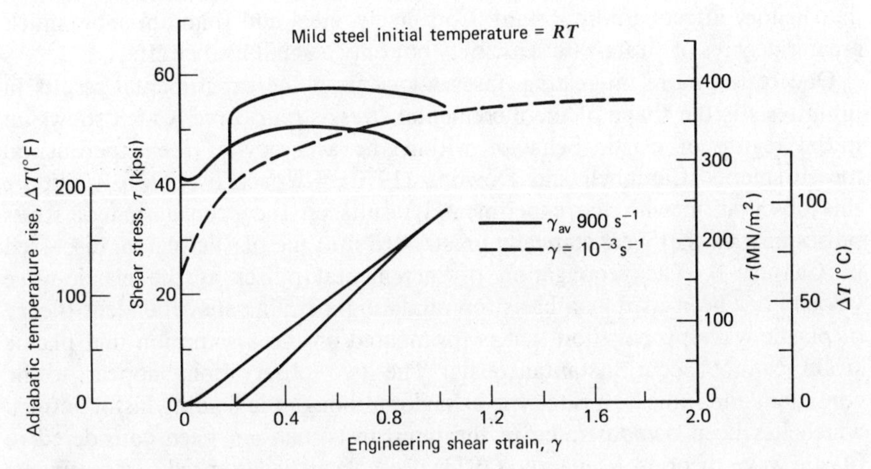

Figure 25 Stress-strain curves for constant-rate and jump tests on mild steel [Eleiche and Campbell (1976b), also in Campbell et al. (1977)].

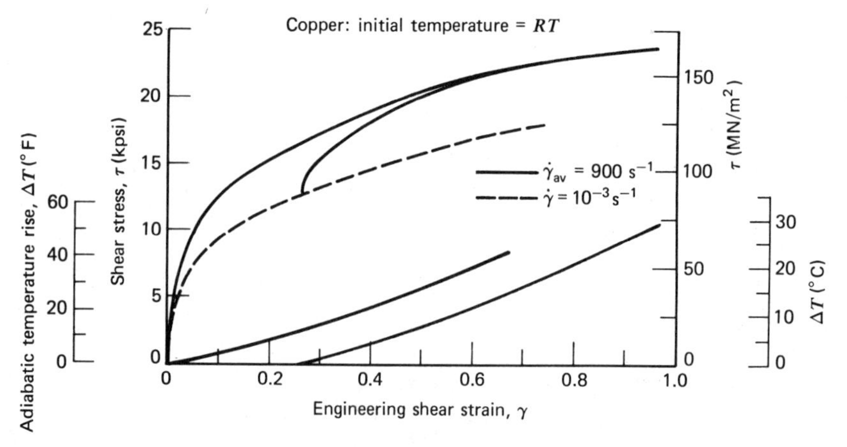

Figure 26 Stress-strain curves for constant-rate and jump tests on copper [Eleiche and Campbell (1976b), also in Campbell et al. (1977)].

difference between the constant rate dynamic test and the jump test is approximately constant for copper and decreases with increasing strain in mild steel. The differences in all cases, however, were too small to explain the differences in flow-stress behavior and indicate that a history effect is present. Senseny et al. (1978) used the explosive loading version of the torsional Hopkinson bar to perform jump tests in aluminum, copper, magnesium, and zinc. These incremental tests also demonstrated a strain-rate history effect in that they provided lower dynamic stresses than those obtained in constant high-rate tests at the same values of strain rate and strain. Duffy (1980) has summarized the results of many investigations. For the *fcc* metals aluminum and copper, the results show that they are not strongly strain-rate sensitive but that history effects are important. Conversely, steel and titanium show much greater degrees of strain-rate sensitivity but only a small history effect.

One of the more interesting observations from the experimental results in jump tests is the shape of the incremental stress-strain curve, which shows an initial region of elastic behavior within the accuracy of the experimental measurements [Campbell and Dowling (1970), Klepaczko (1973)]. Compare this observation with the experimental results on the propagation of stress pulses in rods that were statically prestressed into the plastic region, discussed in Chapter 4. The propagation of incremental pulses at the elastic-wave velocity has been used as a basis for validating a strain rate dependent theory of plastic wave propagation that is formulated on the assumption that plastic strain cannot occur instantaneously. The two observations appear to be consistent and indicate a rate-type behavior although the loading history effect, which has been demonstrated in the jump tests, has not been considered in plastic wave theories. Klepaczko (1973) used plastic wave speed measurements in aluminum, copper, and steel and a rate-independent plastic wave theory to construct the incremental stress-strain curves for these materials. These curves appear to correlate quite well to those obtained in jump test experiments.

Other tests involving strain-rate history effects involve those of Eleiche and Campbell (1976a) who subjected specimens to a sudden reversal in strain rate going from a quasistatic to a dynamic rate and those of Lipkin et al. (1978) who imposed a sudden decrease in strain rate. Results of the latter study show a flow stress that apparently does not decrease immediately following the decrease in strain rate, indicating some type of delay mechanism in OFHC copper. Klepaczko and Duffy (1974) have looked at combined strain-rate and temperature-history effects and conclude that such effects play an important role in the plastic deformation of *fcc* metals in which fading memory effects are also observed.

8.4.4 Constitutive Models

Various forms of constitutive equations have been proposed to describe the dynamic behavior of materials. Some of these that have been employed in the description of wave-propagation phenomena are discussed in Chapter 4. From the results of constant strain-rate experiments on a large number of metals (see Section 8.4.1) it has been found that the variation of flow stress with log strain rate is linear over a wide range of strain rates up to about $10^3\ s^{-1}$. At higher strain rates, many investigations show a linear variation of stress with strain rate that appears as a dramatic increase when plotted against log strain rate. There is considerable controversy over whether these dramatic increases in stress are real or manifestations of test techniques or procedures at high rates of strain where the assumptions in the analysis may be violated. Nonetheless, the modeling of dynamic material behavior does not generally include history effects, which have been shown to be important for many materials. One of the earliest attempts to include history or memory effects in a constitutive model is that of Hart (1970) who introduced the concept of a hardness state that can be related to the dislocation structure generated in the course of plastic deformation. Denoting the hardness parameter by y, a state equation

$$\sigma=\sigma(y,\dot{\varepsilon}) \tag{58}$$

is used to describe polycrystalline metals over a range of strain rates at the lower end of the spectrum where creep and relaxation phenomena occur. Bodner and Partom (1975) introduced an incremental constitutive model to account for history effects. The model is developed in a similar fashion to the semilinear and quasilinear rate theories of plastic wave propagation by decomposing the strain rate into additive elastic and plastic components and relating the elastic component of strain rate to stress rate through Hooke's law. At a given temperature, the plastic strain rate is given in the form

$$\dot{\varepsilon}_p=\dot{\varepsilon}_p(\sigma,Z) \tag{59}$$

where Z is a history-dependent state variable that characterizes the material's resistance to plastic flow. It is assumed that $\dot{\varepsilon}_p$ follows the Prandtl-Reuss flow law of classical plasticity and plastic flow is incompressible, leading to an evolution equation of the general form

$$\dot{Z} = F(J_2, Z) \tag{60}$$

where J_2 is the second invariant of the stress deviator. The state variable Z is assumed to be a function of plastic work, W_p, in the form

$$Z = Z_1 - (Z_1 - Z_0)\exp(-mW_p) \tag{61}$$

and the plastic strain rate for uniaxial stress is given by

$$\dot{\varepsilon}_p = \frac{2}{\sqrt{3}} \frac{\sigma}{|\sigma|} D_0 \exp\left[-\tfrac{1}{2} \left(\frac{Z}{\sigma} \right)^{2n} \left(\frac{n+1}{n} \right) \right] \tag{62}$$

where Z_0, Z_1, m, n, and D_0 are material constants. The evolution equation takes the form

$$\dot{Z} = \frac{\partial Z}{\partial W_p} \frac{dW_p}{dt} \tag{63}$$

for computational purposes. One unique feature of this model is that there is no quasistatic stress-strain curve for a material. Rather, the deformation rate, which is decomposed into an elastic and a plastic portion, is a function of the stress and plastic work. A quasistatic stress-strain curve is thus nothing more than the solution to the boundary value problem of a material loaded in uniaxial tension at a prescribed deformation rate. Bodner and Merzer (1978) used this formulation to model the response of copper to a six-decade change in strain rate over a temperature range from 77° to 523°K. Comparing the results with experimental data of Senseny et al. (1978) shows good correlation between plastic work, temperature, and strain-rate sensitivity through the use of a single internal state variable. Figure 27 shows the comparison between calculated and experimental data on copper at 523°K for constant static and dynamic rates as well as for jump tests. It can be seen that the model provides a reasonably accurate representation of the data and demonstrates the capability to represent not only strain-rate effects, but strain-rate history effects as well.

Another approach to modeling time and history-dependent plasticity is the nonlinear hereditary-type, stress-strain relation. This theory has been widely applied to the time-dependent deformation of polymers. The viscoelastic theory was modified by Rabotnov and Suvorova (1978) to describe the behavior of metals by replacing total strain by plastic strain. The theory explained data of Frantz and Duffy (1972), which showed a drop in stress after yield in incremental strain-rate tests in 1100-0 aluminum and also modeled the experimental data of Klepaczko (1968) where dynamic loading was followed by quasistatic loading. In both cases, excellent agreement was achieved. An alternate approach to describing dynamic plastic behavior of metals is that of Wu and Yip (1980) who developed constitutive equations for strain hardening

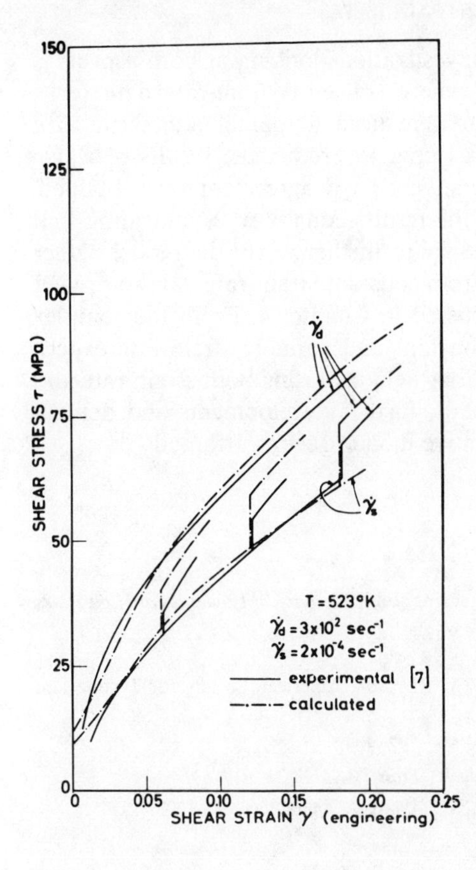

Figure 27 Calculated and experimental static, dynamic and incremental stress-strain curves for copper at 523° K [Bodner and Merzer (1978)].

and softening within the framework of the endochronic theory of viscoplasticity. Their computations compare favorably with experimental data from the literature and also include the capability to model history effects. Perzyna (1966) presents a summary of incremental post-initial yield, stress-strain rate laws, which provides a combined treatment of rheologic and plastic phenomena. The thermodynamic foundations of the theory of viscoplasticity are reviewed by Perzyna (1971) where the theoretical basis for temperature and strain-rate effects is presented from a mathematical viewpoint.

The test of a constitutive model is its ability to explain observed physical phenomena over a wide range of test conditions. Models developed from constant strain-rate or jump-test data should also be capable of explaining observed wave-propagation phenomena. In addition, they should have a physically realistic basis as well as be mathematically tractable. In the large body of literature in dynamic plasticity there has been relatively little attention devoted to the development and use of constitutive models for explaining both high strain-rate experimental data as well as wave-propagation phenomena. Most investigations have tended to focus on one type of problem or the other with the result that constitutive models have been developed separately for each type of investigation. One of the few investigations looking at both aspects is that of Lawson and Nicholas (1972) where constant strain-rate data for a relatively strain-rate sensitive titanium were used to obtain a best fit to a constitutive relation which, in turn, was used to predict the results of wave-propagation experiments. Although relatively good agreement was obtained, other constitutive equations predicted the results equally well, indicating that additional experiments would be necessary to further verify the model. Other uses of constitutive models developed from constant strain-rate data to predict wave-propagation phenomena are discussed in Chapter 4. From the complex behavior that has been observed in constant and variable strain-rate experiments and wave-propagation investigations demonstrating both strain-rate and strain-rate history effects in many metals, further developments and demonstrations of the applicability of constitutive models seem warranted.

REFERENCES

Albertini, C., and Montagnani, M. (1974), in J. Harding (Ed.), *Mechanical Properties at High Rates of Strain*, Institute of Physics, London, p. 22.

Alder, J. F., and Phillips, V. A. (1954), *J. Inst. Met.*, **83**, 80.

ASTM (1974), *Instrumented Impact Testing*, ASTM STP 563, American Society for Testing and Materials.

Bailey, J. A., and Singer, A. R. E. (1963), *J. Inst. Met.*, **92**, 404.

Baker, W. E., and Yew, C. H. (1966), *J. Appl. Mech., Trans. ASME*, **33**, 917.

Barker, L. M., Lundergan, C. D., and Herrmann, W. (1964), *J. Appl. Phys.*, **35** 1203.

Bell, J. F. (1966), *J. Mech. Phys. Solids*, **14**, 309.

Bertholf, L. D., and Karnes, C. H. (1975), *J. Mech. Phys. Solids*, **23**, 1.

Bhushan, B., and Jahsman, W. E. (1978), *Int. J. Solids Struct.*, **14**, 739.

Bitans, K., and Whitton, P. W. (1971), *Proc. Inst. Mech. Eng.*, **185**, 1149.

Bodner, S. R., and Merzer, A. (1978), *J. Eng. Mater. Tech., Trans. ASME*, **100**, 388.

Bodner, S. R., and Partom, Y. (1975), *J. Appl. Mech., Trans. ASME*, **42**, 385.

Bridgman, P. W. (1964), *Studies in Large Plastic Flow and Fracture*, Harvard University Press, Cambridge, MA.

Campbell, J. D., and Dowling, A. R. (1970), *J. Mech. Phys. Solids*, **18**, 43.

Campbell, J. D., Eleiche, A. M., and Tsao, M. C. C. (1977), in R. I. Jafee and B. J. Wilcox (Eds.), *Fundamental Aspects of Structural Alloy Design*, Plenum Press, New York, p. 545.

Campbell, J. D., and Ferguson, W. G. (1970), *Phil. Mag.*, **21**, 63.

Campbell, J. D., and Tsao, M. C. C. (1972), *Quart. J. Mech. Appl. Math.*, **25**, 172.

Chalupnik, J. D., and Ripperger, E. A. (1966), *Exp. Mech.*, **6**, 547.

Chiddister, J. L., and Malvern, L. E. (1963), *Exp. Mech.*, **3**, 81.

Christman, D. R., Isbell, W. M., Babcock, S. G., McMillan, A. R., and Green, S. J. (1971), Report No. DASA 2501-2, MSL 70-23, Vol. II, Arlington, VA.

Clark, D. S., and Duwez, P. E. (1950), *Proc. ASTM*, **50**, 560.

Clausing, D. P. (1969), *J. Mater. JMLSA*, **4**, 566.

Clifton, R. J. (1980), in J. Harding (Ed.), *Mechanical Properties at High Rates of Strain*, Institute of Physics, London, p. 174.

Conn, A. F. (1965), *J. Mech. Phys. Solids*, **13**, 311.

Cooper, R. H., and Campbell, J. D. (1967), *J. Mech. Eng. Sci.*, **9**, 278.

Costin, L. S., Duffy, J., and Freund, L. B. (1977), in G. T. Hahn, and M. F. Kanninen (Eds.), *Fast Fracture and Crack Arrest*, ASTM STP 627, American Society for Testing and Materials, p. 301.

Culver, R. S. (1972), *Exp. Mech.*, **12**, 398.

Curran, D. R., et al. (1979), Annual Report, Army Contract DAAK 11-78-C-0115, SRI International, Menlo Park, CA

Daniel, I. M., La Bedz, R. H., and Liber, T. (1980), *Exp. Mech.*, **21** 71,.

Davies, E. D., and Hunter, S. C. (1963), *J. Mech. Phys. Solids*, **11**, 155.

Davies, R. G., and Magee, C. L. (1975), *J. Eng. Mater. Tech.*, *Trans. ASME*, **97**, 151.

Davies, R. M. (1948), *Phil. Trans. Roy. Soc. Lond. A.*, **240**, 375.

Dowling, A. R., and Harding, J. (1967), *1st Int. Conf. of the Center for High Energy Forming*, Estes Park, University of Denver, Denver, CO.

Dowling, A. R., Harding, J., and Campbell, J. D. (1970), *J. Inst. Met.*, **98**, 215.

Duffy, J. (1974), in J. Harding (Ed.), *Mechanical Properties at High Rates of Strain*, Institute of Physics, London, p. 72.

Duffy, J. (1980), in J. Harding (Ed.), *Mechanical Properties at High Rates of Strain*, Institute of Physics, London, p. 1.

Duffy, J., Campbell, J. D., and Hawley, R. H. (1971), *J. Appl. Mech.*, *Trans. ASME*, **38**, 83.

Edington, J. W. (1969), *Phil. Mag.*, **19**, 1189.

Eleiche, A. M. (1972), Air Force Materials Laboratory Report AFML-TR-72-125, Wright-Patterson AFB, OH.

Eleiche, A. M., and Campbell, J. (1967a), *Exp. Mech.*, **16**, 281.

Eleiche, A. M., and Campbell, J. D. (1976b), Air Force Materials Laboratory Report AFML-TR-76-90, Wright-Patterson AFB, OH.

Eleiche, A. M., and Duffy, J. (1975), *Int. J. Mech. Sci.*, **17**, 85.

Forrestal, M. J., Duggin, B. W., and Butter, R. I. (1980), *J. Appl. Mech.*, *Trans. ASME*, **47**, 17.

Francis, P. H. (1966), *J. Appl. Mech.*, *Trans. ASME*, **33**, 702.

Francis, P. H., and Lindholm, U. S. (1968), *J. Appl. Mech.*, *Trans. ASME*, **35**, 441.

Frantz, R. A., and Duffy, J. (1972), *J. Appl. Mech.*, *Trans. ASME*, **39**, 939.

Fyfe, I. M. (1968), in U. S. Lindholm, (Ed.), *Mechanical Behavior of Materials under Dynamic Loads*, Springer-Verlag, New York, p. 314.

Fyfe, I. M., and Rajendran, A. M. (1980), *J. Mech. Phys. Solids*, **28**, 17.

Goldsmith, W. (1960), *Impact*, Edward Arnold, London.

Gorham, D. A. (1980), in J. Harding (Ed.), *Mechanical Properties at High Rates of Strain*, Institute of Physics, London, p. 16.

Green, S. J., Maiden, C. J., Babcock, S. G., and Schierloh, F. L. (1970), in M. F. Kanninen et al. (Eds.), *Inelastic Behavior of Solids*, McGraw-Hill, New York, p. 521.

Gust, W. H., and Young, D. A. (1980), *Bull. Am. Phys. Soc.*, **25**, 567.

Harding, J. Ed., (1980), *Mechanical Properties at High Rates of Strain*, Institute of Physics, London, No. 47.

Hardin, J., and Huddart, J. (1980), in J. Harding (Ed.), *Mechanical Properties at High Rates of Strain*, Institute of Physics, London, p. 49.

Harding, J., Wood, E. D., and Campbell, J. D. (1960), *J. Mech. Eng. Sci.*, **2**, 88.

Hart, E. W. (1970), *Acta Met.*, **18**, 599.

Hauser, F. E. (1966), *Exp. Mech.*, **6**, 395.

Hauser, F. E., Simmons, J. A., and Dorn, J. E. (1961), in P. G. Shewmon and V. F. Zackay (Eds.), *Response of Metals to High Velocity Deformation*, Interscience, New York, p. 93.

Hayashi, T., and Tanimoto, N. (1978), in K. Kawata and J. Shioiri (Eds.), *High Velocity Deformation of Solids*, Springer-Verlag, Berlin, p. 279.

Hockett, J. E. (1959), *Proc. ASTM*, **59**, 1309.

Hoggatt, C. R., and Recht, R. F. (1969), *Exp. Mech.*, **9**, 441.

Hoggatt, C. R., Orr, W. R., and Recht, R. F. (1967), *1st Int. Conf. of the Center for High Energy Forming*, Estes Park, University of Denver, Denver, CO.

Holt, D. L., Babcock, S. J., Green, S. J., and Maiden, C. J. (1967), *Trans. ASM*, **60**, 152.

Holzer, A. J. (1978), *Int. J. Mech. Sci.*, **20**, 553.

Holzer, A. J. (1979), *J. Eng. Mater. Tech., Trans. ASME*, **101**, 231.

Holzer, A. J., and Brown, R. H. (1979), *J. Eng. Mater. Tech., Trans. ASME*, **101**, 238.

Huffington, N. J., Jr. Ed., (1965), *Behavior of Materials Under Dynamic Loading*, ASME, New York.

Jahsman, W. E. (1971), *J. Appl. Mech., Trans. ASME*, **38**, 75.

Jiang, C. W., and Chen, M. M. (1974), Report No. AMMRC CTR 74-23, Watertown, MA.

Karnes, C. H., and Ripperger, E. A. (1966), *J. Mech. Phys. Solids*, **14**, 167.

Kawata, K., and Shioiri, J. Eds. (1978), *High Velocity Deformation of Solids*, Springer-Verlag, Berlin.

Klepaczko, J. (1967), *Arch. Mech. Stosow.*, **2**, 19.

Klepaczko, J. (1968), *J. Mech. Phys. Solids*, **16**, 155.

Klepaczko, J. (1973), in A. Sawczuk (Ed.), *Symposium on Foundations of Plasticity*, Noordhoff, Amsterdam, p. 451.

Klepaczko, J. (1980), in J. Harding (Ed.), *Mechanical Properties at High Rates of Strain*, Institute of Physics, London, p. 201.

Klepaczko, J., and Duffy, J. (1974), in J. Harding (Ed.), *Mechanical Properties at High Rates of Strain*, Institute of Physics, London, p. 91.

Klepaczko, J., Frantz, R. A., and Duffy, J. (1977), *Pol. Akad. Nauk., Ins. Podstawowych Probl. Techn., Eng. Trans.*, **25**, 3.

Klepaczko, J., and Malinowski, Z. (1978), in K. Kawata and J. Shioiri (Eds.), *High Velocity Deformation of Solids*, Springer-Verlag, Berlin, p. 403.

Kolsky, H. (1949), *Proc. Phys. Soc. B*, **62**, 676.

Krinke, D. C., Barber, J. P., and Nicholas, T. (1978), Air Force Materials Laboratory Report AFML-TR-78-54, Wright-Patterson AFB, OH.

Kumar, A., and Kumble, R. G. (1969), *J. Appl. Phys.*, **40**, 3475.

Lawson, J. E., and Nicholas, T. (1972), *J. Mech. Phys. Solids*, **20**, 65.

Lee, E. H., and Tupper, S. J. (1954), *J. Appl. Mech., Trans. ASME*, **21**, 63.

Lengyel, B., and Mohitpour, M. (1972), *J. Inst. Met.*, **100**, 1.

Leung, E. K. C. (1980), *J. Appl. Mech., Trans. ASME*, **47**, 278.

Lewis, J. L., and Campbell, J. D. (1972), *Exp. Mech.*, **12**, 520.

Lindholm, U. S. (1964), *J. Mech. Phys. Solids*, **12**, 317.

Lindholm, U. S., Ed. (1968a), *Mechanical Behavior of Materials Under Dynamic Loads*, Springer-Verlag, New York.

Lindholm, U. S. (1968b), in U. S. Lindholm (Ed.), *Mechanical Behavior of Materials Under Dynamic Loads*, Springer-Verlag, New York, p. 77.

Lindholm, U. S. (1971), in R. F. Bunshah (Ed.), *Techniques in Metals Research*, Vol. 5, part 1, Interscience, New York.

Lindholm, U. S. (1974), in J. Harding (Ed.), *Mechanical Properties at High Rates of Strain*, Institute of Physics, London, p. 3.

Lindholm, U. S. (1978), in K. Kawata and J. Shioiri (Eds.), *High Velocity Deformation of Solids*, Springer-Verlag, Berlin, p. 26.

Lindholm, U. S., and Bessey, R. L. (1969), Air Force Materials Laboratory Report AFML-TR-69-119, Wright-Patterson AFB, OH.

Lindholm, U. S., Nagy, A., Johnson, G. R., and Hoegfeldt, J. M. (1980), *J. Eng. Mater. Tech., Trans. ASME*, **102**, 376.

Lindholm, U. S., and Yeakley, L. M. (1967), *Exp. Mech.*, **7**, 1.

Lindholm, U. S., and Yeakley, L. M. (1968), *Exp. Mech.*, **8**, 1.

Lindholm, U. S., Yeakley, L. M., and Nagy, A. (1974), *Int. J. Rock Mech. and Mining Sci.*, **11**, 181.

Lipkin, J., Campbell, J. D., and Swearengen, J. C. (1978), *J. Mech. Phys. Solids*, **26**, 251.

Loizou, N., and Sims, R. B. (1953), *J. Mech. Phys. Solids*, **1**, 234.

Maiden, C. J., and Green, S. J. (1966), *J. Appl. Mech., Trans. ASME*, **33**, 496.

Malatynski, M., and Klepaczko, J. (1980), *Int. J. Mech. Sci.*, **22**, 173.

Malvern, L. E. (1980), personal communication.

Meguid, S. A., Campbell, J. D., and Malvern, L. E. (1979), *J. Mech. Phys. Solids*, **27**, 331.

NMAB (1978), Tech. Report NMAB-341, National Materials Advisory Board, National Academy of Sciences, Washington, D. C.

Nicholas, T. (1971), *Exp. Mech.*, **11**, 370.

Nicholas, T. (1973), *J. Appl. Mech., Trans. ASME*, **10**, 277.

Nicholas, T. (1975), Air Force Materials Laboratory Report AFML-TR-75-54, Wright Patterson AFB, OH.

Nicholas, T. (1980a), *Exp. Mech.*, **21**, 177.

Nicholas, T. (1980b), Technical Report AFWAL-TR-80-4053, Wright-Patterson AFB, OH.

Nicholas, T., and Lawson, J. E. (1972), *J. Mech. Phys. Solids*, **20**, 57.

Niordson, F. I. (1965), *Exp. Mech.*, **5**, 29.

Orowan, E. (1950), Technical Report MW/F/22/50, British Iron and Steel Res. Assn.

Papirno, R. P., Mescall, J. F., and Hansen, A. M. (1980), *Proceedings of the Army Symposium on Solid Mechanics—1980*, Technical Report AMMRC MS 80-4, Watertown, MA, p. 367.

Perzyna, P. (1959), in W. Olszak (Ed.), *Non-Homogeneity in Elasticity and Plasticity*, Pergamon, London, p. 431.

Rabotnov, Yu N., and Suvorova, J. V. (1978), *Int. J. Solids Struct.*, **14**, 173.

Raftopoulos, D., and Davids, N. (1967), *AIAA J.*, **5**, 2254.

Recht, R. F. (1978), *Int. J. Eng. Sci.*, **16**, 809.

Ripperger, E. A. (1965), in N. J. Huffington, Jr. (Ed.), *Behavior of Materials Under Dynamic Loading*, ASME, New York, p. 62.

Rohde, R. W., Butcher, B. M., Holland, J. R., and Karnes, C. H. Eds. (1973), *Metallurgical Effects at High Strain Rates*, Plenum Press, New York.

Samanta, S. K. (1971), *J. Mech. Phys. Solids*, **19**, 117.

Senseny, P. E., Duffy, J., and Hawley, R. H. (1978), *J. Appl. Mech., Trans. ASME*, **45**, 60.

Shewmon, P. G., and Zackay, V. F. Eds. (1961), *Response of Metals to High Velocity Deformation*, Interscience, New York.

Shioiri, J., Satoh, K., and Nishimura, K. (1978), in K. Kawata and J. Shioiri (Eds.), *High Velocity Deformation of Solids*, Springer-Verlag, Berlin, p. 50.

Sturges, J. L., Parson, B., and Cole, B. N. (1980), in J. Harding (Ed.), *Mechanical Properties at High Rates of Strain*, Institute of Physics, London, p. 35.

Swift, R. P., and Fyfe, I. M. (1970), *J. Appl. Mech., Trans. ASME*, **37**, 1134.

Tatro, C. A., Scott, R. G., and Taylor, A. R. (1979), Tech. Report UCRL-50057-78, Lawrence Livermore Laboratory, CA.

Taylor, G. I. (1948), *Proc. Roy. Soc. Lond. A.*, **194**, 289.

Walling, H. C., and Forrestal, M. J. (1973), *AIAA J.*, **11**, 1196.

Warnes, R. H., Duffey, T. A., Karpp, R. R., and Carden, A. E. (1980), in M. A. Meyers and L. E. Murr (Eds.), *Proceedings of the International Conference on the Metallurgical Effects of High Strain-Rate Deformation and Fabrication* (in publication).

Wasley, R. J., Hoge, K. G., and Cast, J. C. (1969), *Rev. Sci. Inst.*, **40**, 889.

Watson, H. Jr., and Ripperger, E. A. (1969), *Exp. Mech.*, **9**, 289.

Wesenberg, D. L., and Sagartz, M. J. (1977), *J. Appl. Mech., Trans. ASME*, **44**, 643.

Wiffin, A. C. (1948), *Proc. Roy. Soc. Lond. A.*, **194**, 300.

Wilkins, M. L., and Guinan, M. W. (1973), *J. Appl. Phys.*, **44**, 1200.

Wingrove, A. L. (1971), *J. Phys. E. (Sci. Inst.)*, **4**, 873.

Woodward, R. L., and Brown, R. H. (1975), *Proc. Ins. Mech. Eng.*, **189**, 107.

Wu, H. C., and Yip, M. C. (1980), *Int. J. Solids Struct.*, **16**, 515.

Wulf, G. L., and Richardson, G. T. (1974), *J. Phys. E. (Sci. Inst.)*, **7**, 167.

Yeung Wye Kong, Y. C. T., Parsons, B., and Cole, B. N. (1974), in J. Harding (Ed.), *Mechanical Properties at High Rates of Strain*, Institute of Physics, London, p. 33.

Zaslawsky, M., Tatro, C., and Freeman, T. (1980), Tech. Report UCRL-83484, Lawrence Livermore Laboratory, CA, to be published in *Exp. Mech.*

9

DYNAMIC FRACTURE

Donald R. Curran

The title of this chapter is redundant. All fracture is dynamic; it is a rate process in which material bonds are broken and voids are created in previously intact material. To understand fracture, one must therefore understand both the threshold conditions that trigger this process and the kinetics by which it proceeds.

Mathematical attempts to gain this understanding, and to predict fracture in engineering applications, have taken two different approaches. The first and most successful approach thus far is that pioneered by A. A. Griffith and George Irwin [see, for example, the compendium edited by Harold Liebowitz (1968)], in which a macroscopic crack is treated as a stress-free boundary in a mathematical boundary-value problem. The threshold condition for crack instability is assumed to be a critical-energy density, stress-intensity factor, or other measure of the stress and strain fields at the crack tip. The growth kinetics of this macroscopic crack is also handled as part of the dynamic boundary-value problem.

The boundary-value approach has been extremely successful in predicting failure in cases where the behavior of a single, large crack in fairly brittle material is of prime interest, and the term "fracture mechanics" is commonly used to refer to this discipline. However, in many cases, continuum-fracture mechanics does not apply. A classic example of such a case is the ductile failure of a smooth bar in tension, illustrated in Figure 1. In this case there is no macroscopic crack, and failure occurs by the nucleation, growth, and coalescence of millions of microscopic voids in the necking region of the rod. Continuum-fracture mechanics cannot be used to predict such failures.

Another example is the case of a large process zone around a single large crack. In this case microvoid activity may be observed in a large region around the macrocrack tip, as seen in Figure 2, and continuum-fracture mechanics

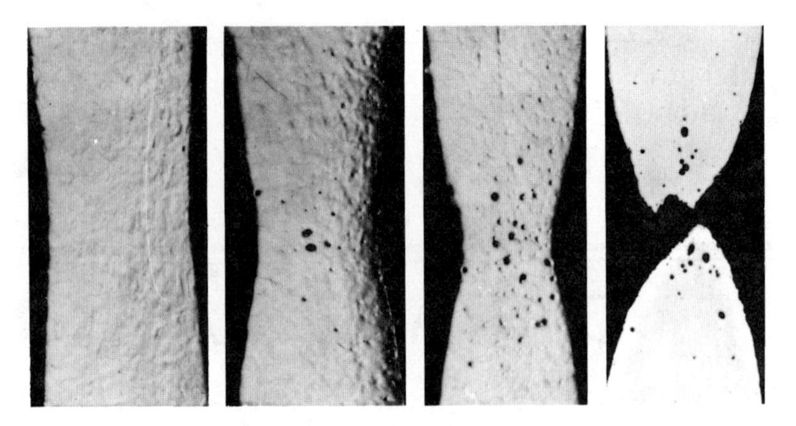

Figure 1 Polished sections through tensile specimens of OFHC copper ($\frac{3}{4}$ hard) at various stages of plastic strain.

cannot be routinely used to predict the failure if the process-zone size becomes on the order of specimen dimensions.

For such cases, treatment of each microcrack individually becomes too difficult, and a second mathematical approach to failure prediction has been developed—namely, to treat some key measures of average microscopic-void behavior as internal-state variables in the constitutive relations for the material [Barbee et al. (1972), Davison and Stevens (1973), Seaman et al. (1976)]. That is, in addition to strain, entropy, and temperature, one also specifies the microvoid concentration and size-distribution functions in a description of the current state of a material particle. In this approach the fracture kinetics involves the microvoid behavior and introduces specific rate dependence into the constitutive relations. In short, the fracture process is built into the stress-strain relationship for the material.

This microstatistical rate-dependent, internal-state variable approach has exciting potential because, in principle, it can be related to microstructural variables and thereby can form a link between fracture mechanics and materials science; moreover it is applicable to problems not easily approached with classical fracture mechanics. However, until quite recently, this potential was largely unrealized. Not only were the constitutive relations describing microstatistical-fracture processes highly nonlinear and therefore resistant to analysis, but also precise data were lacking for the nucleation, growth, and coalescence of the microscopic voids and cracks.

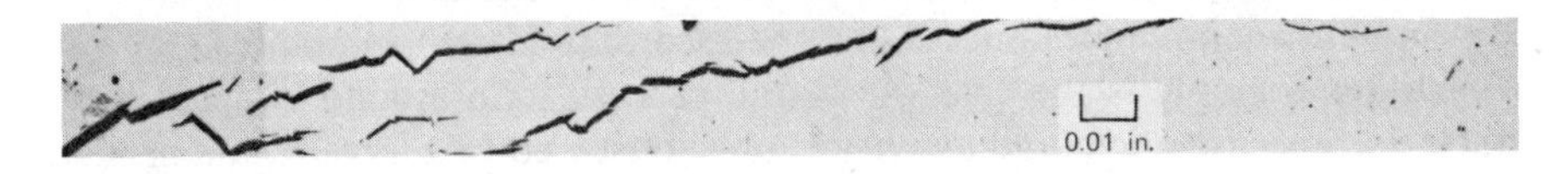

Figure 2 Microcracks near the crack tip in a double-cantilever beam specimen of Fe-3 Si steel.

During the past 10 years, however, significant progress has been made on both fronts. First, the explosive growth of computer modeling has made it possible to computationally simulate highly nonlinear material behavior. Second, the use of controlled impact, explosive, and interrupted quasistatic tests has made is possible to correlate observed microvoids with known stress and strain amplitudes and durations, thus "freezing in" the microvoid kinetics at various stages of development and making it possible to obtain precise kinetics data. Finally, the development of analytical models of microscopic-void processes such as nucleation and growth on grain boundaries and at inclusions has made it possible to combine theory with experimental data and computer modeling to make increasingly realistic descriptions and predictions of material failure under dynamic and quasistatic-loading conditions.

In this chapter we concentrate on reviewing progress in this second, microstatistical internal-state variable approach. We discuss how to construct and test, from experimental data, constitutive-relation models of the nucleation, growth, and coalescence of the microscopic voids and cracks. We then present several examples of how the approach can be used to predict failure in engineering applications.

9.1 EXPERIMENTAL MEASUREMENTS OF MICROVOID KINETICS

In metals there are three commonly observed microscopic-damage modes—ellipsoidal voids (Figure 1), cleavage cracks, and shear bands. To successfully observe the kinetics of each of these damage modes, it is useful to perform dynamic experiments in which material samples at given temperatures are exposed to variable stress or strain amplitudes for variable durations. If the durations can be made short enough compared with the times required for the nucleation, growth, and coalescence of the microscopic voids, cracks, or shear bands, then the damage can be "frozen in" at various stages of development and the kinetics can be inferred.

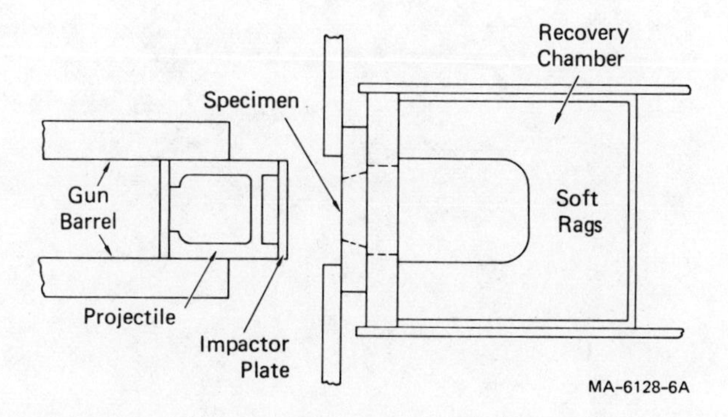

Figure 3 Plate impact experiments for studies of dynamic fracture.

A method commonly used [Barbee et al., 1972; Seaman et al., 1976] to determine the kinetics of microscopic voids and cracks is shown in Figure 3. A gas gun propels a flyer plate to an impact with a flat target plate. As described in detail by Barbee et al. (1972) and Seaman et al. (1976), the resulting stress-wave interactions with the free surfaces of both flyer plate and target plate cause a region of tension to form in the target plate.

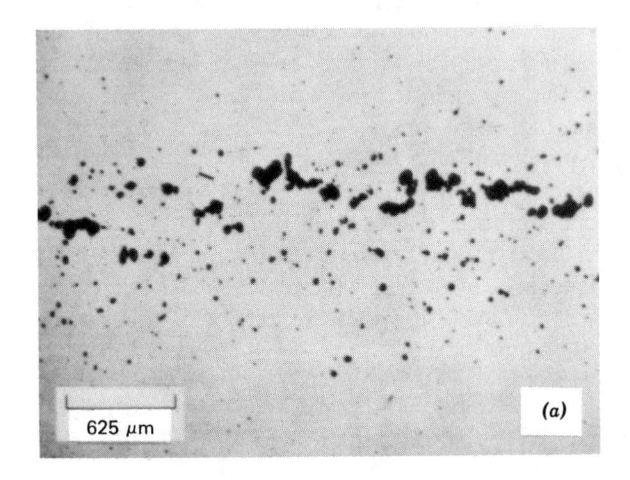

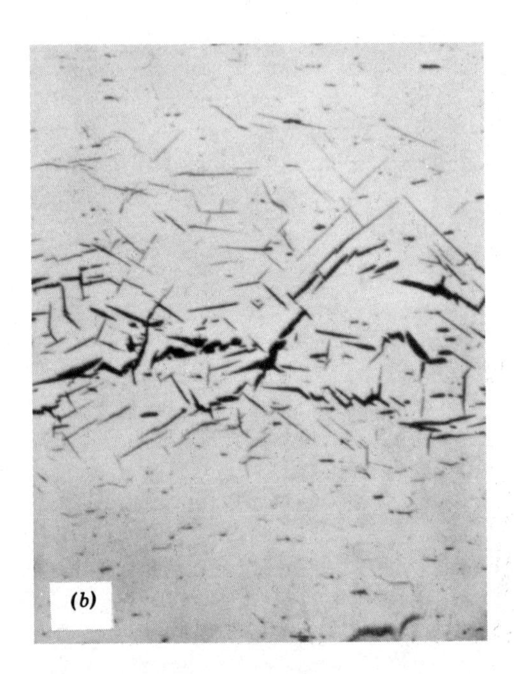

Figure 4 Polished cross sections through plate-impact specimens showing dynamic fracture damage in (a) 1145 aluminum (spherical voids) and (b) Armco iron (planar cracks).

The duration of the tensile stress is greatest in the center of the target and decreases toward the front and rear faces. This causes the damage to be heaviest in the center, as shown in Figure 4. (In Figure 4 the vertical axis corresponds to the direction of impact.)

By performing a series of such experiments at different impact velocities and different plate thicknesses, it is possible to vary the amplitude and duration of the tensile stress pulse over a wide range. After each experiment, the target specimens can be sectioned and polished to reveal the damage, and the voids and cracks can be measured and counted to provide void kinetics data. Figure 5 shows the result of this procedure for Armco Iron. In this example, the number of cracks per unit volume greater than a given crack radius are plotted versus that radius for four different stress durations at a given stress amplitude. As described by Seaman et al. (1976), it is then possible to deduce from these data crack nucleation and growth rates as a function of applied stress.

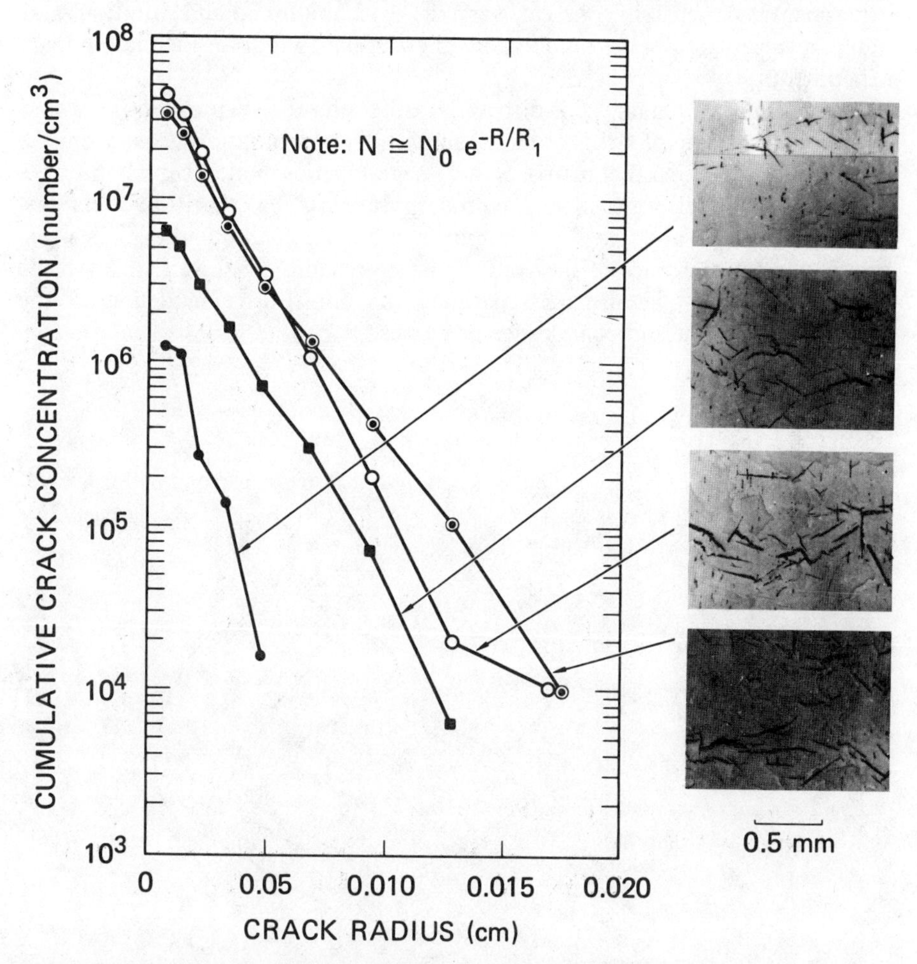

Figure 5 Microcrack size-distribution curves obtained by counting and measuring fracture damage on polished sections through plate-impact specimens of Armco iron.

A similar approach can be taken for shear banding. One experiment that has proved valuable is to detonate an explosive mixture inside a hollow cylinder of the material of interest [Erlich et al. (1980)]. As the cylinder expands, shear bands are nucleated and grown at the inner surface of the cylinder, and these bands eventually cause the cylinder to fragment. However if the cylinder is surrounded by a plastic sheath and massive containment tube, as shown in Figure 6, the expansion of the controlled fragmenting cylinder (CFC) may be arrested at various stages of development. As discussed by Erlich et al. (1980), the shear bands can then be counted and measured to produce cumulative size distributions very like these observed for voids and cracks.

Example data for a 4340 steel cylinder are shown in Figure 7. Similar to the case for the plate-impact experiments, different portions of the cylinder experience different load durations, and therefore receive different damage distributions, as shown in Figure 7. By varying the explosive mix and the experimental dimensions, one can vary the load amplitude and duration, and obtain a variety of kinetics data in a manner similar to that for the plate-impact experiments.

The cumulative damage-size distributions like those shown in Figures 5 and 7, taken for a range of stress or strain amplitudes and durations, thus form the data base from which the microscopic damage-kinetics models can be inferred. The kinetics models must in turn be incorporated in the constitutive relations for the material, and used in computer codes to compute the nucleation, growth, and coalescence of the voids, cracks, or bands to cause failure. In the following sections we discuss separately the constitutive modeling of the nucleation, growth, and coalescence processes.

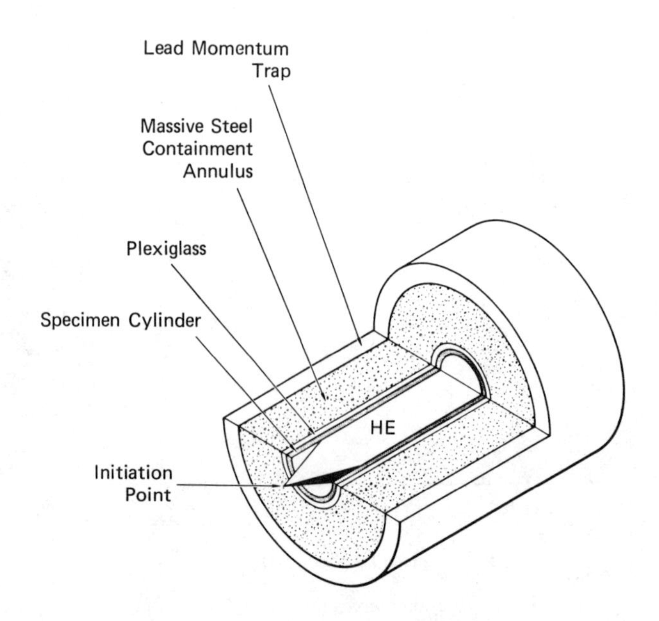

Figure 6 Exploding cylinder experiments for studying shear-band kinetics.

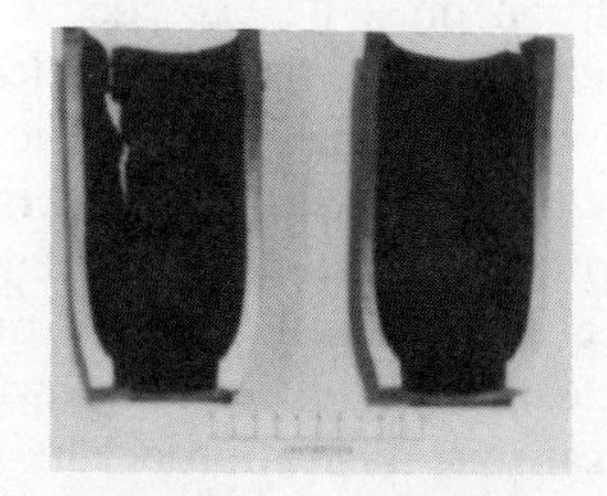

(a) Cylinder Deformation

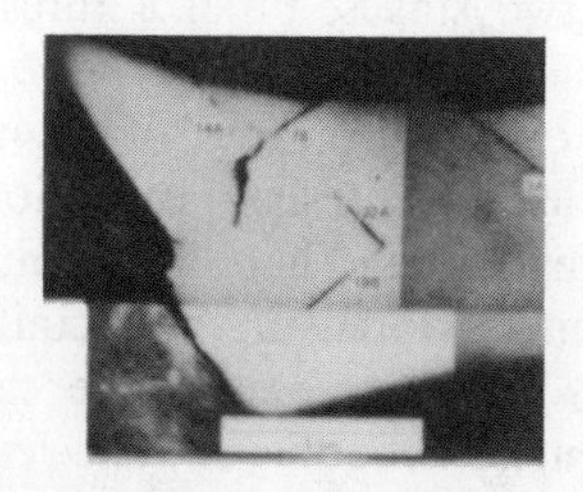

(b) Shear Bands Observed on a Polished Cross Section

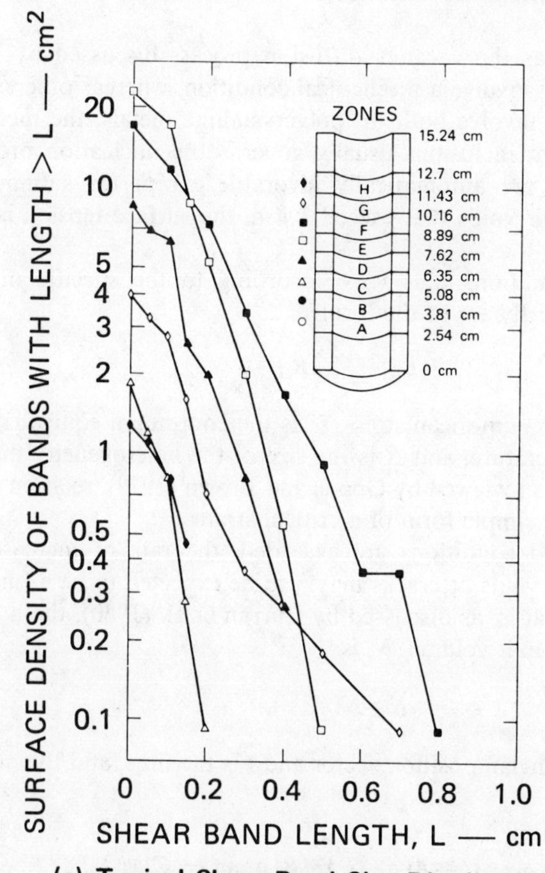

(c) Typical Shear Band Size Distribution

Figure 7 Results of an exploding cylinder experiment on 4340 steel (R_c40).

9.2 CONSTITUTIVE MODELING OF FRACTURE

9.2.1 Processes To Be Modeled

Nucleation. Microdamage nucleates at heterogeneities in the material such as inclusions, grain boundaries, and the like [Shockey et al. (1973)]. Nucleation occurs in two stages. First, a threshold condition must be exceeded before nucleation can begin. Second, once the threshold criterion is met, nucleation will occur over a size range of heterogeneities and at a material-specific rate.

The nucleation threshold condition itself has two parts. First, the energetics must be favorable. That is, the nucleation process must not require more energy than is stored in the vicinity; the free energy must decrease during nucleation. For example, in the case of nucleation by diffusive growth of vacancy clusters in grain boundaries to a critical size, nucleation occurs when the local stresses can overcome the surface tension [Raj and Ashley, (1975)]. Second, the mechanics must be favorable. For example, if nucleation occurs by debonding at an inclusion, the stress across the interface must exceed the bond strength.

Some cases, such as the vacancy diffusion process discussed by Raj and Ashley (1975), do not involve a mechanical condition, whereas others, such as inclusion debonding, involve both. In polycrystalline metals, the mechanical-threshold condition at inclusions usually governs the nucleation process because the energetics are automatically favorable except for submicrometer inclusions. That is, for voids that exceed 1 μm, the surface tension is usually negligible.

The threshold conditions thus vary according to the specific nucleation mechanism, but typically are of the form

$$F(\sigma_m, \bar{\varepsilon}^p, T, R) \geqslant 0 \tag{1}$$

where σ_m is the continuum-mean stress, $\bar{\varepsilon}^p$ is the continuum-equivalent plastic strain, T is the temperature, and R is the size of the heterogeneity that forms the nucleation site. As reviewed by Goods and Brown (1979), relation (1) often takes the particularly simple form of a critical strain.

Once the threshold conditions are exceeded, the rate of increase of the number of nucleated voids or cracks may also be expected to be a function of σ_m, $\bar{\varepsilon}^p$, T, and R. That is, as discussed by Curran et al. (1980), the number of flaws of all sizes per unit volume, N, is

$$N = \hat{N}(\underline{X}, t) \tag{2}$$

where $\underline{X}$ is the Lagrangian position vector and t is the time, and the nucleation rate is given by

$$\dot{N} \equiv \left(\frac{\partial N}{\partial t} \right)_{\underline{X}} = A(\sigma_m, T) + B(\sigma_m)\dot{\sigma}_m + C(\varepsilon_i^p)\dot{\varepsilon}_i^p \tag{3}$$

In this formulation, the dependence on the heterogeneity size R has been suppressed by considering all nucleated cracks or voids of every size.

In (3), the last two terms describe the nucleation due to mechanical debonding of inclusions and the like. The $A(\sigma_m, T)$ term, on the other hand, may be thought of as describing thermal and stress-activated diffusion processes such as vacancy diffusion in grain boundaries [Raj and Ashby (1975); Raj (1978)]. That is,

$$A(\sigma_m, T) = \dot{N}_0 \exp\left(\frac{-\Delta H}{kT}\right) \exp\left[(\sigma_m - \sigma_0)\frac{\Omega}{kT}\right] \tag{4}$$

where $\sigma_m \geqslant \sigma_0$, and where ΔH is the activation energy for the diffusion process, k is Boltzman's constant, Ω is the activation volume, σ_0 is the threshold stress for the diffusion process, and $\dot{N}_0$ is a frequency factor that may also depend on stress, temperature, and the value of N itself.

We may thus view the nucleation-rate process as a competition between diffusion processes that are stress- and temperature-driven and mechanical decohesion processes that are instantaneous on the time scale of interest—that is, when compared to the load duration and microcrack growth times.

Growth. Growth, by definition, is the increase in size of the microscopic cracks or voids, and we must therefore begin a description of growth by first describing the size distributions. As discussed by McClintock (1973) and Batdorf (1979), the inherent flaws in polycrystalline materials tend to be distributed in the form

$$N_g(R) = N_0 \exp\left[-\left(\frac{R}{R_0}\right)^m\right] \tag{5}$$

where $N_g(R)$ is the number of flaws per unit volume with radii greater than R, and where m is 1 or 2, depending on whether the flaws tend to be linear or areal. In practice, we found that (5), with m equal to unity, describes observed flaw-size distributions reasonably well for a variety of materials over the limited size range of interest (i.e., over the size range of nucleation at a few micrometers to a few millimeters).

In applying the microstatistical constitutive-relations approach, we do not wish to describe the growth of individual microcracks or voids. Rather, we wish to describe the evolution of $N_g(R)$ in time. A particularly convenient growth law, and one that ellipsoidal voids in ductile materials have been found under dynamic loads to follow both theoretically and experimentally, is

$$\frac{\dot{R}}{R} = \left[\frac{(\sigma_m - \sigma_{g0})}{4\eta}\right] \tag{6}$$

where $\sigma_m \geqslant \sigma_{g0}$, and where η has the units of viscosity. This viscous growth law has the convenient property of maintaining the exponential form of (5). Indeed, the evolution of the distribution can be followed by simply applying (6) to the characteristic size R_0 in (5).

In more brittle materials, the microcracks may achieve a constant limiting velocity on the order of the Rayleigh wave speed in the material, thus making computer bookkeeping of the evolving size distribution more complicated.

Coalescence and Fragmentation. After the microcracks or voids have grown by an amount comparable to the average void spacing, coalescence must begin. In some extremely ductile materials, the coalescence occurs by direct impingement of ellipsoidal voids. However, in most ductile materials, some sort of plastic localization between voids occurs first (Tvergaard, 1979).

For brittle microcracks and for adiabatic-shear bands, however, direct impingement is the coalescence mechanism [Seaman et al. (1976)]. As coalescence proceeds, the material divides into isolated fragments, and the resulting fragment size distribution is related closely to the crack-size distribution that existed before the onset of coalescence [Seaman et al. (1976)].

Stress Relaxation. As microscopic damage grows, stress-free surface is generated in the material, thus diminishing the amount of stress caused by overall deformation. This stress relaxation occurs by two basically different processes. First, under tension the existing cracks or voids open to accommodate some of the imposed volumetric strain. The volumetric strain of the matrix material can therefore relax elastically, and the associated mean-tensile stress also relaxes. Second, as discussed by Carroll and Holt (1972), the decrease of load-bearing area introduces a correction factor

$$\underline{\sigma}=\underline{\sigma}^s(1-v) \tag{7}$$

where $\underline{\sigma}^s$ is the stress tensor in the matrix material, $\underline{\sigma}$ is the continuum-stress tensor (the average stress obtained by dividing the force on the void-containing material by the area), and v is the relative void volume (shown by Carroll and Holt to be equal to the area fraction of randomly distributed and oriented voids).

In our modeling work for microcracks and cracks under tension [Seaman et al. (1976)], we have found that the first stress-relaxation effect is dominant until the last stages of coalescence. However, for adiabatic shear bands, in which no continuum tensions are present, the second mechanism (7) is dominant.

9.2.2 Computational Approach

To institute the microstatistical constitutive-relations approach in computer modeling of microscopic-damage kinetics, it is necessary that the computer be able to keep records of the $N_g(R)$ function for each material element at successive times, and to modify the strains and stresses accordingly. The modified stresses then govern the material accelerations and, thereby, the incremental deformations over successive time steps. Clearly, finite-difference codes with explicit time-stepping schemes are most suitable for such record keeping.

In the applications to be shown later, a Lagrangian, finite-difference wave-propagation code served as the computational simulation vehicle. The conservation equations for mass, momentum, and energy along with the material-constitutive relations were numerically integrated over successive time steps to obtain the stress, strain, and microdamage histories of each material element.

The complex feedback effect of the developing damage on the stresses makes the resulting constitutive relation nonlinear and history-dependent, thereby illustrating the need for computer simulation.

In Section 9.3 we describe three applications of the above approach to predicting detailed failure histories—namely, explosively fragmented cylinders, explosive fracture of a geologic material, and quasistatic fracture of a metal.

9.3 EXAMPLE APPLICATIONS

9.3.1 Fragmenting Rounds

Many fragmenting munitions form their fragments by intersection of adiabatic-shear bands. A computational model capable of predicting the fragmentation pattern as a function of munitions geometry and the material properties governing shear banding would therefore be very useful. As reviewed by Erlich et al. (1980), constitutive relations have been developed that described the nucleation, growth, and coalescence of shear bands.

Nucleation-Threshold Criteria. As we indicated earlier, the nucleation-threshold criterion consists of two parts, thermomechanical and energetic. The thermomechanical criterion is the condition that must exist for perturbations in the plastic flow at grain boundaries or other heterogeneities to grow into localized bands.

A common explanation for the thermomechanical initiation of adiabatic-shear bands is that they are plastic instabilities that occur when thermal softening overwhelms the work hardening, thus causing the effective hardening modulus to become negative. Rice (1976) and Clifton (1980) have recently summarized the theory of such instabilities. In our constitutive model we therefore use, for the thermomechanical-threshold criterion, the value of equivalent plastic strain at which the thermal softening under adiabatic-loading conditions causes the yield stress to decrease faster than the work hardening increases it. For medium hardness steels, this value of strain is about 20%.

The remaining threshold nucleation criterion to be satisfied is that of favorable energetics. That is, once localization has occurred in the planar bands, enough energy must flow into the band tips to extend or propagate them. Figure 8 is a sketch of a slipping shear-band front. The jog of one side relative to the other is B. If σ_{ij} is the resolved shear stress in the plane of the band, the energetics condition is:

$$(\sigma_{ij})(2R)(B) \geqslant E \tag{8}$$

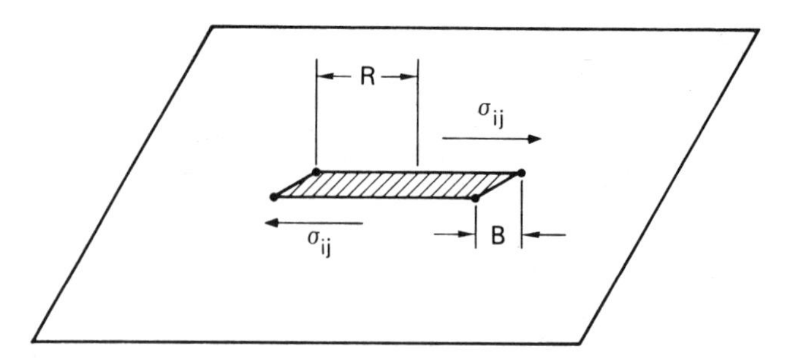

Figure 8 Shear-band geometry.

where E is the energy per unit band front-line length required to remove the shear strength (of the order of the melt energy) inside the process zone (or macrodislocation core).

We next note that E/B is the critical energy per unit area of newly created surface, which we now define as J_{SBC}. Thus, we obtain a fracture mechanics type of critical condition:

$$\sigma_{ij} \geqslant \frac{J_{SBC}}{2R} \tag{9}$$

In short, we may expect a critical value of resolved shear stress that is inversely dependent on band size. At present, this criterion is speculative, and the value of the critical value of σ_{ij} (or, equivalently the critical value of ε_{ij}^{p}) is estimated from the CFC data. In the application to be shown, a critical value of ε_{ij}^{p} was used.

Nucleation Rate. After the nucleation-threshold criteria are fulfilled, our model must specify the rate at which bands appear. This requires that we first specify their size distribution at nucleation and, second, that we specify the rate at which this size distribution is generated.

The computer storage requirements for the shear-band model are decreased by imposing an analytic form on the size distribution. The form we have chosen is the exponential one that fits the shear-band data and is also like that used for ductile and brittle fracture. The number of bands per cubic centimeter with radii greater than R is

$$N_g = N_0 \exp\left(-\frac{R}{R_1}\right) \tag{10}$$

where N_0 is the total number per cubic centimeter, and R_1 is the size parameter of the distribution. The bands are presumed to be circular, with a radius R, and to be in a plane. The bands are assumed to nucleate and grow in six discrete planes, thus approximately representing a complete angular distribution. The analytic size distribution of (10) is used in each angular orientation. Thus, each orientation direction is represented by three numbers, N_0, R_1, and R_B, where R_B is a maximum size of the bands.

The rate of shear band nucleation is described by

$$\frac{dN_{\phi\theta}}{dt} = A_n \left(\frac{d\varepsilon_{\phi\theta}^{PS}}{dt} \right)^n N_{\phi\theta}^m \tag{11}$$

where $N_{\phi\theta}$ and $\varepsilon_{\phi\theta}$ are the number of bands and plastic shear strain in the $\phi\theta$ angular direction, respectively, and A_n, n, and m are experimentally determined constants. A limit on the rate of nucleation is provided by requiring that the total distortion represented by the new bands not exceed the total imposed plastic shear strain. The shear strain associated with nucleation is

$$\Delta\varepsilon^{PS} = \frac{\pi}{2} \int_{R=0}^{\infty} bR^3\, dN \tag{12}$$

where $b = B/R$ is the maximum relative offset across the band and dN is the number nucleated with a radius R. By using (10) to obtain the derivative of N in terms of R, and integrating (12), we obtain

$$\Delta\varepsilon^{PS} = 3\pi b \Delta N R_n^3 \tag{13}$$

where ΔN is the total number per cubic centimeter nucleated, with a size distribution parameter $R_1 = R_n$. Equation (13) is then an upper bound on the number nucleated in any time interval.

In (13), we have incorporated the experimentally observed fact that the total jog B accommodated by a band is proportional to the band radius R. The proportionality constant b is observed in 4340 steel to be about 0.1. The reason for this proportionality is not yet understood, but it appears reasonable because larger bands should exert more leverage on the band tip and create a larger jog.

Growth. Growth or propagation of the shear bands is modeled by a law of viscosity that is suggested, but not confirmed, by the CFC data. That is,

$$\frac{dR_i}{dt} = C_G R_i \frac{d\varepsilon_{\phi\theta}^{PS}}{dt} \tag{14}$$

where C_G is a growth coefficient. When this viscous growth process is applied to a size distribution as in (10), the entire distribution increases in size. The resulting distribution has the same exponential form, N_0 is unaffected, and the change in R_1 is obtained through the use of (14).

Two constraints are applied to the growth law to limit the maximum growth velocity to the transverse sound speed and to require that the total distortion represented by the growth does not exceed the total imposed plastic shear strain. The maximum growth rate occurs at the maximum radius, according to the viscous growth law, and that radius is called R_B. Hence, the velocity limit implies that the maximum radius at the end of the interval, R_{B2}, is

$$R_{B2} \leqslant R_{B1} + V_m \Delta t \tag{15}$$

where V_m is the shear-wave velocity. The increase in the radius parameter is limited by maintaining a constant ratio of R_B/R_1. The distortion limit on growth is derived in the same manner as in (12) and (13). The increase in plastic strain associated with the band growth on the $\phi\theta$ plane is

$$\Delta\varepsilon_{\phi\theta}^{PS} = 3\pi b N_0\left(R_{2\phi\theta}^3 - R_{1\phi\theta}^3\right) \tag{16}$$

where R_2 is the final size-distribution parameter. Hence, on the basis of the imposed distortion, this limit is

$$R_{2\phi\theta}^3 = R_{1\phi\theta}^3 + \frac{\Delta\varepsilon_{\phi\theta}^{PS}}{3\pi b N_{0\phi\theta}} \tag{17}$$

Combining (14), (15), and (17), we find the value of R_2 on the $\phi\theta$ plane to be

$$R_{2\phi\theta} = \text{MIN}\left[\left(R_{B\phi\theta} + V_m\Delta t\right)\frac{R_{1\phi\theta}}{R_{B\phi\theta}},\quad R_{1\phi\theta}\exp\left(C_G\Delta\varepsilon_{\phi\theta}^{PS}\right),\quad \left(R_{1\phi\theta}^3 + \frac{\Delta\varepsilon_{\phi\theta}^{PS}}{3\pi b N_{0\phi\theta}}\right)^{1/3}\right] \tag{18}$$

After calculating growth of the old size distribution and nucleation of a new distribution with size parameter R_n, we combine the two distributions to form a new distribution with the same form as (10). The combined distribution has the parameters N_3 and R_3. The combination has two constraints:

1 The new curve represents the size and number of the largest shear bands ($R = R_B$).
2 The new curve represents the sum of the plastic strain of the nucleated and growing distributions.

Constraint 1 requires that the number of bands at R_B is preserved; hence

$$N_{3B} = N_3\exp\left(\frac{-R_B}{R_3}\right) = N_0\exp\left(\frac{-R_B}{R_2}\right) + \Delta N\exp\left(\frac{-R_B}{R_n}\right) \tag{19}$$

The total plastic strain of the new distribution is

$$\varepsilon^{PS} = 3\pi b N_3 R_3^3 \tag{20}$$

By eliminating N_3 from (20) by using (19), we obtain an expression for R_3:

$$\varepsilon^{PS} = 3\pi b N_{3B}\exp\left(\frac{R_B}{R_3}\right)R_3^3 \tag{21}$$

where N_{3B} is known from (19). A simple algorithm for solving (21) was developed for the program. N_3 is then obtained from (20), and the final size distribution is fully specified.

If the computed total-strain increment accommodated by the shear bands is greater than the total imposed strain, the growth velocities are decreased to the level required to make the shear-band strain increment equal to the total plastic-strain increment.

Coalescence. Constructing a detailed model of the coalescence of the shear bands to form fragments would be a formidable job. However, several simplifications are possible. If we assume that the bands are randomly oriented (sometimes a poor assumption, however), then there will be a statistical tendency for the large bands to intersect to form the large fragments, for intermediate-sized bands to intersect to form intermediate-sized fragments, and so forth. Thus, the fragment-size distribution should reflect the band-size distribution at the onset of coalescence.

At the onset of coalescence, the total plastic strain the bands accommodate in any orientation is

$$\varepsilon^p = \pi b \sum_i N_i R_i^3 \tag{22}$$

This is similar to the Orowan relation for atomic dislocations. When the material breaks into fragments, the volume fraction of fragments is

$$V = T_F \Sigma N_i^f \left(R_i^f \right)^3 = 1 \tag{23}$$

where T_F is a dimensionless volume factor (about 4) and N_i^f and R_i^f are the number and the radius of fragments in the ith-size group.

We assume that the number of fragments is related to the number of bands through the factor β:

$$N_i^f = \beta N_i \tag{24}$$

Chunky fragments usually have six or eight sides that were each formed by a band or crack. Because each band forms a side of two fragments, three or four bands must be associated with each fragment. Therefore, β equals $1/3$ or $1/4$. Similarly, the fragment sizes are related to the band sizes through the factor γ:

$$R_i^f = \gamma R_i \tag{25}$$

Here R_i^f is defined so that the fragment volume is $T_F(R_i^f)^3$, where T_F is about $4\pi/3$. Because the bands forming the fragment sides have about the same area as the sides, γ is approximately equal to 1. Now (23) can be rewritten as a double sum of shear bands over orientations and size groups:

$$V = T_F \beta \gamma^3 \Sigma_{\phi\theta} \left[\sum_i N_{\phi\theta i} R_{\phi\theta i}^3 \right] = 1 \tag{26}$$

A comparison of (26) and (23) shows that damage appears to be related to NR^3 at all times during the calculation. We will use (26) as our definition of complete fragmentation ($V=1$) and let V be the fragmented fraction for values less than full fragmentation. Hence, the quantity $T_F\beta\gamma^3\Sigma_i N_{\phi\theta i}R^3_{\phi\theta i}$ is a measure of the damage associated with the bands in the $\phi\theta$ orientation. When the damage in all orientations sums to unity, the material element is completely fragmented, and the fragment-size distribution is obtained from the final band-size distribution through (24) and (25).

Stress Relaxation. A key part of any constitutive relation damage model is the interaction between the damage and the stresses that are supported by the damaged material. That is, the damage progressively weakens the material, producing a "work-softening" effect and causing the continuum (or average) stresses to relax. In the following sections, we outline our current model of this process, which is part of a complete computational description of the stress-strain relations for a material undergoing shear-band damage.

Basic Assumptions. The major assumption of the stress-relaxation process is that the shear strength of a partially shear-banded material depends jointly on the amount of shear banding and the normal stress across the bands. Quantitatively, the shear strength is

$$S=\frac{l}{h}\sigma_n \tan\phi+\frac{h-l}{h}Y_s \tag{27}$$

where l = the band length

h = the length of the material block in the band direction

σ_n = the normal stress across the band

ϕ = the friction angle

Y_s = the yield strength in shear (one-half the tensile yield strength)

The configuration of the band and stresses is shown in Figure 9. Note that the band reduces the shear stresses S_{12} and S_{21} equally but has no effect on the S_{13} stresses.

Equation (27) is extended to the shear band-size distributions by relabeling the damage density function τ_k, which represents the total area of shear bands per unit volume associated with the kth plane. The plane is fully shear banded for $\tau_k=1$; then (27) is assumed to take the form

$$S_k=\tau_k\sigma_n \tan\phi+(1-\tau_k)Y_s, \qquad S_k\leq Y_s \tag{28}$$

The shear strength is always positive and cannot exceed Y_s.

Central to the use of (28) are the assumptions that

1 The shear band damage can be allocated to planes of interest (such as the X-Y plane), instead of being confined to the actual damage planes in the material.

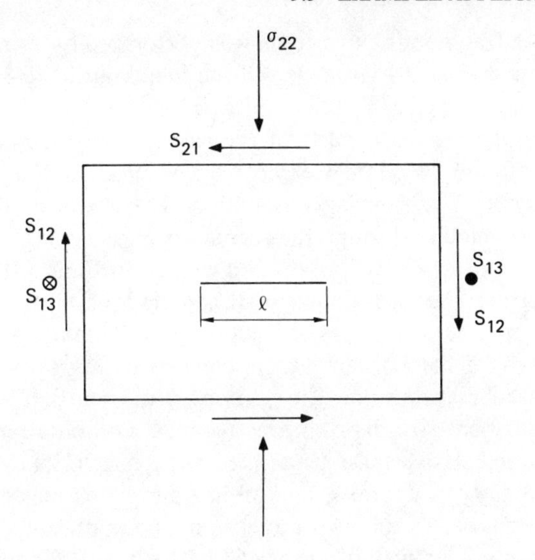

Figure 9 Shear (S) and normal (σ) stress orientation on a block containing a shear band.

2 The yield process can be dealt with on specific planes (a Tresca yield approach), rather than by considering all stresses simultaneously (as in a Mises yield process).

Both these assumptions seem to fit the basic mechanisms mentioned earlier.

In the yield process, we further assume that yielding does not affect the pressure while the material is under compression. Thus the pressure remains a function only of the density and internal energy and is not modified by yielding, except by the heat generated.

These assumptions are the basis of the stress-relaxation model. With these assumptions, the model will provide for an anisotropy of strength so that on planes that are shear banded the shear strength may go to zero, but shear stresses on other planes will retain their full strength. This stress reduction will allow shear bands to propagate through computational cells representing material under high shear.

This stress relaxation model is described in detail in the paper by Erlich et al. (1980).

Application. The application to be described is a case with geometry quite similar to that of the CFC calibration experiments—namely, a self-fragmenting artillery round. We sought to compute the final fragment-size distribution and to compare it with measurements made by Dr. Robert Crowe at the Naval Surface Weapons Center (NSWC) at Dahlgren, Virginia [see Erlich et al. (1980)].

In this calculation, we used a simple form of the model in which the normal stress component $\tau_k \sigma_n \tan\phi$ in (28) was set to zero. The two-dimensional simulation of the fragmenting projectile was performed by using the cell layout shown in the top sketch of Figure 10. All the metal parts were of HF-1 steel, a material for which CFC calibration data were available. No provision was made for the joint in the two parts of the case. The primary explosive in the fuse is CH6 and the secondary explosive was PBXN 106. The simulation included treatment of a running detonation that starts in the primary and sweeps at the C-J velocity through the secondary explosive. The progress of the running detonation is evident from the cell motion in Figure 10, which shows a series of plots from the two-dimensional calculations. The indicators in the figure show the attendant progress of fracturing in the material cells.

Figure 11 shows a comparison of the computed fragment-size distribution with the measured distribution. The computed distribution was obtained by summing the fragment-size distributions from all computational cells and by accounting for the relative mass associated with each cell. Cells that did not fragment were mainly in the nose, and all unfragmented masses were added in as a single large chunk. In this two-dimensional computational simulation, the only shear-band mode activated was at 45° between radial and circumferential

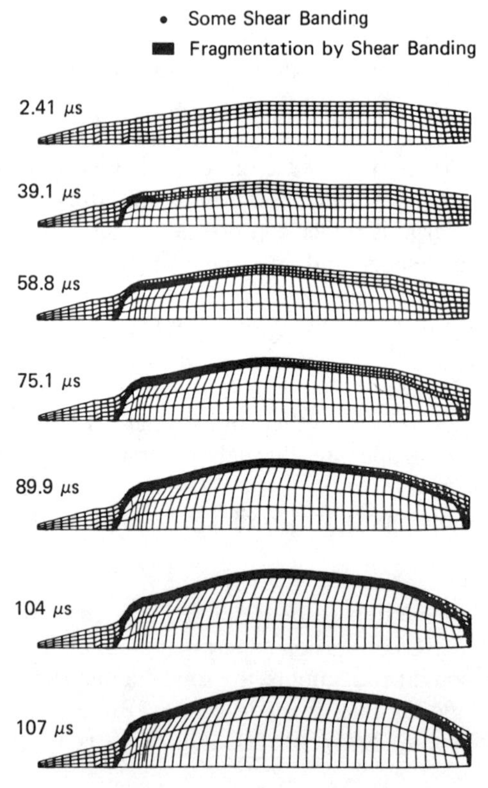

Figure 10 Progress of detonation and shear-band damage computed in the two-dimensional simulation of fragmenting projectile.

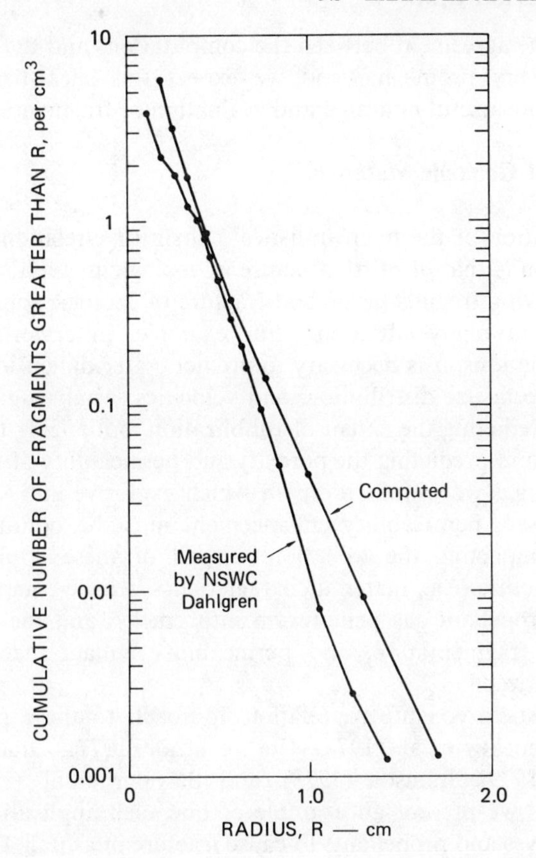

Figure 11 Comparison of computed and observed fragment-size distribution for sample fragmenting projectile.

directions. The duration of detonation was about 65 μs. By 110 μs, nearly all the shear-banding cells were completely fragmented. No shear banding occurred forward of *K*-row 48.

In the model calculations, the shear band nucleation and growth gradually assumed a larger portion of the total plastic strain. The total computed equivalent plastic strain imposed up to complete fragmentation was 60 to 80%; of this, 40% was homogeneous deformation with the remaining 20–40% accommodated by the bands. Experimental measurements indicate that about 35% homogeneous strain occurred in the fragments, which is in agreement with the computations.

In the calculations, we assumed that fragmentation occurs only by shear banding; in some of the experiments, however, microcracking in the outer layer was a factor in the fragmentation process. In future work, the effects of microcracking must be included in the computations.

In summary, the agreement between the computations and the experimental results appears very promising, and we expect this shear-band modeling approach to become useful in design and evaluation of fragmenting munitions.

9.3.2 Fracture of Geologic Materials

A second application of the microstatistical constitutive-relation approach to fracture prediction is that of brittle fracture in geologic materials.

Problems involving fragmentation and fracture of geologic solids, and other materials, occur in many situations. For example, in cratering or earth-penetrating calculations, it is necessary to predict crater dimensions, extent of fragmentation, ejecta-size distribution, and velocities. Analyzing in-situ retort design requires predicting the extent of rubblization, optimizing the fragment-size distribution, and predicting the porosity and permeability of the rubblized zone. In oil- and gas-well stimulations in which explosive and/or propellant techniques are used, permeability enhancement must be optimized without rubblizing or compacting the well bore. In all of these applications, the conditions for fracture (e.g., material characteristics, tensile-compressive stress field, explosive-propellant gas penetration into cracks) and the relationships among fracture, fragmentation, and permeability enhancement should be properly characterized.

The microstatistical constitutive-relation approach to these problems has been taken by Shockey et al. (1974), Curran et al. (1977), Grady and Kipp (1980), Dienes (1979), Johnson (1978), and Barbour et al. (1980). In the following sections we present an example of one such application—namely, the use of explosives and propellants to cause fracture in ashfall TUFF around a borehole [McHugh et al. (1980)].

The computational fracture model used was a subroutine called BFRACT (Brittle Fracture).

BFRACT Model

Size Distribution of Cracks. Just as in the preceding example of shear bands, the BFRACT model treats crack densities (i.e., the number of cracks per unit volume) and orientations (e.g., horizontal or vertical) statistically. Therefore, the fracture parameters in the model do not describe the behavior of individual cracks, but rather the combined behavior of crack arrays. This allows the fracture parameters to be treated as internal-state variables in the continuum with the fracture manifesting itself as a type of work softening.

BFRACT is based on observations of fractures in Armco iron, beryllium, novaculite, and Lexan polycarbonate (a transparent plastic). In impact experiments with these materials, fractures occurred by nucleation and growth of microcracks. We roughly measured penny-shaped cracks on the sectioned surfaces of the samples for length and angular orientation with respect to the direction of loading. The cracks were organized into groups according to size and angle interval. Surface distributions (i.e., number of cracks versus crack radius) were then converted to volumetric distributions with respect to size and angle. For this conversion, the cracks were assumed to be penny-shaped, with the distribution axisymmetric around the direction of impact loading [Seaman et al. (1978)].

Just as in the case for shear bands, these experiments demonstrated that the relationship between cumulative-crack density and crack size is approximately exponential:

$$N_g = N_0 \exp\left(\frac{-R}{R_1}\right) \tag{29}$$

where N_0 is the total number of cracks per cubic centimeter, N_g is the number of cracks with radii greater than R, and R_1 is a constant giving the shape the crack size distribution.

In this model, cracks nucleate at inherent flaws in the material and grow until they intersect other cracks or until the stress vanishes. The extent of fracture or degree of fragmentation resulting from a given loading pulse depends on the number of inherent flaws activated by the stress and on the rate of crack growth. The inherent flaw distribution in the material is specified in the input to BFRACT by the fracture parameters T_3 (the maximum flaw size R_{max} at nucleation). Table 1 sets forth the values of these parameters and other parameters discussed below for ashfall tuff.

Nucleation of Microcracks. Experiments have shown [Seaman et al. (1976)] that cracks nucleate at a rate that depends on the tensile stress, σ, normal to

Table 1 Fracture Properties of tuff Used in BFRACT Calculational Simulations of Laboratory and Gas Frac Field Experiments

Description	Value
T_1: Growth coefficient (1/KPa-s)	-80
T_2: Growth threshold (KPa)	500
T_3: Nucleation-size parameter (m)	1.20×10^{-3}
T_4: Nucleation-rate coefficient (no/m^3)	$3.00 \times 10^{+5}$
T_5: Nucleation threshold (MPa)	-3.50
T_6: Stress-sensitivity factor (MPa)	-20
T_7: Upper-size nucleation cutoff (m)	1.00×10^{-2}
T_8: Spall criterion	1.00
T_9: Coefficient in stress relaxation time constant	1.00
T_{10}: Ratio of number of fragments to number of cracks	0.25
T_{11}: Ratio of fragment radius to radius crack	1.00
T_{12}: ΣNR^3 at coalescence	0.20
T_{13}: Coefficient for calculating fragment volume	1.33
T_{14}: Gas penetration switch	1.00

the plane of the cracks, and that the number of cracks nucleated is governed by the nucleation rate function:

$$\dot{N}=\dot{N}_0 \exp\left[\frac{(\sigma-\sigma_{n0})}{\sigma_1}\right] \tag{30}$$

where $\dot{N}$ is the nucleation rate and $\dot{N}_0$, σ_{n0}, and σ_1 are fracture parameters. $\dot{N}_0=T_4$ is the threshold nucleation rate, $\sigma_{n0}=T_5$ is the nucleation-threshold stress, and $\sigma_1=T_6$ governs the sensitivity of nucleation rate to stress level. For problems in which shear cracking is important, (30) has been replaced by an equation describing a nucleation rate proportional to a shear strain rate [McHugh et al. (1980)].

Growth of Cracks. In BFRACT, crack growth can be caused by a combination of a tensile stress normal to the orientation, σ, and/or a fluid or gas pressure on the internal crack surfaces, P_c. The relationship for crack growth is:

$$\frac{dR}{dt}=T_1(\sigma+P_c-\sigma_{g0})R \tag{31}$$

where T_1 is the growth coefficient ($=1/4\eta$ where η is an effective viscosity at the crack tip); $\sigma_{g0}=T_2$ is the growth threshold stress; and R is the crack radius. The parameter T_{14} is a "switch" in the model that determines whether the calculation is performed with fluid penetration. If $T_{14}=0$, no fluid penetration takes place (i.e., $P_c=0$); otherwise fluid penetration occurs. The growth velocity, $\dot{R}$, is allowed to increase by (31) to a limiting value of the Rayleigh wave velocity, or one-third the longitudinal sound speed in a geologic material such as tuff.

In some materials of geologic interest, such as novaculite and oil shale, flaw activation is assumed to occur according to Griffith-Irwin fracture-mechanics concepts, and hence the normal stress will activate flaws larger than a critical radius R^* but leave smaller flaws unaffected [Shockey et al. (1974)]. That is, we used Sneddon's relation [see Liebowitz (1968)] for the critical size of a penny-shaped crack subjected to a uniform tensile stress normal to the plane of the cracks in an infinite medium:

$$R^*=\pi\frac{K_{IC}^2}{4\sigma^2} \tag{32}$$

where K_{IC}, the fracture toughness, is a material property describing the resistance of the material to crack propagation. If σ_{g0} is the critical stress for crack growth, then

$$\sigma_{g0}=K_{IC}\sqrt{\frac{\pi}{4R^*}} \tag{33}$$

and hence σ_{g0} depends on crack size. The size distribution of inherent flaws in the rock sample thus governs the number of flaws activated by the normal tensile stress using (29), (30), and (33). The primary effect of using (33) for σ_{g0}

instead of a constant value for the growth threshold stress is to skew the calculated crack-size distributions toward larger cracks.

Stress Relaxation Due to Crack Interaction. The BFRACT model provides for a natural relaxation or reduction of stresses with developing damage. Each increment of strain $\Delta\varepsilon^T$, applied normal to the plane of fracture is separated into an elastic-plastic portion taken by the intact material and a fracture portion taken in crack opening:

$$\Delta\varepsilon^T = \Delta\varepsilon^S + \Delta\varepsilon^c \tag{34}$$

where $\Delta\varepsilon^c$ is the relative void volume of the crack. For our calculations of fracture in materials with high-crack densities and small cracks, the crack opening is assumed to be in equilibrium with the stress. Thus the opening time of the crack and engulfment time by stress waves were considered to be smaller than the computational time step.

For the rock fracture problems in the present study, there were so few cracks that we assumed that the crack opening did not instantly relax the stresses according to (34). To delay the effect of crack opening, we used a relaxation time τ_n equal to the time for stress waves to run between the cracks:

$$\tau_n = \frac{N_0^{-1/3}}{C_l} \tag{35}$$

Here N_0 is the fracture density and C_l is the longitudinal sound speed. This relaxation time was used to compute an effective crack-opening strain, $\Delta\varepsilon_n^c$, which approaches the instantaneous opening $\Delta\varepsilon_n^{ci}$ gradually as follows:

$$\Delta\varepsilon_n^c = \Delta\varepsilon_n^{ci} - \left(\Delta\varepsilon_n^{ci} - \Delta\varepsilon_{n-1}^c\right)\exp\left(-T_9\frac{\Delta t_n}{\tau_n}\right) \tag{36}$$

where $\Delta\varepsilon_{n-1}^c$ is the effective opening at the beginning of the time increment Δt, and T_9 is a dimensionless constant taken to be unity. With (36) governing crack opening, the material softening caused by damage is delayed until the material has been unloaded by stress-relief waves originating at the fractures.

Spallation. BFRACT defines "spallation" as complete fragmentation and separation in a computational cell. If the fracture density is large enough, the fractures are assumed to coalesce and form a spall plane. The coalescence and fragmentation algorithm used is nearly identical to the one described earlier for shear banding. The stress on this plane is set to zero if no gas penetration takes place; otherwise, the stress is set to P_c (the gas pressure in the crack). BFRACT also provides for consolidation following spallation [Seaman et al. (1980)].

The criterion for spall is that $\tau_p \geqslant T_8$, where τ_p is the fragmented fraction:

$$\tau_p = \beta\gamma^3\tau_z T_F \tag{37}$$

where $\beta = T_{10}$ is the ratio of number of fragments to number of cracks (set to 0.25); $\gamma = T_{11}$ is the ratio of fragment size to crack size (set to 1.00); $\tau_z = \Sigma NR^3$

(where the summation, Σ, is over the total crack density N and the crack size R); and $T_F = T_{13}$ is a proportionality constant defined such that the volume of a fragment is $T_F \gamma^3 R^3$.

The spallation algorithms calculate the stresses and increment the spall-opening strain. The stress state is obtained from the usual stress-strain relations for an elastic-plastic material and from the free-surface condition in the direction of spall.

Laboratory Experiments to Calibrate the Model. The BFRACT model parameters listed in Table 1 were obtained from laboratory experiments quite similar to those described earlier to determine the shear-banding model parameters. First, the size-distribution parameters R_1 and R_{max} (T_3 and T_7 in Table 1) were obtained by microscopic examination of core material. Second, high explosive was detonated in thick-walled cylinders of material taken from cores. The cylinders were 12-in.-diameter by 12-in.-long cores of tuff in which $\frac{1}{2}$-in. diameter boreholes were drilled along the axes. The experiments were designed to separate the effects of the tensile stresses produced by the explosively generated shock wave from gas penetration into the fractures. One set of experiments was conducted with in-material gauges to record the stresses and particle velocities during the test (for correlation with the calculations), and one set was conducted with only borehole pressure gauges so that the fracture would not be affected by in-material gauge planes (for correlation with the calculational simulations of fracture and determination of the BFRACT parameters).

The BFRACT computational subroutine assumes that the fractures are nucleated only by the tensile stresses or shear strains from the shock wave. (These fractures can then be grown by both the tensile stresses in the material and the gas pressure in the cracks.) It was originally planned, therefore, that the BFRACT parameters would be derived from experiments in which only the tensile stresses from the shock wave cause crack extension to separate effects produced by the tensile stresses and by gas penetration. To determine the effects produced by the tensile stresses acting without gas penetration, we used thin (6 mil) steel sheaths in three experiments to line the borehole and contain the explosive detonation products. Pressure gauges between the sheath and the tuff monitored the stress input to the tuff for use in the calculations. For comparison purposes and for determining the distribution of fractures from gas penetration, an experiment was also conducted with a glass liner. The glass liner protected the explosive from the water in the tuff, but fractured after detonation of the explosive and allowed the explosive gases to enter the fractures.

The explosive source in each case was PETN*, packed into plastic straws, centered in the borehole, and detonated using a bridgewire configuration. The charges were stemmed in each core by a Plexiglas plug epoxied into the borehole, and a $\frac{3}{4}$-in.-thick Plexiglas plate was epoxied over the borehole.

*Pentaerythritol tetranitrate.

During each experiment, the sample was inserted into a solenoid jacket in a pressure vessel. (The solenoid provided the magnetic field required for the operation of the particle velocity gauges.) The vessel was filled with lightweight oil, sealed, and a 1000-psi confining pressure was applied to the sample using compressed nitrogen to pressurize the oil.

After the PETN was detonated, the confining pressure was relieved. Then the sample was removed from the pressure vessel, sectioned, and photographed. Also, the gauge records from the experiments were converted from voltage-time histories to stress-time or particle velocity-time histories.

The observed damage distributions were then compared with computed and measured stress and strain histories in a manner similar to that described for the shear bands, to derive first estimates of the BFRACT parameters. Iterative computational simulations of the experiments were then performed to refine the parameters and improve the agreement between computed and observed damage levels. The final values of the BFRACT parameters are those shown in Table 1.

Application to Field Experiments. In 1978 Sandia Laboratories, Albuquerque, NM performed a series of field experiments in ashfall tuff at the Nevada Test Site [Schmidt et al. (1979) (1980)].

Briefly, Sandia performed three tests using the Gas-Frac fracturing technique: GF1 (using a "slow" burning propellant), GF2 ("intermediate"), and GF3 ("fast"). After each experiment, the region surrounding the borehole was excavated to reveal the fracture patterns. Although variations in fracturing occurred along the length of the boreholes, it was possible to obtain semiquantitative data for ranking the three tests in terms of the fracture densities, lengths, and orientations.

The number of cracks and their lengths in each test were determined using the same method as in the laboratory springing experiments. The volume within which the cracks were nucleated was also determined in the same way—that is, by using the pressure-source length and performing calculations to determine the furthest extent of the zone of tensile stresses. Difficulties in observing the fracture pattern resulted in an estimated factor of 2 uncertainty in the measured fracture densities. Nonetheless, the three different borehole-loading histories produced clearly different fracture patterns, and therefore provided a challenging test for the BFRACT model.

Computational simulations of the above three field experiments were performed with BFRACT, using the laboratory-derived parameters of Table 1. In these simulations, both the tensile stress wave and internal gas-pressure algorithms were used in BFRACT. Several simulations were performed to evaluate the effect of the gas pressure gradient in the cracks (which was not measured in these tests) on fracturing; the material and fracture property input was not varied during these simulations. Although the computed crack lengths are influenced by assumptions about the gas pressure gradient, the crack densities are not. Consequently, these field tests provided a rigorous test of BFRACT's ability to predict fracture using laboratory-derived data.

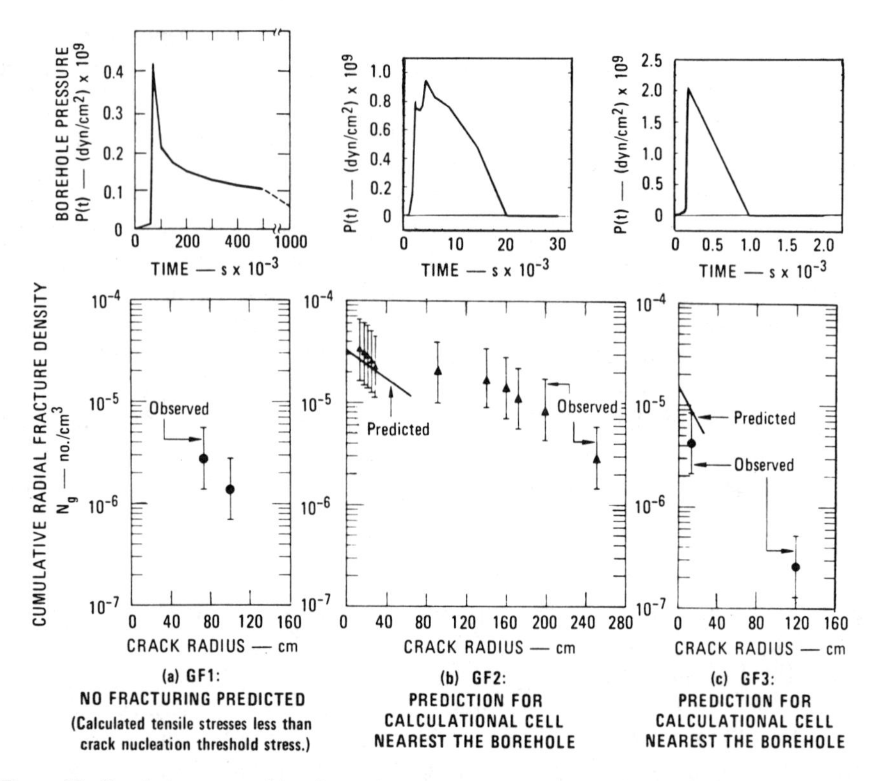

Figure 12 Borehole-pressure histories and comparison of predicted and observed fracture distribution in gas-frac field experiments.

Figure 12 compares computed fracture distributions with those observed. In Gas-Frac Test simulation GF1, no fracturing was computed to occur because the computed tensile stresses never exceeded the crack-nucleation threshold. Possible reasons for this discrepancy in GF1, which produced only two large cracks and had an almost quasistatic loading rate, are

1 The predicted tensile stresses may be incorrect because the in-situ stresses actually varied from about 5.4 MPa (780 psi) to 10.3 MPa (1500 psi), rather than exactly equaling the assumed 6.9 MPa (1000 psi.)

2 The nucleation threshold [derived for a higher rate loading: 2.06×10^8 MPa/s (i.e., 30,000 psi/μs)] was too large for this lower loading-rate experiment.

3 Large-scale flaws (i.e., joints) were activated.

4 Shear cracking occurred.

In simulating GF2, a linear gas-pressure gradient was applied at the time of cracking. The gas pressure was equal to the pressure at the borehole and decayed to zero at the calculational cell with cracking furthest from the borehole. Figure 12 compares replicated fracture distributions for GF2 with the observations. The total computed fracture density (N_0) is about equal to that observed, and the slope of the computed distribution is about two-thirds of that observed. The computed maximum crack length (R_{max}) is a factor of 4 less than that observed.

In the simulation of GF3, a linear gas-pressure gradient was also used; Figure 12 compares the computed fracture distributions with those observed. The decrease in fracture densities is caused by yielding around the borehole and the attendant suppression of fracturing. The predicted ranges in N_0 and R_1 overlap the observations; the calculations tend to predict larger values (by a factor of 2 in some cases) of N_0 and R_1 than those observed.

In summary, BFRACT correctly ranked the field experiments according to number and extent of fractures. GF1 was predicted to have no fractures because the slow burning propellant never produced high enough circumferential tensile stresses near the borehole to activate the tensile fracture process. In fact, two fractures were observed, probably because of late-time hydrofracturing. GF2, using an intermediate-rate propellant, was correctly predicted to produce the most damage because the borehole pressure was high enough to produce tensile stresses sufficient to activate the fracturing, but was low enough to avoid plastic deformation around the borehole with associated elimination of the circumferential tensile stresses. GF3, using the fast propellant, was correctly predicted to produce fewer fractures because plastic flow was produced in this case.

The main discrepancy between predictions and observations was that the predicted maximum crack sizes were roughly a factor of 4 too small. The reason for this discrepancy is that the current form of BFRACT does not permit further crack extension in a computational cell after full fragmentation (spall) has been attained. This constraint is realistic when gas has not penetrated the cracks, but is unrealistic when the cracks can be extended by gas pressurization. In ongoing work we are removing this unrealistic constraint on crack extension and expect to obtain improved agreement with data from field experiments.

Thus, it appears that the microstatistical constitutive relation approach is very promising in the application to brittle fracturing of geologic materials, even in the case where relatively few cracks are produced, thereby making the statistical assumptions somewhat dubious.

9.3.3 Quasistatic Ductile Fracture of Metals

The microstatistical constitutive-relation approach has long been used to predict dynamic fracture in metals caused by impact-, explosive-, or radiation-induced loads. In fact, such problems were perhaps the first application of the technique [Barbee et al. (1972)]. Recently, however, the technique has also been used to successfully predict the details of quasistatic failure of ductile metals.

In the following, we summarize the results of applying the microstatistical constitutive-relation approach to the quasistatic fracture of A533B-pressure vessel steel [Shockey et al. (1979)].

The phenomenology of tensile failure in A533B steel at reactor operating temperature (561°K) was investigated by examining fracture surfaces and polished sections of fractured and partially fractured Charpy and round-bar tensile specimens tested at upper shelf temperatures (355°K). Fracture surfaces showed ductile hemispherical dimples in a bimodal size distribution, and polished cross sections showed spherical voids. We concluded that tensile fracture occurred by nucleation, growth, and coalescence of voids.

Nucleation takes place predominantly at MnS and Al_2O_3 inclusions, which are typically 5 to 10 μm in diameter, and grow plastically and independently of one another as the tensile strain is increased, maintaining a roughly spherical shape. As adjacent growing voids approach one another, a population of much smaller voids nucleates at the 0.2-μm-diameter Fe_3C particles, which are homogeneously dispersed throughout the material along the interfaces of the bainite laths. The submicrometer voids form on rather well-defined surfaces connecting the larger voids and grow to radii of about a micrometer before coalescing with neighboring voids by plastic impingement. Thus, the mechanism of coalescence of larger voids is the nucleation, growth, and coalescence of smaller voids.

To obtain quantitative expressions describing the microfailure process, we measured the size distributions of inclusions, statistical void populations, and fracture-surface dimples and correlated these measurements with the imposed strain field. The photograph in Figure 13 shows the voids intersecting the plane of polish through a round-bar tensile specimen. The voids were counted and measured. These surface data were converted via a statistical transformation [Seaman et al. (1978)] to obtain void radius per unit volume for each zone indicated in the figure. The cumulative void-size distributions regularly increase in size and number with proximity to the fracture plane from the inclusion-size distribution to the dimple-size distribution. No voids were observed in regions of plastic strain less than 11%, suggesting that 11%, the strain at the onset of necking, is also the threshold strain for nucleation. Void growth rates were deduced from these data by drawing (horizontal) lines of constant void number across the distribution lines and plotting against strain the points where the lines intersect the size distribution curves.

The ductile fracture model consists of analytical expressions for void nucleation, growth, and coalescence and an algorithm for gradually reducing the specimen strength as the voids develop. The model was constructed to describe as closely as possible experimental observations and measurements.

Void nucleation was presumed to occur by decohesion of the matrix from inclusions. Thus, material is assumed to contain no voids until it experiences a strain of 11%, at which time it acquires a population of voids identical to the inclusion-size distribution (Figure 3); as for shear bands and brittle cracks, this

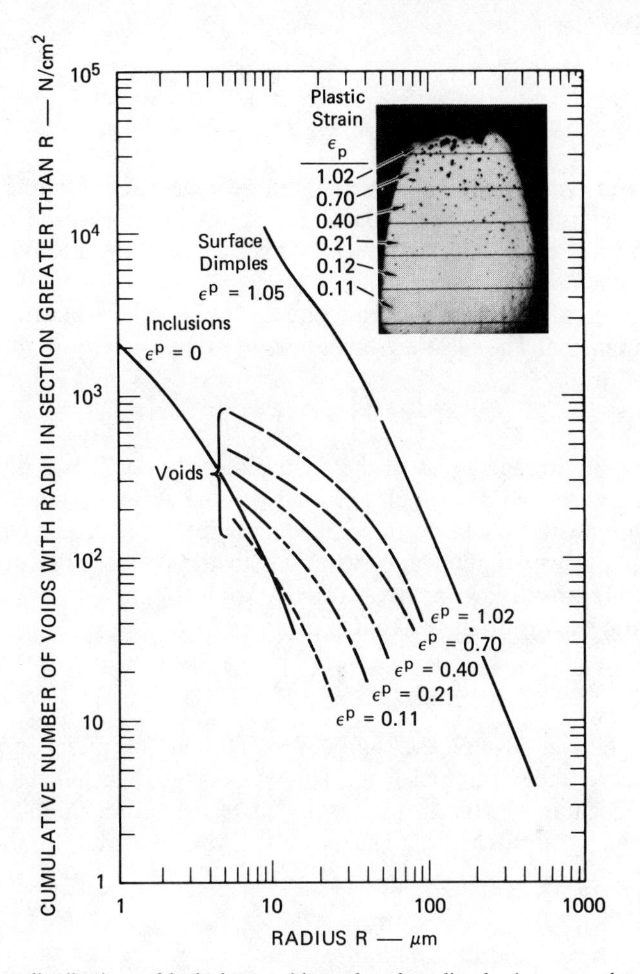

Figure 13 Size distributions of inclusions, voids, and surface dimples in a smooth tensile bar of A533B steel, heat CDB, at various plastic strains, ε^p.

distribution was fitted to the exponential form:

$$N_g = N_0 \exp\left(\frac{-R}{R_n}\right) \tag{38}$$

where N_g is the number of inclusions (voids) with radii greater than R, N_0 is the total number per cm^3, and R_n is the characteristic size parameter of the distribution. The voids are presumed to be distributed uniformly throughout each computational cell, so that the failing material can be treated as homogeneous and isotropic. The experimental observations indicate nucleation is primarily a function of plastic shear strain $\bar{\varepsilon}^P$ and suggest the following bilinear nucleation model:

$$N_0 = N_1(\bar{\varepsilon}^P - \varepsilon_1), \qquad \varepsilon_1 < \bar{\varepsilon}^P < \varepsilon_2,$$

$$N_0 = N_1(\varepsilon_2 - \varepsilon_1) + N_2(\bar{\varepsilon}^P - \varepsilon_2), \qquad \varepsilon_2 < \bar{\varepsilon}^P \tag{39}$$

where N_1 and N_2 are material-specific void densities, and ε_1 and ε_2 are strain thresholds determined from the data. The results of the tensile tests suggest values of 5×10^6 cm^{-3} and 1×10^5 cm^{-3} for N_1 and N_2, respectively, and 0.11 and 0.13 for ε_1 and ε_2.

Once nucleated, voids grow gradually by plastic flow, elastic strain, and thermal expansion. The plastic growth law

$$V_v = V_0 \exp\left[-T_1 \frac{P_s}{Y}(\bar{\varepsilon}^P - \bar{\varepsilon}_0^P)\right] \tag{40}$$

where P_s is the average stress in the solid material and T_1 is a dimensionless coefficient determined from a plot of void volume versus strain, was obtained from measurements on a series of notched and unnotched tensile bars [Shockey et al. (1979)]. This equation is very similar to the theoretically derived Rice-Tracey (1969) growth law for stress triaxialities less than 1.

The elastic expansion of the void is given by

$$\Delta V_v = -V_v \Delta P_s \frac{(1 - 3K/4G)}{K} \tag{41}$$

which was derived from Love's (1906) expression for radial motion of an infinite sphere under external load. This expansion is included to provide a physically accurate stiffness for porous material, although its influence is usually small. The thermal expansion of the void is simply the ratio of the temperature factors

$$1 + \frac{\Gamma P_{s0} E}{K} \tag{42}$$

at the beginning and end of the strain increment, where Γ is the Grüneisen ratio, E is internal energy, and ρ_{s0} is the initial solid density.

Combining the expressions for growth by plastic flow, elastic strain, and thermal expansion yields the following expression for the final void volume at the end of a strain increment:

$$V_{v3} = V_{v1}\left[\frac{1 + \Gamma\rho_{s01}E/K}{1 + \Gamma\rho_{s03}E/K}\right]\left[1 - \frac{\Delta P_s}{K}\left(1 + \frac{3K}{4G}\right)\right]\exp\left[-\frac{T_1 P_s}{Y}\Delta\bar{\varepsilon}^P\right] \tag{43}$$

As voids form and grow in the material, the load-bearing capacity of the specimen decreases. This reduction in strength is an important consequence of void development and is accounted for in the failure model in a manner similar to that described earlier for brittle cracks.

A detailed model of the void coalescence was not attempted. Instead, we assumed that void growth continues until a relative void volume of about 0.01, indicated by the round-bar tension experiments, is reached, at which point coalescence and specimen failure occur.

The model was written as a subroutine DFRACTS, incorporated into the Lagrangian, two-dimensional stress wave-propagation computer code TROTT [Seaman and Curran (1978)] and used to compute the failure behavior in a smooth and in a severely notched tensile bar. Comparisons of measured and computed force-displacement curves and relative void volume as a function of location in the specimens were made to test the reliability of the model.

Iterative simulations were made first of a smooth tensile-bar test, varying the stress-strain curve and the void-growth coefficient until the computed load-deflection curve and void-size distributions agreed with observation. Then a simulation of a circumferentially notched* round-bar tensile test (which produces a significantly higher stress triaxiality and hence a different environment for void development) was performed, using the stress-strain relation and void-growth coefficient indicated by the smooth-bar iterative simulations. Good agreement was obtained between computed and measured load-deflection curves and rupture strains.

To test the ability of the model to predict failure for cases when standard fracture toughness parameters apply, we performed a computational simulation of the center-cracked panel shown in Figure 14a. The panel dimensions and crack length were chosen to give plane-strain conditions and J-controlled initiation and growth; that is, the half-crack length, ligament, and panel thickness exceeded 200 J_{I_c}/σ_0 [Shih et al. (1978); McMeeking and Parks (1977)]. For economy, a complete simulation using the coarse computational grid of Figure 14b was performed first to establish the boundary motions of a small butterfly-shaped region at the crack tip. This smaller region was then subdivided into much smaller cells, Figure 14c, and the failure behavior of this region was computed by applying the boundary motions indicated in the first simulation. These motions, governed by the equations for the conservation of momentum, lead to strain changes in each cell, from which stresses are computed with a stress-strain relation that includes standard elastic, and work-hardening relations as well as the provisions for nucleation and growth of voids. Figure 15 shows the calculated plastic-strain field and the process zone at the macrocrack tip within which the void activity occurs.

Table 2 presents the computed values for three fracture-toughness parameters at the point where the void volume in one cell† has reached 1% (the onset of macrocrack growth) and compares the values with values reported by Shih et al. (1978, 1979). The fracture resistance predicted by the microfracture model is only slightly above the range of values measured experimentally, demonstrating that fracture-toughness parameters can be computed from microvoid-kinetics models.

*The specimen had a 1.27-cm-diameter gross section and a 0.897-cm-diameter net section. The notch had a 60° included angle with 0.025-cm root radius.

†The cell size was chosen to be a few inclusion spacings; thus, the constitutive model contains a material-specific scale size.

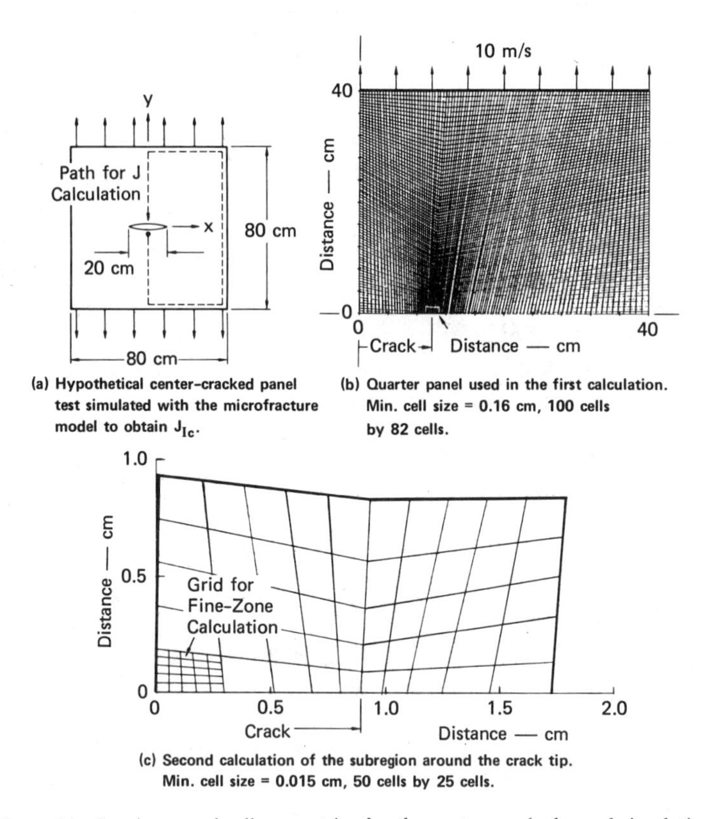

(a) Hypothetical center-cracked panel test simulated with the microfracture model to obtain J_{Ic}.

(b) Quarter panel used in the first calculation. Min. cell size = 0.16 cm, 100 cells by 82 cells.

(c) Second calculation of the subregion around the crack tip. Min. cell size = 0.015 cm, 50 cells by 25 cells.

Figure 14 Specimen and cell geometries for the center-cracked panel simulation.

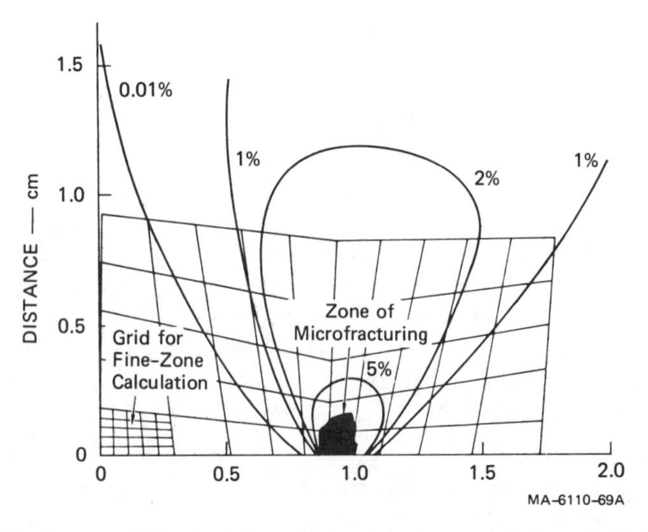

Figure 15 Contours of plastic strain and fracture in center-cracked panel simulation.

Table 2 Comparison of Computed and Measured Toughness Parameters

	Present Computation	Experimental Measurements
Fracture toughness, J_{I_c} (MJ/m^2)	0.45	0.23–0.43
Crack opening, δ_c (cm)	0.084	0.035–0.072
Crack opening half-angle (degrees)	54	[a]

[a] Measured values of this quantity are quite high and not well defined during the initial stages of crack extension. A constant value of 11 to 17° is approached after the crack has advanced several millimeters.

9.4 DISCUSSION

When large numbers of microscopic voids, cracks, or shear bands are produced during failure of a solid, it appears possible to use the microstatistical constitutive-relations approach to describe the failure history of the material. For the many dynamic-loading conditions and for the quasistatic-loading conditions to which classical fracture mechanics does not apply, this approach seems especially promising. A particular advantage is the potential for relating failure kinetics more closely to microstructural variables.

However, the examples of the constitutive-relations approach discussed show that such models are by necessity complex. Thus, increased understanding comes at the expense of increased complexity. On the other hand, the increased understanding also allows us to use simplified approaches with more confidence when they are justified. For example, in cases where the designer wishes to avoid damage altogether, only the threshold-nucleation criteria are of interest.

In summary, fracture is the result of a variety of microscopic rate processes, but each of these processes is accessible to analysis and modeling. These models, together with the use of modern computers, show promise for making fracture an understood and predictable event.

REFERENCES

Barbee, T. W., Jr., Seaman, L., Crewdson, R., and Curran, D. (1972), *J. Mater.*, **7**, 3, 393.

Barbour, T. G., Maxwell, D. E., and Young, C. (1980), "Numerical Model Developments for Stimulation Technologies in the Eastern Gas Shales Project," Science Applications, Inc., for DOE/METC.

Batdorf, S. B. (1975), *Nuclear Eng. Des.*, **35**, 349.

Carroll, M., and Holt, A. C. (February 1972), *J. Appl. Phys.*, **42**, 759.

Clifton, R. J. (1980), in "Materials Response to Ultra-High Loading Rates," W. Herrmann, Ed., National Materials Advisory Board, Report No. NMAB-356.

Curran, D. R., Shockey, D. A., Seaman, L., and Austin, M. (1977), "Mechanisms and Models of Cratering in Earth Media," in *Proceedings of the Symposium on Planetary Cratering Mechanics —Impact and Explosion Cratering*, Pergamon Press, New York, p. 1057.

Curran, D. R., Seaman, L., and Shockey, D. A. (1980), "Dynamic Fracture," in *Proceedings of the UCLA Short Course on Impact Dynamics*, University of California, Los Angeles.

Davison, L., and Stevens, A. L. (February 1973), *J. Appl. Phys.*, **44**, 2, 668.

Dienes, J. K. (1979), Los Alamos Scientific Laboratory Annual Report, LA-8104-PR.

Erlich, D. C., Seaman, L., and Shockey, D. A. (1980), Final Report for U.S. Ballistic Research Laboratory, Contract DAA DOS-76-C-0762.

Goods, S. H., and Brown, L. M. (1979), *Acta. Met.*, **27**, 1.

Grady, D. E., and Kipp, M. E. (1980), *Int. J. Rock Mech. Min. Sci. Geom. Abstr.*, **17**, 147.

Johnson, J. N. (1978), Los Alamos Scientific Laboratory Annual Report, LA-8104-PR.

Liebowitz, H., Ed. (1968), *Fracture*, Vols. I–VII, Academic Press, New York.

Love, A. E. H. (1906), *A Treatise on the Mathematical Theory of Elasticity*, Cambridge University Press, London, p. 142.

McClintock, F. A. (1973), *Fracture Mechanics of Ceramics*, Vol. 1, Plenum Press, New York.

McHugh, S. L., DeCarli, P. S., and Keogh, D. D. (1980), SRI Quarterly Reports 1 and 2 for Contract No. DE-AC21-79MC11577, for Dept. of Energy, Morgantown Energy Technology Center.

McMeeking, R., and Parks, D. M. (1977), Paper presented at ASTM Symposium on Elastic-Plastic Fracture, Atlanta, GA.

Raj, R. (1978), *Acta. Met.*, **26**, 995.

Raj, R., and Ashby, M. F. (1975), *Acta. Met.*, **23**, 653.

Rice, J. R. (1976), *Theoretical and Applied Mechanics*, North-Holland, p. 207.

Rice, J. R., and Tracey, D. M. (1969), *J. Mech. Phys. Solids*, **17**, 202.

Schmidt, R., Worpinski, N., and Northrup, D. (1979), Proc. 3rd Eastern Gas Shales Symposium, METC/AP-79/6, U.S. Dept. of Energy, Morgantown Energy Technology Center.

Schmidt, R. A., Worpinski, N. R., and Cooper, P. W. (May 1980), Paper presented at Society of Petroleum Engineers Meeting, Pittsburgh, PA.

Seaman, L., and Curran, D. R. (1978), "TROTT Computer Program for Two-Dimensional Stress Wave Propagation," SRI International Final Report for U.S. Army Ballistic Research Laboratory, Aberdeen Proving Ground, MD.

Seaman, L., Curran, D. R., and Shockey, D. A. (1976), *J. Appl. Phys.*, **47**, 4814.

Seaman, L., Curran, D. R., and Crewdson, R. C. (1978), *J. Appl. Phys.*, **49**, 10, 5221.

Seaman, L. (1980), SRI Report PLTR 001-80.

Shih, C. F., et al. (1978), EPRI Special Report NP-701-SR.

Shih, C. F. (1979), Paper presented at OECD-CSNI Specialist Meeting, Washington University, St. Louis, MO, Sept. 25–27.

Shockey, D. A., Curran, D. R., and Seaman, L. (1973), in *Metallurgical Effects at High Strain Rates*, Plenum Press, New York, p. 473.

Shockey, D. A., Curran, D. R., Seaman, L., Rosenberg, J. T., and Petersen, C. F. (1974), *Int. J. Rock Mech. Sci. Geom. Abstr.*, **11**, 303.

Shockey, D. A., Seaman, L., and Curran, D. R. (1979), in *Nonlinear and Dynamic Fracture Mechanics*, AMD, Vol. 35, ASME, New York.

Tvergaard, V. (1979), Technical University of Denmark Report No. 159.

10

NUMERICAL SIMULATION OF IMPACT PHENOMENA

Jonas A. Zukas

Analytical models, although limited in scope, are quite useful for developing an appreciation for the dominant physical phenomena occurring in a given impact situation and for sorting experimental data. They may even be useful in making predictions, provided care is taken not to violate the simplifying assumptions introduced in their derivation or exceed the data base from which their empirical constants are derived. If a complete solution to impact situations is necessary, recourse must be made to numerical simulation. This is especially true for oblique impacts or situations where a three-dimensional stress state is dominant, for there are virtually no models that can deal with such complexity. Two- and three-dimensional computer codes obviate the need for various simplifications and are capable of treating complex geometries and loading states. However, their accuracy and utility is limited by the material descriptions embodied in their constitutive equations. Excellent results have been obtained for situations where material behavior is well understood and characterized [Bertholf, et al. (1975)].

Numerical simulations of high-velocity impact phenomena in two dimensions have been performed routinely for a number of years. Current interest centers on three-dimensional simulations. The range of problems addressed is fairly wide, including computations in the hypervelocity regime to determine structural configurations capable of protecting spacecraft against meteorite impact and the study of erosion and fracture of missile and space-vehicle heat shields during reentry. The bulk of the effort has been on military problems, namely the penetration and perforation of solids and structures subjected to kinetic energy (inert) missile and shaped charge attack as well as the reverse problem of the design of armors against such threats. In geophysics, computations complement the study of materials under very high pressures and provide historical details for formation of craters produced by meteor impact [Roddy et al. (1977)]. Industrial problems addressable computationally include explosive forming, explosive welding, shock synthesis of materials, mining, and massive earth removal.

A compact description of the computational process is shown in Figure 1. The three stages listed may be incorporated in a single computer program, or code, or may exist as three distinct codes. At any rate, some sort of automatic generation capability exists to provide a detailed computational mesh for the geometry of interest from an abbreviated description provided by the user. This information is coupled to a description of the materials making up the geometric bodies by specifying appropriate parameters for the equation of state, the stress-strain relationship used by the code in both the elastic and plastic regimes, and the failure criteria to be used. A description of boundary and initial conditions ends this stage of the process. The preprocessor, as this stage is commonly called, prepares the information in a form usable by the next part, the main processor, and also prints or interactively displays the initial geometry and conditions for user verification. The current crop of codes (those developed within the last few years) tend to allow the user, via a graphics terminal, to view the results, either complete or partial, of the preprocessor and modify the computational grid, initial conditions, boundary conditions, and material description.

The conservation laws for mass, momentum and energy, coupled to an equation of state for determination of pressures, a constitutive relationship, a

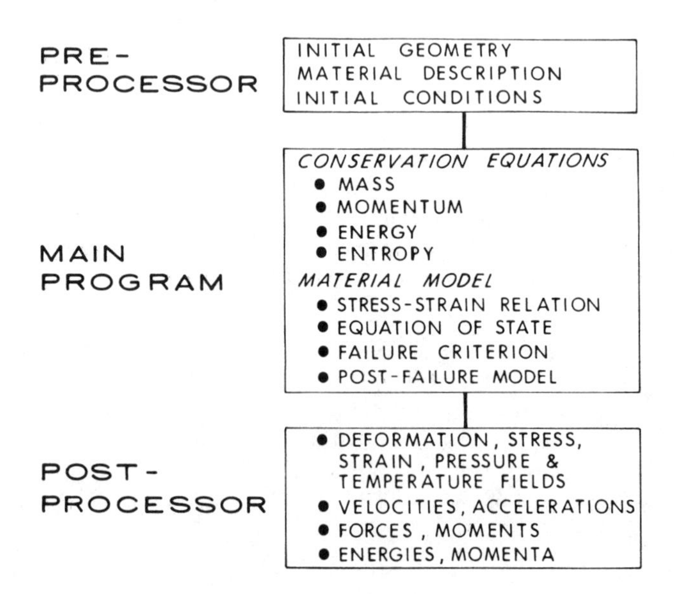

Figure 1 The computational process.

failure criterion and a post-failure model are cast into finite-difference or finite-element form and integrated in time in the next phase, or main processor, using the information generated by the preprocessor. These computations are of necessity quite long and demanding of computer storage and except for a very few problems never run to completion in one pass. Hence, provision is almost always made for a restart capability, so that computations may be resumed after interruption for physical or administrative reasons.

Results of the computations, or output, is generally massive. Codes typically produce full-field descriptions of physical quantities and material conditions (intact, failed, plastically deformed) throughout the problem as a function of time. Output listings of two- and three-dimensional codes can run into hundreds of pages and are impossible to read (indeed, sometimes to carry). Recourse is therefore made to post-processors, computer programs that prepare graphical displays of the items of interest, that is, deformation, velocity, temperature, energy fields at given times, and time histories of variables of interest, and so forth. The degree of sophistication of these plot packages varies considerably from code to code. They are also very dependent not only on the machine on which the code is installed but also on the installation. Transferring both code and plot package from one installation to another can be a nontrivial task. It is not unusual for such transfers to require several man-months of full-time effort.

The decision to use such codes should not be made lightly for in no sense can the current crop of programs be treated as "black boxes"! As a rule, at least three to six months of effort is required before new users can run practical problems on existing codes. During the learning period, frequent contact with the code developers or persons experienced in their use will be required. Even after the essentials have been mastered, pathological situations will arise that will require guidance from experienced users and may require code modification. Aside from the man-months expended, computational costs will be nonnegligible. However, the judicious combination of computer simulations and experiments (which will be accompanied by efforts to characterize candidate materials at high strain rates if such data is not already available) can lead to considerable improvements in engineering design with reduced manpower and computer costs.

Some examples of current computer-code capabilities are presented in Section 10.7. Throughout this chapter, it should be kept in mind that computational results are directly related to the quality of the material model in the code—the better the description of material behavior at the strain rates encountered experimentally and of its failure modes at those strain rates—the better the computational results. Improper materials characterization leads not only to quantitatively incorrect results but frequently to descriptions that are qualitatively incorrect. Imperfect understanding of this situation has frequently led to "...an undesirable iterative procedure of matching imperfectly understood experiments with theoretical computations based on incomplete models" [Mescall (1974)]. Fortunately, this is currently an area of intense activity so that it is not unreasonable to expect that within the next few years improved understanding of the dynamic behavior of materials at ultra-high loading rates will lead to improvement in code quantitative predictions.

This chapter provides a brief glimpse of various aspects of numerical simulation of the dynamics of impacting continuous bodies. In keeping with the spirit of the book, the primary emphasis is on methods capable of treating loading conditions sufficiently intense to produce large deformations and possible penetration. Such problems are characterized by loading and response times measured in microseconds. Deformations resulting from impact or impulsive loading will be severe and highly localized and will be governed by the constitution and properties of the materials within the affected region as opposed to the global characteristics of the structure in which they are contained. If failure occurs, it will be due to the interaction of stress waves with boundaries, material interfaces, and each other. This is in marked contrast to situations involving the response of structures to time-dependent excitations. Here, loading and response times are in the millisecond regime. The entire structure, or a significant portion of it, responds to the low-frequency components of the externally applied excitation so that the vibration characteristics of the structure become extremely important. A sizable literature exists on numerical treatment of transient structural response. Reference is made to this area when needed to point out similarities and differences in analysis of the dynamic deformation of materials and structures to intense impulsive loading.

10.1 DISCRETIZATION METHODS

It is necessary in a computer analysis to replace a continuous physical system by a discretized system. In the discretization process, the continuum is replaced by a computational mesh. The discretization techniques most commonly used are the finite-difference and finite-element methods.

Historically, the finite-difference codes were first developed as programs to treat hypervelocity-impact situations. Later a material-strength model, usually a simple-elastic, perfectly plastic or rigid-plastic model, was tacked on to the codes to treat the later stages of hypervelocity situations or penetration problems in the ordnance-velocity regime. Finite-element methods began at the opposite end of the loading-rate spectrum, being originally used to approximate the behavior of arbitrary structures and structural systems subjected to static loadings. Recently, finite-element programs capable of treating problems in wave propagation, large plastic flow, and fluid flow have appeared and are taking their place with the finite-difference codes.

In the finite-difference method, spatial and time grids are constructed by replacing derivatives in the governing equations of continuum dynamics with difference approximations. Standard techniques for solution of large equation sets are employed to obtain spatial solutions. Solutions in time are obtained by integration.

The finite-element approach is an outgrowth of structural-analysis techniques. Here, instead of manipulating the governing equations into differential-equation form and then attempting a numerical solution, the discretization procedure is employed from the very start. The procedure [Zienkiewicz (1971)] consists of:

1 Dividing the continuum by means of imaginary lines into a finite number of regions, or elements that are assumed to interact only at a discrete number of points called nodes. The displacements at these nodal points are the basic unknowns of the problem.
2 A set of functions is postulated to define displacements at any point within an element in terms of the nodal displacement.
3 These displacement functions now define a state of strain within each element. These, together with constitutive properties, then define the state of stress.
4 A system of forces concentrated at the nodes that equilibrate external loads is determined. This procedure results in a stiffness relationship-equation relating internal loads, external loads, and nodal displacements. Individual element (local) data are then assembled into global arrays and solutions for nodal displacements are obtained with conventional techniques for large systems of algebraic equations. Element strains are then determined from the nodal displacements.

A common property of both the finite-difference and finite-element methods is the local separation of the spatial dependence from the time dependence of the dependent variable. This permits separate treatment of the space and time grids. Characteristics of the spatial discretization schemes are discussed in some detail in several papers [Herrmann, Hicks, and Young (1971), Walsh (1972), Herrmann and Hicks (1973), Herrmann (1975), Herrmann, Bertholf, and Thompson (1975), Chang (1977), Belytschko (1977), Argyris et al. (1979)]. The following is therefore a brief summary of conclusions contained therein.

It has been shown [Belytschko (1977)] that the discrete forms of the equations of motion of the finite-element method are equivalent to those of the finite-difference method for a number of cases. Thus, since there is no basic mathematical difference between the two methods they should have the same degree of accuracy in numerical computations. The main differences lie not in the methods themselves but in the data-management structure of the computer programs that implement them.

Finite-element codes have a distinct advantage in treating irregular geometries and variations in mesh size and type. This is because in the finite-element method, the equations of motion are formulated through nodal forces for each element and do not depend on the shape of the neighboring mesh. In the finite-difference method, equations of motion are expressed directly in terms of the pressure gradients of the neighboring meshes. This is not inherently a problem, but the difference equations must be formulated separately for irregular regions and boundaries.

Another major difference occurs in numbering of meshes. In finite-difference programs, the regularity of the mesh implicitly establishes the connectivity information. In finite-element programs, mesh connectivity is explicitly stored, a feature that facilitates automatic generation of complex mesh systems. This limitation can be overcome for finite-difference methods, for example, Liszka and Orkisz (1977), but versatility is generally achieved at the expense of large computer storage and central processor (CPU) time.

10.2 MESH DESCRIPTIONS

The bulk of the computer codes used on a production basis for impact studies fall into two categories: Lagrangian and Eulerian. Each has peculiar advantages and disadvantages for various problem classes. Various hybrid schemes have also been developed over the years (coupled Eulerian-Lagrangian codes, calculations with convected coordinate systems), but by and large these have not found their way into the mainstream of numerical impact studies. The review papers by Herrmann (1975) and Noh (1976) are recommended for further information on hybrid methods.

10.2.1 Lagrangian Methods

Lagrangian codes follow the motion of fixed elements of mass. The computational grid is fixed in the material and distorts with it. Lagrangian methods have several advantages:

1 The codes are conceptually straightforward since the equations of mass, momentum, and energy conservation are simpler because of the lack of convective terms to represent mass flow in the coordinate frame. Since fewer computations are required per cycle, they should, in theory, be computationally faster.
2 Material interfaces and free surfaces are stationary in the material coordinate frame, hence allowing sharp definition and straightforward treatment of boundary conditions. It should be noted though that fairly complex logic is required to define behavior at material interfaces, that is, opening and closing of voids, frictional effects. While such logic enhances the generality and applicability of the codes, a price is paid in the number of additional computations required per cycle, thus slowing overall run time.
3 Some constitutive equations require time histories of material behavior. In the Lagrangian method, this is easily and accurately accounted for.

A typical example of a Lagrangian grid is shown in Figure 2. As already mentioned, the Lagrangian computational grid distorts with the material.

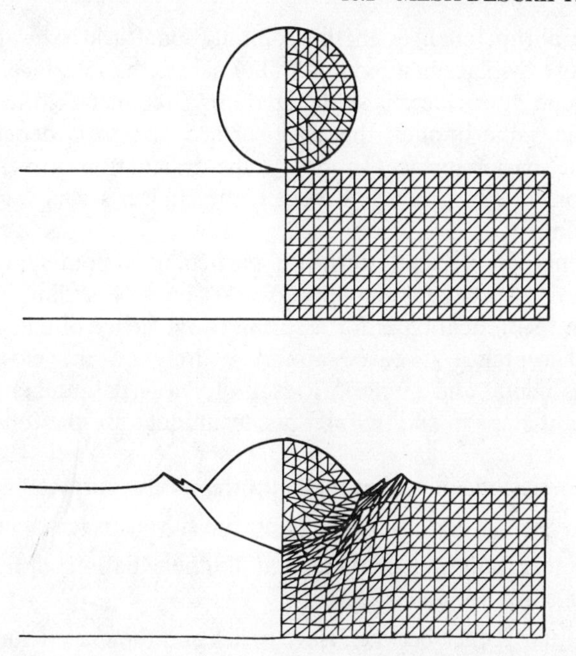

Figure 2 Lagrangian computational grid.

Inaccuracies in the numerical approximations grow when cells become significantly distorted due to shear or fold over themselves resulting in negative masses. These problems can be overcome to some extent through the use of sliding interfaces and rezoning.

Sliding interfaces are appropriate for problems involving materials that can be expected to slide over one another. Situations involving the interactions of gases and fluids with solid walls, penetration of targets by projectiles, and contact between colliding bodies require the use of sliding interfaces. They prove useful also in regions where very large shears and fractures can be expected to develop. Most sliding-interface methods are based on the decomposition of acceleration and velocity into components normal and tangential to the interface. Motions in the normal direction are continuous when materials are in contact but are independent when they are separated. Tangential motions are independent when materials are separated or the interface is frictionless, but are modified if there is contact and a finite-frictional force is present. On sliding interfaces, frictional forces ranging between zero and infinity may be specified. Materials at either side of an interface may separate if a specified criterion is exceeded and may collide again if previously separated.

A special case is the situation where two surfaces are constrained to remain in contact, either for the duration of the computation or until some criteria specified by the analyst is satisfied. Such "tied" sliding interfaces are useful in regions where abrupt changes in the computational grid size are introduced since they insure displacement compatibility across the interface.

One of the sliding surfaces is designated a master surface, the other a slave surface. As the name implies, the slave-surface motion is dependent on the behavior of the master surface. In theory, the designation of master and slave surfaces is arbitrary. In practice, however, the choice is very highly problem- and code-dependent.

The experience of the analyst with a particular computer program is frequently the key to successful computations. At present, sliding-interface treatments lack the theoretical basis for assuring convergence of calculations using them. Their acceptance is based almost entirely on the close correlation between experimental and numerical results for a large number of situations.

The important steps in sliding-interface techniques are the following:

1 Identify a series of nodes that make up the master surface.

2 Identify a series of nodes that make up the slave surface.

3 For each integration time increment, apply the equations of motion to both master and slave nodes.

4 Check for interference between slave nodes and the master surface. Various approaches have been employed to determine interference. Typically a spherical or triangular region defined by several master nodes is chosen to search for penetration by a slave node. This search is performed for each free slave node not already on the master surface. Often, for convenience, a transformation of coordinates is made from the global system to a local system orthogonal to the search area. If local penetration of the master surface by a slave node has been found, several different approaches may be employed to remedy the situation. A linear spring may be inserted in the global stiffness matrix once a slave nodes comes infinitesimally close to a master region or penetrates it. The linear spring does not affect sliding and is not activated until penetration occurs. Once it does, a restoring force acts to move the offending node onto the master surface [Hallquist (1979)]. A simpler approach involves placing the slave node on the master surface in a direction normal to the appropriate master triangular plane [Johnson et al. (1978)]. In another variation, all slave nodes not on the master surface can be checked at each time step to see if penetration will occur. If so, the time step is scaled back in such a manner that the potential intruder just reaches the master surface at the end of the time step. At the beginning of the next time step, the slave node is constrained to slide on the master segment it impacted [Hallquist (1976, 1977, 1978)].

5 Once penetrating slave nodes are returned to the master slave nodes, momentum balance is typically invoked. Frictional forces are applied, if appropriate. If tensile forces exist at the interface, slave nodes are released and voids permitted to form. If necessary, a transformation is made back to global coordinates and velocities for master and slave nodes determined.

6 Steps 3–5 are repeated for each computational cycle (time step).

When mesh distortions become very large in Lagrangian computations, truncation errors rise to unacceptably high values. Quadrilateral meshes become entangled and may take on shapes resulting in negative volumes, thus rendering computations useless. To some extent this can be overcome with different computational grids. Triangular and tetrahedral elements provide greater resistance to distortion and have proven useful in impact calculations [Jonas and Zukas (1978), Kimsey (1980)] provided care is taken in selecting their orientation in a computational grid [Nagtegaal et al. (1974)]. Improper grouping of elements can lead to undesirable oscillations in stress and pressures, even for low-distortion situations. However, for very large distortions, such elements have a tendency to lock up and produce unrealistically high pressures [Wilkins (1980)]. A further difficulty arises if an explicit scheme is used to integrate the equations of motion. For explicit methods, the time step size is controlled by the smallest mesh dimension. As distortions increase, the time step is progressively reduced and approaches zero for large distortions, rendering computations economically impractical.

When such difficulties occur, recourse must be made to rezoning. A new grid is overlaid on the old one and the rezone program maps mesh quantities of the old grid onto the new grid such that conservation of mass, momentum, and total energy and the constitutive relationship are satisfied. Rezone techniques have been used quite successfully in one-dimensional codes, especially to increase definition in regions where physical quantities vary rapidly. For two- and three-dimensional codes, however, rezoning is a costly and complex operation. Difficulties arise in averaging internal-state variables representing material memory, since a given new mesh may cover several old meshes each of which has experienced a somewhat different history. Several rezoning operations may be required before a successfully rezoned grid is obtained and the final result is very much a function of the experience of the operator performing the rezoning. Frequent rezones make the mesh effectively semi-Eulerian [Belytschko (1976)] with all the advantages and limitations of Eulerian methods—large distortions are accounted for, but material history and boundaries may be diffused. This last point can be crucial if advanced failure models are employed that make use of material history to initiate or propagate cracks and voids in materials.

Mesh distortions can be detected by a variety of tests. One of the simplest is reduction of the time step to an unacceptable level, at which point a restart file is written. Rezoning may be performed manually by an experienced analyst or done automatically with special purpose programs. Even the most complicated and sophisticated rezone routines have been disappointing for the two-dimensional case, with the apparent exception of the TOODY and DYNA codes.

10.2.2 Eulerian Methods

For problems in which large distortions predominate or where mixing of materials initially separated occurs, an Eulerian description of material behav-

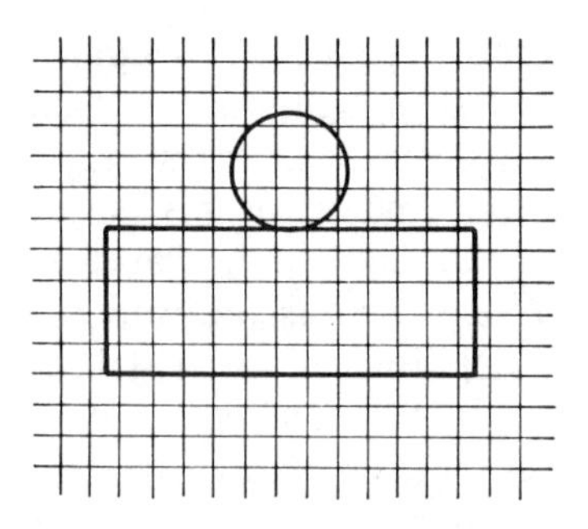

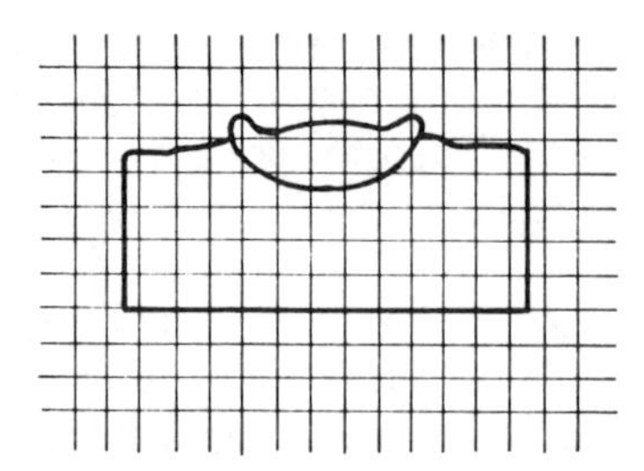

Figure 3 Eulerian computational grid.

ior is necessary. In the Eulerian approach, the computational grid is fixed in space while material passes through it (Figure 3). Material can be represented as either discrete points or a continuum. A number of approaches are available in existing Eulerian codes. In one, the material transport terms are included in the difference algorithm for the partial derivatives from the outset, for example, Roach (1972). In another, at each time step a Lagrangian calculation is performed, followed by a separate calculation of the convective transport terms, which in effect resembles a rezone of the distorted Lagrangian grid back to the undistorted Eulerian grid, for example, Hageman et al. (1975), Durrett and Matuska (1978).

Unless special provisions are made for locating material surfaces and interfaces, these will diffuse rapidly throughout the computational grid. A number of techniques have been invented. In one technique, the actual positions of surfaces and interfaces are calculated by introducing massless Lagrangian tracer particles. Such particles can also be used to provide material-history data but this introduces considerable computational complexity and expense. Diffusion may also be limited by resorting to preferential transport of materials. In some techniques, care is taken to identify the materials being transported from the donor mesh to the acceptor mesh and to transport only one material until it is exhausted in the donor mesh. Material surfaces and interfaces are thereby defined only to within one mesh width, but they do not progressively diffuse.

10.2.3 Hybrid Methods

Since no one computational technique can handle all situations in impact dynamics, hybrid methods have been developed. Many variations exist. The most common techniques involve use of an Eulerian method at the beginning of a computation when gross flow and distortions are expected, the mapping to a Lagrangian grid for completion of the calculations. Alternative approaches include mixed Eulerian and Lagrangian computations in parallel. Convected coordinate methods have also been used. Computer programs based on hybrid methods tend to be written for specific applications and are not easily adapted to other problems. Further discussions and references are given by Herrmann (1975), Belytschko (1976), and Belytschko et al. (1978).

10.3 ARTIFICIAL VISCOSITY

A characteristic of high-amplitude wave propagation in solids is the presence of shock waves. Even if such waves are not introduced by initial or boundary conditions, they may arise spontaneously in the body by the steepening of compressive waves due to the nonlinear response of the material. While shock waves do not exist as mathematical discontinuities in real solids, shock widths several orders of magnitude smaller than body dimensions have been observed experimentally.

Difficulties are encountered in numerical solutions to impact problems when shock waves are present. One approach to treating shock waves is to use the Rankine-Hugoniot jump conditions (see Chapter 1). These equations, together with the equation of state in characteristic form could be solved to determine the thermodynamic quantities at the shock as well as the shock velocity and position [Walsh (1972)]. However, this involves an iterative solution to complicated nonlinear equations and is not practical for other than one-dimensional problems where the technique is usually described as "shock fitting" [Herrmann (1975), Karpp and Chou (1972), Herrmann and Hicks (1973)].

In multidimensional calculations, shock fitting is computationally complex and expensive and does not solve all problems since shocks can occur spontaneously within the material. Consequently, virtually all existing computer codes, even those that use differencing schemes (such as the Lax-Wendroff method) with built-in dissipation make use of an artificial dissipative term. This is commonly added to the pressure term of the original differential equations, so that the pressure P is replaced by $P+q$ where q is a function of the spatial derivatives of velocity and has the form of a viscosity. The concept was originally proposed by von Neumann and Richtmyer (1950) for one-dimensional shock propagation in inviscid fluids and had the form

$$q=(k_2)^2\rho(\Delta x)^2\left(\frac{d\dot{x}}{dx}\right)^2 \tag{1}$$

where x is the coordinate in the direction of motion, Δx is the grid spacing, ρ the density and k_2 a constant often taken to have a value of ≈ 2. The quantity q is zero for $d\dot{x}/dx \geqslant 0$.

Equation (1) can be rewritten in the form of a diffusion equation

$$q = \alpha_1 \frac{d\dot{x}}{dx} \tag{2}$$

where
$$\alpha_1 = (k_2)^2 \rho (\Delta x)^2 \left| \frac{d\dot{x}}{dx} \right| \tag{3}$$

can be thought of as a diffusion coefficient. The diffusion coefficient is scaled to the grid spacing of a problem. The term q does not represent a real viscosity but rather serves to smear the shock over a region of constant width two to three times Δx. Because q is quadratic in the velocity derivative, it disappears rapidly away from the shock, so that the Rankine-Hugoniot conditions must be satisfied across the shock region.

For computations involving steep stress gradients in solids where sound speeds exist even at zero pressure, dissipation over larger distances than that provided by (1) is frequently required. Hence it is almost universal practice to use an artificial viscosity that has both linear and quadratic terms

$$q = k_1 \rho c \Delta x \left| \frac{d\dot{x}}{dx} \right| - k_2^2 \rho (\Delta x)^2 \left| \frac{d\dot{x}}{dx} \right| \frac{d\dot{x}}{dx} \tag{4}$$

where c is the local sound speed and k_1 and k_2 are constants. Their values are problem-dependent. Typically, k_1 ranges between 0.1–0.5, and $k_2 = 2$.

To extend the von Neumann-Richtmyer idea to two and three dimensions it is necessary to obtain an expression for the rate of strain given by the velocity gradient in (1) and to determine a characteristic grid length. A comprehensive discussion of artificial viscosity in multidimensional calculations is given by Wilkins (1980). Expressions for strain rate and characteristic length are to be found therein, together with examples of artificial-viscosity effects in grid stabilization. A comprehensive discussion of artificial-viscosity effects has also been given by Noh (1976).

The introduction of damping, either through the numerical scheme itself or explicitly through artificial viscosity always results in some distortion of the solution. Care is always necessary in the use of any numerical method for nonlinear wave-propagation problems to be sure that, on the one hand, the high-frequency content is adequately represented and shock waves are not spread to the point where the solution is meaningless while on the other hand spurious numerical noise does not override the solution. Since these effects are mesh-size dependent, comparison of computations with different mesh sizes is necessary whenever new problems are begun to insure that the solution is not affected to an unacceptable degree by numerical artifacts [Herrmann (1975)].

10.4 TIME INTEGRATION

Situations involving impact or impulsive loading excite wide-spectrum frequencies in the affected structure. If the response of the structure is controlled by a relatively small number of low-frequency modes, the problem is said to be in the *structural dynamics* category. In *wave-propagation* problems, on the other hand, the high-frequency modes dominate the response throughout the time of interest. Here we consider high frequencies to be on the order of, or greater, than the characteristic acoustic frequencies while low frequencies would typically be an order of magnitude less than acoustic frequencies. In practice, combinations of these conditions are frequently encountered. In many low-velocity impact situations, in blast of explosive loading, in accidental collisions, for example, initial high-frequency transients gradually decay to the steady state or free vibration regime. In certain cases involving dynamic instabilities (e.g., flutter), bounded responses can suddenly grow without limit as critical conditions are attained.

Situations wherein some portions of a structure are mechanically stiffer than other portions create difficulties for numerical integration of the equations of motion. If a time step is chosen small enough to accurately treat the rapidly varying components, then it will be excessively small for the remaining components, resulting in excessive computation effort and round-off error in the low-frequency components. On the other hand, if the time step is adjusted to the slowly varying components, instability may result and the high-frequency components will not be accurately resolved.

Time-integration routines are the heart of most structural dynamics programs. Hence the subject has been extensively studied [e.g., Goudreau and Taylor (1972), Nickell (1973), Jensen (1974), Geradin (1975), Belytschko (1974, 1975, 1976), Tillerson (1975), Belytschko et al. (1976, 1979), Fellippa (1977), Underwood and Park (1977), Hughes et al. (1978), Felippa and Park (1978a, b, 1979), Belytschko and Mullen (1978), Wright (1979), and Argyris et al. (1979)]. Broadly, it may be said that implicit integration methods are most effective for structural dynamics problems while explicit integration methods are best for wave-propagation problems. The choice of integration method is still very much problem-dependent and in many cases (e.g., fluid-structure interaction) it is advantageous to use both methods in different parts of the problem [Belytschko and Mullen (1978)].

Methods for integrating the discretized equations of continuum mechanics are called explicit if displacements at some time $t+\Delta t$ in the computational cycle are independent of the accelerations at that time. The second order central-difference algorithm is one of the most widely used tools in integration schemes. For purposes of discussion, take the equations of motion to be in the form

$$M\ddot{u}+Ku=F(t,u) \tag{5}$$

where M is the mass matrix, either lumped or consistent, K a stiffness matrix, u

the displacement vector, $\ddot{u}$ the acceleration and F the load vector that includes mechanical loads, thermal loads, and pseudo forces due to both material and geometric nonlinearities. Express velocities and accelerations by difference equivalents in time

$$\dot{u}\left(t+\frac{\Delta t}{2}\right)=\frac{1}{\Delta t}\left[u(t+\Delta t)-u(t)\right] \tag{6}$$

$$\ddot{u}(t)=\frac{1}{\Delta t}\left[\dot{u}\left(t+\frac{\Delta t}{2}\right)-\dot{u}\left(t-\frac{\Delta t}{2}\right)\right] \tag{7}$$

Replacing the acceleration term in (5) by (7) and using (6) to put things in terms of displacements yields a recurrence relation to be solved at each time

$$Mu(t+\Delta t)=(\Delta t)^2F(t)+\left[2M-(\Delta t)^2K\right]u(t)-Mu(t-\Delta t) \tag{8}$$

At any time step, the velocities and displacements are known. The rate of deformation and strain can be computed from the strain-displacement relations and the stresses at that time step are found from the constitutive relationship. The equation of motion is then used to find the accelerations which, together with velocities, are stepped forward in time to find new displacements and the entire procedure is repeated once again.

The computed response may become unstable (grow without bound) in explicit integrations unless care is taken to restrict the size of the time step. This problem has been studied rigorously for linear problems by Courant, Friedricks, and Lewy (1928) who found that in explicit integration the computation will be stable if the time step Δt satisfies the relation

$$\Delta t \leqslant \frac{2}{\omega} \tag{9}$$

where ω is the highest natural frequency of the mesh. No rigorous stability criterion has been determined for nonlinear problems but it is customary to determine the time step from

$$\Delta t=\frac{kl}{c} \tag{10}$$

where l is the minimum mesh dimension in the computational grid, c the sonic velocity, and k a factor chosen to be less than unity, generally between 0.6–0.9.

In an implicit scheme, the displacements at any time $t+\Delta t$ cannot be obtained without a knowledge of the accelerations at the same time. The relationships among velocity, displacement, and accelerations must be combined with the equations of motion and the resulting set of simultaneous equations solved for the displacements. The resulting nonlinear equations are generally solved by some kind of linearization method. Among the most popular methods are predictor-corrector schemes, trapezoidal techniques such as Houbolt's method, the Wilson method and the Newmark β method [Tillerson (1975), Belytschko (1977), and Argyris et al. (1979)].

Many implicit methods have been shown to be unconditionally stable. However, the price of stability is the need to solve a set of equations at each time step. The local truncation error of most implicit and explicit schemes is of order Δt^3. While this is insignificant for explicit schemes, it is a matter of concern for implicit methods where the time step is so much larger.

The Newmark β method [Newmark (1959)] seems to be among the most widely used of the many implicit methods available and appears to give excellent results for a wide class of problems. In this scheme, displacements and velocities are expressed as

$$\dot{u}(t+\Delta t)=\dot{u}(t)+\Delta t[(1-\gamma)\ddot{u}(t)+\gamma\ddot{u}(t+\Delta t)] \tag{11}$$

$$u(t+\Delta t)=u(t)+(\Delta t)\dot{u}(t)+(\Delta t)^2[(\tfrac{1}{2}-\beta)\ddot{u}(t)+\ddot{u}(t+\Delta t)] \tag{12}$$

The parameter γ may be thought of as a generalized acceleration parameter. It is generally selected based on the experiences of users. If γ is not set to $\frac{1}{2}$, spurious damping is automatically introduced into the response. This feature may be used in much the same fashion as artificial viscosity is used in shock-propagation situations and similarly requires a fair amount of experience for proper application. Generally, it is recommended that the method be used with $\gamma=\frac{1}{2}$ [Tillerson (1975)].

The Newmark formulation has been shown to be equivalent to several other methods depending on the choice of the β parameter. If

$\beta=\frac{1}{4}$—constant average acceleration during a time step
$\beta=\frac{1}{6}$—linear acceleration
$\beta=\frac{1}{8}$—step-function variation of acceleration
$\beta=0$—explicit, same as second-order, central-difference formulation

Another popular implicit scheme attributed to Houbolt (1950) is obtained using the following difference formula for the acceleration vector

$$\ddot{u}(t+\Delta)=\frac{1}{(\Delta t)^2}[2u(t+\Delta t)-5u(t)+4u(t-\Delta t)-u(t-2\Delta t)] \tag{13}$$

This difference expression is equivalent to fitting a third order curve through four discrete points in time or alternatively to assuming $\dddot{u}$ is zero. Tillerson and Stricklin (1970) reviewed nine different integration techniques for the nonlinear dynamic analysis of shells of revolution. The Houbolt method was judged superior from the standpoints of efficiency and practicality in situations where the low-frequency response of the structures was significant. Goudreau and Taylor (1972) critically evaluated the Newmark, Wilson, and Houbolt methods for electrodynamics problems, including wave propagation. Newmark's method was found superior to the other two.

Problems involving high-velocity impact (0.5–2 km/s) result in a strong initial shock wave (alternatively, a large stress or velocity gradient) which can lead to material failure and must be accurately resolved. This demands fine spatial as well as temporal resolution and results in very small time steps and a large number of computational cycles.

Typically, for design problems, computations must be run to tens or hundreds of wave-transit times across the characteristic length dimension of the problem. This places a severe burden on computer resources and spurred investigation of implicit integration schemes.

As a general rule, it has been found [Herrmann and Hicks (1973), Herrmann et al. (1975), Herrmann (1977)] that, for wave-propagation problems, the time step for implicit methods must be about the same as that for explicit methods to satisfy accuracy requirements. Since implicit methods require considerably more computations per cycle than explicit integrations, their use has generally been limited to problems where the details of wave propagation are not as significant as the overall response of the material.

10.5 MATERIAL MODELS

The predictive capability of current computer codes for impact studies is dependent on the material model and properties that are used in calculations. Constitutive relations applicable to stress wave-propagation situations have been extensively studied. Formulations for metal plasticity have been reviewed by Nolle (1974), Herrmann (1976), and by two National Materials Advisory Board Committees (1978, 1980). Reviews for porous compaction have been given by Herrmann (1971) and Seaman (1976), for viscoelastic materials by Herrmann and Nunziato (1972), Schuler et al. (1973), and Nunziato et al. (1974) among others. Dispersion in composites has been reviewed by Bedford et al. (1976), Moon et al. (1976), and Dvorak (1978). While data and models for the response and failure of concrete under static loading is plentiful, virtually none exist for its dynamic response to high-pressure loading. Examples of computational procedures for concrete structures subjected to impact can be found in the reports of Gupta and Seaman (1978) and Osborn (1981). Much of the current work involves the behavior of metals and metallic structures subjected to high rate loading. The following remarks will therefore briefly summarize the situation in this area.

It is common in existing production codes for the study of high-velocity impact phenomena to divide the deformation behavior of metals into volumetric and shear (deviatoric) parts. Metals undergo plastic yielding at modest levels of deviator flow stress. This is usually taken to be independent of pressure, so that the volumetric behavior can be treated independently of the shear behavior.

The volumetric behavior is described in terms of an equation of state relating pressure, volume, and some thermal parameter, usually the internal energy or temperature. The Mie-Gruneisen equation of state is frequently used, although some codes allow a choice of equations. In the newer, modular codes, equations of state can readily be changed.

For solid-solid impacts in the 0.5–2 km/s velocity regime, only moderate pressures (300–500 kb) are generated and these decay rapidly to values comparable to the strength of the material. Hence, the equation of state in impact calculations is of secondary importance. Considerable data exists for the current crop of equations in various compilations for most metals of interest [Kohn (1969), Van Thiel (1977)] and additional data can be readily obtained. Consequently, the state of the art in equation of state is adequate for most present needs.

An incremental elastic-plastic formulation is used to describe the shear response of metals in present finite-difference and finite-element codes. The plasticity descriptions are usually based on an assumed decomposition of the velocity strain tensor, $\dot{e}$, into elastic and plastic parts

$$\underset{\sim}{\dot{e}} = \underset{\sim}{\dot{e}}^e + \underset{\sim}{\dot{e}}^p \tag{14}$$

together with incompressibility of the plastic part

$$\dot{e}^p_{11} + \dot{e}^p_{22} + \dot{e}^p_{33} = 0 \tag{15}$$

Stress and strain tensors are divided into volumetric and deviator parts. The volumetric parts, namely the pressure and volume, are determined through the equation of state. The von Mises yield criterion and the Prandtl-Reuss incremental theory are typically used to describe plastic behavior. Since strains in problems involving penetrations are large, questions have been raised regarding the validity of this approach and alternatives proposed [Green and Naghdi (1965), Lee (1970), and Clifton (1979)].

Plasticity models for computations have been reviewed by Armen (1979). Generally, the plasticity models in wave-propagation production codes are relatively simple elastic, perfectly plastic descriptions following the method formulated by Wilkins (1964). Minor modifications have been introduced to this basic description by allowing the yield stress to vary with the amount of plastic work, temperature, strain rate, or some combination thereof, for example, Johnson et al. (1978)].

Herrmann and Lawrence (1978) have looked at material models that have been successfully used to describe experimental observations of stress wave propagation in metals, polymers, composites, and porous materials. These were used in a series of one-dimensional finite-difference calculations of stress pulses propagating in a semiinfinite medium. They, as well as the NMAB Committee on Materials Response to Ultra-High Loading Rates (1980), find that for metals, a perfect plasticity approach can serve as a first order approximation for many high-strength alloys used in high-velocity (0.5–2 km/s) impact situations provided care is taken to select an average dynamic value of flow stress. Excellent results have been obtained with this approach [Bertholf et al. (1975), Wilkins and Guinan (1973), and Norris et al. (1977)] and it is appealing from the point of view that the degree of dynamic material characterization required is quite low. Also, many high-strength alloys show little variation in flow stress with strain rate and relatively low rates of strain hardening.

A priori determination of an appropriate dynamic flow stress is another matter however. Until very recently such information was not generally available. In a few cases, a dynamic yield stress can be estimated from static test data but in general the understanding of micromechanical deformation mechanisms at very high strain rates is so limited that estimation of dynamic properties from static data is hazardous. This approach is best used in conjunction with dynamic material property tests.

No one dynamic property test technique can provide information over the range of stresses, strains, strain rates, and temperatures encountered in high-velocity impact. Several relatively simple techniques exist however which, despite their limitations, provide useful data for numerical computations. Methods for dynamic characterization of materials have been reviewed by Lindholm (1971, 1980). Several methods commonly used from which a substantial body of data exists are described in Chapter 8.

10.6 COMPUTER RESOURCE REQUIREMENTS

Questions of accuracy and resolution required for dynamic stress-wave solutions are discussed by Herrmann (1977). Such solutions are characterized by a very high frequency content and adequate numerical representation requires a large number of meshes in areas where large stress gradients propagate. Two-dimensional computations are routinely done with 4000–10,000 meshes or elements. Adequate resolution in three-dimensional problems is even more difficult to achieve. For practical problems 20,000–50,000 meshes are not uncommon. For high-velocity impact situations, adequate information for design purposes (projectile velocity and orientation, extent of deformation in projectile and target, energy deposition in the target) was obtained for an oblique impact situation with 25,000 elements [Jonas and Zukas (1978)] while the same calculation with 12,000 elements underpredicted penetration depth by 40%. For problems involving steep gradients or advanced failure models, even greater resolution may be required. Since most practical problems are run with variable meshes to conserve CPU time and computer storage, there is no a priori method for determining an optimum grid for a given computation, although guidelines for educated guesses exist [Herrmann and Hicks (1973), Herrmann (1977)]. Experienced code users can generally arrive at acceptable grids within a few iterations.

Computing costs tend to be quite high for production problems. Running times for one-dimensional codes are measured in minutes. For two-dimensional codes with some 5–6000 meshes or elements, running times of several hours on CDC 7600 class machines are typical. Three-dimensional problems will typically run from 4–10 hours or longer. Three-dimensional problems also place severe limitations on computer storage. It is virtually impossible to run a three-dimensional impact problem totally in-core. Thus, to permit adequate resolution, most codes have provisions for keeping only a small portion of the grid in core and the remaining information on a mass memory device. There is still an upper limit on resolution, though. If the number of meshes or elements is too great, the bulk of the total computer time will be spent in data transfer and very little in advancing computations.

As expensive as such calculations may be, they are often less costly than full-scale experiments. Indeed, in certain situations, experimentation may not be possible and reliance must be made on computations. Additionally, unless the software is reasonably efficient in permitting automated mesh generation and graphical representation of results, the major portion of the total cost will be associated with manpower charges incurred by analysts performing numerical studies. An illuminating example is shown in the article by Herrmann (1977).

10.7 EXAMPLES OF CURRENT CODE CAPABILITIES

It is not possible to review here the spectrum of calculations that have been done with existing codes. Rather, representative examples selected from the literature are presented which illustrate problems tractable by existing codes.

10.7.1 Hypervelocity Impact of a Nylon Sphere With a Steel Plate

We begin with an example of a two-dimensional computation, using both Eulerian and Lagrangian codes, of a hypervelocity situation involving the impact of a 9.53 mm diameter sphere of nylon against a 12.7 mm thick armor-steel plate at a velocity of 5.182 km/s. This example was part of an experimental program initiated by the Mobility Equipment Research and Development Center to study spallation in armor steels. Hypervelocity experiments for projectiles weighing between 0.5 and 5 g were conducted at the Naval Research Laboratory light-gas gun facility. Metallurgical examination of the impacted specimens for the nylon sphere problem was done at the Stanford Research Institute by Shockey et al. (1975) while numerical simulation was performed by Bertholf et al. (1975) at Sandia Laboratories. This situation is an excellent one for study since it brings out the complexities involved in the physical response of materials under high-velocity impact situations and the excellent results that can be achieved computationally when meticulous care is taken to model the dominant physical processes occurring in the experiment.

Material response to high-rate, high-pressure loading as encountered in hypervelocity-impact situations is quite complex. The most noticeable characteristics of impacted target plates are the large crater that forms at the impact site and the spall damage that occurs near the rear surface for finite-thickness targets. There are other, subtle, material changes—phase changes, shear banding, fragmentation—that strongly influence the observed cratering and spallation. These tend to be highly material dependent and need to be accounted for in predictive relationships or analytical models on a case by case basis.

Steel is a particularly interesting material for impact studies for it can exhibit many physical phenomena. Steel undergoes both pressure-induced and temperature-induced phase changes, can fracture in either a ductile or brittle manner, and is susceptible to shear instabilities.

The metallographic analysis of the impact situation indicated the presence of an α (bcc)–ε (hcp) polymorphic phase change in the armor steel used for the study (Brinell hardness 351 or approximately 38 on the Rockwell *C* scale). Profuse shear banding was also found to be present in the subcrater region. The patterns of the shear bands and associated cracks suggested strongly to Shockey et al. (1975) that crater formation, at least in the late stages, occurred by the joining of the cracks to create fragments that were then ejected. Fracture damage to the back of the target was assumed as a result of nucleation, growth, and coalescence of microfractures followed by a widening of the resulting macrocrack and subsequent scabbing of the back surface by a shearing process that does not appear to be adiabatic.

The numerical simulation was performed with the Eulerian CSQ code and the TOODY/TOOREZ Lagrangian codes. The first problem was to accurately specify the constitutive behavior of both the nylon and steel over a wide stress range. The most probable material effects [Bertholf et al. (1975)] for the nylon were deemed to be its:

1 Viscoelastic nature.
2 21 GPa polymorphic phase change.
3 Melting transition.
4 High-temperature decomposition.

For the steel:

1 Elastic-plastic response, including work hardening and rate dependence.
2 A failure description (including failure criteria, spall stress, cumulative damage, and other effects such as adiabatic shear and fragmentation).
3 The 13 GPa-polymorphic phase change.

Since determination of complete constitutive descriptions for both materials would be prohibitively expensive, even if possible, Bertholf and his coworkers compared experimental results with numerical solutions obtained by making a number of simplifying assumptions concerning the response of both the nylon sphere and steel plate to determine which effects were dominant. It was finally decided to perform the analysis under the assumptions that:

1 The nylon has no shear strength and undergoes only a simple liquid-vapor transition.
2 The steel behaves as an elastic, perfectly plastic solid with a yield point of 1.2 GPa. The polymorphic phase change was assumed to occur at 13 GPa.

3 A simple, maximum-normal stress-failure criterion was used, neglecting the effects of cumulative damage and time-dependent fracture. The spall strength for the steel was assumed to be 5.3 GPa.

Computations with the above assumptions were then performed with both the CSQ and TOODY codes and compared with post-impact experimental results at 20 μs. The results are shown in Figure 4. The calculations demonstrated the need for a very fine grid in the simulation, since a Lagrangian calculation similar to that shown in Figure 4*c* with half as many meshes in each direction gave no indication of back-surface spallation. Doubling the mesh size increases computation time by a factor of 8 but in this situation it was vital for correct modeling of the impact event. Differences in results were also noted due to the way failure was modeled in each of the codes. In the Eulerian code, the pressure in a failed cell is immediately set to zero once the failure criterion is exceeded whereas in the Lagrangian code a spall plane is introduced (a zero-thickness gap with free surfaces on both sides) at the time and place of failure and the magnitude of the hydrostatic tension is restricted to a specified level, so that the stress is reduced over a finite time.

It is also evident from Figures 4*a* and *b* that the effect of the phase change in steel is very significant. When the phase change is included in the calculations, the loading portion of the stress wave consists of two parts as the wave

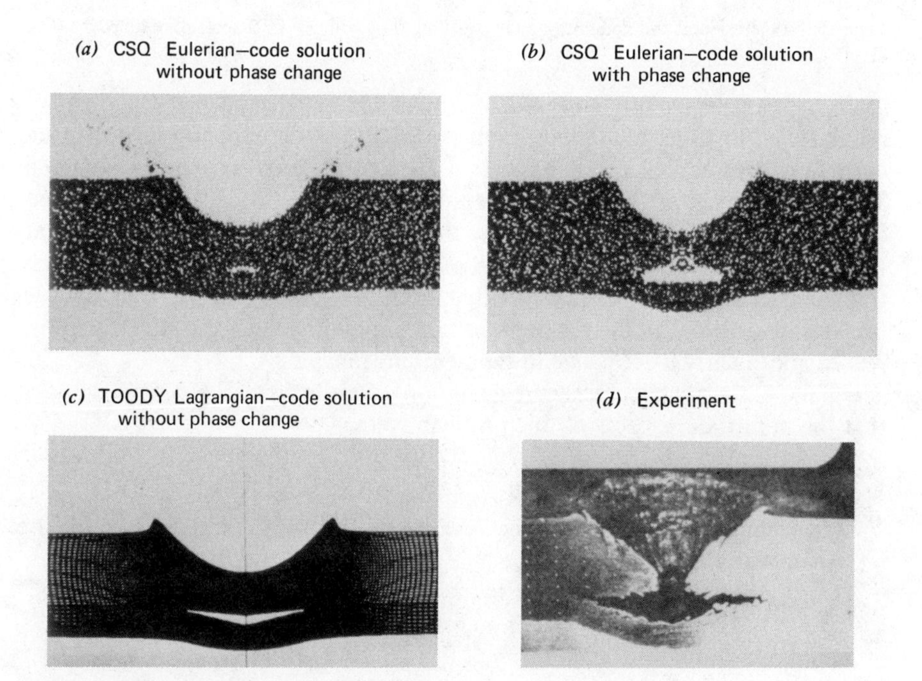

Figure 4 Comparison of computational and experimental results at 20 μs with 5.3 GPa spall strength [Bertholf et al. (1975)].

(*a*) CSQ Eulerian–code solution with 13GPa phase transition

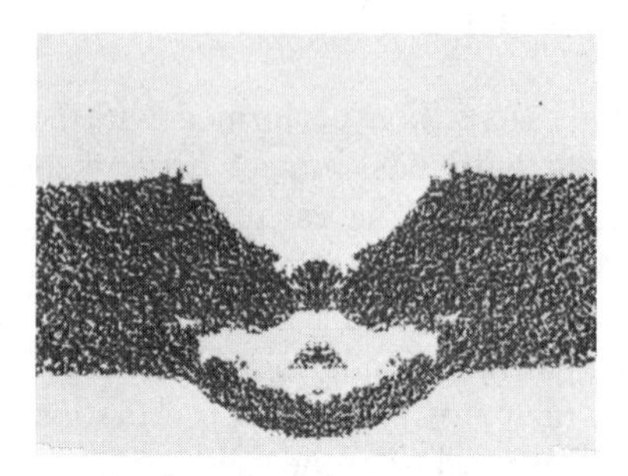

(*c*) TOODY Lagrangian–code solution with 13GPa phase transition

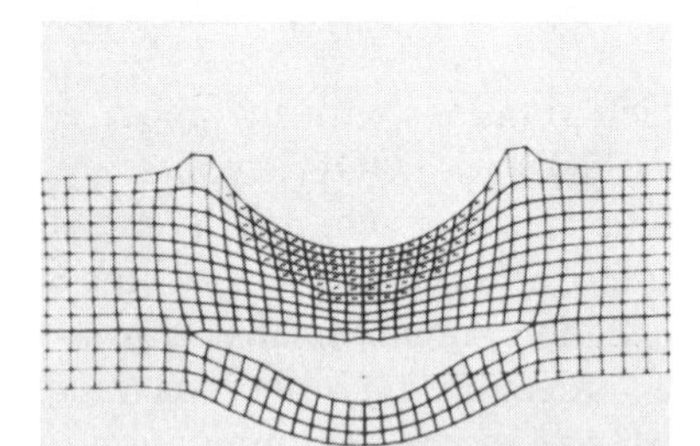

(*b*) CSQ Eulerian–code solution without 13GPa phase transition

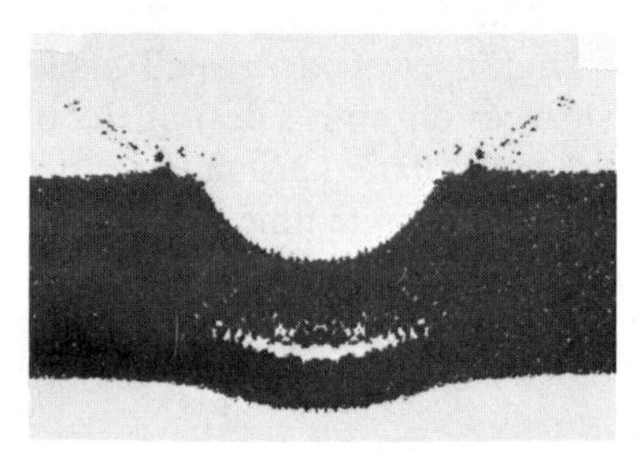

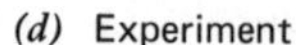

(*d*) Experiment

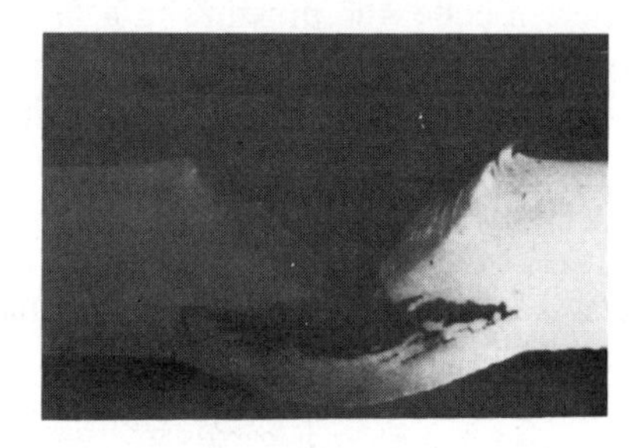

Figure 5 Experimental and computational results at 20 μs with 3.8 GPa spall strength [Bertholf et al. (1975)].

separates at 13 GPa. More important is the difference observed as unloading occurs. A rarefaction shock is present when the phase change is included. These differences in the stress-wave structure not only cause a cylindrical-conical failure to occur directly below the crater, but also result in a propagating pulse that is nearly square for the case with a phase change and nearly triangular without. The square pulse, upon reflection from a free surface, transfers essentially all the momentum to the spall layer, whereas the triangular pulse is not nearly as effective in momentum transfer.

On the basis of this series of calculations, Bertholf et al. (1975) concluded that the important aspects of this problem were:

1 Sufficient numerical resolution.
2 An adequate description of the 13 GPa-polymorphic phase change in the target material.
3 A good numerical model of the failure process.
4 The correct spall strength of the steel plate.

Being reasonably confident of all but the last item, they undertook a series of flyer-plate experiments to verify the spall strength. The tests indicated that a spall strength of 3.8 GPa was a more realistic value for the steel material at hand. The calculations were repeated with that value and the results are shown in Figure 5.

Agreement with experiment is excellent, especially for the Lagrangian results where the crater diameter, crater depth, spall layer thickness, spall length, spall bulge, cylindrical-conical failure, and the area that undergoes a phase change all show nearly 1-1 correspondence with experiment. Approximately 12 hours of CDC 6600 time were required for the numerical solutions. Only a few man-hours were required for the Eulerian calculation. Considerably more man-hours were required for the Lagrangian calculation since frequent rezoning was required along the way.

10.7.2 Long Rod Impact

To illustrate current capabilities of finite-difference and finite-element codes in dealing with ballistic-impact situations at high obliquity, a representative calculation is presented and compared to experimental data. The information is taken from Jonas and Zukas (1978).

The penetration of a staballoy (depleted uranium) rod into a rolled-homogeneous armor (RHA) plate at 60° was computed with both the plane-strain version of the HELP code (a two-dimensional Eulerian finite-difference calculation) and the EPIC3 code (a three-dimensional Lagrangian finite-element calculation). Three-dimensional codes have been successfully used on various problems but they make severe demands on computer storage and are quite costly, though this latter aspect may become a minor problem with the advent of the parallel processors. Hence, in the past, plane-strain approximations have been used to obtain at least a qualitative appreciation of the behavior of rod and target under oblique impact conditions, which are clearly of great practical importance and sufficiently different from the normal impact case due to the added complexities of severe bending and asymmetric loading to warrant separate consideration. Two-dimensional plane-strain calculations are straightforward enough, relatively inexpensive and provide some interesting information. At sufficiently early times, they can even be quantitatively correct. It must be recognized however that when oblique impact of an ogival projectile is treated as the impact of an infinitely long wedge (Figure 6) important physical phenomena are being neglected not the least of which are the out-of-plane motions leading to lateral stress relaxations. Useful qualitative information about the early stages of an oblique impact can be obtained from plane-strain solutions. Their utility degrades with increasing time after impact, however, so that for late times, when important aspects of penetrator and target response are being determined, plane-strain solutions can be speculative at best. The utility of such calculations for high-velocity impacts has been examined by Zukas et al. (1979).

As an example, consider the deformation field around an ogival (frictionless) penetrator. In the plane-strain approximation, this would be modeled as an

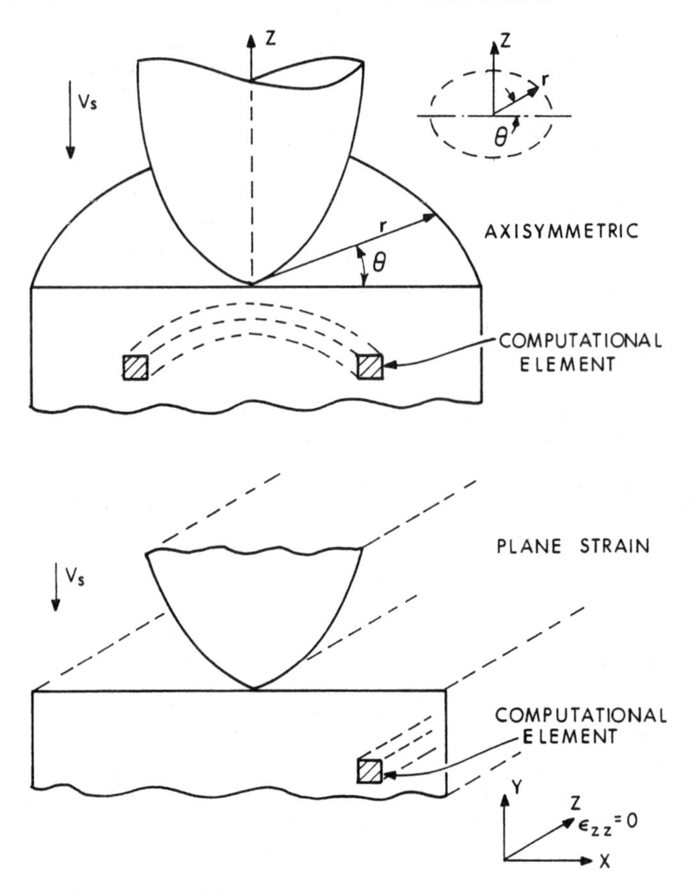

Figure 6 Computational elements for axisymmetric and plane-strain calculations.

infinitely long wedge. A deformation front (the boundary between elastic regions and regions of permanent plastic deformation) which may or may not be attached to the apex of the wedge, moves with the wedge as it penetrates. As this front moves across an element of rectangular cross section running parallel to the wedge, it will distort the cross section in shear and start it translating laterally, possibly with a small vertical motion. Sectional distortion will continue until the sides of the element have been made parallel to the sides of the wedge, at which point penetration would proceed without any further distortion of that element. It should be subjected only to translation as there is now no geometric requirement for further distortion, that is, steady-state deformation will always occur in the mode requiring the smallest energy input.

Now consider the deformation surrounding a frictionless ogive. Again the deformation front moves with the ogive. However, the element under study is now a toroid of rectangular cross section that is concentric with the ogive. When the front passes the element, the cross section is distorted and the element is translated radially. Because of this radial displacement, the cross-sectional area of the element must be reduced to satisfy continuity. Thus, so long as penetration continues, that element will be both displaced and distorted. This is a fundamental difference between the plane-strain approximation and the exact (axisymmetric) computation of penetration and is sufficient reason to expect different energy-displacement relationships for the two modes. The requirement for continuing distortion of all material within the plastic-elastic boundary clearly suggests that axisymmetric penetration should be a higher energy deformation mode than plane-strain (wedge-like) penetration.

Figures 7 and 8 show the comparison between plane-strain computational results and radiographs obtained at the Phermex facility of Los Alamos at 12

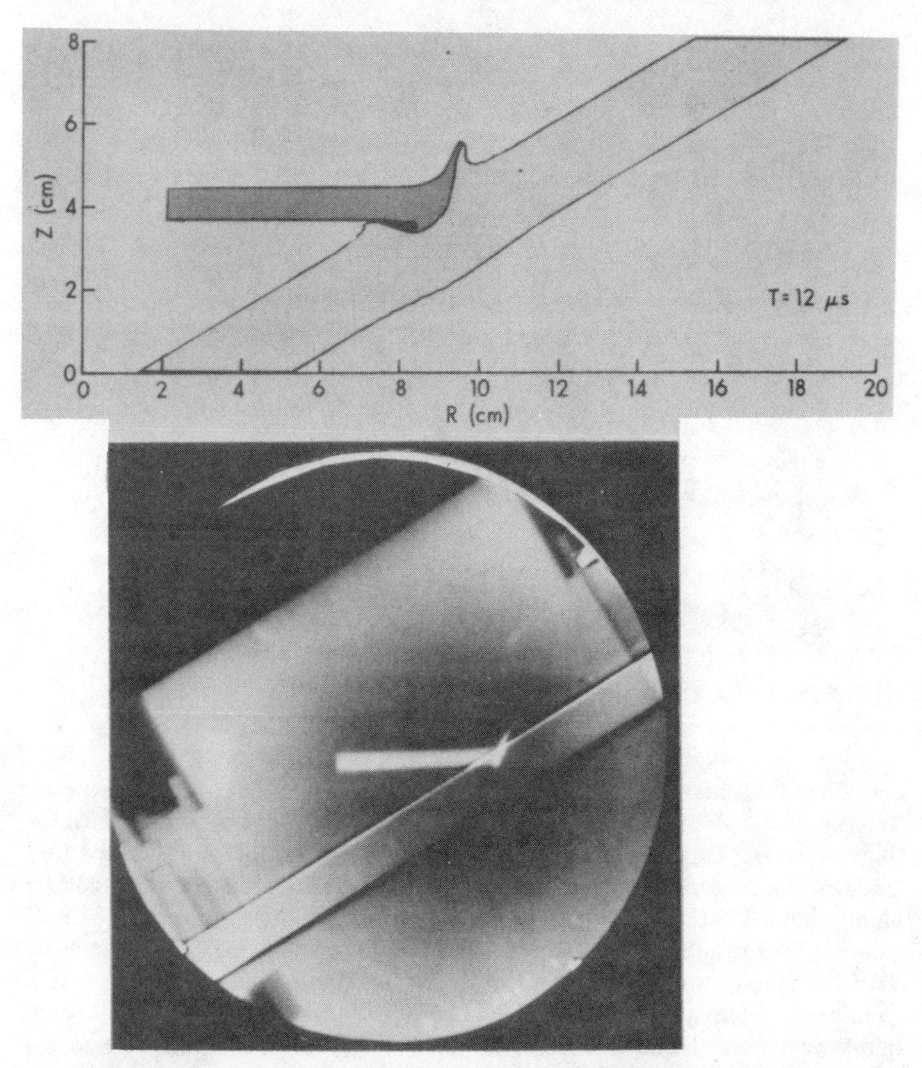

Figure 7 Plane-strain and experimental results at 12 μs [Jonas and Zukas (1978)].

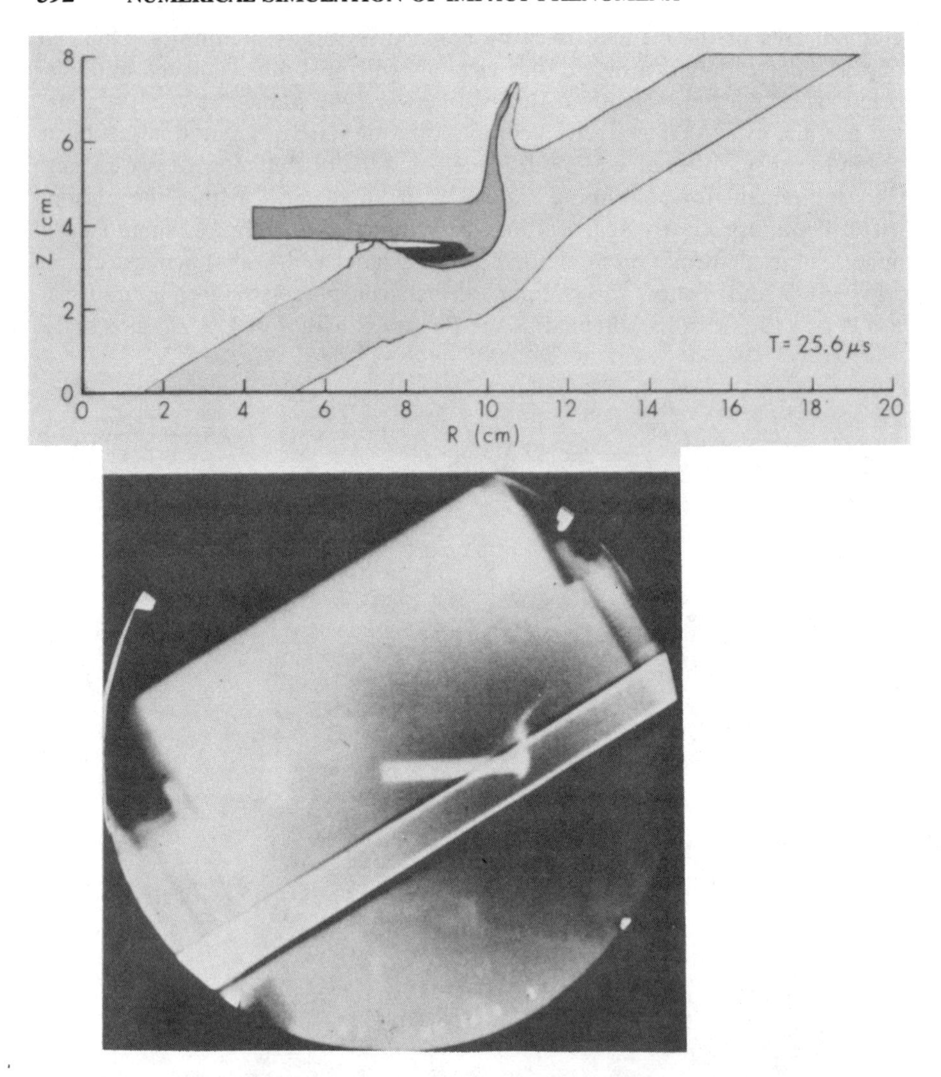

Figure 8 Plane-strain and experimental results at 25.6 μs [Jonas and Zukas (1978)].

and 25.6 μs. (Phermex is essentially a 6 MeV x-ray source capable of shining through 6 or more in. of steel. It is invaluable for penetration studies since for the first time it is possible to obtain information about penetrator deformation and orientation *within* the target rather than having only initial and post-mortem data obtained with standard 150–300 KeV x-ray facilities.) Figures 9 and 10 show EPIC3 results for the same situation. On the whole, agreement between both computations and experiment is remarkably good. The plane-strain computation was performed on a UNIVAC 1108 computer with some 4300 mesh points and required 5 hours of computing time. The three-dimensional

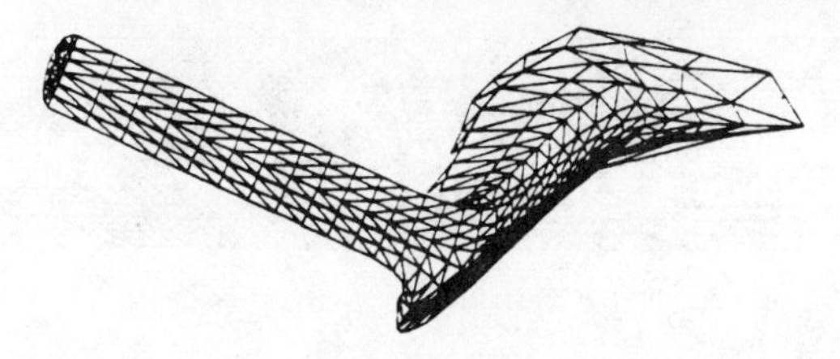

Figure 9 EPIC-3 results for rod deformation at 25.6 μs [Jonas and Zukas (1978)].

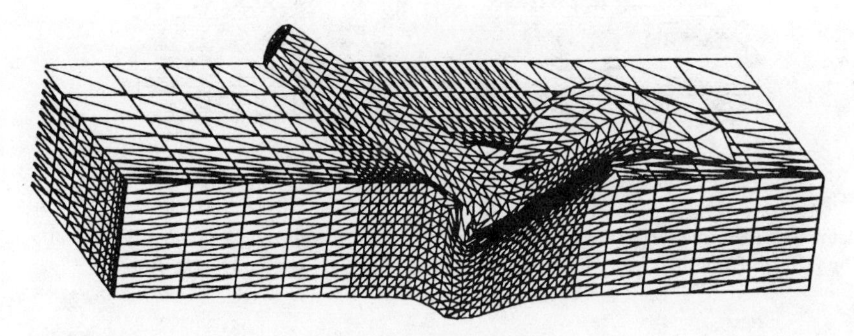

Figure 10 EPIC-3 predicted target and rod deformation at 25.6 μs [Jonas and Zukas (1978)].

calculation used 4900 nodal points and some 23,000 elements. It was performed on a CDC 7600 and required some 7 hours.

10.7.3 Sphere Ricochet

In certain impact situations it becomes necessary to account for the effects of friction on the motion of colliding bodies. Figures 11 and 12 show isometric plots of EPIC-3 results for the case of an SAE 52100 steel sphere, 0.635 cm in diameter, striking a 2024-T3510 aluminum plate, 0.635 cm thick at an obliquity of 60° from the plate normal at a velocity of 720 m/s. Experimental data for this problem is given by Backman and Finnegan (1976). The problem was first addressed computationally by G. R. Johnson (1977) with the early version of EPIC-3 using a frictionless sliding interface. Despite a very coarse grid, good agreement was obtained for the deformed profile of the target plate. The present calculation was done with a finer grid (5202 nodes and 24576 elements) with the current version of EPIC-3. A computation was also made with a grid consisting of 14768 nodes and 75168 elements, but results did not differ appreciably from the identical case with the 24576-element configuration.

The variation of projectile velocity with time and sliding friction coefficient is shown in Figure 13. For the frictionless interface, the computed residual velocity differs by some 38% from the value of 303 m/s reported by Backman and Finnegan. Progressive increase in the surface friction coefficient reduces the residual velocity of the sphere (and imparts a correspondingly greater spin

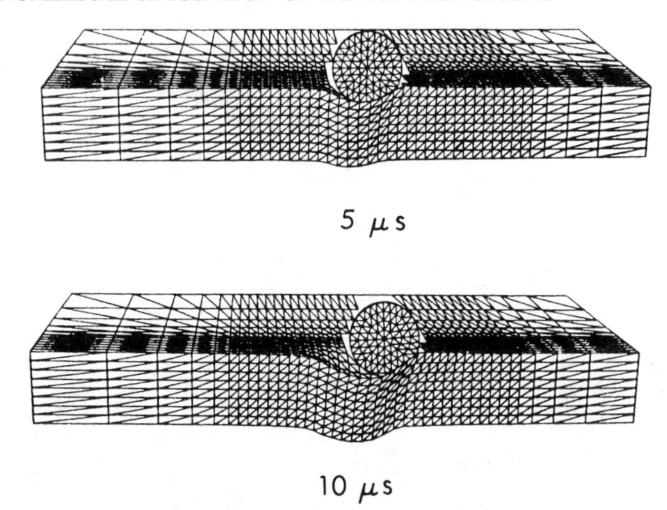

Figure 11 Deformation profiles for oblique impact of steel sphere into aluminum target at 5 and 10 μs [Zukas (1980)].

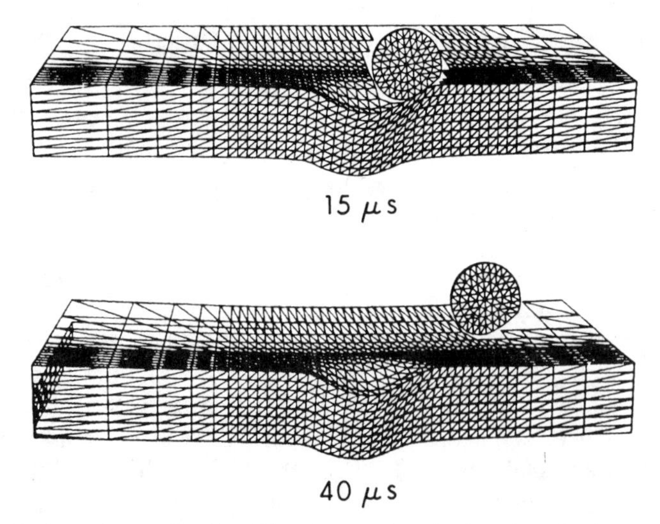

Figure 12 Deformation profiles for oblique impact of steel sphere into aluminum target at 15 and 40 μs [Zukas (1980)].

to it). At 25% friction, the computed residual velocity of 318 m/s differs by less than 5% from that reported experimentally. The deformation pattern in the target is little affected by the frictional interface and is substantially the same for all cases.

In actuality, surface-friction effects would be a function of the relative velocity of the impacting bodies. It is unlikely to be significant at ordnance-

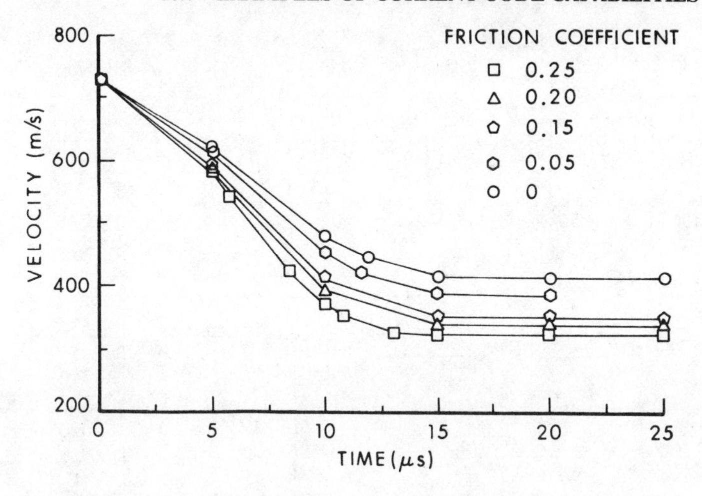

Figure 13 Velocity histories for steel-sphere impact [Zukas (1980)].

velocity impacts where very high pressures are generated at the impact interface producing a thin layer of material that acts as a fluid. At lower velocities and for drawing and punching problems frictional effects can be significant. Estimates of frictional coefficients are not easily obtained, however. In a recent paper, Ghosh (1977) determined the coefficient of friction for sheets of various materials being struck by a hemispherical steel punch. Average values of the coefficient of friction were calculated for a number of materials under various conditions of lubrication. The aluminum alloys considered were 2036-T4, 3003-0 and 5182-0 aluminum. The calculated coefficient of friction for these alloys ranged from 0.22 to 0.41 for the dry state and from 0.07 to 0.27 for test pieces coated with Teflon and polyethylene. With these results in mind and the close agreement with experiment for projectile-residual velocity and target deformed shape, it is reasonable to conclude that EPIC-3 results are a quite reasonable approximation of the principal features of the impact event.

All the computations were done on a CDC 7600 computer. Typical running times for all but the fine grid problem were just under 2 hours.

10.7.4 Spaced-Plate Impact

Figure 14 illustrates the capability of the HULL code to treat the computationally difficult problem of spaced-plate perforation. The calculation involves the impact of a copper rod, 13 cm long and 2.54 cm in diameter into a target consisting of 1 cm rolled-homogeneous armor (RHA) a 100-cm air gap and a second RHA plate 10 cm thick at a velocity of 3 km/s. Figure 14 is a tracing of density contours from more detailed plots furnished by Ms. C. Westmoreland, formerly of the Air Force Armament Laboratory. The calcula-

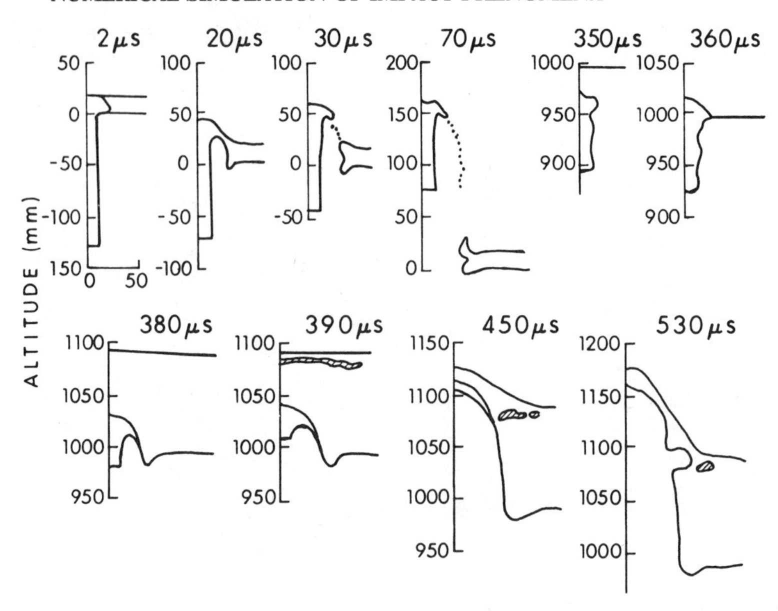

Figure 14 HULL code results of spaced-plate calculation.

tion was done by D. Matuska while with AFATL. The rod defeats the first target plate by about 20 μs. The plate actually separates in the calculation allowing the rod to travel through the air gap. Because of the energy imparted to the rod by this impact with the first plate, it continues to shorten while traveling through the air, shedding material from the nose (impact end) until 160 μs. The rod in effect runs into itself as the rear of the rod overtakes the more slowly moving front section. Some 5 cm of length are lost due to impact with the first plate. Velocity equilibrium is achieved at 160 μs and the rod moves in rigid body fashion until 360 μs at which time it impacts the second plate. By 390 μs, voids are formed across a large plane near the back surface due to reflection of the compressive wave. By 450 μs, the penetrator has been completely consumed and simply lines the inside of the impact crater. Continued liner motion and target momentum have closed most of the void plane. However, a small section remains and provides a stress concentration point for tearing material from the target plate at 530 μs, The calculation correctly models phenomena observed experimentally and illustrates the capability of a relatively simple failure criterion and post-failure model to simulate real events.

Other spaced-target calculations have been done by Bertholf et al. (1977, 1979) who used the TOODY and CSQ codes in the plane-strain modes to compute various aspects of long rod penetration of spaced steel plates at obliquity.

10.7.5 Hydrodynamic Ram

Hydrodynamic ram refers to the high pressures that are developed when a fluid reservoir is penetrated by a kinetic energy (*KE*) projectile. Hydrodynamic ram in aircraft fuel cells can damage structural components or rupture tank walls which in turn can lead to fuel starvation, fire, and explosion.

The hydrodynamic-ram event is generally considered to consist of a shock phase, a drag phase, a cavitation phase, and an exit phase. The shock phase occurs during initial impact with the fluid at which time the projectile impulsively accelerates the fluid and generates an intense pressure field bounded by a hemispherical shock wave. This shock wave expands radially away from the impact point and may produce petaling of the entrance panel. As the projectile traverses the fluid it transfers a portion of its momentum to the fluid as it is decelerated due to viscous drag. If the projectile tumbles in the fluid, a significantly larger portion of the projectile's momentum will be transferred to the fluid. The radial velocities imparted to the fluid during the drag phase lead to the formation of a cavity behind the penetrator. This is often termed the cavitation phase. As the fluid seeks to regain its undisturbed condition, the cavity will oscillate. The time interval during which the exit panel of the fluid cell is perforated by the *KE* projectile is referred to as the exit phase.

Kimsey (1980) has employed the EPIC-2 code to study the impact of an S7 steel rod into a cylindrical fuel-cell simulator. The rod was a right-circular cylinder with a hemispherical nose, weighed 50 g, had a length-to-diameter ratio of 3 and was assumed to strike the fluid-filled container normally at a velocity of 909 m/s. The outer diameter of the container was taken to be 50.8 cm, its depth was 15.2 cm and the wall thickness (2024-T3 aluminum) was 1.8 mm. The fluid was simulated as water. Sliding was permitted between the projectile and the water as well as between the water and the interior portions of the entrance and exit panels of the simulator. 2943 nodes and 5424 triangular elements were used for the simulation.

Deformation and pressure profiles at 40 and 180 μs after impact are shown in Figure 15. The impulsive acceleration of the fluid during the shock phase (≈ 10 μs after impact) generates peak pressures of ≈ 280 MPa, which decay rapidly to ≈ 14 MPa and persist at about that level during the drag phase. Petaling of the entrance panel due to the action of the water is also evident. By 180 μs, the exit panel has been sufficiently loaded to initiate bulging prior to perforation. The entrance panel has been deflected considerably and an additional cavity between the entrance panel and the water has formed. The cavity formed behind the water was formed by permitting total failure of elements that exceed an equivalent strain of 2.5.

No experimental data for this case was found. However, Kimsey was able to compare the predicted residual velocity from EPIC-2 with that determined from an empirical relationship derived on the basis of a number of hydrodynamic-ram experiments. The two values differed by 4%.

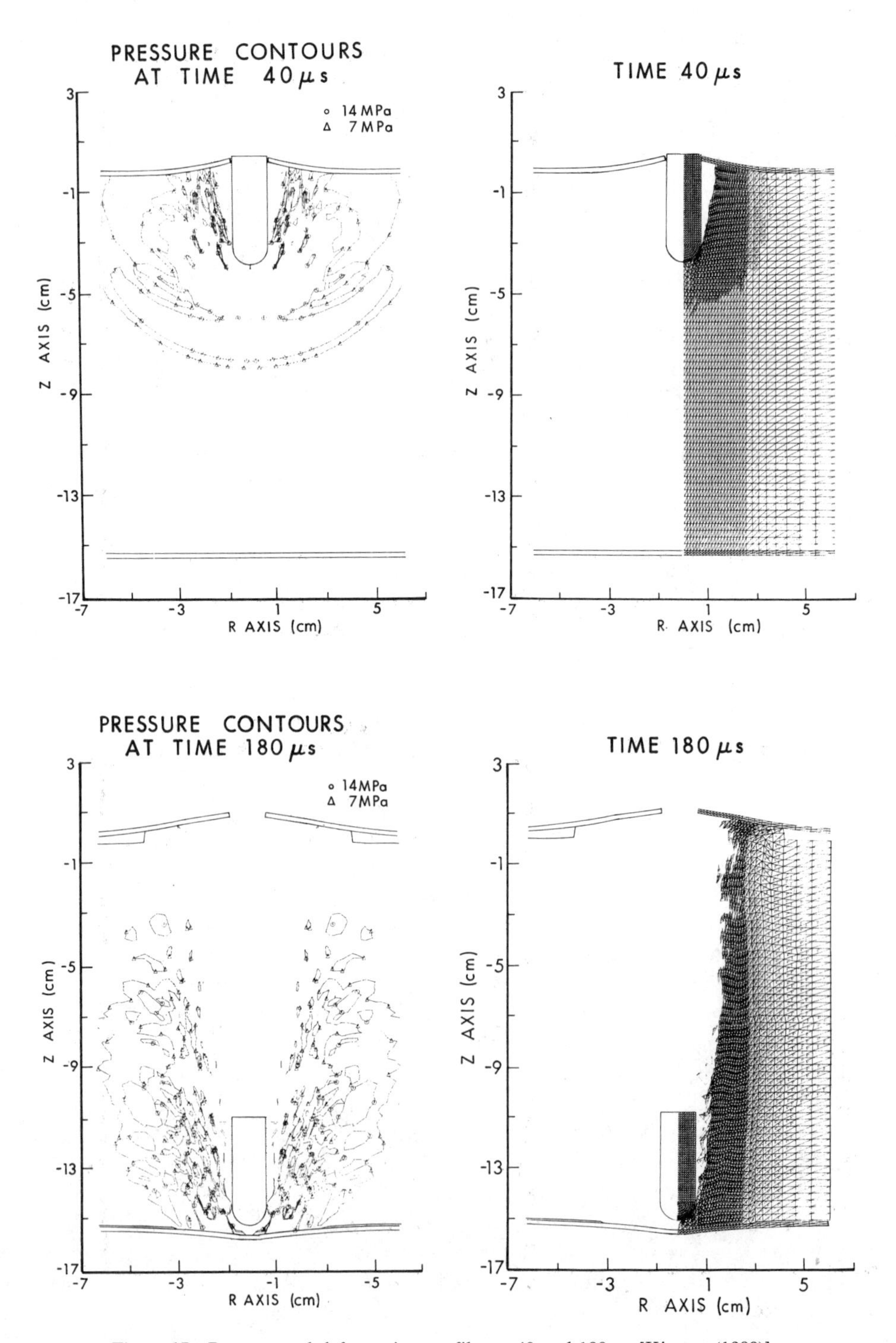

Figure 15 Pressure and deformation profiles at 40 and 180 μs [Kimsey (1980)].

10.7.6 Impact Involving Anisotropic Media

Walsh and Sedgwick (1974) and Sedgwick et al. (1976) studied normal-incidence hypervelocity impacts into isotropic and anisotropic materials. Their studies were intended to assist in the development of erosion-resistant missile and space-vehicle materials by determining the effects of various material parameters on mass loss and material damage. Also studied were the effects of prestressing and porosity on the target materials.

The target materials considered were copper impregnated tungsten (Cu-W), 6061 aluminum, ATJ-S graphite, and MAR-M-200 steel. The projectiles were mostly spheres of glass, ice and water, although computations were also made with the Air Force Standard Rock and rods. Striking velocities were 1.524, 3.048, and 6.096 km/s.

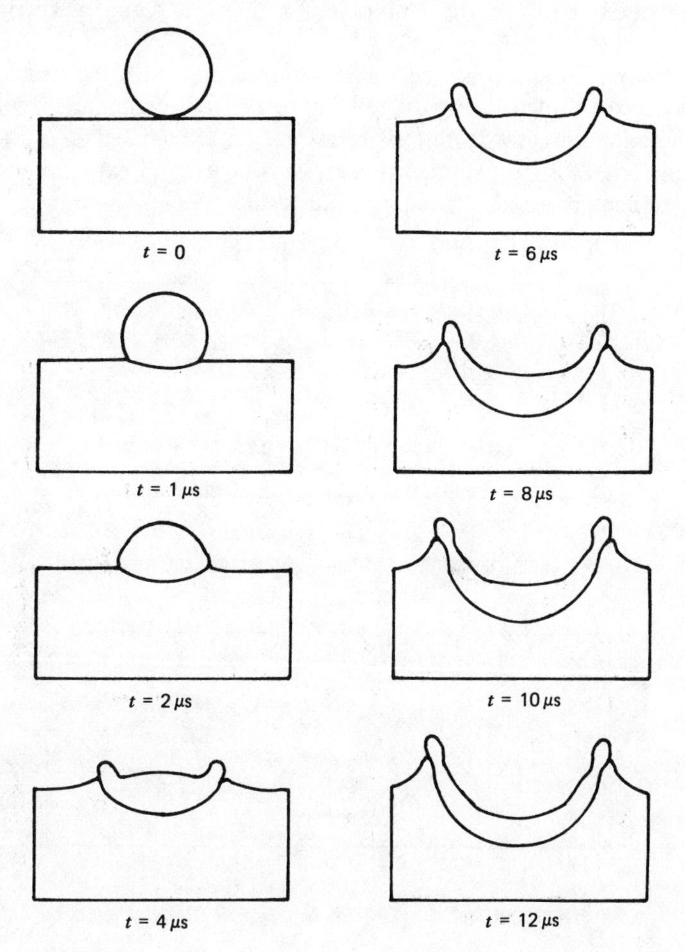

Figure 16 Projectile-target deformation for glass-Cu-W impact at 6.096 km/s [Sedgwick et al. (1976)].

The parametric study with HELP predicted greatest improvements in erosion resistance for increases in failure threshold and for prestressing while little added resistance resulted from either shear modulus or shear yield-strength variation. For some of the impact conditions, experimental data was available. For others, code predictions for penetration depth were compared with an empirical formula of the form

$$\frac{P}{D}=K\left(\frac{\rho_p}{\rho_t}\right)^{\alpha_1}\left(\frac{V}{c_t}\right)^{\alpha_2}\left(\frac{\sigma_{yt}}{\rho_t c_t^2}\right)^{\alpha_3} \tag{16}$$

where $K, \alpha_1, \alpha_2, \alpha_3$ are parameters derived from a fit of the form to experimental data, P was the penetration depth, D projectile diameter, ρ density, V striking velocity, c sound speed, σ_y yield strength, and the subscripts p and t refer to projectile and target, respectively. For striking velocities above 3.048 km/s, agreement between the empirical formula and code predictions was good.

The sequence of events for hypervelocity impact of a spherical striker against a thick target are shown in Figure 16. Figure 17 shows comparison of HELP code-predicted crater profile and experimental data from W. Gray of Martin Marietta in Orlando, FL. Agreement between theory and experiment is quite good. Agreement is good as well for the results shown in Figure 18 for a

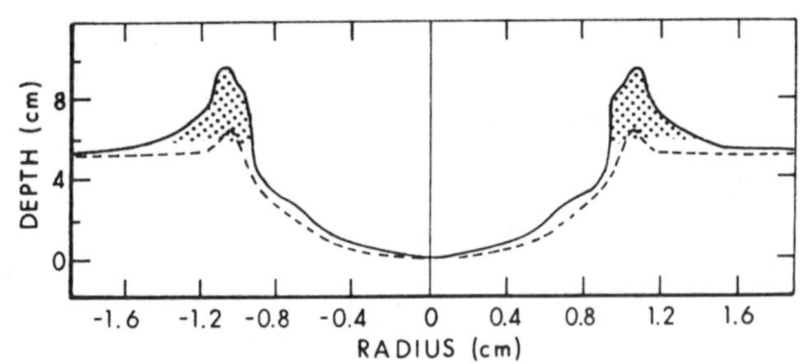

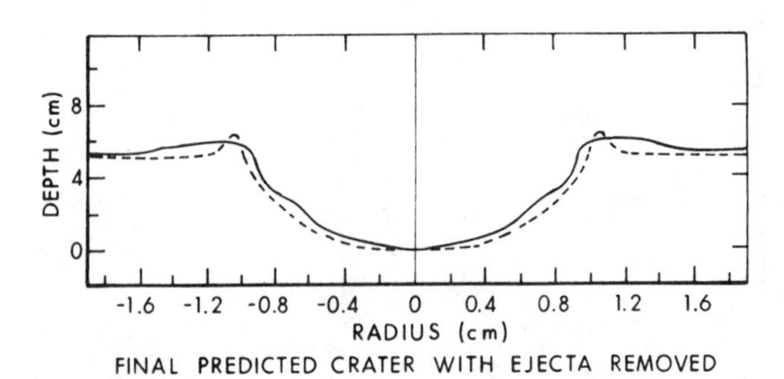

Figure 17 HELP code calculation and experimental results for ice-aluminum impact at 3.048 km/s [Sedgwick et al. (1976)].

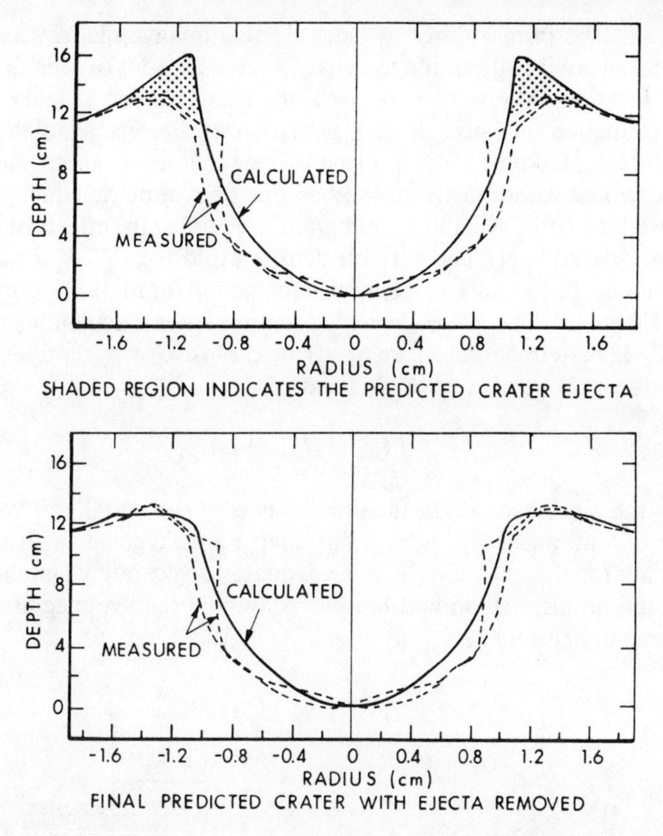

Figure 18 HELP code calculation and experimental results for glass-aluminum impact at 3.048 km/s [Sedgwick et al. (1976)].

glass-aluminum impact. Comparison of predicted crater size and mass loss for ATJ-S graphite impacts with data provided by L. Rubin of Aerospace Corporation were not as good as the above examples, although mass loss predictions fell within the bounds of experimental data. The discrepancy was attributed to inadequate modeling of ATJ-S behavior. In the course of this work, an anisotropic material model was developed and incorporated in HELP. Anisotropic models are also incorporated in TOODY [Swegle and Hicks (1979)] and EPIC 3 [G. R. Johnson et al. (1980)].

10.7.7 Explosive-Metal Interaction

G. R. Johnson (1981) has investigated computationally explosive devices that accelerate a metal liner known as a self-forging fragment. Unlike a conical-shaped charge that forms a high-velocity jet consisting of multiple particles, a self-forging fragment remains essentially intact and eventually forms a relatively rigid body as it travels at a constant velocity after being accelerated by the explosive. The primary uses for such devices involve military and mining applications where it is desirable to deliver a large amount of kinetic energy to a distant location. Many such devices are axisymmetric (Figure 19). The dynamic formation of such devices has been accurately simulated in two dimensions [e.g. Hallquist (1980)]. Little is known however about the effect of three-dimensional variations from the baseline axisymmetric condition. G. R. Johnson used the EPIC-3 computer program to examine the effects of off-center detonation, tapered liners, and variable density explosive.

The dynamic formations of four different configurations (Figure 19) are shown in Figure 20. The finite-element model for these calculations used 1288 nodes and 4392 tetrahedral elements. The explosive was represented by a Gamma Law equation of state of the form

$$P=F(\gamma-1)E_s(1+\mu) \tag{17}$$

where F is the burn fraction, E_s the internal energy per initial unit volume, γ a material constant, and $\mu=\rho/\rho_0-1$. The coarse grid, together with the simple Gamma Law for the explosive, is not adequate to accurately simulate all the details of the problem. It should however show the relative magnitude of the various three-dimensional effects.

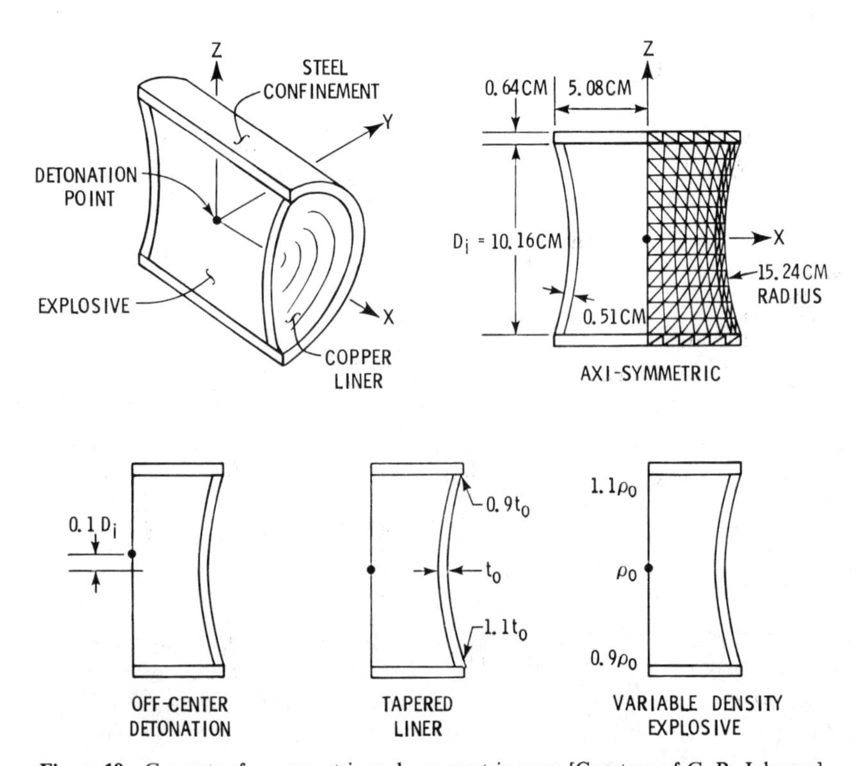

Figure 19 Geometry for symmetric and asymmetric cases [Courtesy of G. R. Johnson].

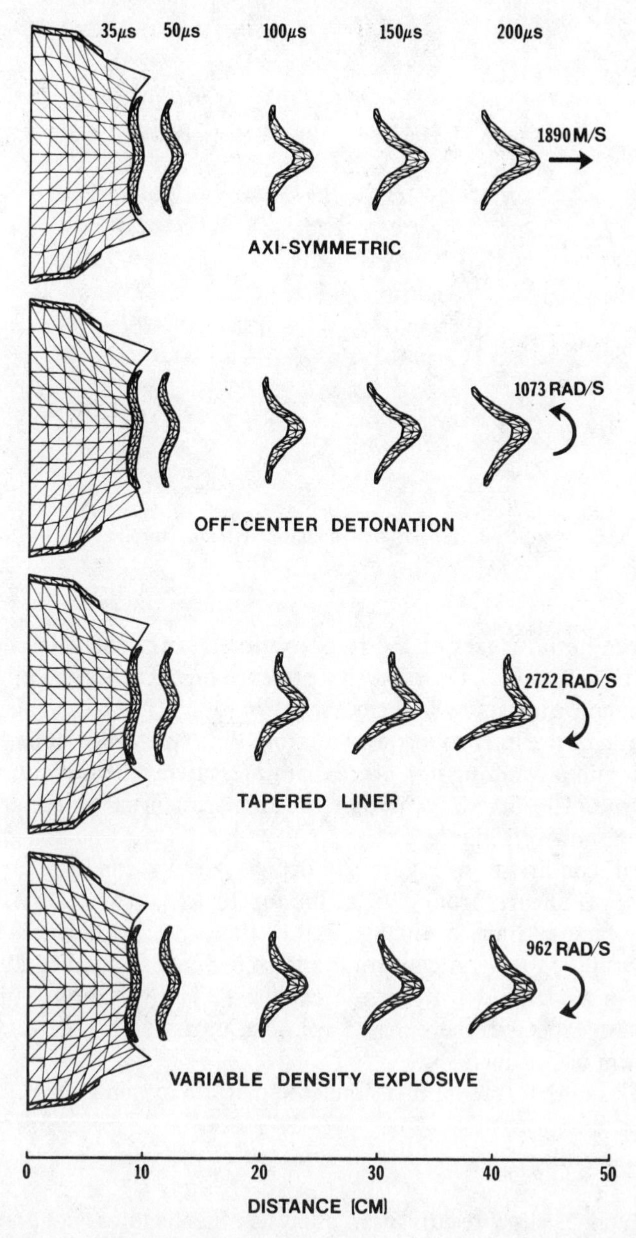

Figure 20 Cross-sectional views of dynamic formations [Courtesy of G. R. Johnson].

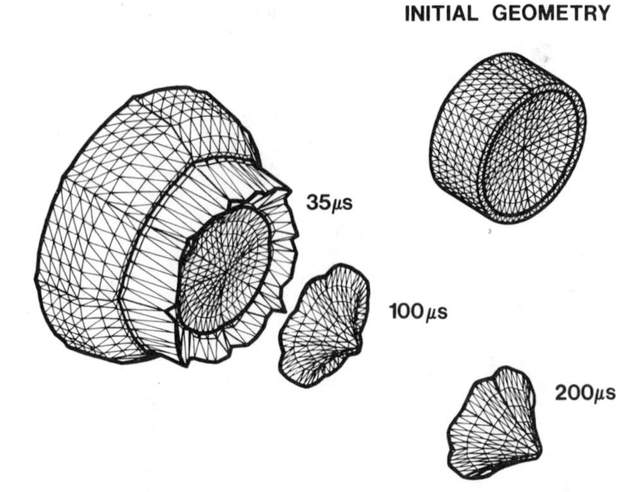

Figure 21 Three-dimensional view of axisymmetric dynamic formation [Courtesy of G. R. Johnson].

During the initial stages of the response for an off-center detonation (1.06 cm above the centerline), the explosive pressure first reaches the upper portion of the liner and imparts to it a clockwise rotational velocity. This causes the upper portion of the liner to move away from the explosive at a faster rate than the lower portion, resulting in a decrease in pressure at the top and an increase at the bottom. The net result is that the final rotational velocity is counterclockwise as shown in Figure 20.

The third condition in Figure 20 depicts results for a liner where the thickness varies linearly from $0.9t_0$ at the top to $1.1t_0$ at the bottom. The total mass of the tapered liner is equal to that of the constant thickness liners. This condition results in a significant rotational velocity. The final condition shows the effects of variable density in the explosive. The variation is indicated in Figure 19. As expected, the denser explosive imparts a higher velocity to the upper portion of the liner.

Figure 21 shows a three-dimensional view of the axisymmetric condition.

10.7.8 Cone Collapse

Figures 22 and 23 show results of an analysis of a steel nose cone by Hallquist (1979b). The nose cone was designed to limit the resultant force transmitted to the aft section on impact. The computation was performed with the DYNA3D finite-element computer code. The mass of the aft section was simulated by using a material with artificially high density (131,477 gm/cm^3) in the top rows of elements.

Figure 23 shows deformed shapes at 3000 μs intervals. The peak deformation is reached at 15 ms and the nose cone begins to rebound. Computational

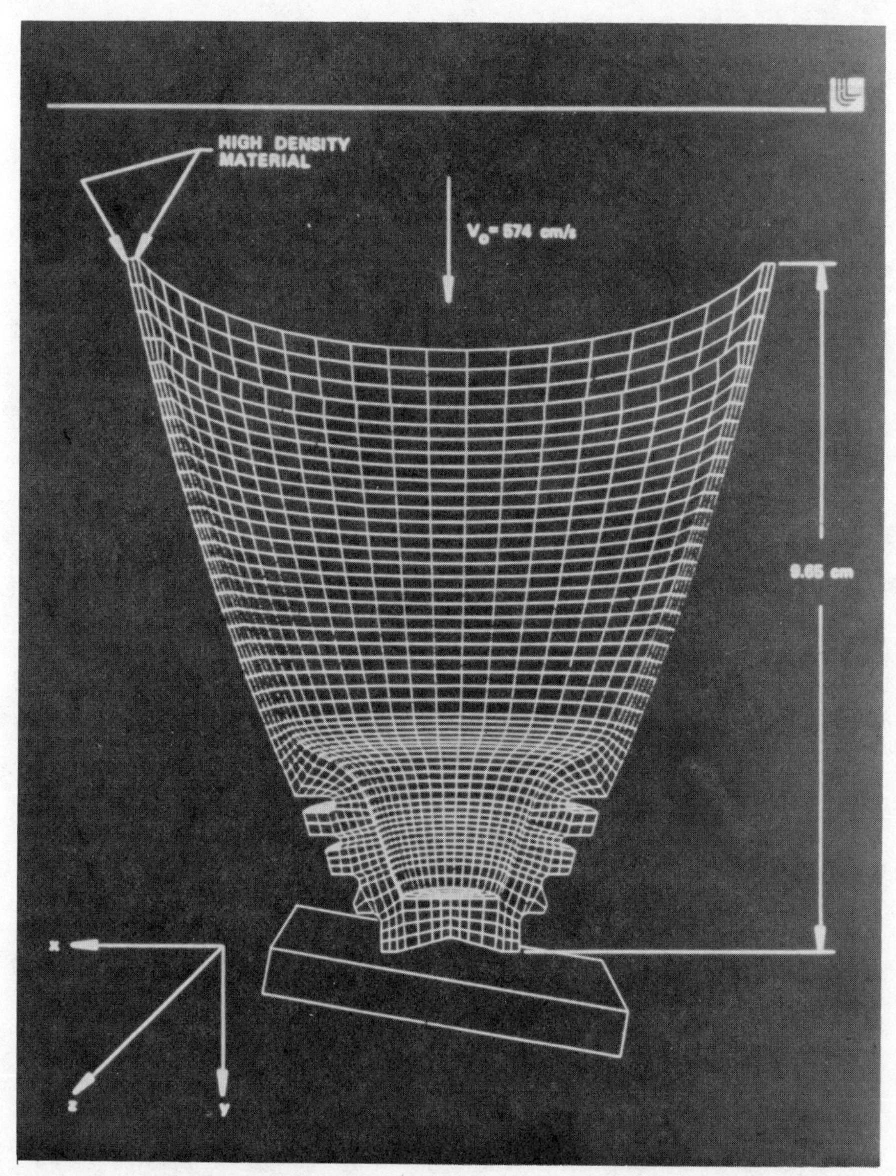

Figure 22 Initial conditions for nose-cone impact [Hallquist (1979)].

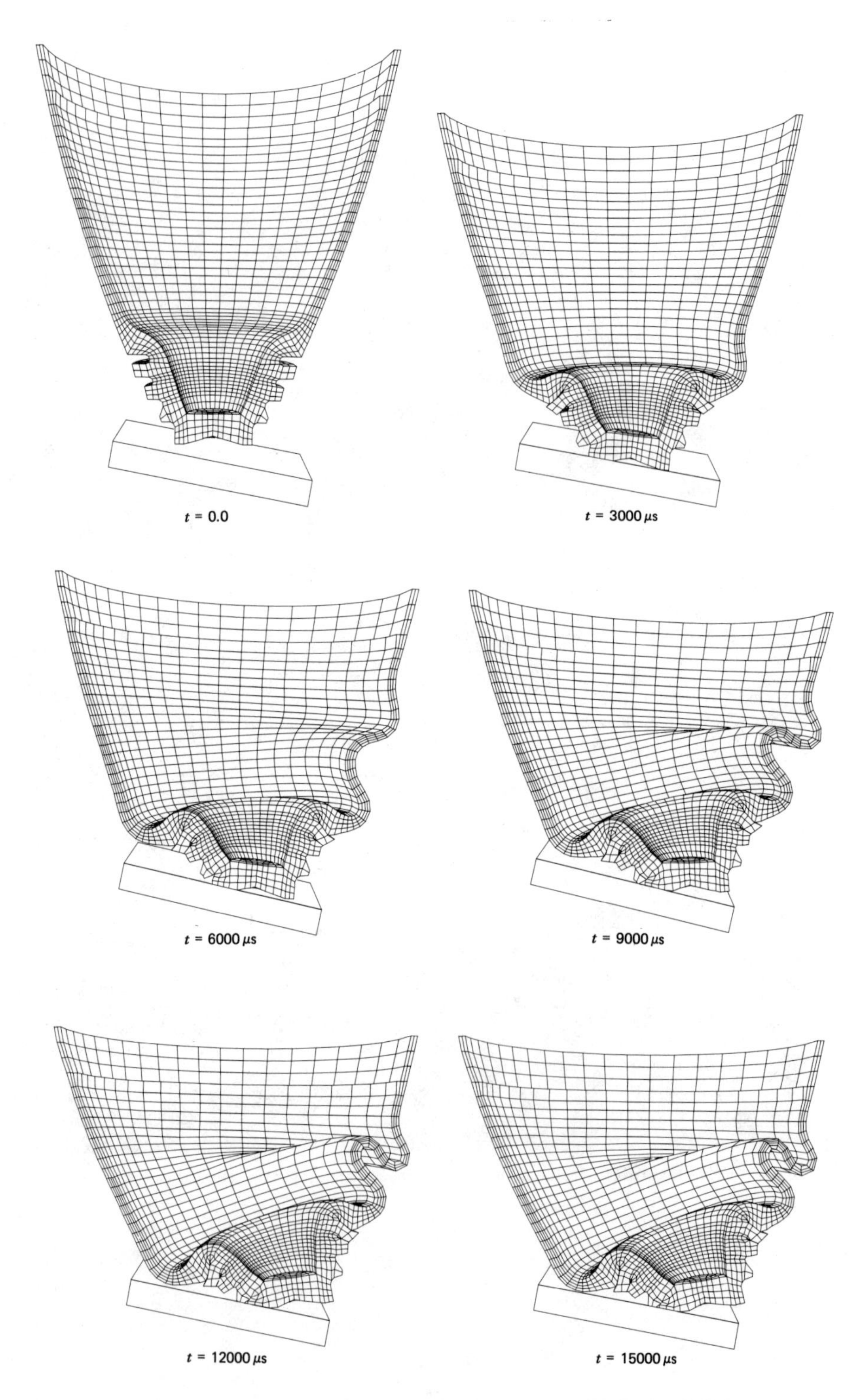

Figure 23 Deformation history for nose-cone impact [Hallquist (1979)].

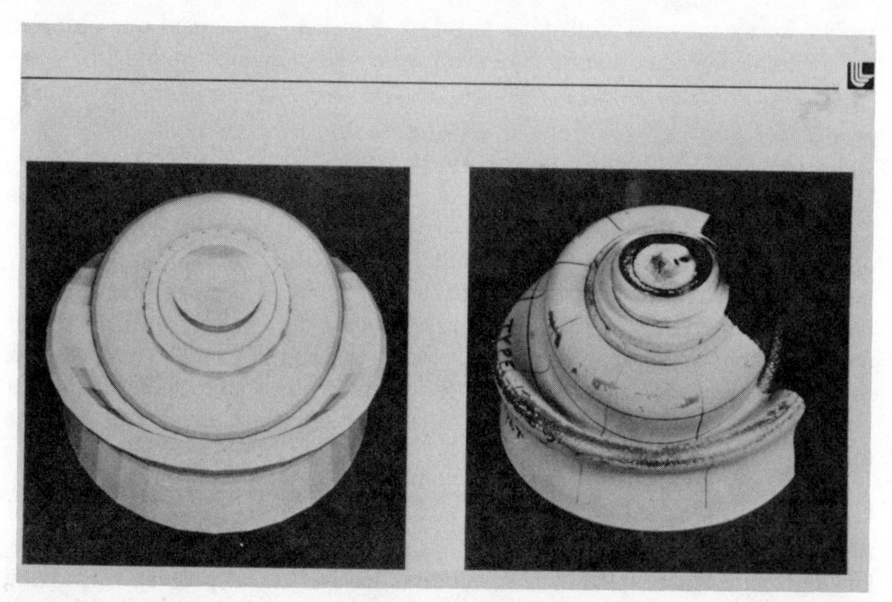

Figure 24 Computed (left) and recovered (right) deformed cones [Courtesy of J. O. Hallquist].

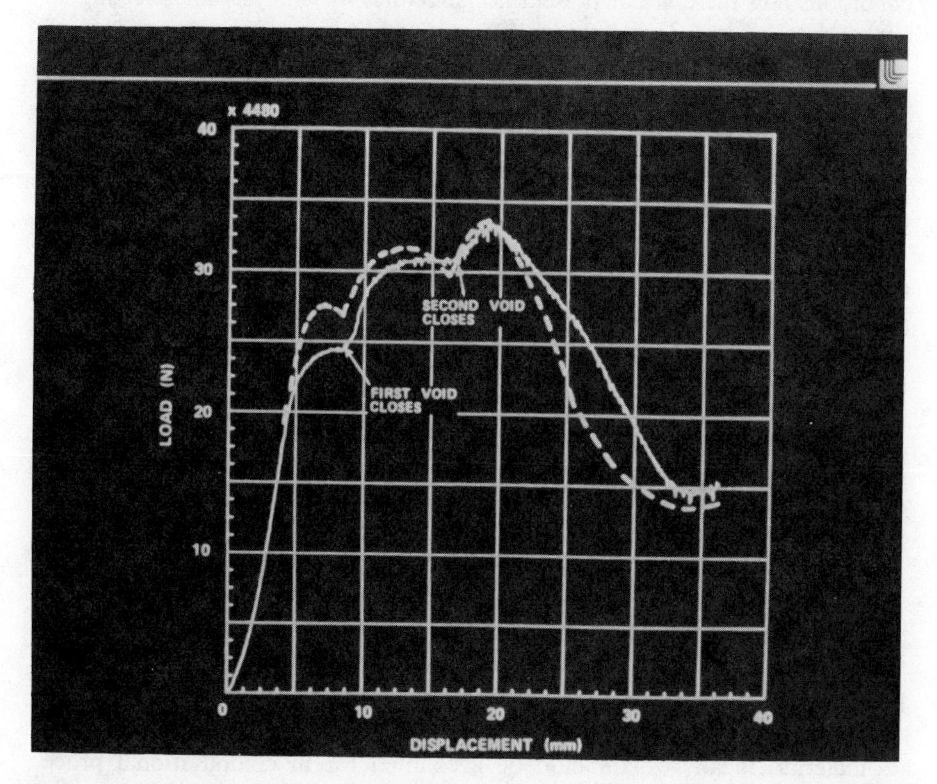

Figure 25 Computed (dashed) and experimental (solid) load-displacement curves [Hallquist (1979)].

results were compared with static test data. Excellent agreement was obtained for the final shape and the force-deflection behavior (Figures 24 and 25).

10.8 COMPUTATIONAL FAILURE MODELS

The most serious limitation to extensive use of computational techniques arises not from their cost or complexity (ballistic experiments are neither less costly nor less complex) but from the inadequacy of the models describing material failure. Under dynamic loading, failure can occur by a variety of mechanisms dependent on the material constitution and the state of stress, temperature, rate of loading, and a number of other variables. The methods and results of quasistatic fracture studies are of little use in situations involving high-rate loading.

Simple, empirical failure models of varying degrees of complexity exist. Some have been applied successfully in high-velocity impact calculations. For the most part, though, failure criteria and models are of an ad-hoc nature, lacking a micromechanical basis to comprehensively treat problems involving brittle, ductile, and shear failure. Different criteria apply for different impact conditions and there are at present no guidelines for analysts for selecting an appropriate failure criterion under different conditions.

Computational failure models for impact loading situations are discussed in a review article by Seaman (1975) and in some depth in the report by the National Materials Advisory Board Committee on Materials Response to Ultra-High Loading Rates (1980). Hence, the following remarks are intended to summarize the current state of affairs in numerically simulating material failure. The cited references should be consulted for additional details.

It is now generally accepted that material failure under impact loading is a time-dependent phenomenon. Experimental observations lead to the following conclusions about the general features of materials failure [Seaman (1975)]:

1 A range of damage is possible; there is no instantaneous jump from undamaged to fully separated material.
2 Damage grows as a function of time and the applied stress. Hence, a single field variable of continuum mechanics (stress, strain, plastic work, etc.) at any time cannot be expected to characterize the dynamic fracture process. At least some time-integral quantity must be used to represent the dynamic strength. Of course, if the intensity of the applied load is so severe that the damage occurs in negligibly short time, reasonable results may be expected from instantaneous failure criteria.
3 As the level of damage increases, the material is weakened and its stiffness reduced. This changes the character of the wave propagation through the material. Ideally, this should be accounted for in computational procedures.

4 Even incipient damage levels are important since voids or cracks, difficult though they may be to observe, may seriously weaken a structure.

There are a number of failure descriptions used in current production codes. They involve an initiation criterion and some description of the post-failure behavior.

The simplest of the initiation criteria are based on instantaneous values of a field variable, such as pressure, stress, strain, plastic work, or some combination thereof. Once the criterion has been satisfied at a given location, failure is considered to occur instantaneously. The post-failure response can be described in a number of ways. The failed material may be entirely removed from the computation or be described by a modified constitutive function that describes a weakened material.

One of the simplest of such models, for example, is the pressure cutoff. When the hydrostatic pressure reaches a critical tensile (or zero value), failure is assumed to occur instantaneously. Further expansion occurs at that fixed value of pressure. If subsequently recompression occurs, compressive pressures are again allowed. More complicated versions of this basic approach have been implemented. Excellent results have been obtained with critical stress criteria for hypervelocity-impact situations where dynamic tensile-spall situations exist [e.g. Bertholf et al. (1975)].

Time-dependent initiation criteria represent the next level of sophistication and have been successfully applied in several situations. One of the earliest is attributed to Tuler and Butcher (1968). Failure is assumed to occur instantaneously when a critical value of the damage K, defined by the integral

$$K = \int_0^t (\sigma - \sigma_0)^\lambda \, dt \tag{18}$$

is reached. Here $\sigma(t)$ is a tensile stress pulse of arbitrary shape, σ_0 a threshold stress level below which no significant damage will occur regardless of stress duration, and λ is considered a material-dependent parameter chosen to fit experimental data. The parameters λ, σ_0, and K_{critical} are taken to be material parameters that must be determined from separate experiments.

Criteria in which the damage accumulation is a function of the extent of damage as well as field variables have also been devised [Davison et al. (1977)]. Here the damage accumulation function is taken to be a function of strain, temperature, and the current damage level

$$K = f(\varepsilon, T, K) \tag{19}$$

and the post-failure description includes progressive weakening of the material as the damage increases.

In an attempt to include micromechanical behavior in a continuum-damage model, researchers at SRI International [Seaman and Shockey (1975), Seaman et al. (1976), Erlich et al. (1980)] have developed models for ductile, brittle, and shear failure. For ductile failure damage is initiated when the average stress exceeds a tensile-pressure criterion. Brittle fracture is initiated when the maximum normal stress exceeds a tensile threshold. Shear banding begins when the maximum plastic-shear strain exceeds a critical value. After initiation, voids, cracks, or shear bands nucleate and grow according to experimentally determined rate equations. These equations are determined from experiments in which samples are exposed to pulse loads of varying amplitudes and durations. The samples are then sectioned and examined metallographically to measure the microcrack, void, or shear band size and orientation distributions as a function of the imposed stress and strain histories. Calculations are made by storing the crack, void, or shear band size and orientation distributions for each location. These distributions serve as internal-state variables. As the damage accumulates, the stresses are relaxed by amounts and in directions governed by the damage distribution functions. The brittle fracture and shear-banding models have been extended to predict fragment-size distributions on complete failure, based on very simple descriptions of the coalescence of individual flaws.

It is clearly established that the material descriptions that most affect computational results are the flow stress and the failure model. This latter aspect is especially evident in the computations performed by Gupta and Misey (1980). They used the EPIC-2 finite-element Lagrangian code and the HELP finite-difference Eulerian code to study surface strains in long rods during penetration. Computational results with EPIC-2 in which failure was not allowed to occur showed little resemblance to experimentally determined strain-time records. When a failure criterion based on effective plastic strain was included, good agreement was obtained between computed and experimentally determined strain-time records. Computations with the HELP code used a failure criterion based on a minimum allowable density ratio. In general, better agreement with experimental results was obtained with the finite-element code using an effective plastic-strain failure criterion than with the finite-difference code with failure based on a density ratio.

For lack of any substantive theoretical guidance, most computations are performed with the simplest available failure criteria. These generally require the least amount of material characterization and are easily implemented in computer codes. Too often, the failure criterion is used as an adjustable parameter to bring into coincidence computational and experimental results, with the ensuing argument that the close agreement is proof of the applicability of the particular model and data chosen. Such arguments can and should be dismissed out of hand. This is most easily done for situations where the failure model has no micromechanical basis. But in general the tendency to treat material descriptors as adjustable parameters is to be discouraged.

Problems of accounting for dynamic material behavior and fracture are difficult but not insurmountable, even in view of our present limited knowledge. A practical approach is suggested in the NMAB report. It is suggested that the iterative procedure of successive refinements involving computations with existing relatively simple failure descriptions, dynamic materials characterization employing relatively simple and standardized techniques such as those described in the preceding section, and ballistic-test firings may produce useful results for design purposes in many applications. The report suggests that "...rough computations, using simple material models with published or even estimated material properties, may be used in conjunction with exploratory test firings to scope an initial design. Comparison of test data with the predictions may reveal discrepancies which suggest refinements in the computations or material models, and the need for some dynamic material property measurements. Once reasonable agreement has been achieved, another round of computations may then be performed to refine the design. Test firings of this design might use more detailed diagnostic instrumentation. This sequence is iterated, including successively more detail in computational models, material property tests and ordnance test firings, until a satisfactory design is achieved. In this procedure, unnecessarily detailed computations, material property studies or test firings are minimized; only the details necessary to achieve a satisfactory design are included."

The ultimate solution to the problem requires development of micromechanical failure theories through theoretical and experimental research. It is not likely that this will occur even within the next decade. In the interim, the iterative method appears to be a practical approach for rational and cost-effective design of materials and systems that must withstand high-intensity impact loading.

10.9 PROGNOSIS

A summary of the state of events is in order before hazarding a few opinions on developments that may be anticipated in the near future.

1 The study of impact processes encompasses a wide range of materials responses that cross traditional academic boundaries. Because of the complexity of the subject, no comprehensive solution to impact problems in all velocity regimes exists. Analytical models have very restricted application because of the simplifying assumptions employed in their derivation. The bulk of the work in this area is experimental in nature.

2 Many of the high-velocity tests that are performed have specific objectives in mind. Data is gathered to satisfy those objectives. Because of this limited scope, such experiments add little to our knowledge of impact processes. The information collected from routine ballistic tests concerns the initial and final states of striker and target (too often, the latter information is obtained from post-mortem examinations). Time and cost constraints do not generally permit acquisition of a data base to permit either development of unambiguous approximate analytical models or confirmation or rejection of the various material descriptions used in present computer codes. Techniques such as instrumented impact tests and cineradiography

provide time-resolved information on stress, strain, and deformation states in colliding materials that is of great value in assessing mechanisms governing the impact process and developing models. At present, though, these methods are still in their infancy and quite expensive.

3 Numerical techniques offer the hope of obtaining a complete solution to impact problems. Computational techniques have advanced to the point where extremely difficult situations can be analyzed quickly and cost-effectively— typically— one-dimensional problems in tens of minutes to a few hours, and three-dimensional problems in a few hours to tens of hours.

4 Within broad limits, the geometry of a given problem poses no barrier to computations. In situations where material failure is not a major problem, computations with simple material models for elastic-plastic behavior produce excellent qualitative and quantitative results, especially when care is taken to characterize the materials involved under dynamic conditions appropriate to the problem. A number of standard methods are available for such characterization and a small but growing data base for the behavior of technologically significant alloys at high strain rates already exists.

5 Hypervelocity-impact situations readily lend themselves to computations. Here, the principal factor in characterizing material behavior is the equation of state. Excellent work in this area has been done over the past three decades and accurate equation of state data now exists for a wide variety of materials and loading conditions. Good materia characterization here leads to good qualitative and quantitative results.

6 The description of dynamic material behavior, *especially material failure under high loading rates*, remains the greatest single limitation on the accuracy and utility of computer codes for solid-solid impact situations.

7 Despite lingering uncertainties about the short-duration loading response and failure characteristics of materials, practical solutions of an ad hoc nature for difficult problems involving material failure can be obtained by the iterative use of computations, dynamic material characterization and ballistic testing.

In short then, a clever analyst with access to a large computer, dynamic materials data, and a ballistic-test facility can obtain an engineering solution to almost any problem involving the intense impulsive loading of a material or structure, even through the solution may be of an ad hoc nature. What refinements does the future hold for this state of affairs?

The probable situation with regard to computing is somewhat easier to predict. The trend in computer hardware has been toward increased capability at ever-decreasing cost. Three-dimensional impact simulations involving penetration or perforation of solids can now be run in 4–10 hours of CPU time on a CDC 7600 computer. The CRAY-1 computer, now operational at a number of sites, is estimated to be 7–10 times faster than the CDC 7600 for comparable problems. Further developments of large mainframes should result in even greater increases in speed and memory capacity. Thus, hardware developments alone should bring computing times for three-dimensional penetration problems to under 1 hour of CPU time within the next five years.

In addition, there is considerable activity in developing numerical schemes that are both fast and accurate for problems involving large stress and velocity gradients. Hallquist's work at Lawrence Livermore Laboratory has already been mentioned. Hicks and Madsen (1979) at Sandia Laboratories has developed a two-dimensional, wave-propagation code using shock-fitting techniques. Hicks has also made promising progress in three-dimensional impact calculations. In many impact situations, there is very little activity in a large part of the mesh. Typically, only 5% of the mesh has significant stress or velocity gradients that must be adequately resolved. In explicit schemes, these regions determine the size of time step. This is then used for the entire grid. Using operator-splitting and implicit techniques, Hicks was able to use a big time step in a large portion of the grid. In regions of intense activity, subcycling was employed to pick up the necessary time resolution. These methods are still experimental in nature. However, the running time for a three-dimensional impact problem using the above approach was approximately one-twentieth of that of an an explicit scheme [Herrmann (1977)]. Other methods are under development that would use both implicit and explicit methods. Explicit integration would be used in the early stages of impact where stress or velocity gradients need to be resolved. As the activity then decayed and high-frequency components were no longer significant, implicit schemes with their larger time step would then be used to obtain the overall dynamic response of the structure. The combination of improved hardware and innovative software development should make three-dimensional computations routine within the next decade since their cost will become an insignificant portion of the overall project costs.

Development of software for pre and post-processing of computational results will also be accelerated as CPU costs decline. Development of interactive graphics software for automatic mesh generation and graphical analysis of computational results is an expensive proposition but constitutes a one-time cost, well justified by savings of an analyst's time. Software that places an unnecessary burden on an analyst's time is being gradually replaced. Situations that were common only a few years ago where a week or more was required to prepare input, and several weeks to analyze the output of a 2-hour computer run can no longer be logically or economically justified.

The minicomputer will play a significant role in three-dimensional impact computations. Minicomputers with their associated mass storage and graphics capability can be used as hosts for elaborate pre- and post-processing packages for production codes. Once meshes have been generated interactively, computations may be done in batch on a mainframe and the results returned to the minicomputer for post-processing. Alternatively, the minicomputer itself may be used for the computations. Although slower than current mainframes, a minicomputer dedicated to a particular code can readily be justified on the basis of cost and manpower savings. Many of the codes described in Chapter 11 are already operational on minicomputers.

The ability to routinely perform three-dimensional impact calculations at moderate cost will fuel a demand for high strain-rate materials data and better descriptions of materials failure at high loading rates. This should soon result in a substantial data bank of dynamic properties (at strain rates up to $10^4\ s^{-1}$) for many materials, similar to the situation that now exists with equation of state data. Development of failure models with a firm micromechanical basis will remain a formidable task. Impressive work is being done at a number of research centers along this line and it is not unreasonable to expect that first-order models suitable for engineering applications for metals under impact loading may be available within a decade, even though they are likely to be semiempirical in nature.

Testing will remain a labor-intensive undertaking. Little reduction in cost can be expected here, but with the availability of advanced instrumentation techniques and the use of minicomputers for automated data acquisition, reduction, and reporting, both the quality and quantity of test data should be improved considerably. This in turn will spur development of approximate analytical techniques and refinement of material descriptions in computer codes.

No one method—computations, ballistic testing, dynamic materials characterization—will by itself lead to fuller understanding of the mechanisms that govern the behavior of materials and structures subjected to intensive impulsive loading. The judicious combination of all three can, however, lead to significant advances of our understanding of short-time response phenomena and result in designs capable of withstanding such severe environments.

REFERENCES

_____ (1978), *Response of Materials and Metallic Structures to Dynamic Loading*, National Materials Advisory Board, NMAB-341, Washington, D.C.

_____ (1980), *Materials Response to Ultra-High Loading Rates*, National Materials Advisory Board, NMAB-356, Washington, D.C.

Argyris, J. H. et al. (1979), *Comp. Methods in Appl. Mech. Eng.*, **17/18**, 341.

Armen, H. (1979), *Comp. Struct.*, **10**, 161.

Backman, M., and Finnegan, S. (1976), Naval Weapons Center, NWC-TP-5844.

Bedford, A., Drumheller, D. S., and Sutherland, H. J. (1976), in S. Nemat-Nasser (Ed.), *Mechanics Today*, Vol. 3, Pergamon, New York.

Belytschko, T. (1974), in W. Pilkey et al. (Eds.), *Structural Mechanics Computer Programs*, University Press of Virginia, Charlottesville, Va.

Belytschko, T. (1975), in W. Pilkey and B. Pilkey (Eds.), *Shock and Vibration Computer Programs: Reviews and Summaries*, Shock and Vibration Information Center, Washington, D.C.

Belytschko, T. (1976), *Nuc. Eng. Des.*, **37**, 23.

Belytschko, T. (1977), in R. F. Hartung (Ed.), *Computing in Applied Mechanics*, AMD-18, ASME, New York.

Belytschko, T., Chiapetta, R. L., and Bartel, H. D. (1976), *Int. J. Num. Methods Eng.*, **10**, 579.

Belytschko, T. B., Kennedy, J. M., and Schoeberle, J. M., ASME Paper 78-PVP-60, *ASME/CSME Pressure Vessel and Piping Conference*, Montreal, Canada, June 1978.

Belytschko, T., and Mullen, R. (1978), *Int. J. Num. Methods Eng.* **12**, 1575.

Belytschko, T., Yen, H-J, and Mullen, R., (1979), *Comp. Methods Appl. Mech. Eng.*, **17/18**, 259.

Bertholf, L. D. et al. (1975), *J. Appl. Phys.*, **46**, 3776.

Bertholf, L. D., Kipp, M. E., and Brown, W. T. (1977), Ballistic Research Laboratory, BRL-CR-333.

Bertholf, L. D., et al. (1979), Ballistic Research Laboratory, ARBRL-CR-00391.

Chang, Y. W. (1977), in T. A. Jaeger and B. A. Boley (Eds.) *Proc. 4th International Conference on Structural Mechanics in Reactor Technology*, paper E1/1, North-Holland, Amsterdam.

Clifton, R. J. (1979), in J. Harding (Ed.), *Mechanical Properties at High Rates of Strain*, Conference Series 47, Institute of Physics, Bristol.

Davison, L., Stevens, A. L., and Kipp, M. E. (1977), *J. Mech. Phys. Solids*, **25**, 11.

Durrett, R. E., and Matuska, D. A. (1978), Air Force Armament Laboratory, AFATL-TR-78-125.

Dvorak, G. J. Ed. (1978), *Research Workshop on Mechanics of Composite Materials*, Army Research Office and National Science Foundation, Durham, N.C.

Erlich, D. C. et al. (1980), Ballistic Research Laboratory, ARBRL-CR-00416.

Felippa, C. A. (1977), in A. Wexler (Ed.), *Large Engineering Systems*, Pergamon, Oxford.

Felippa, C. A., and Park, K. C. (1978a), *J. Appl. Mech., Trans ASME*, **45**, 595.

Felippa, C. A., and Park, K. C. (1978b), *J. Appl. Mech., Trans ASME*, **4**, 603.

Felippa, C. A., and Park, K. C. (1979), *Comp. Methods Appl. Mech. Eng.*, **17/18**, 277.

Geradin, M. (1975), *Shock Vib. Digest* **7**, 9, 3.

Ghosh, A. K. (1977), *Int. J. Mech. Sci.*, **19**, 457.

Goudreau, G. L., and Taylor, R. L. (1972), *Comp. Methods Appl. Mech. Eng.* **2**, 69.

Green, A. E., and Naghdi, P. M. (1965), *Arch. Rat. Mech. Anal.* **18**.

Gupta, A. D., and Misey, J. J. (1980) AIAA Paper 80-0403, presented at the AIAA 18th Aerospace Sciences Meeting, Jan. 14–16, 1980, Pasadena, CA.

Gupta, Y. M., and Seaman, L. (1978), Electric Power Research Institute, NP-1217.

Hageman, L. J. et al. (1975), Systems, Science, and Software, SSS-R-75-2.

Hallquist, J. O. (1976), Lawrence Livermore Laboratory, UCRL-52066.

Hallquist, J. O. (1977), Preprint 2956, *ASCE Fall Convention and Exhibit*, San Francisco, CA.

Hallquist, J. O. (1978), in K. C. Park and D. K. Gartling (Eds.), *Computational Techniques for Interface Problems*, AMD-30, ASME, New York.

Hallquist, J. O. (1979a), Lawrence Livermore Laboratory, UCRL-52678.

Hallquist, J. O. (1979b), Lawrence Livermore Laboratory, UCID-17268, Rev. 1.

Hallquist, J. O. (1980), Lawrence Livermore Laboratory, UCID-18756.

Herrmann, W. (1975) in B. Pilkey and W. Pilkey (Eds.), *Shock and Vibration Computer Programs, Reviews and Summaries*, Shock & Vibration Information Center, Washington, D.C.

Herrmann, W. (1976), in E. Varley (Ed.), *Propagation of Shock Waves in Solids*, ASME, New York.

Herrmann, W. (1977), in T. C. T. Ting, R. J. Clifton, and T. Belytschko (Eds.), *Nonlinear Waves in Solids*, National Science Foundation, Washington, D.C.

Herrmann, W., Bertholf, L. C., and Thompson, S. L. (1975), in J. T. Oden (Ed.), *Computational Methods for Stress Wave Propagation in Nonlinear Solid Mechanics*, Springer Verlag, New York.

Herrmann, W., and Hicks, D. L. (1973), in R. W. Rhode, et al. (Eds.), *Metallurgical Effects at High Strain Rates*, Plenum Press, New York.

Herrmann, W., and Lawrence, R. J. (1978), *J. Eng. Mat. Technol.*, **100**, 84.

Herrmann, W., and Nunziato, J. (1972) in P. C. Chou, and A. K. Hopkins (Eds.), *Dynamic Response of Materials to Intense Impulsive Loading*, U.S. Govt. Printing Office, Washington, D.C.

Hicks, D. L., and Madsen, M. M. (1979) Sandia Laboratories, SAND-78-1806.

Houbolt, J. C. (1950), *J. Aero. Sci.* **17**, 540.

Hughes, T. J. R., Pister, K. S., and Taylor, R. L. (1978), *Comp. Methods Appl. Mech. Eng.*, **17/18**, 159.

Jensen, P. S. (1974), *Comp. Structures*, **4**, 615.

Johnson, G. R. (1977), in *Proc. 3rd International Symp. on Ballistics*, Karlsruhe, German.

Johnson, G. R., Colby, D. D., and Vavrich, D. J. (1978), Air Force Armament Laboratory, AFATL-TR-78-81.

Johnson, G. R., Vavrich, D. J., and Colby, D. D. (1980), Ballistic Research Laboratory, AR-BRL-CR-00429.

Johnson, G. R. (1981), *J. Appl. Mech., Trans ASME*, **103**, 30.

Jonas, G. H., and Zukas, J. A. (1978), *Int. J. Eng. Sci.*, **16**, 879.

Karpp, R., and Chou, P. C. (1972), in P. C. Chou and A. K. Hopkins (Eds.), *Dynamic Response of Materials to Intense Impulsive Loading*, U.S. Govt. Printing Office, Washington, D.C.

Kimsey, K. D. (1980), in *Proc.* 1980 *Summer Computer Simulation Conference*, Seattle, WA., Aug. 25–27, 1980, Simulation Councils Inc., La Jolla, Calif.

Kohn, B. J. (1969), Air Force Weapons Laboratory, AFWL-TR-69-38 (AD852300).

Lee, E. H. (1970), in J. J. Burke and V. Weiss (Eds.), *Shock Waves and the Mechanical Properties of Solids*, Syracuse University Press, Syracuse, N.Y.

Lindholm, U.S. (1971), in R. F. Bunshah, (Ed.), *Techniques of Metals Research*, Vol. 5, Part 1, Wiley, New York.

Lindholm, U.S. (1980), in J. R. Vinson (Ed.), *Emerging Technologies in Aerospace Structures, Design, Structural Dynamics and Materials*, ASME, New York.

Liszka, T., and Orkisz, J. (1977), in T. A. Jaeger and B. A. Boley (Eds.), *Proc. 4th International Conference on Structural Mechanics in Reactor Technology*, North-Holland, Amsterdam.

Moon, F. C., Kim, B. S., and Fang-Landau, S. R. (1976), Natl. Aeronautics and Space Admin., NASA-CR-134999.

Nagtegaal, J. C., Parks, D. M., and Rice, I. R. (1974), *Comput. Methods Appl. Mech. Eng.*, **4**, 153.

Newmark, N. M. (1959), *Proc. ASCE, J. Eng. Mech. Div.* **85**, EM3, 67.

Nickell, R. E. (1973), *Proc. ASCE, J. Eng. Mech. Div.*, **99**, EM2, 303.

Noh, W. F. (1976), Lawrence Livermore Laboratory, UCRL-52112.

Nolle, H. (1974), *Int. Met. Rev.*, **19**, 223.

Osborn (1981), Ballistic Research Laboratory, ARBRL-CR-00456.

Roach, P. J. (1972), *Computational Fluid Mechanics*, Hermosa Publishers, Albuquerque, NM.

Roddy, D. J., Peppin, R. O., and Merrill, R. B., Eds. (1977), *Impact and Explosion Cratering*, Pergamon Press, New York.

Seaman, L. (1975), in W. Pilkey and B. Pilkey (Eds.), *Shock and Vibration Computer Programs: Reviews and Summaries*, Shock & Vibration Information Center, Washington, D.C.

Seaman, L., Curran, D. R., and Shockey, D. A. (1976), *J. Appl. Phys.*, **47**, 4814.

Seaman, L., and Shockey, D. A. (1975), Army Materials and Mechanics Research Center, AMMRC-CTR-75-2.

Segwick, R. T. et al. (1976), Ballistic Research Laboratory, BRL-CR-322.

Shockey, D. A., Curran, D. R., and DeCarli, P. S. (1975), *J. Appl. Phys.*, **46**, 3766.

Spilker, R. L., and Witmer, E. A. (1977) in T. A. Jaeger and B. A. Boley (Eds.), *Proc. 4th International Conference on Structural Mech. in Reactor Technology*, North-Holland, Amsterdam, paper M7/3.

Tillerson, J. R. (1975), *Shock and Vib. Digest*, **7**, 4, 2.

Tillerson, J. R., and Stricklin, J. A. (1970), Texas A & M University, unnumbered report for NASA (NASA-CR-108639).

Tuler, R. F., and Butcher, B. M. (1968), *Int. J. Fracture Mech.*, **4**, 431.

Underwood, P., Park, K. C. (1977), in T. A. Jaeger and B. A. Boley (Eds.), *Proc. 4th International Conference on Structural Mechanics in Reactor Technology*, North-Holland, 1977, paper M5/2.

Van Thiel, M. (1977), Lawrence Livermore Laboratory, UCRL-50108, vols. 1–3.

von Neumann, J., and Richtmyer, R. D. (1950), *J. Appl. Phys.*, **21**, 232.

Walsh, R. T. (1972), in P. C. Chou and A. K. Hopkins (Eds.), *Dynamic Response of Materials to Intense Impulsive Loading*, U.S. Govt. Printing Office, Washington, D.C.

Walsh, J. M., and Sedgwick, R. T. (1974), Systems, Science and Software, Topical Report SSS-R-74-2399.

Wilkins, M. L. (1964), in B. Adler et al. (Eds.), *Methods in Computational Physics*, Academic Press, New York.

Wilkins, M. L. (1980), *J. Comp. Phys.*, **36**, 281.

Wright, J. P. (1979), *Comp. Structures*, **10**, 235.

Zukas, J. A., Jonas, G. H., and Misey, J. J. (1979), Ballistic Research Laboratory, ARBRL-MR-2969.

11

THREE-DIMENSIONAL COMPUTER CODES FOR HIGH VELOCITY IMPACT SIMULATION

Jonas A. Zukas

Computer codes have matured considerably since their initial development some 25 years ago [Evans and Harlow (1957)]. They now serve as valuable tools in studies of materials and structures subjected to intense impulsive loading. Until quite recently, only two-dimensional codes were available. It was generally felt that three-dimensional computations were impractical from the standpoint of both cost and computer-storage capability. Hence, only problems involving normal impact resulting in waves propagating in one or two dimensions could be studied. Such situations, however, are the exception rather than the rule in the real world. Recourse could always be made to plane-strain approximations to at least qualitatively simulate select aspects of oblique impacts or nonsymmetric loading conditions. However, it has now been established that while plane-strain solutions can provide good qualitative (and sometimes quantitative) information for very early times after impact for problems involving long, slender strikers [Norris et al. (1977)], they cannot be relied on for even qualitative information for times long in comparison to the wave-transit time across the characteristic dimension of the problem [Zukas et al. (1979)].

The advent of third- and fourth-generation computers such as the CDC 7600 and CRAY machines has made three-dimensional computations attractive. Consequently, a number of computer programs capable of solving the fully three-dimensional equations of continuum physics are being used at various laboratories on a production basis. More are being developed and tested. A review of the characteristics of existing three-dimensional production codes to assist potential users in making a selection is therefore appropriate.

The material in this chapter is taken from a report by Zukas, Jonas, Kimsey, Misey, and Sherrick [Zukas et al. (1981)]. Any such review is at least somewhat subjective and strongly influenced by the past experience of the authors in applying two- and three-dimensional computer codes to problems involving high-velocity impact and penetration of solids. It is the view of the authors that good documentation is vital to successful implementation and use of any computer program not written by the user. Equally important is an expansive and flexible mesh-generation capability and a post-processor capable of synthesizing the hundreds of pages of numbers generated by such codes into *useful* graphics. Without comprehensive pre and post-processing packages, situations will readily arise that require a week or more of an analyst's time to prepare input and several weeks to analyze the results of a computation that required 2 hours of CPU time. It is not possible to support experimental programs under such conditions, yet the optimum use of codes is in close conjunction with impact testing and dynamic material characterization. Neither, in our view, can such a situation be economically justified, for computer charges are rapidly decreasing while labor costs continue to rise. Finally, it is both ineffective and shortsighted to require an imperfect human to perform mundane and repetitive tasks that a machine can do rapidly and efficiently. Optimum benefit is gained from codes when the analyst is free to concentrate on the problem and the computer on repetitive details.

The emphasis in this chapter is on three-dimensional codes that can be used on a production basis or will shortly be available for such use. Many are an outgrowth of their two-dimensional forerunners. Details of these codes will be mentioned only when needed for completeness. Previous reviews of one- and two-dimensional codes for wave propagation and impact have been prepared by Mescall (1974), von Riesemann et al. (1974), Herrmann (1975), and Belytschko (1975). These should be consulted for details. Codes for static and dynamic structural analysis will not be considered. The recent review by Noor (1980) and references therein contain extensive information on such codes.

11.1 GENERAL CODE CHARACTERISTICS

The computer programs to be reviewed here were principally developed to solve problems characterized by:

1. The presence of shock waves (alternatively, steep stress or velocity gradients).
2. Localized materials response (as contrasted to global structural response).
3. Loading and response times in the submillisecond regime.

As a result, they share a number of common characteristics.

With exceptions to be noted, such codes use an explicit central-difference time-integration algorithm. The time step, Δt, is controlled by the Courant stability condition

$$\Delta t \leqslant \frac{kl}{c} \tag{1}$$

where l is a characteristic mesh dimension in the grid, c is the local sound speed (which varies with pressure), and k is a user-specified constant, usually called a stability factor, which must be less than unity. Generally, k is chosen to be between 0.6–0.8.

Metals, until recently, were the materials of primary interest. Hence, the material model in most codes decouples the effects of volumetric and shear behavior. The hydrostatic component of stress is taken from an equation of state. For impacts where the striking velocity is below the sonic velocity, some variant of the Mie-Gruneisen equation of state is commonly used. Beyond this range (hypervelocity impact), the Tillotson equation of state is favored. The constitutive model for elastic-plastic behavior is typically an incremental form, employing a von Mises yield criterion and providing for some form of hardening, thermal softening, or strain-rate effects. By way of illustration, the equations used in the HEMP3D Lagrangian code are given in Appendix A while the equations for the Eulerian HULL code are given in Appendix B. Representative equation of state formulations for metals, explosives, and gases are given in Appendix C. The relations in the appendices are representative of those in most production codes. By and large, the basic physics contained in these codes can be written down on the back of a business envelope and is contained in two or three subroutines. The remainder of the coding, typically upward of 10,000 FORTRAN statements, is devoted to assorted bookkeeping —input/output operations, slide line logic, material transport, and the like.

Because of interest in porous materials, various geological materials, and concrete, many of the newer three-dimensional codes incorporate material models for crushable solids or foam. Figure 1, taken from the DYNA3D manual, shows a typical pressure-volume relationship for such models. The notation V and V_0 refer to the current and initial volumes of a computational element, respectively.

The greatest uncertainty in code computations comes in the description of material failure. Micromechanically based failure models applicable in the high-pressure regime for multiaxial-transient loading situations are generally not available. Those under development, namely the SRI International nucleation and growth models [Seaman et al. (1976)], require an inordinate degree of material characterization. Hence, failure models for codes tend to be quite simple for several reasons:

1 Since material failure will be incorrectly modeled, it should be done as simply and cheaply as possible.

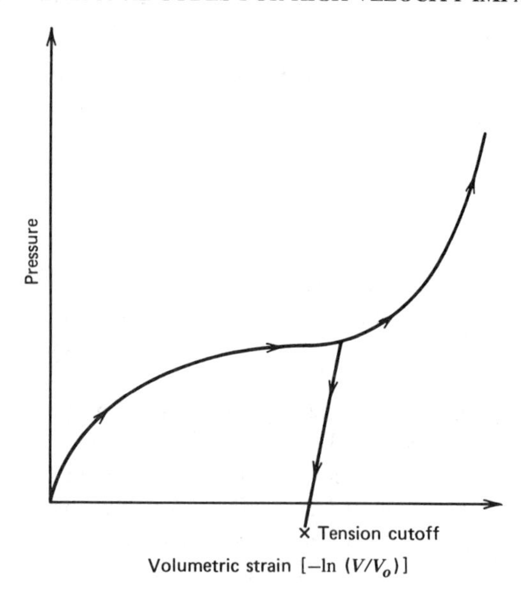

Figure 1 Characteristic volumetric strain versus pressure curve for soil and crushable materials.

2 The material characterization required (and therefore cost) is kept to a minimum.

3 In many cases, the simple models produce results that are in substantial agreement with experimental data. Precious few guidelines exist to permit a selection of a failure model a priori for any given encounter situation.

All the codes use an artificial viscosity to smear-shock fronts over several mesh widths in the calculation. At least linear and quadratic terms are available in all the codes. The generalized viscosity term, which is added to the pressure computed from the equation of state, is of the form

$$q=\begin{cases} c_1 \rho l a|\dot{\varepsilon}|-c_0 \rho l^2 \dot{\varepsilon}|\dot{\varepsilon}|, & \dot{\varepsilon}<0 \\ 0, & \dot{\varepsilon}\geqslant 0 \end{cases} \tag{2}$$

where c_0 and c_1 are usually user-defined constants, l is a characteristic grid length, a is the local sound speed, $\dot{\varepsilon}$ is the volumetric strain rate, and ρ is the density. Expressions for strain rate and characteristic length for three-dimensional computations are given by Wilkins (1980). A lucid discussion of the artificial viscosity concept is to be found in the report by Noh (1976).

The formulation of q in (2) principally serves to suppress spurious oscillations behind a steep compression wave. With quadrilateral or cubical grid, a linear instability can occur in two or three dimensions commonly referred to as hourglass distortion. Artificial viscosity formulations for grid stabilizations differ. The formulation employed in HEMP3D is discussed in the article by Wilkins (1980).

Virtually all the three-dimensional codes employ a rectangular Cartesian coordinate system, a restart capability, and are written in FORTRAN.

A shortcoming of almost all the codes to be discussed is the paucity of documentation. Most three-dimensional codes were developed to satisfy demands of immediate users and little thought was given to providing the extensive documentation typically associated with commercial structural-analysis codes such as MARC and NASTRAN. An intimate association with developers and users is therefore necessary for effective implementation and use of such codes. Typically, some six months of full-time commitment to one of these codes is required before they can be used with any degree of confidence. A one-man, one-code philosophy is heartily recommended.

Production codes available as of December 1980 are now reviewed. Some comments on current developments in this field conclude the chapter.

11.2 LAGRANGIAN CODES

The primary advantage of Lagrangian codes is their ability to accurately track material boundaries and interfaces. They are ideally suited for constitutive models that require time histories of material behavior. A major portion of any Lagrangian code is the logic that defines the behavior of material interfaces. Materials at an interface are designated as slave and master and procedures are built into programs to permit contact, separation, and sliding, with or without friction, between master and slave surfaces. Generally, versatility in sliding-interface treatment is achieved at the expense of execution time. Lagrangian grids distort with the material flow. Once distortions become severe, a new mesh must be overlaid on the old one and the computations continued. Automatic rezoning programs exist for two-dimensional Lagrangian codes. None exist for their three-dimensional counterparts.

11.2.1 HEMP3D

Developers: M. L. Wilkins and Associates
Lawrence Livermore National Laboratories
Livermore, CA

Documentation: Wilkins et al. (1975)

Discretization: The equations of Appendix A are discretized using a second-order accurate finite-difference formulation analogous to that in the two-dimensional HEMP code [Wilkins (1964), Giroux (1973)].

Material Description:

1. *Material Model*-see Appendix A.
2. *Equation of State*-HEMP3D employs the Gamma law, JWL, and a Mie-Gruneisen type of equation of state.

3. *Failure criteria/post-failure models*-the code documentation gives no information about treatment of failed cells in HEMP3D. Wilkins et al. (1975) state that models for ductile and brittle fracture are available. The code has been applied on a number of static and dynamic problems involving fracture [Wilkins (1977), Chen and Wilkins (1976), Chen (1978)].

Code Characteristics:

1. *Sliding surfaces*-intersecting sliding surfaces are permitted with separation and closing dependent on forces acting on the interface. The sliding-surface logic is not described in the documentation.
2. *Preprocessing capability*-a semiautomatic zoning program, VMESH3D, is available and is described in the documentation. It permits users to define blocks and automatically subdivides these into finer zones. The blocks may then be assembled to form the final mesh. VMESH3D is derived from the MESH3D program of Hutula and Zeiler (1972) and is implemented on the LLNL CDC 7600.
3. *Post-processing capability*-an interactive program, VPIX3D is used for verification of input data and interpretation of output. VPIX3D generates pictures of three-dimensional objects in perspective with hidden lines removed. The program is interactively capable of rotating, translating, and scaling an object in the field of view. Output devices include teletypewriter, FR80 color monitor, and television-monitor display system. Three-dimensional pictures may be displayed as line drawings or halftone pictures with shading keyed to a selected flow-field variable. A complete description of program capabilities is given in the documentation.
4. *Comments*-the basic equations and differencing procedures for HEMP3D are well described in the cited reference, but little or no information is provided on sliding-surface logic, core-storage requirements, and other program limitations. Early versions of HEMP3D were implemented on the CDC 7600. Later versions, with sliding logic, are said to be running on the STAR and CRAY computers. Training programs for potential users of HEMP and HEMP3D are available at LLNL. Attendance at such training prior to acquisition of either code is highly recommended by LLNL.

11.2.2 EPIC-3

Developer: G. R. Johnson
Honeywell, Inc.
Defense Systems Division
Hopkins, MN

Documentation: Johnson et al. (1978, 1979, 1980)
Johnson (1977a, b)

Discretization: Tetrahedral finite elements are employed in EPIC-3. Tetrahedral and triangular elements offer inherently greater resistance to distortion than cubic or quadrilateral elements. Lagrangian calculations can therefore be run much further with them. A penalty for this behavior is paid because of the introduction of artificial stiffness and some asymmetries into calculations. This can be somewhat overcome with finer zoning as well as arrangement of the triangles in a crossed configuration (four triangles per quadrilateral). The added stiffness of the tetrahedral elements proves to be a boon for high-velocity impact calculations and very good correlation with experiments has been obtained [Jonas and Zukas (1978), Zukas (1980), Misey et al. (1980)]. For low-velocity contact-impact problems, these aspects of triangles and tetrahedra may prove undesirable.

Material Description:

1. *Material model*-for isotropic materials, the basic equations are similar to those in Appendix A. Strain-rate and temperature effects are accounted for through

$$\sigma_D = \left[\sigma(\varepsilon) + c_1 \ln(\bar{\dot{\varepsilon}})\right]\left[c_2 + c_3 T\right] \tag{3}$$

   where σ_D is the dynamic stress, $\sigma(\varepsilon)$ a strain-dependent effective stress, $\bar{\dot{\varepsilon}}$ the equivalent strain rate, T the temperature, and $c_1 - c_3$ are user-supplied constants. Temperature is determined from

$$T = T_0 + \frac{E_s}{c_s \rho_0} \tag{4}$$

   where T_0 is the initial temperature in an element, E_s is the internal energy per unit original volume, c_s is the specific heat, and ρ_0 is the initial density.

   The current version [Johnson et al. (1980)] also has an anisotropic model for strength and pressure. The formulation follows that of Valliapan (1972) for elastoplastic analysis of orthotropic materials. The method is a piecewise linear theory based on the von Mises yield criterion proposed by Hill (1948, 1950). Anisotropic work-hardening follows the formulation derived by Hu (1956).

2. *Equation of state*-for isotropic materials, the Mie-Gruneisen, JWL, and Gamma law options are provided. For orthotropic materials, an equation of the form

$$P = P^* + K_2 \mu^2 + K_3 \mu^3 \tag{5}$$

   where

$$\mu = \frac{\rho}{\rho_0} - 1$$

$$P^* = \Sigma(\Delta P) = -\Sigma\left(\frac{\Delta\sigma_x + \Delta\sigma_y + \Delta\sigma_z}{3}\right) \tag{6}$$

is used where $\Delta\sigma_\alpha$ ($\alpha = x, y, z$) are stress increments that are derived from computed strain fields through the elastic constants.

3. *Failure criteria/post-failure model*-element failure may be initiated by exceeding user-specified values for volumetric strain, equivalent true plastic strain, or some combination of the two. A tensile-pressure cutoff may also be specified. Elements may fail partially or totally. Partial failure implies that an element can carry neither tensile nor shear forces. Its behavior becomes similar to that of a fluid since compressive loads only can be tolerated. A totally failed element can carry no loading whatsoever and is ignored in the computational loop.

Code Characteristics:

1. *Sliding surfaces*-an extensive slide-line logic, well documented in Johnson et al. (1980, 1978) permits sliding with or without friction, void formation and collapse, tunneling, and erosion of material designated as slave. Intersecting sliding planes can be treated with the current logic. The number of sliding surfaces is currently set at five. This can be changed by altering the appropriate dimension statement. Nothing in the interface logic inhibits the number of sliding planes that can be used.
2. *Preprocessing Capability*-EPIC-3 has a preprocessor that permits rapid, automatic generation of rectangular and circular plates, hollow or solid uniform or tapered rods, hollow or solid spheres, and hollow or solid hemispherical, ogival, and conical nose shapes.
3. *Post-processing capability*-an extensive post-processing capability exists with the EPIC-3 post processor. Plots can be obtained on CALCOMP plotters or Tektronix graphics terminals. The post-processor is compatible with the Tektronix PLOT 10 software package. Geometry plots in several planes, perspective plots, velocity vectors, and contour plots of pressure, normal, and shear stresses, equivalent plastic stress and strain, temperature, internal energy, and plastic work are readily obtained. For long rods, axial, bending, and shear loads may also be plotted. Time-history plots may also be obtained for all energies, angular momenta and velocities, system center of gravity positions, momenta and velocities, nodal positions, velocities and accelerations, element pressures, stresses, and temperatures.
4. *Comments*-EPIC-3 has been implemented on the UNIVAC 1108, Honeywell 6080, CDC 6600, CDC 7600, and the CRAY computers. With the exception of subroutine ASTRESS (the anisotropic model), the current version is vectorized [Johnson et al. (1978)]. Nodal data may, at the user's option, be core-contained or buffered to mass storage. EPIC-3 is highly modular and portable. Typical computation times on CDC 6600 machines is 6–9 ms/node/cycle while on CDC 7600s the typical value is 1.5 ms/node/cycle. Use of the anisotropic model or generalized search routine for the sliding-surface logic requires considerably more time.

11.2.3 DYNA3D

Developer: J. O. Hallquist
Lawrence Livermore National Laboratories
Livermore, CA

Documentation: Hallquist (1979a, 1976, 1977, 1978a, b)
Goudreau and Hallquist (1979)

Discretization: The present version employs both the HEMP difference equations and a Popov-Galerkin finite-element formulation using four-, six-, or eight-node quadrilateral solid elements. Future versions of DYNA3D will contain only the finite-element option since it has proven to be more efficient than the differencing scheme.

Material Description:

1. *Material model*-the current version of DYNA3D solves finite-difference or finite-element analogs of equations similar to those in Appendix A. The formulation is for elastoplastic work-hardening materials undergoing finite strains and rotations. A crushable foam model permits treatment of concrete and geological materials.
2. *Equation of state*-currently DYNA3D contains a one-term equation of state relating pressure and compression through the bulk modulus. Additional options are to be provided in future versions.
3. *Failure criteria/post-failure model*-failure models currently include a pressure cutoff for tensile-fracture simulation. Failure can also be based on volumetric strain. Failed elements cannot carry tensile or shear loads but may carry compressive hydrostatic stress if reloaded.

Code Characteristics:

1. *Sliding surfaces*-DYNA3D has an extensive slide-line capability that is very well documented in the references cited. As is characteristic of Lagrangian codes, a master-slave logic is employed that permits sliding with and without friction (Coulomb friction model). Void opening (separation of master and slave surfaces), and closing is automatically treated, requiring no user intervention. Provision is also made for sliding without separation and for tied sliding (neither separation nor sliding). Tied sliding permits a sudden transition in zoning while insuring that displacement compatibility is maintained along an interface between slave nodes and corresponding master segments. It is also useful in situations where two surfaces in contact must remain in contact with no relative motion until some criteria, determined by the analyst, are satisfied. Rigid barriers are also permitted in DYNA3D. There is no limit in the logic on the number, type, and orientation of interfaces.
2. *Preprocessing capability*-a mesh generator is currently being developed for DYNA3D.

3. *Post-processing capability*-The DYNAP post-processor is used to obtain time-history plots for DYNA3D calculations. It also generates Lagrange and Almansi strain measures, strain rates, rigid body displacements, velocities, and accelerations for each material in a DYNA3D calculation. The GRAPE post-processor [Brown (1980)] is used to generate deformation plots. It is compatible with most graphics display devices as well as the DICOMED COM system.
4. *Comments*-the current version of DYNA3D runs on both the CDC 7600 and CRAY computers at LLNL. It is currently core-contained, permitting 10 K elements on the 7600 and 30 K on the CRAY. The GRAPE post-processor is not operational on the CRAY. Execution speeds are impressive. The finite-element option runs at 4.67 CPU min/10^6 mesh cycles on the CDC 7600 (0.67 on the CRAY). The finite-difference option requires 9.65 CPU min/10^6 mesh cycles on the 7600 (5.00 on CRAY). The finite-difference option is not vectorized on the CRAY. Because some of the I/O routines are programmed in machine language for execution under the CDC NOS system, users may need to rewrite some programs when running on other machines. DYNA3D is under constant development. Potential users should contact Dr. John Hallquist for an update on DYNA3D capabilities.

11.3 EULERIAN CODES

Eulerian codes are ideally suited for treating large distortions. Eulerian calculations can be done either by including the convective derivatives explicitly in the discretization scheme or by performing a two-step operation, the first phase being essentially a Lagrangian calculation and the second or transport phase amounting to a rezoning of the computational grid to its original conditions. Production codes tend to adopt the latter approach. To inhibit diffusion of material interfaces and boundaries, Lagrangian features such as massless tracer particles are added across which material is not permitted to diffuse (HELP), or a preferential mass-transport scheme may be used (HULL).

Rezoning in Eulerian calculations consists of a translation of the computational mesh to insure that fine zoning is located in regions with the greatest activity. This rezoning can be continuous or user initiated.

Eulerian computations do not require explicit sliding-surface specification. Depending on the discretization scheme, they may not require explicit artificial viscosity.

11.3.1 HULL

Developers: D. A. Matuska and R. E. Durrett
Orlando Technology, Inc.
Shalimar, FL

Documentation: Durrett and Matuska (1978)
Gaby (1978)

Several variants of HULL exist. The code was first written in 1971 as a purely hydrodynamic code and rewritten in 1976 to permit computations involving elastic-plastic flow. In 1979, additional development was begun to link HULL with the three-dimensional Lagrangian EPIC-3 code. As a result of this effort, additional documentation was prepared and should begin to appear by mid-1981 in the form of joint Army Ballistic Research Laboratory-Air Force Armament Laboratory technical reports.

Discretization: A second-order accurate finite-difference scheme is employed in HULL. The differencing method is explained in the documentation.

Material Description:

1. *Material model*-see Appendix B.
2. *Equation of state*-HULL has an extensive equation of state capability. Solids and liquids are modeled with the Mie-Gruneisen equation of state, the Gamma law equation being used after vaporization. HULL also contains a plain concrete equation of state model that is valid in a 1–300 kbar pressure range. The model is constructed from Hugoniot data generated by Gregson (1972) and static yield-strength data from Chinn and Zimmerman (1965). The aggregate used in Gregson's work had an average diameter of 3.2 mm and an average initial density of 2.185 g/cc. The concrete consisted of 18% voids, 25% granite aggregate, 20–25% quartz grains, and 25–30% cement paste by volume. The unloading path allows the concrete to return to its solid density when loaded past the point where all voids are filled. The yield strength of the concrete increases as the confining pressure increases. The concrete equation of state has been used successfully in penetration calculations into concrete at velocities from a few to over 300 m/s. A JWL and various ad hoc equation-of-state formulations may also be selected to model a large number of explosives.
3. *Failure criteria/post-failure model*-failure initiation criteria in HULL are based on maximum principal stress or strain. When these variables exceed user-specified limits for a material, stress components are set to zero and that region is not permitted to support future tensile or shear loads. A triaxial failure criterion for metals [Hancock and Mackenzie (1976)] has recently been implemented. Once failure has occurred, a small but numerically significant quantity of void (or a low-pressure gas region) is inserted into the zone. Under shear or tensile loading, the void may grow to occupy the entire zone or propagate into adjacent zones. Recompression of failed zones is not inhibited.

Code Characteristics:

1. *Rezoning*-a rezoner that permits a continuous translation of the computational mesh at a continuous velocity in any one or all of the coordinate directions is implemented in three-dimensions.

2. *Preprocessing capability*-the KEEL preprocessor defines the initial conditions, mesh coordinates, and material property specifications for three components of velocity, density, internal energy, mass, and volume of each computational cell. KEEL input is divided into three sections—initial conditions, mesh generation, and material generation. Material properties for over 35 solids, liquids, and gases can be generated automatically by specification of a material number. The KEEL mesh generator permits construction of a fine computational mesh in the region of most interest and gradual increase of zone sizes away from that region. KEEL includes a number of geometric configurations for insertion of materials, tracer particles and stations for collection of flow-field data. The three-dimensional geometric definitions include boxes, wedges, pyramids, cylinders, spheres, and assorted curves, to name but a few.
3. *Post-processing capability*-graphical output for HULL is generated by the PULL program that is capable of producing three types of plots:

   1 A flow field variable (density, pressure, energy, etc.) versus a spatial coordinate.

   2 Up to 20 contours of a flow field variable.

   3 Plots of vector quantities indicating direction and magnitude.

   Information collected at flow-field stations is also plotted by PULL. Time histories of pressure, density, and impulse may be obtained from purely hydrodynamic calculations. If strength is also included, plots of radial motion, axial motion, individual stress, and stress-deviator components, second-stress invariant, principal stresses, and mass flux as functions of time may be obtained. The PULL processor is a very comprehensive graphics package and can produce a large number of contour plots of various field variables and derived quantities. Unfortunately, the quality of PULL graphics leaves much to be desired. PULL is currently being revised under an Air Force contract and the new version, it is hoped, will make PULL output compatible with state-of-the-art, computer-graphics technology. PULL supports the Stromberg Carlson SC4020 plotter, META system, and CALCOMP hardware.
4. *Comments*-HULL is a system of computer programs capable of running cylindrical or cartesian coordinate calculations in two spatial dimensions or cartesian calculations in three dimensions. It is a multi-material (10) code that employs a diffusion limiter to preserve material interfaces. The HULL system of codes is composed of:

   1 KEEL—an initial condition and mesh generator.

   2 HULL—the finite difference code.

   3 PULL—the graphics package.

4 BOW—the dump tape and problem library.
5 PLANK—the alternate input generator for SAIL.
6 CONTROL—an interactive-program status monitor.
7 SAIL—the system file manager and executive processor.

All codes in the HULL system including SAIL itself are maintained on a single SAIL program library file. Compilation of any program element of HULL entails execution of SAIL to process the program library, select the proper subroutines and options desired for a particular application and produce a compilable file. This processing activity is done to minimize storage requirements. HULL calculations are performed with four mesh planes contained in-core, with other planes being buffered in from a secondary mass-storage device. Versions of HULL are available that run on CDC (6400, 6500, 6600, 7600, CYBER 176), IBM (360, 370), and CRAY mainframes with the CDC SCOPE 2.0, 3.4, and IBM OS/VS 2 operating systems. Recently HULL has been installed on a DEC VAX. HULL is probably the most complex system of codes available for the study of the behavior of solids subjected to intense impulsive loading. An offsetting advantage is the existence of a large network of HULL users. Thus, as modifications and improvements are made, all members of the network reap the benefits.

11.3.2 TRIOIL/TRIDORF

Developer: W. E. Johnson
Computer Code Consultants
Los Alamos, NM

Documentation: TRIOIL—W. E. Johnson (1967)
Reed and Henderson (1975)
TRIDORF—W. E. Johnson (1976, 1977)

The documentation on these codes is minimal and outdated. Potential users are urged to contact W. E. Johnson directly for the current status of either code.

Discretization: Both codes use finite differences for temporal and spatial discretization. In 1975, TRIOIL was reconfigured to run on the ILLIAC IV computer and renamed TRIOIL IV. The code uses operator splitting, is limited to two materials, permits variable zone sizes, reflective and transmissive boundaries, and a maximum of $64 \times 70 \times 70$ cells in the respective coordinate directions. A complete computational cycle consists of three calculational sweeps over the grid, one in each coordinate direction. The sweep directions are permuted to account for the six possible combinations of the three coordinate directions to avoid preferential mass flow.

Material Description:

1. *Material model*-the TRIOIL code solves the hydrodynamic equations (Appendix B equations with all stress terms set to zero) in three dimensions. TRIDORF is a two-material version of TRIOIL and includes a rigid plastic-strength formulation. The yield function in simple shear is taken to be a function of pressure and energy in the form

$$Y=(Y_0+\alpha P)\left(1-\frac{Q}{Q_m}\right) \tag{7}$$

where

Y_0 = the yield strength
α = constant
P = hydrostatic pressure
Q_m = energy to melt
Q = internal energy

The rigid plastic equations in TRIDORF are only applied to those zones that have $\rho/\rho_0>1$ or $\rho/\rho_0<1$ and $Q>Q_i$ (user-specifiable parameter). Those cells in which the original velocity has not changed by at least 5% are bypassed for the strength calculation.

2. *Equation of state*-the Tillotson equation of state is used in both codes.
3. *Failure criteria/post-failure models*-unknown.

Code Characteristics:

1. *Rezoning*-a rezoning feature permits motion and expansion of the grid to maintain fine zoning in regions of maximum activity.
2. *Preprocessing capability*-both codes have the capability to generate material-geometric packages for shapes such as cones, simple parallelipipeds, cylinders, and others. These may be combined to generate complex geometries.
3. *Post-processing capability*-both codes employ graphics packages that are proprietary to Los Alamos Scientific Laboratories and Lawrence Livermore National Laboratories.
4. *Comments*-both TRIOIL and TRIDORF are core-contained. They have been run on CDC 7600, UNIVAC 1108, the VAX minicomputer, and CRAY. The CRAY versions are not vectorized.

11.3.3 METRIC

Developers: L. J. Hageman and E. P. Lee
Systems, Science and Software, Inc.
La Jolla, CA

Documentation: Hageman et al. (1976)
Hageman and Lee (1976)

Discretization: A first-order finite-difference method is employed in METRIC, which is the three-dimensional version of the HELP code. The discretized equations and material transport scheme are well-documented in the above references.

Material Description:

1. *Material model*-the basic formulation follows that of Appendix B. The strength model uses a linear-elastic constitutive relation for isotropic materials in the form

$$\Delta s_{ij} = 2G\Delta\varepsilon_{ij}^{e} \tag{8}$$

where G is the shear modulus, $\Delta\varepsilon_{ij}^{e}$ the elastic strain-increment components, and Δs_{ij} the incremental deviatoric stresses. The von Mises yield criterion is used in the form

$$s_{ij}s_{ij} \leqslant 2Y^2$$

where

$$Y = \left[Y_0 + Y_1\left(\frac{\rho}{\rho_0} - 1\right) + Y_2\left(\frac{\rho}{\rho_0} - 1\right)^2\right]\max\left(0, 1 - \frac{E}{E_{\text{melt}}}\right) \tag{9}$$

where Y is the yield strength in shear, ρ_0 is the original density, ρ is the current density, E is the specific internal energy, and E_{melt} is the specific internal energy corresponding to the material's melting temperature. Y_0, Y_1, and Y_2 and E_{melt} are treated as material constants. Once plastic flow has occurred, the plastic flow rule is analogous to that given in Appendix B.

There is no limitation on the number of materials that may be treated.

2. *Equation of state*-in METRIC, the Tillotson equation of state is used for initially condensed material (Appendix C). It is modified to give a smooth transition between condensed and expanded states. The blended portion of the equation of state has the form

$$P = \frac{(E - E_s)P_E + (E_s' - E)P_c}{E_s' - E_s} \tag{10}$$

where E_s is the specific internal energy required to bring the material to vapor temperature and E_s' is regarded as a material constant. For multimaterial cells, an iteration is used to determine the cell pressure and densities of the various materials within the cell. The values of the various equation of state constants for 19 materials are built into the code.

3. *Failure criteria/post-failure model*-the material failure criterion in METRIC is based on a maximum allowable distension of the material $(\rho/\rho_0)_{\min}$, which is related to the material's hydrodynamic tensile strength and to its melting point. It is assumed that the hydrodynamic tensile strength approaches zero as the specific internal energy ap-

proaches E_{melt}, and also that the material has its maximum hydrodynamic tensile strength when it is cold, that is, when E is zero.

Failed materials carry no shear or tensile loads. Recompression is permitted.

Code Characteristics:

1. *Rezoner*-none.
2. *Preprocessing capability*-material-geometric packages for shapes such as cylinders, spheres, hemispheres, simple parallelipeds, and cones may be combined to create complex geometries. Variable zoning of the mesh is possible. Initial conditions—velocities, energy, density—for each computational cell can be automatically specified. Material property tables for 19 solids and four high explosives are built into the code and may be selected by specifying a single number associated with individual materials.
3. *Post-processing capability*-printer maps and CALCOMP plots may be generated from METRIC output files.
4. *Comments*-material interfaces and free surfaces are maintained by a special transport prescription, which is described in some detail in the documentation cited. The number of material packages in a calculation is limited only by the dimensions of the material arrays. Cell quantities for 15 rows are kept in core, the remaining information being buffered from mass memory. All grid boundaries in METRIC can be reflective or transmittive. A rigid boundary can also be generated.

11.3.4 K3

Developers: J. May, P. Snow, D. Williams, and W. Windholz
Kaman Sciences Corp.
Colorado Springs, CO

Documentation: May et al. (1980)
Thompson (1979, 1975)

Discretization: The finite-difference method is employed. Discretization and transport logic are well-documented in the basic reference.

Material Description:

1. *Material model*-K3 is a three-dimensional extension of Sandia's CSQ II code and uses the standard Eulerian equations (see Appendix B).
2. *Equation of state*-the equations of state in CSQ II are employed here as well.
3. *Failure criteria/post-failure model*-failure is based on either pressure or maximum principal stress. If the user-specified parameter for those quantities is exceeded, a void is introduced into the material.

Code Characteristics:

1. *Rezoning*-none as yet.
2. *Preprocessing capability*-programs PREK3 and KSGEN generate the necessary geometric packages and control cards for start and restart tapes. Geometric shapes available for material generation include a box, tetrahedron, ellipsoid, a general solid with six faces, and conical frustrum. These may be combined to form arbitrary shapes.
3. *Post-processing capability*-program K3PLT is the graphics program. At present, the graphics package is specific for the BMDATC color system described by Kirk and Robertson (1976). The plot package edits the restart dump and produces dot plots of pressure and density.
4. *Comments*-the program is currently configured to run on a CDC 7600. It is limited to a maximum of 80 lines in any one coordinate direction, 10 materials, and 40 variables per cell. A sample problem with extremely coarse grid has been run of a steel ball striking a plate of steel and aluminum. No run times are indicated. The documentation indicates that further development can be expected. Familiarity with the CSQ code is helpful.

11.4 HYBRID CODES

11.4.1 CELFE

Developers: C. H. Lee and Associates
Lockheed Missiles and Space Company
Huntsville Research & Engineering Center
Huntsville, AL

Discretization: The designers of CELFE view impact as a problem in the structural dynamics category. The CELFE procedure is divided into three parts. The first part uses Eulerian finite elements to simulate the dynamic behavior in the impact zone. The second part is a classical finite-element procedure for analyzing the dynamic response of a Lagrangian zone. The third part consists of interfacing procedures to integrate the first two separate parts. The code may be used in-core or coupled with a structural analysis program such as NASTRAN. For the Eulerian region, a finite-element algorithm based on the theorem of weak solutions was used to insure that the Rankine-Hugoniot jump conditions are automatically satisfied. Brick-type finite elements are used including isoparametric, linear 8-node, quadratic 20-node, and cubic 32-node elements. Time integration is accomplished through a modified Lax-Wendroff scheme [Richtmyer and Morton (1976)].

Material Description:

1. *Material model*-the Eulerian formulation is similar to that of other codes. Next to the Eulerian region is a transition region described with a moving coordinate system. The relative velocity varies in such a way that for each time it becomes identical to the local velocity within the Eulerian zone and decreases monotonically through the moving boundary through the zone and vanishes at the Lagrangian zone. The Lagrangian equations are typical for small-deflection structural dynamics. The formulation employs the Jaumann stress-rate tensor, includes the von Mises yield criterion, and a Prandtl-Reuss flow rule. Provision is made for treating orthotropic materials.
2. *Equation of state*-the Tillotson and Los Alamos equations of state are used.
3. *Failure criteria/post-failure model*-the Chamis criterion [Lee (1978a)] is used to model failure in orthotropic composites. Characteristic of structural analysis, the onset of plasticity is considered a failure criterion for isotropic materials.

Code Characteristics:

1. *Sliding surfaces*-sliding with (Coulomb law) and without friction is permitted.
2. *Rezoning*-none.
3. *Preprocessing capability*-none
4. *Post-processing capability*-if one exists, it is not described in the documentation.
5. *Comments*-the code at present can accommodate rigid boundaries, free surfaces, symmetry planes, and interfaces. Preliminary tests with simple problems indicate need for additional work. Running times are also slow—360 s/cycle with core-contained CELFE, 1800 s/cycle with a CELFE-NASTRAN link. The slow-running times for the CELFE-NASTRAN calculation are attributed to inefficient interfacing routines.

11.5 CURRENT DEVELOPMENTS

A number of three-dimensional codes are under development and should soon be available. Hallquist at LLNL had developed an implicit code, NIKE3D, for large-deformation dynamic analysis of structures subjected to severe impact loading. According to Hallquist (1980), NIKE3D will have many of the features of NIKE2D [Hallquist (1979b)], its two-dimensional analog. NIKE2D includes material models for isotropic and orthotropic elastic, thermoelastoplastic and linear-viscoelastic behavior. A soil and crushable-foam model is also included. The finite-element discretization is accomplished with an incremental-iterative numerical algorithm based on a Jaumann stress-rate formulation with rotational terms treated explicitly. The Newmark β method is used for time integration. Provision is made for lumped or consistent mass matrices, but global mass and stiffness matrices are never assembled. A post processor computes contours for nearly 100 different quantities. The manual for NIKE3D is currently being reviewed. The documentation and code are expected to be available sometime in 1981.

Many impact situations are not well simulated with either Eulerian or Lagrangian codes alone. A combination of both is frequently desired and to this end, a link between the three-dimensional Eulerian HULL codes and the three-dimensional Lagrangian EPIC-3 code has been developed. The motivating problem for the linkup is the study of the deformation and breakup of projectiles penetrating spaced plates. However, the techniques developed are not limited to that specific problem.

A Lagrangian calculation for this problem could in principle be performed with EPIC-3. However, during penetration of relatively thick plates (those for which the ratio of target thickness to projectile diameter exceeds 2) the grid becomes badly distorted, shortening the time step. Consequently it becomes uneconomical to run the calculation to completion. An Eulerian code such as HULL could be used, but large numbers of zones would be required since the entire volume of effects would need to be included in the computational solution space and fine zoning would be required to properly define the details of the projectile structure. Such a calculation would require a considerable amount of computer time on even the fastest machinery available.

The first phase of the HULL/EPIC link has been completed and documentation is being prepared. As presently configured, the HULL three-dimensional code provides a simulation of penetration of the first target plate. It carries enough of the projectile to insure faithful reproduction of the velocities imparted to the projectile by contact with the target. Lagrangian stations are located along the projectile surface and over several cross sections near its nose. The HULL calculation stops when contact with target plate ceases—typically 20 to 30 μs after impact. The EPIC-3 calculation then takes over. It is run with a finely gridded representation of only the projectile from the time of initial impact using nose and surface velocities from the HULL calculation. Any material sheared from the nose area of the projectile is dropped from the EPIC-3 calculation at the appropriate time. This is accomplished by dropping nodes from the calculation and switching to a new cross section of HULL loading points. When contact with the target plate ceases, the nodes using HULL velocity input are changed to free boundaries and the projectile continues its journey to the next plate.

A three-dimensional Eulerian code, JOY, is being developed by Richard Couch at LLNL. Plans include development of a link to the Lagrangian DYNA3D code. The code uses cartesian coordinates and has available a number of equation-of-state options. The constitutive model includes the effects of work hardening, pressure hardening, thermal softening, and melting. The ability to generate complex geometries from simple shapes (arbitrary planes, spheres, and conic sections) is included. Post-processing is presently limited to edit prints and planar interface and contour plots. This will no doubt be expanded in the future. The code has been tested on simple problems involving hydrodynamic flow. Further document development and testing is expected throughout 1981. The code will also be documented during this period.

THREEDY is a three-dimensional Lagrangian code for the study of wave propagation in gases, liquids, and solids, developed by D. L. Hicks and others at Sandia [Hicks et al. (1974, 1975, 1976), Hicks and Madsen (1976, 1977)]. It is used as a research code to examine various algorithms and numerical techniques for problems involving shock-wave propagation and as such is not available for distribution. THREEDY makes use of operator splitting techniques, wave fitting, and movable meshes. Implicit differences with subcycling is used to avoid the Courant time-step restriction. However, for accuracy, time-step restrictions are employed that prevent volumes and energies from changing too much in a subcycle. An explicit option is included to bypass the implicit scheme if the Courant restriction happens to be satisfied by the accuracy criterion.

Several derivatives of the two-dimensional HEMP and TOODY codes are also available. A three-dimensional code is included in the STEALTH family of codes, developed by R. Hofman at SAI for the Electric Power Research Institute [Hofman (1978, 1980)]. Problems in the structural dynamics category (millisecond loading and response times) have been run with 2D options in STEALTH. No three-dimensional results have been published thus far.

Physics International of San Leandro, CA, markets a series of one-, two-, and three-dimensional codes (Eulerian, Lagrangian, coupled Eulerian-Lagrangian) under the acronym PISCES. The three-dimensional code, PISCES 3DE, is a first-order finite-difference Eulerian code using explicit time integration. The code is formulated to treat multimaterial problems involving gas and fluid flow. The capability to treat materials with strength has been added but is not documented in the user's manuals [Trigg et al., DDPLOT]. Letters and bulletins are periodically issued to bonafide users.

A technical data sheet for PISCES 3DE indicates that the constitutive models are based on the Von Mises and Mohr-Coulomb laws with strain-hardening and strain-rate effects. Failure criteria are based on maximum tensile pressure, minimum density, and maximum effective strain. The equation of state by library includes the Gamma law, Tillotson and JWL equations among others. The code uses free-format input with automatic coordinate generation (boxes, spheres, cylinders, and half-spaces). Contours, vectors, spatial, and time histories and deformed grids may be plotted with DD PLOT. The PISCES codes are installed on the CDC CYBERNET system and on the CRAY1. PISCES 3DE is written in FORTRAN IV and is fully vectorized.

11.6 SUMMARY

In practice, nonnormal incidence and asymmetries in materials and structures undergoing impact are the rule rather than the exception. Advances in the development of computer hardware and decades of research in numerical simulation of the partial differential equations of continuum physics have combined to provide analytical tools that make three-dimensional impact studies both feasible and cost-effective. As computer capabilities continue to grow and the cost of computations continues to decline, it is reasonable to expect that within the next five years the cost of such computations will drop to the point where they will be performed as a matter of routine.

The judicious combination of impact experiments, dynamic materials characterization, and three-dimensional numerical simulations holds tremendous promise for solving formidable technological problems (e.g. design of reactor-containment structures, space vehicles, personnel armor systems, and vehicles of all types) affecting all facets of society. Computations supplement the meager information currently attainable from impact experiments [Zukas (1980)]. Despite their current high cost, three-dimensional simulations are often less expensive than field experiments. Indeed, in certain situations (e.g. study of reactor-core disruptions) experimentation may not be possible and reliance must be placed on computations. In general, the combined analytical-experimental approach leads to improved designs at significant cost reduction over testing alone. An illuminating example in this regard is given by Herrmann (1977).

Three-dimensional codes are of great utility now and will prove even more useful in the future. But before rushing to acquire one or more of the existing production codes, potential users would do well to recall several points made thus far.

Acquisition, implementation, and use of a three-dimensional code for wave propagation and impact studies is a nontrivial undertaking. Because of the meager documentation available with most codes, close association with developers and experienced users is mandatory. Some six man-months are required to develop the necessary familiarity to run such codes with some feeling of security.

A comprehensive and flexible pre and post-processing capability is also essential if such codes are to be used on a production basis. If this capability is inadequate for the planned applications, additional one-time investments in computer graphics hardware and software is advisable.

Three-dimensional computer codes are quite demanding of computer resources. While one- and two-dimensional calculations can be done quite inexpensively on fourth-generation computers, running times for three-dimensional codes are still measured in terms of hours to tens of hours. Hence, some thought should be given to acquiring a two- as well as three-dimensional code. Preliminary two-dimensional calculations can be used to scope out problems, decide questions of temporal and spatial resolution and indicate physical parameters that most affect computational results and therefore need to be modeled most accurately. Given such advanced information, a full three-dimensional computation with the appropriate material model and resolution can then be undertaken to obtain design information at minimum cost.

The greatest limitation on code utility remains the constitutive model in the code, especially as it relates to material failure. Thus, modularity in codes is a highly desirable feature so that various aspects of the material description such as the equation of state, the strength model, the failure criteria, and post-failure models can be readily changed by users. Since the utility of computations is directly related to the quality of the material model, access to dynamic materials data and awareness of current developments in characterization of material behavior and failure under high rate loading conditions will insure effective use of an expensive but very powerful analytical tool.

APPENDIX A HEMP3D EQUATIONS

The conservation equations, equations of motion, and the constitutive relations listed below are solved by finite differences in the HEMP3D code [Wilkins et al. (1975)].

Conservation Equations

$$\dot{M}=0 \tag{1}$$

$$\dot{E}=-(P+q)\dot{V}+V[s_{11}\dot{\varepsilon}_{11}+s_{22}\dot{\varepsilon}_{22}+s_{33}\dot{\varepsilon}_{33}+s_{12}\dot{\varepsilon}_{12}+s_{13}\dot{\varepsilon}_{13}+s_{23}\dot{\varepsilon}_{23}] \tag{2}$$

M = mass element

E = internal energy per original volume

V = volume

P = hydrostatic pressure

q = artificial viscosity pressure

$(\dot{\ })$ = time derivative

Equations of Motion

$$\rho\ddot{u}_j=\sigma_{ij,i}(i, j=1,2,3) \tag{3}$$

u_j = displacement components

ρ = current density

Constitutive Relations

1 Stress deviators.

$$\dot{s}_{ij} = \begin{cases} 2G\left[\dot{\varepsilon}_{ij} - \dfrac{1}{3}\dfrac{\dot{V}}{V}\right] & (i=j) \\ G\dot{\varepsilon}_{ij} & (i\neq j) \end{cases} \tag{4}$$

G = shear modulus
ε_{ij} = strain components
s_{ij} = stress deviators
$\dot{V}/V = u_{i,i}$ from continuity

2 Velocity strains.

$$\begin{aligned} \dot{\varepsilon}_{ii} &= u_{i,i} \text{(no sum)} \\ \dot{\varepsilon}_{ij} &= \dot{u}_{i,j} + \dot{u}_{j,i}\ (i\neq j) \end{aligned} \tag{5}$$

3 Total stress.

$$\sigma_{ij} = s_{ij} - \delta_{ij}(P+q) \tag{6}$$

4 von Mises yield criterion.

$$s_{ij}s_{ij} \leqslant \tfrac{2}{3}Y^2 \tag{7}$$

where

$Y = a(b+\bar{\varepsilon}^p)^c$
$\bar{\varepsilon}^p$ = effective plastic strain
a, b, c = flow stress constants

Artificial Viscosity

$$q = \frac{\rho_0}{V}\left[c_0^2 l^2 \left(\frac{\dot{V}}{V}\right)^2 + c_1 la\frac{\dot{V}}{V}\right] \tag{8}$$

ρ_0 = initial density
l = characteristic length
a = characteristic velocity
c_0, c_1 = constants

Note: The current version of HEMP3D used ds/dt, the strain rate in the direction of acceleration [Wilkins (1980)], in the artificial viscosity expression in place of $\dot{V}/V$.

Hydrostatic Pressure

$$P=a(\eta-1)+b(\eta-1)^2+c(\eta-1)^3+d\eta E \tag{9}$$

$$\eta=\frac{\rho}{\rho_0}$$

a, b, c, d = equation-of-state constants

APPENDIX B HULL EQUATIONS

The partial differential equations describing the dynamic behavior of nonconducting linear elastic-plastic materials are solved in finite-difference form in the HULL code [Durrett and Matuska (1978)].

Conservation Equations

$$\dot{\rho}+\rho\dot{u}_{i,i}=0 \tag{1}$$

$$\rho\dot{E}-(\sigma_{ij}\dot{u}_j)_{,i}=-\rho\dot{u}_j g_j \tag{2}$$

where

g_j = body force due to gravity
$E=I+\frac{1}{2}\dot{u}_j\dot{u}_j$ — total specific energy
I = specific internal energy

Equations of Motion

$$\rho\ddot{u}_j=\sigma_{ij,i}-\rho g_j \tag{3}$$

Constitutive Relations

1 Stress deviators.

$$\dot{s}_{ij}=2G\left(\dot{\varepsilon}_{ij}-\tfrac{1}{3}\delta_{ij}\dot{u}_{k,k}\right) \tag{4}$$

2 Velocity strains.

$$\dot{\varepsilon}_{ij}=\tfrac{1}{2}(\dot{u}_{i,j}+\dot{u}_{j,i}) \tag{5}$$

3 Total stress.

$$\sigma_{ij}=s_{ij}-\delta_{ij}(P+q) \tag{6}$$

4 von Mises yield criterion.

$$s_{ij}s_{ij}\leqslant\tfrac{2}{3}Y^2 \tag{7}$$

Y = yield stress
δ_{ij} = Kronecker delta

When the yield stress is exceeded, the stress deviators are brought back to the yield surface by

$$s_{ij}^{p}=s_{ij}\left[\frac{2Y^2}{3s_{ij}s_{ij}}\right]^{1/2} \tag{8}$$

Artificial Viscosity

See (2) of Chapter 11.

APPENDIX C EQUATION-OF-STATE FORMULATIONS

Mie-Gruneisen

$$P=P_H\left(1-\frac{\Gamma\mu}{2}\right)+\Gamma\rho(E-E_0) \tag{1}$$

$$P_H=\begin{cases} K_1\mu+K_2\mu^2+K_3\mu^3 & \text{if } \mu\geqslant 0 \\ K_1\mu & \text{if } \mu<0 \end{cases} \tag{2}$$

$\mu=\frac{\rho}{\rho_0}-1$

Γ = Gruneisen parameter

ρ = current density

ρ_0 = initial density

E = internal energy per unit mass

E_0 = internal energy per unit mass at ambient density and pressure

K_1, K_2, K_3 = constants

Tillotson

For $\rho>\rho_0$ and $0\leqslant E\leqslant E_s$

$$P=P_\pi+A\mu+B\mu^2 \tag{3}$$

$$P_\pi=E\rho\left[a+\frac{b}{E/(E_0\eta^2)+1}\right] \tag{4}$$

For $\rho<\rho_0$ with $E>E_s$

$$P=aE\rho\left[\frac{bE\rho}{\left[1+E/(E_0\eta^2)\right]}+A\mu e^{-\beta(1/\eta-1)}\right]e^{-\alpha(1/\eta-1)^2} \tag{5}$$

where

$$\eta = \rho/\rho_0$$
$$\mu = \eta - 1$$
$$E_s = \text{sublimation energy}$$
$a, b, A, B, E_0, \alpha, \beta$ = parameters that depend on the material

Los Alamos

$$P = \begin{cases} \dfrac{A\mu + \rho_0 E(B + \rho_0 EC)}{\rho_0 E + \phi_0} & \text{if } \mu \geqslant 0 \\ \dfrac{\mu A_1 + \rho_0 E(B_0 + \mu B_1 + \rho_0 EC)}{\rho_0 E + \phi_0} & \text{if } \mu \leqslant 0 \end{cases} \tag{6}$$

where

$$A = A_1 + \mu A_2$$
$$B = B_0 + \mu(B_1 + \mu B_2)$$
$$C = C_0 + \mu C_1$$
$A_1, A_2, B_0, B_1, B_2, C_0, C_1, \phi_0$ = constants dependent on the material.

For explosives, the following equations of state are typical in computer codes for impact studies.

Wilkins

$$P = a\eta^\alpha + b\left(1 - \frac{\omega}{R}\eta\right)e^{-R/\eta} + \omega\eta E \tag{7}$$

a, b, α, ω, R = constants

Gamma Law

$$P = (\gamma - 1)(1 + \mu)E \tag{8}$$

where

γ = material constant

JWL (Jones, Wilkins, Lee)

$$P = A\left(1 - \frac{\omega\eta}{R_1}\right)e^{-R_1/\eta} + B\left(1 - \frac{\omega\eta}{R_2}\right)e^{-R_2/\eta} + \omega\eta E \tag{9}$$

where

A, B, ω, R_1, R_2 = constants

REFERENCES

______(1972), DD-PLOT User's Manual, Physics International, San Leandro, CA.

______ (1978), PISCES 3DE, Technical Data Sheet 211, Physics International, San Leandro, CA.

Belytschko, T. (1975), in W. Pilkey and B. Pilkey (Eds.), *Shock and Vibration Computer Programs: Reviews and Summaries*, Shock and Vibration Information Center, Washington, D.C.

Brown, B. (1980), Lawrence Livermore Laboratory, UCID-18507.

Chamis, C. C. (1969), National Aeronautics and Space Administration, NASA TN D 5367.

Chen, Y. M., and Wilkins, M. L. (1976), *Int. J. Fract.*, **12**, 607.

Chen, Y. M. (1978), *Eng. Fract. Mech.*, **10**, 699.

Chinn, J., and Zimmerman, R. M. (1965), Air Force Weapons Laboratory, AFWL TR 64-163.

Durrett, R. E., and Matuska, D. A. (1978), Air Force Armament Laboratory, AFATL-TR-78-125.

Evans, M. W., and Harlow, F. H. (1957), Los Alamos Scientific Laboratory, LA-2139.

Gaby, L. P. (1978), Air Force Weapons Laboratory, C4-C-4041.

Giroux, D. (1973), Lawrence Livermore Laboratory, UCRL-51079, Rev. 1.

Goudreau, G. L., and Hallquist, J. O. (1979), in *Proc. 5th International Seminar on Computational Aspects of the Finite Element Method*, Berlin, W. Germany (also Lawrence Livermore National Laboratory Reprint UCRL-82858).

Gregson, V. R., Jr. (1972), Defense Nuclear Agency, DNA 2797F.

Hageman, L. J., and Lee E. P. (1976), Ballistic Research Laboratory, BRL-CR-305.

Hageman, L. J., Waddell, J. L., and Herrmann, R. G. (1976), Systems, Science and Software, SSS-R-76-2973.

Hallquist, J. O. (1976), Lawrence Livermore Laboratory, UCRL-52066.

Hallquist, J. O. (1977), *Proc. ASCE Fall Convention and Exhibit*, San Francisco, CA, Preprint 2956.

Hallquist, J. O. (1978a), in K. C. Park and D. K. Gartling (Eds.), *Computational Techniques for Interface Problems*, AMD-30, ASME, New York.

Hallquist, J. O. (1978b), Lawrence Livermore Laboratory, UCRL-52429.

Hallquist, J. O. (1979a), Lawrence Livermore Laboratory, UCID-17268, Rev. 1.

Hallquist, J. O. (1979b), Lawrence Livermore Laboratory, UCRL-52678.

Hallquist, J. O. (1980), private communication.

Hancock, J. W., and Mackenzie, A. C. (1976), *J. Mech. Phys. Solids*, **24**, 147.

Herrmann, W. (1975), in W. Pilkey and B. Pilkey (Eds.), *Shock and Vibration Computer Programs: Review and Summaries*, Shock and Vibration Information Center, Washington, D.C.

Hicks, D. L., Lauson, H. S., and Madsen, M. M. (1974), in *Proc. CUBE Symp.*, Lawrence Livermore Laboratory, CONF-741001.

Hicks, D. L., Lauson, H. S., and Madsen, M. M. (1975), Sandia Laboratory, SAND-75-0350.

Hicks, D. L., Lauson, H. S., and Madsen, M. M. (1976), Sandia Laboratory, SAND-75-0578.

Hicks, D. L., and Madsen, M. M. (1976), Sandia Laboratory, SAND-76-0436.

Hicks, D. L., and Madsen, M. M. (1977), Sandia Laboratory, SAND-76-0744.

Hill, R. (1948), *Proc. Roy. Soc. Lond., Ser. A*, **193**, 281.

Hill, R. (1950), *The Mathematical Theory of Plasticity*, Clarendon Press, Oxford.

Hofmann, R. (1978), Electric Power Research Institute, EPRI NP 176-1.

Hofmann, R. (1980) in H. G. McComb, Jr., and A. K. Noor (Eds.), *Research in Nonlinear Structural and Solid Mechanics*, National Aeronautics and Space Admin., NASA Conf. Pub. 2147.

Hu, L. W. (1956), *J. Appl. Mech., Trans. ASME*, **23**, 444.

Hutula, D. N., and Zeiler, S. M. (1972), Westinghouse Electric Corp., Atomic Power Division, WAPD-TM-1079.

Johnson, G. R. (1977a), *Proc. Third International Symp. on Ballistics*, Karlsruhe, W. Germany.

Johnson, G. R. (1977b), Ballistic Research Laboratory, BRL-CR-343.

Johnson, G. R., Colby, D. D., and Vavrick, D. J. (1978), Air Force Armament Laboratory, AFATL-TR-78-81.

Johnson, W. E. (1967), General Atomic, GAMD 7310.

Johnson, W. E. (1976), Computer Code Consultants, CCC-976.

Johnson, W. E. (1977), Ballistic Research Laboratory, BRL-CR-338.

Johnson, G. R., Colby, D. D., and Vavrick, D. J. (1979), *Int. J. Num. Methods Eng.*, **14**, 1865.

Johnson, G. R., Vavrick, D. J., and Colby, D. D. (1980), Ballistic Research Laboratory, ARBRL-CR-00429.

Jonas, G. H., and Zukas, J. A. (1978), *Int. J. Eng. Sci.*, **16**, 879.

Kirk, R. W., and Robertson, F. E. (1976), Systems Development Corp., TM-HU-211/000/00.

Lee, C. H. (1978a), National Aeronautics and Space Administration, NASA-CR-159395 (N79-29832).

Lee, C. H. (1978b), National Aeronautics and Space Administration, NASA-CR-159396 (N79-29833).

May, J., Snow, P., Williams, D., and Windholz, W. (1980), Kaman Sciences Corp. K-80-226(R).

Mescall, J. F. (1974), in W. Pilkey et al. (Eds.), *Structural Mechanics Computer Programs*, University of Virginia Press, Charlottesville.

Misey, J. J., Gupta, A. D., and Wortman, J. M. (1980), in J. R. Vinson (Ed.), *Emerging Technologies in Aerospace Structures, Design, Structural Dynamics and Materials*, ASME, New York.

Noh, W. F. (1976), Lawrence Livermore Laboratory, UCRL-52112.

Noor, A. K. (1980), *Comput. Struct.* **13**, 425.

Norris, D. M., Scudder, J. K., McMaster, W. H., and Wilkins, M. L. (1977), in *Proc. High Density Alloy Penetrator Materials Conf.*, AMMRC-SP-77-3.

Osborn, J. J., and Matuska, D. A. (1978), Air Force Armament Laboratory, AFATL-TR-78-24.

Reed, L. L., and Henderson, D. R. (1975), Defense Nuclear Agency, DNA 3865T (AD-A027091).

Richtmyer, R. D., and Morton, K. W. (1976), *Difference Methods for Initial Value Problems*, 2nd ed., Interscience, New York.

Seaman, L., Curran, D. R., and Shockey, D. A. (1976), *J. Appl. Phys.*, **47**, 4814.

Thompson, S. L. (1975), Sandia Laboratories, SAND-74-0122.

Thompson, S. L. (1979), Sandia Laboratories, SAND-77-1339.

Trigg, M., Hancock, H., Cowler, M., and Abbott, K., *PISCES 3DELK User's Manual*, Physics International, San Leandro, CA, undated.

Valliapan, S. (1972), *Arch. Mech. Stos.*, **24**, 465.

Von Riesemann, W. A., Stricklin, J. A., and Haisler, W. E. (1974), in W. Pilkey et al. (Eds.), *Structural Mechanics Computer Programs*, University of Virginia Press, Charlottesville.

Wilkins, M. L. (1964), in B. Alder, S. Fernback, and M. Rotenberg (Eds.), *Methods in Computational Physics*, Vol. 3, Academic Press, New York.

Wilkins, M. L. (1977), in *Proc. Conf. on Fracture Mechanics and Technology*, Noordhoff, Leyden.

Wilkins, M. L. (1980), *J. Comp. Phys.*, **36**, 281.

Wilkins, M. L., Blum, R. E., Cronshagen, E., and Grantham, P. (1975), Lawrence Livermore Laboratory, UCRL-51574, Rev. 1.

Wilkins, M. L., French, S. J., and Sorem, M. (1970), in *Proc. 2nd International Conf. Num. Methods in Fluid Dynamics*, Berkeley, CA, Springer-Verlag, New York.

Wilkins, M. L., and Reaugh, J. E. (1980), Lawrence Livermore Laboratory, Preprint UCRL-84793.

Zukas, J. A. (1980), in J. R. Vinson (Ed.), *Emerging Technologies in Aerospace Structures, Design, Structural Dynamics and Materials*, ASME, New York.

Zukas, J. A., Jonas, G. H., and Misey, J. J. (1979), Ballistic Research Laboratory, ARBRL-MR-02969.

Zukas, J. A., Jonas, G. H., Kimsey, K. D., Misey, J. J., and Sherrick, T. M. (1981), Ballistic Research Laboratory, to appear.

INDEX